Advances in Intelligent Systems and Computing

Volume 1066

The series "Advances in Intelligent Systems and Computing" contains publications on theory, applications, and design methods of Intelligent Systems and Intelligent Computing. Virtually all disciplines such as engineering, natural sciences, computer and information science, ICT, economics, business, e-commerce, environment, healthcare, life science are covered. The list of topics spans all the areas of modern intelligent systems and computing such as: computational intelligence, soft computing including neural networks, fuzzy systems, evolutionary computing and the fusion of these paradigms, social intelligence, ambient intelligence, computational neuroscience, artificial life, virtual worlds and society, cognitive science and systems, Perception and Vision, DNA and immune based systems, self-organizing and adaptive systems, e-Learning and teaching, human-centered and human-centric computing, recommender systems, intelligent control, robotics and mechatronics including human-machine teaming, knowledge-based paradigms, learning paradigms, machine ethics, intelligent data analysis, knowledge management, intelligent agents, intelligent decision making and support, intelligent network security, trust management, interactive entertainment, Web intelligence and multimedia.

The publications within "Advances in Intelligent Systems and Computing" are primarily proceedings of important conferences, symposia and congresses. They cover significant recent developments in the field, both of a foundational and applicable character. An important characteristic feature of the series is the short publication time and world-wide distribution. This permits a rapid and broad dissemination of research results.

**** Indexing: The books of this series are submitted to ISI Proceedings, EI-Compendex, DBLP, SCOPUS, Google Scholar and Springerlink ****

More information about this series at http://www.springer.com/series/11156

Miguel Botto-Tobar · Joffre León-Acurio ·
Angela Díaz Cadena · Práxedes Montiel Díaz
Editors

Advances in Emerging Trends and Technologies

Volume 1

Editors
Miguel Botto-Tobar
Eindhoven University of Technology
Eindhoven, Noord-Brabant, The Netherlands

Joffre León-Acurio
Universidad Técnica de Babahoyo
Babahoyo, Ecuador

Angela Díaz Cadena
Universitat de Valencia
Valencia, Valencia, Spain

Práxedes Montiel Díaz
Centro de Investigación y Desarrollo Profesional
Babahoyo, Los Rios Province, Ecuador

ISSN 2194-5357 ISSN 2194-5365 (electronic)
Advances in Intelligent Systems and Computing
ISBN 978-3-030-32021-8 ISBN 978-3-030-32022-5 (eBook)
https://doi.org/10.1007/978-3-030-32022-5

This Springer imprint is published by the registered company Springer Nature Switzerland AG
The registered company address is: Gewerbestrasse 11, 6330 Cham, Switzerland

Contents

Architecture for Demand Prediction for Production Optimization: A Case Study

Inabel Karina Mazón Quinde[1], Sang Guun Yoo[1,2(✉)], Rubén Arroyo[1], and Geovanny Raura[1]

[1] Departamento de Ciencias de la Computación, Universidad de las Fuerzas Armadas ESPE, Sangolquí, Ecuador
kmazon1991@gmail.com,
{yysang, rdarroyo, jgraura}@espe.edu.ec

[2] Departamento de Informática y Ciencias de la Computación, Escuela Politécnica Nacional, Quito, Ecuador
sang.yoo@epn.edu.ec

Abstract. A proper product demand projection is an aspect that can be decisive for the competitiveness and survival of companies. However, in most of cases, this process is carried out based on empirical knowledge of the marketing personnel generating a high level of error in the results. To solve this problem, this paper presents a production planning architecture based on demand analysis by using business intelligence architecture and analytical algorithms. The proposed architecture has been validated by means of a case study which results indicate that the effectiveness increases from 25% to more than 85%. We believe that the proposed model may be applicable in other entities.

Keywords: Production planning · Consumer demand · Advanced analytics · Manufacturing company

1 Introduction

A company requires an accurate vision of consumer demand to put in place proper production, inventory, distribution and purchase plans in order to survive in the highly competitive market. Large companies around the world such as Ford, Yanbal International, Western Union and Adidas have implemented advanced analytics projects, related to the product demand of their organizations. The experiences of these companies are considered as success stories since they have allowed managers to take decisions in a more precise and scientific manner. However, in most of cases, the architecture of their solutions (including tools and analytical models) are not revealed to the scientific and technological community due to confidentiality issues; this situation makes difficult for other companies to learn from those success experiences. In this situation, this work intends to share our experience in developing a demand projection solution for a real company in the manufacturing sector. In this work, have proposed our own implementation architecture that could serve as a basis for data analysis in other entities of the manufacturing sector.

M. Botto-Tobar et al. (Eds.): ICAETT 2019, AISC 1066, pp. 1–11, 2020.
https://doi.org/10.1007/978-3-030-32022-5_1

The rest of the paper is organized as follows. First, a literature review is presented in Sect. 2. Then, the proposed architecture for product demand projection is detailed and validated in Sects. 3 and 4. Finally, Sect. 5 concludes the present paper.

2 Literature Review

Given the diversity of tools and algorithms found in the field of business intelligence (BI), for the presented literature review, we have decided to use the strategy of triangulation of information sources. For this process, we have considered the following parameters: (1) To review studies conducted by Gartner Inc. to understand the strengths and weaknesses of BI solutions as well as their market trends, (2) To review analytical algorithms used in the industry based on KDnuggets (a specialized portal in business analytics, big data, etc.), and (3) To contrast KDnuggets study with scientific articles published on international journals and conferences.

Study Made by Gartner: Among the different studies carried out by the Company, we have analyzed Gartner's Magic Quadrant 2018 for Data Science and Machine Learning Platforms. In this study, Gartner classifies companies that provide data science and machine learning solutions in 4 groups i.e. leaders, visionaries, challengers and visionaries [1, 2, 5]. Since the first group incorporate those companies that have obtained the highest score in their capacity of vision and execution, we have only analyzed the tools provided for those companies i.e. KNIME, Alteryx, SAS, RapidMiner, and H2O.ai.

Studies of Algorithms and Statistical Methods: KDnuggets is a recognized website in the field of Business Analytics, Big Data, Data Mining, Data Science and Machine Learning [3]. The latest KDnuggets survey (conducted on 844 professionals) identified the main algorithms used by data scientists. KDnuggets asked what methods or algorithms the professionals have used in the last 12 months for real applications. Such study indicated that industry-level data scientists are more likely to use time series, visualization, statistics, regression, and random forests. Same study indicated that the governmental sector is more likely to use visualization, PCA and time series, and academic researchers are more likely to use PCA and deep Learning [4].

Triangulation Method: Under the triangulation method approach, the present work have carried out a review of the literature to contrast the KDnuggets' research. The inclusion criteria defined for the literature search were the followings: (1) Only works (books, articles and papers) from 2009 were included, (2) Papers must refer to completed decision-making projects and they should describe information on analytic architectures focused on predictions. On the other hand, the exclusion criteria defined for the present systematic literature review were the followings: (1) Papers that are not related to predictive analytics and papers that only refer to the study of analytical algorithms were omitted, and (3) Content in languages different to English and Spanish were also excluded. The final search string selected after performing several tests was: ("Prediction" or "Business Analytics" or "Business Intelligence Decision support systems") and ("Companies" or "industries" or "Manufacturing") or ("Real-time systems").

To perform the search, the SCOPUS digital database was used because it covers a wide range of publications in the field of science and engineering, and maintains a complete and consistent database. The application of the search chain found 213 scientific documents (filtering the information created since 2009 and selecting only content from journals and conferences). From the result, the filtering process was carried out, giving as a result 11 primary studies, which were classified into 4 categories: comparison of data extraction algorithms, studies in the area of medicine, education and industries (see Table 1).

Table 1. Literature review in application of data mining algorithms

Category	Reference	Methodology
Comparison of data mining algorithms	[6]	Time series, lineal regression
	[7]	C4.5, k-Means, SVM, apriori, EM, PageRank, AdaBoost, kNN, Naive Bayes and time series
	[8]	Time series
Studies of data mining algorithms in the area of medicine	[9]	Time series
Studies and evaluation of data mining algorithms in the education area	[10]	Multilayer perception, Naïve Bayes, SMO, J48, REPTree
	[11]	Naive Bayes algorithm
	[12]	Lineal regression
Studies of data mining algorithms in the industry	[13]	Time series – ARIMA
	[14]	Box-Jenkins' time series (ARIMA) and multivariate regression
	[15]	Time series, lineal regression and decision tree
	[16]	Time series and lineal regression

According to the executed analysis, we could verify that the most frequently used methods in the industry area are time series and linear regressions. This analysis coincides with the analysis carried out by KDnuggets indicating that the best options for industrial data analysis are time series and regressions.

3 Proposed Architecture for Demand Projection of Current Demand

This section explains the details of the proposed architecture for product demand projection. For testing the proposed architecture, we have made a case study with real data of an important manufacturing company in Ecuador. Name of the selected company has been omitted because of the confidentiality agreement.

Through a previous works review, we could notice that there are several researches regarding advanced analytics and product demand projection focused on different areas such as retail, insurance, health, transport, among others. However, we could not find a detailed study focused on the manufacturing sector. In this sense, we believe that this work will deliver an important contribution to those people that are planning the development of product demand projection projects. The present project will contribute with the presentation of a proposal of improvement of product projection solution for the manufacturing sector; it will propose an architecture and prototype model for analyzing the projection of the demand that considers the factors of Ecuadorian reality.

In this section, we first explain how the tools and algorithms used in the proposed architecture were selected. Then, we explain the decision making solution of the case study company before implementing the proposed architecture. Finally, we explain the details of the proposed architecture for analyzing product demand projection and its results in the company.

3.1 Selection of Analytical Tools

For the selection of analytical tools and algorithms, a comparative analysis was executed. This analysis was divided into two parts: (1) evaluation of analytical tools and (2) evaluation of analytical algorithms. In addition, the result of the analysis was contrasted with the literature review executed in Sect. 2.

Evaluation of Analytical Tools: The first filter was made considering tools and algorithms used in the global market. In addition, success stories of analytical architectures exposed by different providers were also considered. Results of this analysis are summarized below.

Based on the Gartner study published in February 2018, the main analytical tools of the market were: KNIME (K), Alteryx (A), SAS Enterprise Miner (S) and RapidMiner (R). This list was used to carry out a focus group to determine the most important characteristics that an analytical tool should have. The results of the evaluation are shown in Fig. 1 which indicates that Alteryx received 17 points, Rapidminer 14 points, SAS Enterprise 13 points and Knime 7 points. In conclusion, the focus group recommended the use of the Alteryx tool Software for real industrial implementations because it contains the most important features for an advanced analysis. The detail of the assessment of each of the characteristics is shown in Table 2.

Important: The focus groups executed in this work had the participation of 7 international experts on implementing real solutions of data analytics and big data analysis (10 years of experience on average).

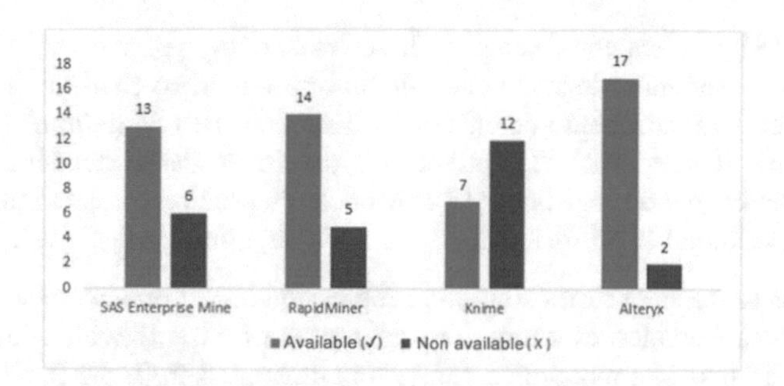

Fig. 1. Results of the focus group evaluating analytical tools

Table 2. Evaluation details of analytical tools

Features	Tools			
	S	R	K	A
Free license	X	✓	✓	X
Multiplatform	✓	✓	✓	X
Can combine models	✓	✓	X	✓
Incorporated module of descriptive techniques (grouping)	✓	✓	✓	✓
Built-in modules of predictive techniques (classification and others)	✓	✓	X	✓
Friendly interface	✓	✓	X	✓
Allows data visualization	✓	✓	✓	✓
Flexibility	X	✓	✓	✓
Easy to configure	X	✓	X	✓
Easy to install	✓	✓	✓	✓
Easy data conversion and filters	✓	✓	X	✓
Integration with R	X	✓	X	✓
Customization of built-in prediction modules	X	X	X	✓
Creation of macros, modules and the possibility of reusing them	X	X	✓	✓
Processing of large volumes of information	✓	X	X	✓
Support in Ecuador (where the case study company is located)	✓	X	X	✓
Success stories exposed in their official pages	✓	X	X	✓
Validation of the model	✓	✓	X	✓

• S = SAS, R = RpidMiner, K = Kmine, A = Asteryx

Evaluation of Analytical Algorithms: Based on the analysis made in the previous section, it was possible to demonstrate that the mostly used algorithms by data scientists for industrial implementations were time series and linear regression. For this reason, a review of relevant studies using those algorithms was carried out. Here are some cases found and the most important results.

- Cases using Time series: "Análisis de series de tiempo en el pronóstico de la demanda de almacenamiento de productos perecederos" [17], "Modelos de series de tiempo aplicados a los expedientes de la Comisión de Derechos Humanos del Distrito

Federal" [18], "Aplicación de análisis de series de tiempo para modelar y pronosticar visitas a departamentos de emergencia en un centro médico en el sur de Taiwán" [9], "Predicción del rendimiento de cultivos utilizando series de tiempo" [16].
- Cases using Regressions: "La predicción del rendimiento académico: regresión lineal versus regresión logística" [12], "Modelo de predicción electoral: el caso de la elección municipal 2015 de León de los Aldama, Guanajuato" [19].

Once success cases of times series and regression algorithms were searched, a new focus group was executed to know the preferences of experts with wide experiences between algorithms. The mentioned analysis is illustrated in Table 3.

Table 3. Analytical algorithms analysis by means of a focus group

Algorithm	Selection of the focus group
Time series	✓ (Unanimity)
Regression	X

Important: The focus groups executed in this work had the participation of 7 international experts on implementing real solutions of data analytics and big data analysis (10 years of experience on average).

Through the evaluation process, it was possible to demonstrate that there are more implementations made using the Time Series than the regressions. In addition, it could be understood that experts prefer the use of time series over regression (see Table 3). For these reasons, time series was selected as the algorithm to be used in the proposed architecture.

3.2 Current Decision-Making Architecture of the Case Study Company

The company selected for the case study has a hybrid decision-making architecture with Microsoft Dynamics AX 2012 ERP as the main repository of information. It creates OLAP cubes in SQL Server for information consolidation and analyze them using QlikView business intelligence tool (see Fig. 2). Although, the current architecture is stable, there are important limitations such as longtime development, limited information visualization, low performance of the database, and lack of implemented modules. Additionally, although the Microsoft Dynamics ERP has a forecast theme incorporated, it has not produced satisfactory results because the processing is very slow (it takes up to 5 days to get information) and is not as effective as expected. For this reason, the company has opted to perform the demand projection through calculations in Excel file according to the criteria of the marketing personnel.

3.3 Proposed Decision-Making Architecture

For the architecture definition, we have relied on the business analysis of the company and its need to improve the product demand projection process. The most suitable and easiest architecture has been proposed to cover their requirements. Due to information

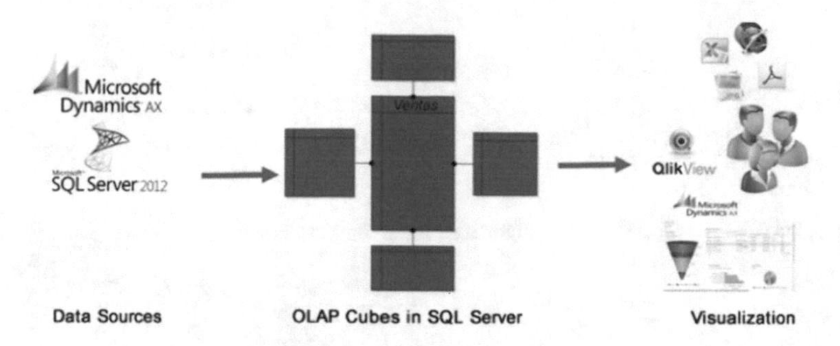

Fig. 2. Current architecture of information management of the case study company

centralization and analysis issues, we decided to adapt the demand projection process to the current business intelligence solution, making improvements and recommendations in each of the layers. Although many variants and elements can be found within the business intelligence architectures implemented in companies, most of them include a series of main components. Here are a brief explanation of those components.

- **Information Sources:** Relational databases, excel sheets, flat files, and other sources that store organizational information that feeds the data warehouse.
- **ETL Process:** Before storing the data in the data warehouse, they must be transformed, cleaned, filtered and redefined.
- **Data warehouse with the Metadata or Data Dictionary:** It seeks to store the data in a way that maximizes its flexibility, ease of access and management.
- **OLAP Engine:** This component provides the capacity of calculation, queries, planning functions, forecasting, and analysis of scenarios in large volumes of data.
- **Visualization tools:** This component allows the visualization of data and analysis results.

Once the components of a BI system are understood, it is fundamental to define all the layers of the architecture, with a special focus on the part where the demand projection module is involved. In this sense, the proposed architecture for the case study company is shown below (see Fig. 3). Our proposal is comprised of 5 layers: (1) data sources, (2) data quality, (3) data storage, (4) advanced analytics, and (5) information exploration and visualization.

The first layer corresponds to the data sources (core of the BI solution). This layer includes the Microsoft Dynamics AX 2012 and SQL Server 2012 ERP solutions which store all information collected from the different mobile devices and platforms of the company.

The second layer focuses on the quality of the data. This layer was implemented using Alteryx software which is an advanced data quality and analysis tool used by large retailers and manufacturers [20].

The third layer corresponds to the data warehouse where the business information is centralized. Due to the large size of data, we have implemented an analytical database to have a summarized and centralized resource for the business analysis execution. For data warehouse, we have used Exasol since it is a database with massive parallel

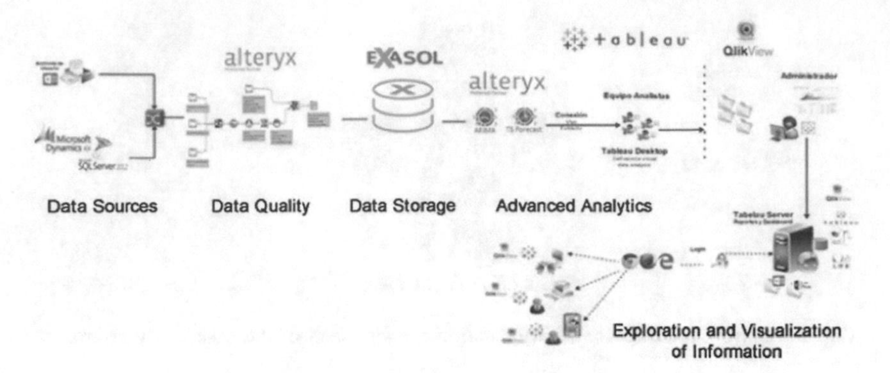

Fig. 3. The proposed BI architecture

process (MPP) feature of low cost and high performance that will allow to achieve an extensible data analysis framework. Additionally, Exasol was selected since it is a free analytical database compatible with Alteryx Software. From this warehouse, all the business analysis will be carried out, in order to decrease the performance overhead of Microsoft SQL Server database.

The fourth layer corresponds to the data mining and advanced analytics processes. This stage was carried out with Alteryx Software because, besides being useful for data quality, it is a powerful software for advanced analytics and it was the best evaluated according to the research carried out in the Sects. 2 and 3 of the present work. In this layer, Time Series algorithm (specifically ARIMA) was implemented to the sales prediction process.

The fifth layer corresponds to the exploration and visualization of the information. In this part, reports and data analysis were done with Tableau Software; this software leads the business intelligence market for 6 consecutive years as mentioned by Gartner [21]. In addition of Tableau Software, we have added the possibility of using the current implementation of QlikView, forming a hybrid implementation of business intelligence exploration and visualization.

4 Validation of the Proposed Solution

In this section, a validation process of the proposed solution is carried out using different data mining techniques and the use of Alteryx Software components. It is important to mention that, following the confidentiality agreement made with the case study company, we have limited the visualization of data used for the analysis.

For the evaluation, results precision was calculated to know the percentage of successes of the proposed solution. We also have compared with the results of the current solution of the company. The formula used to calculate the accuracy are as follows:

$$\text{Prediction accuracy } (\%) = (1 - \text{Absolute Error}) * 100$$
$$= (1 - ((\text{abs}(\text{real sales} - \text{forecast}))/\text{real sales})) * 100$$

First, we have executed a supervised learning process using the Alteryx software. We have used 80% of the data for the learning process and 20% for the validation of the model. Twelve months were predicted for this evaluation and the results were visualized using the Alteryx software. The result obtained is shown below (see Table 4). The average accuracy obtained from the demand projection for 8 products (sold by the company) within the 12 months was 54%. In this analysis, we can conclude that the generated model got a huge improvement over the 25% of precision that current solution of the company has in the prediction of sales.

Table 4. Prediction accuracy calculation

Feature	SKU								Total
	1	2	3	4	5	6	7	8	
Forecast	80	50	10	35	45	19	9	52	300
Sales	45	90	55	180	67	35	49	31	552
Absolute Error	35	40	45	145	22	16	40	21	252
MAPE (%)	78	44	82	81	33	46	80	68	46
Precision (%)	22	56	18	19	67	54	18	32	54

Additionally, a new experiment was carried out using the statistical model validation feature of Alteryx Software where 50% of the sales records have been taken for learning process and the remaining one for model validation. In this experiment, a better result were gotten. With the validation of Alteryx through a sample of 50% of random records, an error of 11.8% is presented, which gives us a model accuracy of 88.2% (see Fig. 4).

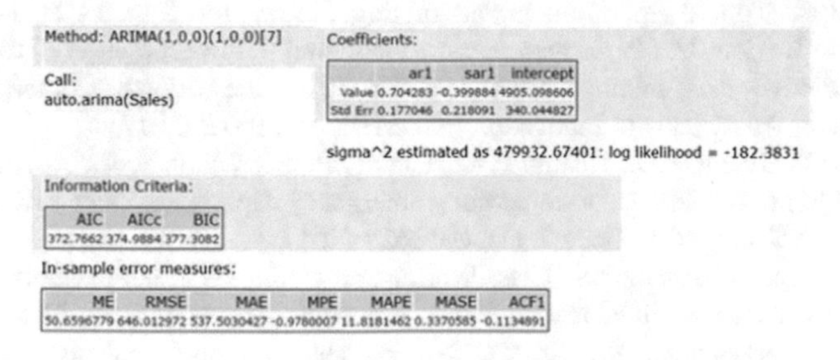

Fig. 4. Summary of ARIMA model results

5 Conclusions

In the present paper, we have proposed an architecture for product demand projection analysis and we have demonstrated its efficiency by a real implementation in a real company. The result showed an average accuracy of 88.2% (through a validation of Alteryx Software) which delivered a huge improvement over the 25% accuracy reported by the current solution of the company. Although the proposed architecture has been validated through a case study in only one company, we can dare to indicate that it is a promising result that can be extended as a prototype architecture in other companies of the manufacturing sector.

References

1. SIAG Consulting: Qué es exactamente el famoso cuadrante mágico de Gartner? Solo pienso en TIC - Consultoría tecnológica, 07 November 2016. http://www.solopiensoentic.com/cuadrante-magico-de-gartner/. Accessed 18 Mar 2018
2. Piatetsky, G.: Gainers and losers in gartner 2018 magic quadrant for data science and machine learning platforms, Feburary 2018. https://www.kdnuggets.com/2018/02/gartner-2018-mq-data-science-machine-learning-changes.html. Accessed 18 Mar 2018
3. Piatetsky-Shapiro, G.: Acerca de KDnuggets, analytics, big data, data mining y data science leader (2018). https://www.kdnuggets.com/about/index.html. Accessed 20 Mar 2018
4. Gregory Piatetsky, Kd.: Algoritmos principales y métodos utilizados por los científicos de datos, September (2016). https://www.kdnuggets.com/2016/09/poll-algorithms-used-data-scientists.html?lipi=urn%3Ali%3Apage%3Ad_flagship3_pulse_read%3Bhrjw5C3BQvOXA9P0xAjKdw%3D%3D. Accessed 20 Mar 2018
5. Gartner: Magic quadrant for data science and machine-learning platforms, Feburary 2018. https://www.gartner.com/doc/3860063/magic-quadrant-data-science-machinelearning. Accessed 18 Mar 2018
6. Wu, L., Yan, J.Y., Fan, Y.J.: Data mining algorithms and statistical analysis for sales data forecast. In: 2012 Fifth International Joint Conference on Computational Sciences and Optimization, pp. 577–581 (2012)
7. Wu, X., et al.: Top 10 algorithms in data mining. Knowl. Inf. Syst. **14**(1), 1–37 (2008)
8. Tanwar, H., Kakkar, M.: Performance comparison and future estimation of time series data using predictive data mining techniques. In: 2017 International Conference on Data Management, Analytics and Innovation (ICDMAI), pp. 9–12 (2017)
9. Juang, W.-C., Huang, S.-J., Huang, F.-D., Cheng, P.-W., Wann, S.-R.: Application of time series analysis in modelling and forecasting emergency department visits in a medical centre in Southern Taiwan. BMJ Open **7**(11), e018628 (2017)
10. Kaur, P., Singh, M., Josan, G.S.: Classification and prediction based data mining algorithms to predict slow learners in education sector. Procedia Comput. Sci. **57**, 500–508 (2015)
11. Saranya, S., Ayyappan, R., Kumar, N.: Student progress analysis and educational institutional growth prognosis using data mining. J. Eng. Sci. Res. Technol. **3**, 1982–1987 (2014)
12. Alvarado, J., Jiménez, A.: La predicción del rendimiento académico: regresión lineal versus regresión logística. Univ. Complut. Madr. **12**, 5 (2010)
13. Francis, H., Kusiak, A.: Prediction of engine demand with a data-driven approach. Procedia Comput. Sci. **103**, 28–35 (2017)

14. Alon, I., Qi, M., Sadowski, R.J.: Forecasting aggregate retail sales: a comparison of artificial neural networks and traditional methods. J. Retail. Consum. Serv. **8**(3), 147–156 (2001)
15. Ngai, E.W.T., Hu, Y., Wong, Y.H., Chen, Y., Sun, X.: The application of data mining techniques in financial fraud detection: a classification framework and an academic review of literature. Decis. Support Syst. **50**(3), 559–569 (2011)
16. Choudhury, A., Jones, J.: Crop yield prediction using time series models. J. Econ. Econ. Educ. Res. **15**, 53 (2014)
17. Contreras Juárez, A., Atziry Zuñiga, C., Martínez Flores, J.L., Sánchez Partida, D.: Análisis de series de tiempo en el pronóstico de la demanda de almacenamiento de productos perecederos. Estud. Gerenciales **32**(141), 387–396 (2016)
18. Quintana, M.J.G., Jiménez, S.A.M.: Modelos de series de tiempo aplicados a los expedientes de la Comisión de Derechos Humanos del Distrito Federal. Econ. Inf. **398**, 89–99 (2016)
19. López, J.A., López, M.A.A.: Modelo de predicción electoral: el caso de la elección municipal 2015 de León de los Aldama, Guanajuato1. Estud. Políticos **35**, 87–101 (2015)
20. Alteryx: About Alteryx | Alteryx (2018). https://www.alteryx.com/about-us. Accessed 25 Jan 2018
21. Gartner: Tableau lidera cuadrante mágico Gartner de BI 2018. Microsystem, 28 February 2018. https://www.microsystem.cl/tableau-lidera-gartner-bi-2018/. Accessed 11 Mar 2018

Administration and Management Platform of Electricity Consumption for Home Appliances Based on IoT

Fabián Cuzme-Rodríguez[1(✉)], Ricardo Madera-Rosero[1], Mauricio Domínguez-Limaico[1], Daniel Jaramillo-Vinueza[1], Carlos Pupiales-Yépez[1,2], and Luis Suárez-Zambrano[1]

[1] Carrera de Ingeniería en Telecomunicaciones, Universidad Técnica del Norte, Av. 17 de Julio 5-21 y Gral. José María Córdova, 100105 Ibarra, Ecuador
fgcuzme@utn.edu.ec

[2] Universidad Politécnica de Cataluña, Barcelona, Spain

Abstract. This paper proposes the design of a management platform that contributes to the responsible consumption of electricity in a home environment through a HAN (Home Area Network) network. The platform is composed of two subsystems, the first one, an electronic module in charge for sampling and sending data, Internet connection, and other functionalities that are detailed in the document; and the second, the management software that manipulates initial parameters of measurement, storage and visualization of the information sent by the first subsystem, allowing an interaction through a web page. For the validation of the platform a record of the electrical consumption of some common devices present in a household is taken, whose purpose is to detect an irregular consumption of any equipment associated with the electronic device, handling alerts in case of a power consumption greater than the normal.

Keywords: Smart grid · IoT · HAN

1 Introduction

Human activity contributes every day to the increase of the greenhouse effect, which seriously affects global climate change, and in turn, can dangerously affect natural ecosystems and society [1]. Yvo de Boer, former secretary of the United Nations Framework Convention about Climate Change said: “Energy efficiency is the most promising way to reduce greenhouse gas emissions in the short term,” the efficiency is to achieve optimal use of energy, maintaining the same level of life quality. Likewise, it’s considered that energy consumption will be the main problem of humanity in the next 50 years, as affirmed by the Nobel Prize winner Richard E. Smalle in 2003 [2].

In a home you can get to waste between 7% and 11% of electrical energy, this phenomenon occurs because most home appliances and electronic devices work with low voltage and direct current, however, the current supplied in the electrical network is high and voltage alternating, for this reason a transformer is required to acquire the adequate levels for each device, because the transformer works all the time, even if the

M. Botto-Tobar et al. (Eds.): ICAETT 2019, AISC 1066, pp. 12–22, 2020.
https://doi.org/10.1007/978-3-030-32022-5_2

equipment is off, there is an additional and unnecessary consumption of electricity, called consumption in standby mode. To avoid this waste of electrical energy the devices should be disconnected when they are not being used, however, execute this task manually can be tedious for the user [3].

In this sense, there is another common problem, which is the use of home appliances with more than 10 years old, which increases the value of the electricity bill until 20%, since they require more electricity to operate, this additional value depends on the time of use and maintenance provided to the equipment. This is due mainly to the fact that the user is not aware of the real electrical consumption of their devices since he has no way to know it because he doesn't have a system that allows him to constantly monitor them [4].

This proposal seeks to provide an adequate solution for the benefit of energy efficiency, by providing tools to control the electrical consumption of equipment in a home, and in turn, avoid wasting energy, thus favoring the user's economy and conservation of the planet.

1.1 Smart Grids

Electricity went from being a technological novelty that aroused curiosity and admiration among the people of large cities, to become, half a century later, a multifaceted source of energy (light, strength, heat), which was increasingly identified with the progress and modern life [5]. Later, with the advent of information technology in the following decades, a new paradigm known as smart grids was created, defined according to the European Union, as: "an electrical network that can efficiently integrate the behavior and actions of all the users connected to it (generators, consumers or both), to guarantee an economically efficient energy system, sustainable with low losses, and high levels of quality and security of supply".

On the other hand, the Department of Energy of the United States (DOE) carried out an analysis of the systems of an electrical network and determined the areas where improvements must be made to be considered as a smart grid, in this sense the following is proposed: Reliability, security, economy, efficiency, respect for the environment and safety [6].

1.2 The Multi-layer System Architecture of a Smart Grid

This kind of network is considered an interactive platform, which is composed of layers that are shown in Fig. 1. This architecture describes the interaction between different protocols to be used in the deployment of a smart grid.

1.3 Applications in Local Networks

There are several types of networks that are classified according to their coverage range and data rate as shown in Fig. 2, among which is the home area network (HAN), building area network (BAN) and industrial area network (IAN); neighborhood area network (NAN) and field area network (FAN); and wide area network (WAN).

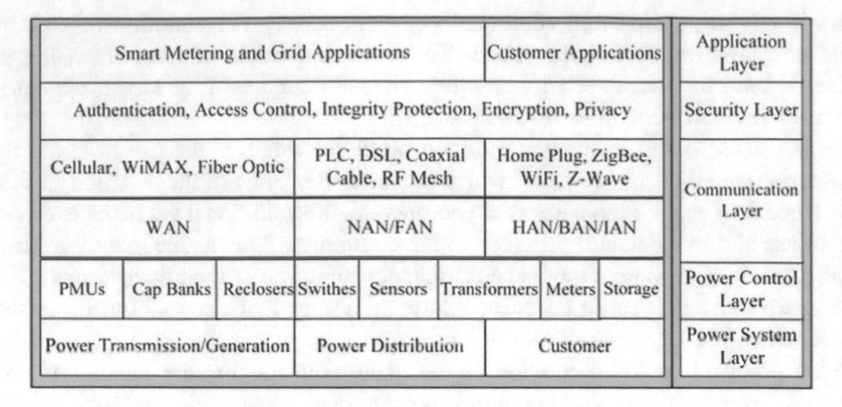

Fig. 1. The multi-layer system architecture of a smart grid

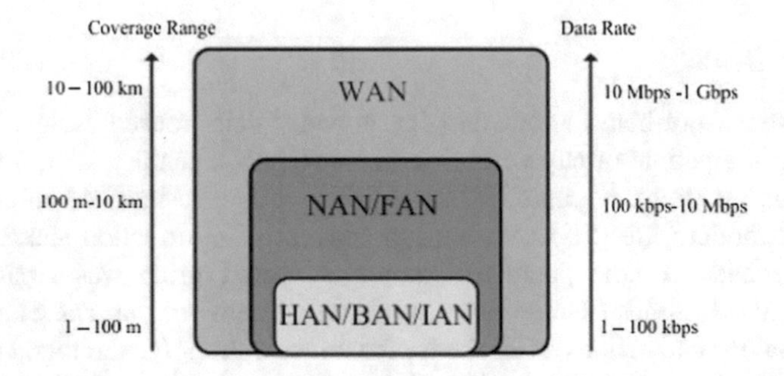

Fig. 2. Classification of communication networks according to coverage range and data rate

The local networks HAN, BAN and IAN are at the client end in the network architecture. These networks support communications between electrical and electronic devices, electric vehicles, and other equipment at the customer's premises.

HAN provides communications for home appliances and equipment that can send and receive signals for an intelligent meter, in-home displays (IHDS) and/or home energy management systems (HEM). There are applications such as residential automation, optimal adjustment of the thermostat for thermal zones of the home, control, and management of electric charges.

BAN and IAN are used for commercial and industrial customers, focus on building automation, heating, ventilating and air conditioning (HVAC), among other industrial energy management applications [7].

Among the most common communication technologies for local networks HAN, BAN and IAN, we have the following: Power Line Communications (PLC), Ethernet, Z-Wave, Bluetooth, ZigBee, WiFi and Wireless Mesh [8].

1.4 IoT Communication Protocols

At present, the Internet-based information architecture facilitates the exchange of goods and services between the equipment connected to the network. From this derives a paradigm known as IoT (Internet of Things), which results from a combination of sensors and actuators that provide and receive information, through bidirectional networks that transmit data to be used by multiple services for users [9]. IoT devices cover all types of everyday elements capable of sending data to the Internet, current trends indicate that in the future everything will be managed through IoT platforms [10].

Among the most important IoT protocols, there are HTTP/2, Websockets, and MQTT (Message Queue Telemetry Transport). The underlying qualifier between the three protocols is the efficiency in which they use resources such as bandwidth, battery, and the versatility of their characteristics. Therefore, the best option is MQTT, since neither HTTP nor WebSockets was designed specifically taking into account machine-to-machine communication [10].

2 Materials and Methods

The development methodology used in this project is known as the V Model. This methodology has four levels of development: In the first level the requirements and specifications are analyzed; in level 2 the functional characteristics of the system are established; level 3 defines the hardware and software components and level 4 the implementation and testing [11].

In addition, the project is based on the ISO/IEC/IEEE 29148: 2011 standard, which contains provisions for processes and products related to the requirements engineering for software systems, products and services along the life cycle [12].

Figure 3 shows the proposed Platform, which is composed of two subsystems: Electronic Data Acquisition Module (EDAM) and Administration and Management Software through Internet (AMS).

2.1 EDAM Subsystem

EDAM is located on a conventional outlet and it's responsible for measuring the electrical current consumed by a device connected to it, this module has an ACS712 current sensor that registers the electrical variable and a microprocessor that sends the data obtained through wifi to the administration and management software; in addition a power control that allows to cut and reactivate the electrical energy in the outlet.

The hardware components of this subsystem were determined by the ISO/IEC/IEEE 29148 standard, consisting of Arduino Nano, WiFi ESP-018, ACS-712, RTC3231, Relay 1 channel, Relay 2 Channels, Outlet, Plug and Metalic box. The connection diagram of electronic components was made using the Fritzing software, which is shown in Fig. 4.

The EDAM is energized by the outlet through a 5v DC source; the relay module is connected directly to the digital pins D4 and D5 of the Arduino; and in series with the current sensors, which are connected to analog pins A0 and A1 of the Arduino

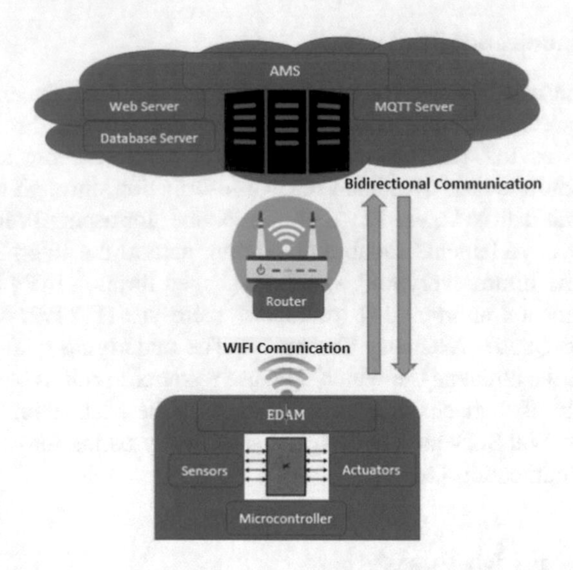

Fig. 3. General diagram of the administration and management platform

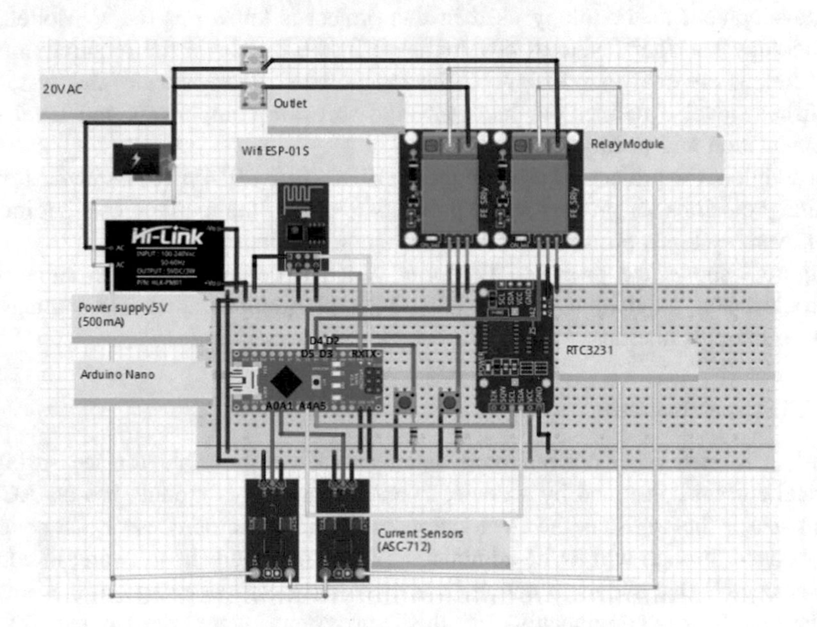

Fig. 4. Connection diagram of electronic components

respectively. It's very important to have two independent circuits in the outlet so that the measurements of each connected device can be differentiated. The real-time clock module keeps the current time in its memory and contains a button stack to keep the record in case of no power.

In addition, the WiFi module communicates through a serial link with the controller board and acts as a bridge between the EDAM and AMS. When the electronic module does not have to send data, at times out of use, it will enter a state of low consumption, where its power consumption is practically zero, however, a pushbutton is added to pin D3 of the Arduino, if it's necessary to activate the EDAM in this state; and another button on D2 to restart the microcontroller.

2.2 AMS Subsystem

AMS consists of a virtual machine hosted on the Internet (cloud computing), which receives the data sent by the EDAM and displays it on a web page for monitoring. In case of exceeding the assigned consumption threshold, it issues an alert on the website and sends an email to the user. Also, through a web interface, it can control the electrical flow to the EDAM and establish the initial measurement and automation parameters.

AMS is developed on infrastructure as a service (IaaS) provided by Amazon Web Services (AWS), in the free version. The software elements necessary for the development of the AMS are selected based on the ISO/IEC/IEEE 29148 standard and are shown in Table 1.

Table 1. Administration and management of software components

Request	Selection
IaaS provider	Amazon Web Services (AWS)
Operative system	Red Hat Enterprise Linux (RHEL)
IoT protocol	Message Queuing Telemetry Transport (MQTT)

AMS has Red Hat Enterprise Linux (RHEL) as an operating system, and a XAMPP server is installed and configured on it, which is an independent free code server that integrates some services such as web server, database server, PHP and Perl language interpreters, FTP server, among other modules. In the same way, a broker is required for the MQTT protocol, which works as a central node and it's responsible for network management and message transmission among the clients (subsystems), this broker is implemented in the same instance of AWS.

On the other hand, the website is based on HTML, PHP, CSS, and Javascript. Figure 5 presents the measurements registered and displayed to the user through the web page of the system; for this, the data is obtained from the electrical consumption of three devices that consume a large amount of energy, which is: microwave oven, stereo, and clothes iron. In the left part of Fig. 5 the consumption in watts of the associated devices is shown, in the lower part the dates when the measurements were taken are indicated, and in the upper part, there is the line-time that allows filtering results.

The platform allows parameterizing the input of the devices that will be monitored, through the options New equipment and Shutdown lapse (see Fig. 6). When entering new equipment, the referential nominal power consumption in watts must also be recorded, that is, the normal consumption power that is given by the manufacturer, and the measurement frequency given in minutes. On the other hand, the shutdown lapse refers to the time interval in which the electronic module remains off, interrupting the flow of current to the equipment, that is, the time in which no person makes use of the equipment, this allows obtain an energy saving per consumption in standby mode, or in case that one or several devices are kept on unnecessarily by the user's carelessness. In the test, a shutdown lapse has been defined from 00:10 to 6:10.

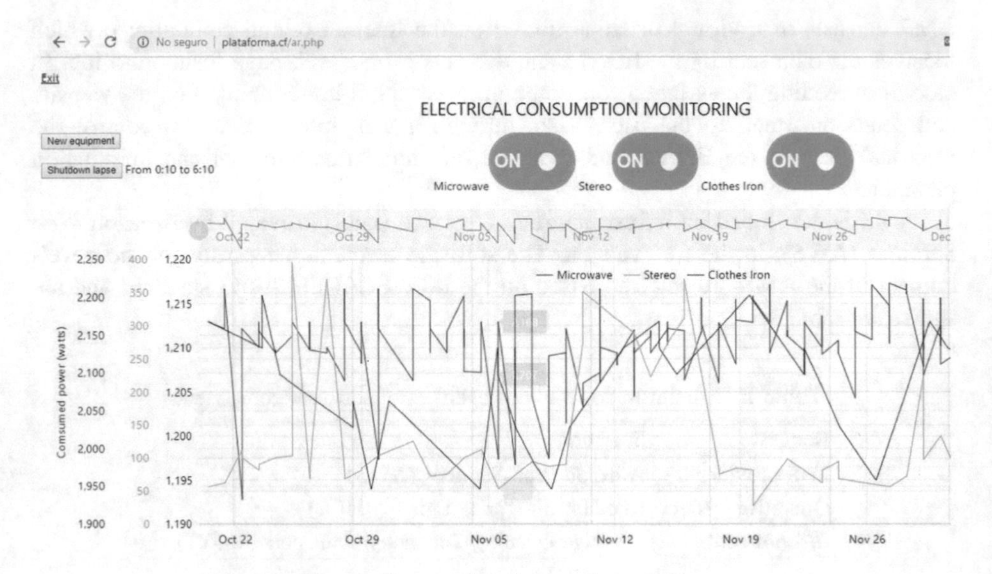

Fig. 5. Data registration on the website

New equipment

Microwave	1450	watts	Measuring frequency (minutes)	
Stereo	600	watts	Measuring frequency (minutes)	
Clothes Iron	1200	watts	Measuring frequency (minutes)	

Accept

Shutdown lapse From 0:10 to 6:10

0	:	10	Shutdown time
6	:	10	On time

Accept

Fig. 6. Initial parameterization of equipment

Finally, the web interface has icons of manual on and off, and the event logs in case of excessive power consumption in any of the registered equipment (see Fig. 7).

Fig. 7. Flow control icons of electric current and event history

Besides, if a registered device exceeds the registered power consumption, the platform will issue an alert to the user, where a message is displayed on the web page and a programmed email is sent (see Fig. 8).

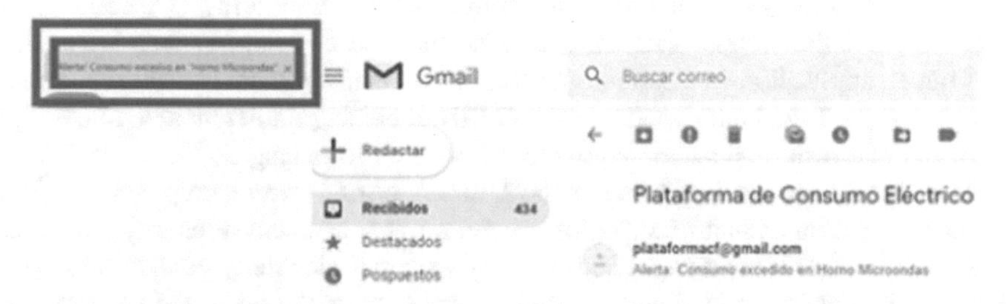

Fig. 8. Alert actions in case of excessive electrical consumption

3 Results and Discussion

Tests were executed for two months, for which an internal meter was placed in a home for the record of weekly electric consumption values. In Fig. 9a the values obtained without the implementation of the platform along the first month are shown, likewise, once the platform was implemented, the electric consumption values are shown in Fig. 9b along the second month.

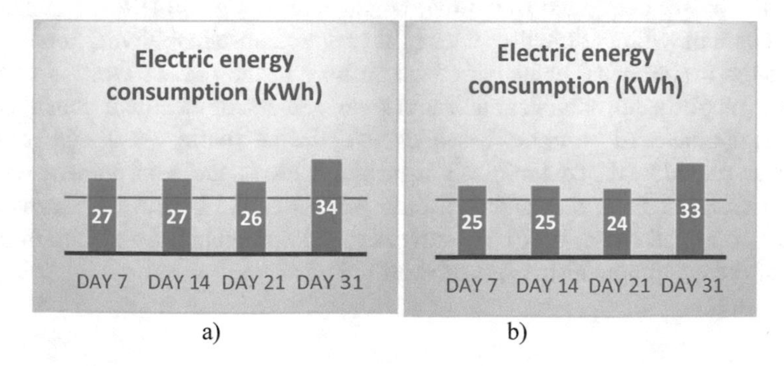

Fig. 9. Electric consumption for one month. (a) without platform implementation (b) with platform implementation

According to the results obtained, the consumption of global electric power was reduced from 114 KWh to 107 KWh, which represents a saving of 6.14% per month, which could be improved if this system is used in other additional equipment, thus increasing this saving up to 13.16%, for example, if it is used with 6 devices. On the other hand, the platform allowed to identify that the *Microwave Oven* equipment consumes 40% more energy than stipulated by the manufacturer, which indicates a possible failure in its operation.

Similar works have been developed, for example, at the Polytechnic University of Catalonia, where the monitoring and analysis of energy consumption is carried out on one of its buildings, using the PowerStudio and SIRENA management software [13], which differs from the proposed platform since its monitoring is done at the scale of buildings or sections of the building

Another example is carried out at the University of the Armed Forces (ESPE) [14], which consists of a system of visualization and mobile monitoring of electricity consumption; while the present platform allows access to it through any web browser, not only from an application on a cell phone; and also contains remote control functions for turning on and off the home appliances, apart from being parameterizable allowing the user of the platform to enter the equipment he wants to manage.

Moreover, the work developed at the University of Navarra proposes the design and implementation of smart meters for the remote management of energy resources at the household level, however, it focuses on the overall electricity consumption of the home, that is, not Monitors devices individually, such as the proposed platform [15].

The solution proposed in this development, unlike the ones mentioned above, is to individually manage each of the household electrical and electronic devices, considering that if a device presents an abnormal consumption, the application detects it and informs the user through an alert so that the respective measures are taken. Furthermore, by integrating it together with a digital meter, it allows knowing the overall consumption of the house, which facilitates the user to establish good practices for the use of devices that improve the optimization of energy consumption.

The administration and management platform implemented will serve as a baseline for the analysis of electricity consumption in the home through data obtained by the platform, which allows graphing curves of daily consumption by electrical device in the home and its trends; In addition to the energy consumed, in the future it is intended to study the power demanded over time, which makes it possible to implement measures to continue with the trend of energy saving and sustainability at home.

Although the principal objective of the authors in the present work is to test and evaluate a platform for administration and management of electrical consumption of household appliances in a home based on IoT; in the future, additional research is required to provide safe communication mechanisms in the environment where the EDAM operates, establishing authentication protocols and lightweight cryptographic algorithms [16]; Likewise, it is necessary to apply vulnerability verification methodologies of the ASM administration software [17].

4 Conclusions

The implementation of this platform allows visualizing data of the real energy consumption of several electrical and electronic devices in a home, in this way the user can discern between a value of energy consumption in normal ranges and a value of excessive energy consumption due to irregularities present in their equipment.

Bidirectional communication between EDAM and the AMS is done in real time, due to the adequate integration of the hardware and software components selected through the ISO/IEC/IEEE 29148 Standard and the use of the MQTT communication protocol, ideal for IoT applications, since it uses very few processing resources and bandwidth.

The tests established during two months with three home appliances, allowed to determine a minimum saving in electricity consumption of around 6%, percentage that can be increased as more household devices are added to the management of electricity consumption; this saving is attributed to the process of automatic disconnection of electrical and electronic devices, according to the schedules programmed through the AMS. This energy and economic savings can be more significant if it's implemented on a large scale.

The administration and management platform can be implemented in larger physical infrastructures, such as computer labs, companies, buildings, among others; where a greater energy saving is expected, when managing a greater quantity of equipment. However, for industrial equipment that consumes a very high amount of electrical energy, the initial requirements of the platform must be redefined.

References

1. Ojea, L.: IMÁGENES Y DATOS: Así nos afecta el cambio climático, Huelva (2018)
2. Huang, E.: Generando ahorros a través de una gestión energética eficiente (2013)
3. Arroyo García, D.: Consumos fantasma residenciales. eltelégrafo (2016). https://www.eltelegrafo.com.ec/noticias/punto/1/consumos-fantasma-residenciales. Accessed 21 Mar 2019
4. Pérez, T.: Electrodomésticos antiguos incrementar el costo de luz | Novedades Quintana Roo. SIPSE.com (2015). https://sipse.com/novedades/electromesticos-viejos-suben-el-costo-de-luz-149966.html. Accessed 21 Mar 2019
5. Tafunell, X.: La revolución eléctrica en América latina, pp. 1–56, November 2015
6. Falvo, M.C., Martirano, L., Sbordone, D., Bocci, E.: Technologies for smart grids: a brief review. In: 2013 12th International Conference on Environment and Electrical Engineering, pp. 369–375 (2013)
7. Kuzlu, M., Pipattanasomporn, M., Rahman, S.: Communication network requirements for major smart grid applications in HAN, NAN and WAN. Comput. Networks **67**, 74–88 (2014)
8. Ananthakrishnan, R., Du, S., Lu, X.: Routing protocols for power line communications (plc). US 8,958,356 B2 (2015)
9. Salazar, J., Silvestre, Y.S.: Internet de las cosas (2016)

10. Singh, R.: Internet of Things: battle of the protocols (HTTP vs. Websockets vs. MQTT). Linkedin (2017). https://www.linkedin.com/pulse/internet-things-http-vs-websockets-mqtt-ronak-singh-cspo. Accessed 21 Mar 2019
11. Rodríguez, J.: Metodología de desarrollo de sotware. El Modelo en V o de Cuatro Niveles (2008). http://www.iiia.csic.es/udt/es/blog/jrodriguez/2008/metodologia-desarrollo-sotware-modelo-en-v-o-cuatro-niveles. Accessed 21 Mar 2019
12. ISO/IEC/IEEE 29148 - SEBoK https://www.sebokwiki.org/wiki/ISO/IEC/IEEE_29148. Accessed 21 Mar 2019
13. Álvarez Pérez, R.: Monitorización y análisis del consumo energético de la ETSEIB MEMORIA Autor: Escola Tècnica Superior d'Enginyeria Industrial de Barcelona. Universidad Politécnica de Cataluña (2018)
14. Chamba, C.: Diseño e implementación de un sistema de visualización y monitoreo móvil del consumo eléctrico a través de una red zigbee con módulo xbee Quito (2015)
15. Oyarzun, A.L., Álvarez, J.M.: Monitorización del consumo eléctrico de un hogar: procesado de datos mediante Raspberry Pi Trabajo Fin de Grado. Universidad Pública de Navarra (2015)
16. Cuzme-Rodríguez, F., et al.: Access network improvement for a WLAN based on 802.1X and CAPsMAN protocols intelligence for embedded systems view project access network improvement for a WLAN based on 802.1X and capsman protocols, pp. 43–49 (2018)
17. Cuzme-Rodríguez, F., León-Gudiño, M., Suárez-Zambrano, L., Domínguez-Limaico, M.: Offensive security: ethical hacking methodology on the web, vol. 884 (2019)

State of the Art Determination of Risk Management in the Implantation Process of Computing Systems

Felipe Ortiz[1(✉)], Mauricio Dávila[1,2], Marisa Panizzi[1,2], and Rodolfo Bertone[3]

[1] Information Systems Engineering. Graduate School, Universidad Tecnológica Nacional. Regional Buenos Aires, Av. Castro Barros 91 (C1178AAA), Caba, Argentina
ortizfd@gmail.com, davilamr.80@gmail.com, marisapanizzi@outlook.com

[2] Department of Information Systems Engineering, Facultad Regional Buenos Aires, Universidad Tecnológica Nacional, Av. Medrano 951 (C1179AAQ), Caba, Argentina

[3] Institute of Computing Research- III-LIDI. School of Computing, Universidad Nacional de La Plata (UNLP), 50 and 120, La Plata, Argentina
pbertone@lidi.unlp.edu.ar

Abstract. Risk management is a vital step for software development projects to be successful. A systematic mapping study (SMS) to systematize the empirical evidence of methodologies, methods and standards that deal with risk management in software development projects is presented. Subsequently, a characteristics-based analysis (DESMET method) is conducted to evidence the degree of compliance of the analyzed methodologies regarding the risks for the computing systems implantation process. Finally, conclusions and future works are presented.

Keywords: Computing systems implantation · Risk management · State of the art · Systematic mapping study

1 Introduction

Software projects are high-risk activities that result in variable performance results [1]. A risk is the probability that a loss may occur. The measurement of the probability and severity of adverse effects in the development, acquisition, maintenance, etc. of a system are the technical risks of the software. In a software development project, the loss may consist of the reduction of the software product's quality [2].

Industry surveys suggest that only a quarter of the software projects are fully successful (that is, they are completed as scheduled, budgeted and specified), and thousands of millions of dollars are lost annually due to failures or to projects that do not yield the expected benefits [1, 3]. In this sense, using metrics is a way to reduce the subjectivity bias since they could provide stakeholders with better knowledge, control and improvement of the risk management processes [4].

M. Botto-Tobar et al. (Eds.): ICAETT 2019, AISC 1066, pp. 23–32, 2020.
https://doi.org/10.1007/978-3-030-32022-5_3

Risk management is a vital step for software development projects to be successful. Pressures from competition, regulatory changes and the evolution of techniques may force the administrators of software projects to alter plans and strategies during the execution of a project. Changes in users' requirements, the emergence of new tools and technologies, the constant security threats and changes in the payroll add more pressure to the software project team and affect decision-making [5].

The absence of risk management, communication and comprehension of the requirements are the main factors related to the low rate of success in software projects [6]. In the context of this research line, the implantation process is a set of required activities and tasks that allow the transfer of the completed software product to the environment where it will be used by the user community [7].

Several factors can affect software development projects, such as: lack of changes in priorities, changes in project objectives, inaccurate requirements, inappropriate planning, etc., but according to [1], one of the most important factors are unmanaged risks.

Despite several studies and published experiences on risk management, the software industry, in general, does not seem to follow a model to analyze and control risks through the development of its products [8]. According to [9], two approaches can be identified in software projects. The traditional is reactive in nature and addresses the generic problems for all software projects in a systemic manner. The risk-oriented approach, however, is proactive since it seeks to identify and manage unique aspects of a specific project before they impact the project.

Software risk management is a crucial part of successful project management, but it is often not well implemented in real-world software projects [10]. According to [11], managing risks through a predictable software development process, provides a foundation that allows the software to be generated consistently for the end user, and at a lower cost.

While Software Engineering has evolved and proposed a set of methodologies or standards for software development and quality models, they do not support the implantation process of computing systems in a comprehensive fashion [12]. The analysis of the implantation process of computing systems performed by Panizzi et al. [7] shows either the absence of elements or weaknesses in the process under study.

Section 2 deals with the systematic mapping study of the literature that answers the research question-problem: is it possible to strengthen the implantation process through the identification of a set of risks and procedures for their mitigation?

2 Development of the Literature Systematic Mapping

This SMS followed the guidelines proposed by Marcela in [13] complying with the process proposed by Bárbara Ann Kitchenham in [14]. It is composed of three stages: planning (Sect. 2.1), review (Sect. 2.2) and the results report (Sect. 2.3).

2.1 SMS Planning

Table 1 summarizes the tasks performed in the "Review planning activity".

Table 1. Tasks of the review planning activities

Tasks	Description
Identify the need for the study	Determine which methodologies, methods and standards dealing with risk management in software development projects are mostly studied by the scientific community
Pose research questions (RQ)	RQ1: Which are the most used methodologies, methods and standards that deal with risk management in software projects? RQ2: What type of contribution regarding risk management in software projects was made at the academic level? RQ3: Are all the main processes in building a software product taken into account by methodologies, methods and standards dealing with risk management in software development projects?
Define research terms (RT)	RT1: Software, RT2: management, RT3: mitigation, RT4: software project, RT5: risk, RT6: software engineering, RT7: risk assessment, RT8: implementation
Specify Search strings	(("Software Project") AND ("Risk")) OR (("Software") AND ("Risk") AND ("Management")) OR (("Software Engineering") AND ("Risk Assessment")) OR (("Management") AND ("Risk") AND ("Implementation")) OR (("Software") AND ("Risk") AND ("Mitigation"))
Define criteria for the selection of studies	Inclusion criteria: • Articles answering the research questions • Studies containing target terms in the title and/or in the abstract • Articles published as of 2008 • Articles in Spanish and English Exclusion criteria: • Unavailable studies • Articles that do not match the inclusion criteria
Establish data source	IEEE Explore, Elsevier Science, ACM Digital Library Google Academic, SEDICI[a]
Type of publication	Publications in conferences and journals
Protocol validation	The review protocol was validated by thesis advisors for the Master's degree program

[a]SEDICI: Digital repository of Universidad Nacional de La Plata, website: http://sedici.unlp.edu.ar/

However, search strings defined are generic for the construction of the state of art of this work. They are considered relevant for the development of SMS whose purpose is to identify the methodologies or standards that address risk management in software projects.

2.2 Review

The number of articles chosen in each research source is presented in Table 2 and the extraction scheme is shown in Table 3.

Table 2. Articles published in each of the search sources

Search sources	Relevant articles	Dismissed articles	Primary articles
IEEE Explore	10	4	6
ACM Digital Library	4	2	2
Google Academic	95	15	80
SEDICI	5	2	3
Science Direct	18	9	9

Table 3. Extraction scheme

Extracted data	
Dimension	Categories
General	ID (Article registration identifier), search string, year of publication, title, author/s, source, search source, type of publication (in a conference or a journal), country, continent, key words, apa citation, number of citations, problem, proposal and results
Context	Main processes: requirements, development, deployment or implantation, architecture and design, maintenance, tests Support and organizational processes: verification and validation, configuration management, risk management [15]
Contribution	Metrics, tool, model, method, process, good practice
Purpose	Evaluate, validate, propose a solution, report an experience, expert opinion
Solutions	Different existing solutions in relation to managing, mitigating risks. It is important to point out that the inclusion of a new category in the dimension was performed as it appeared in relevant studies

2.3 Results Report

A quantitative summary is presented and then an analysis of the primary articles obtained is performed in order to answer each of the research questions.

2.3.1 Quantitative Summary of the SMS

Graph 1 shows the number of articles according to their source of publication, where 42% corresponds to conferences and 58% to journals. Graph 2 presents the number of articles per search source.

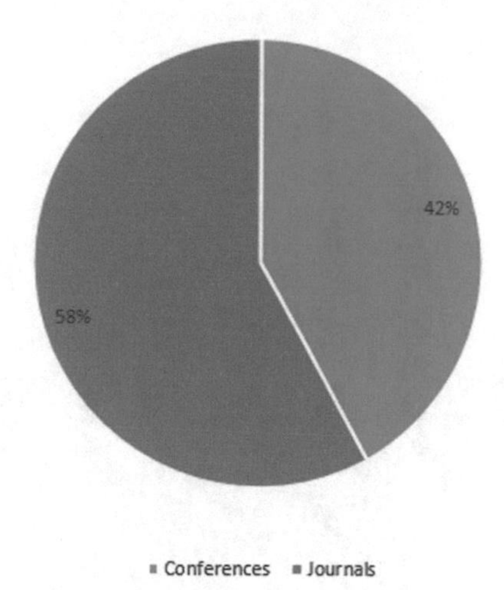

Graph 1. Percentage distribution of primary articles per type of publication

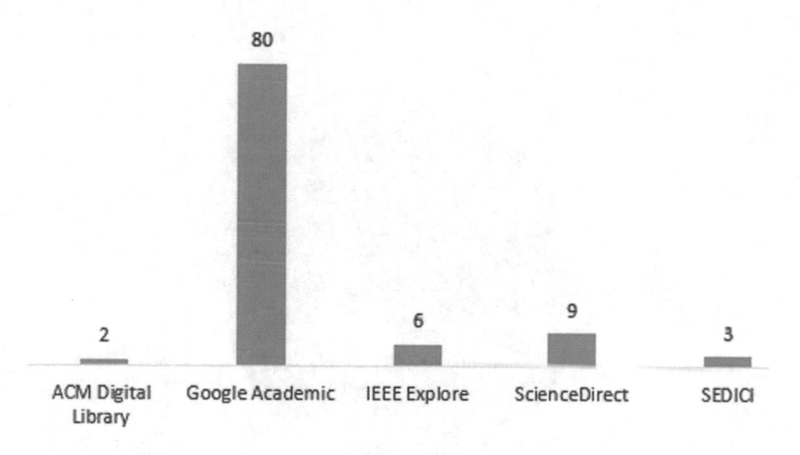

Graph 2. Number of primary articles per search source

2.3.2 Answers to the Research Questions (RQ)

Graph 3 presents the answer to RQ 1; the most frequently used methodology is CMMI (28 references) followed by PMBOK (26 references). These are followed by Software Risk Evaluation (23 references) and Artificial Intelligence techniques with 11 references. Then, Risk Management Frameworks (6 references), Prince2 (5 references), AS/NZS 4360 (3 references), ISO 31000 (3 references), ISO 12207 (2 references), Risk IT (2 references) and Magerit (1 references). Graph 4 shows the answer to RQ2, the types of contributions made by the primary articles.

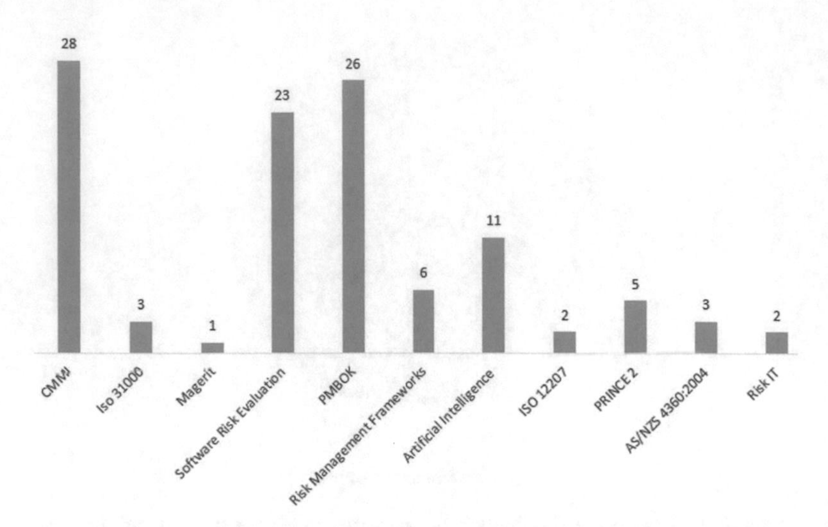

Graph 3. Number of references to methodologies, methods and standards dealing with risk management

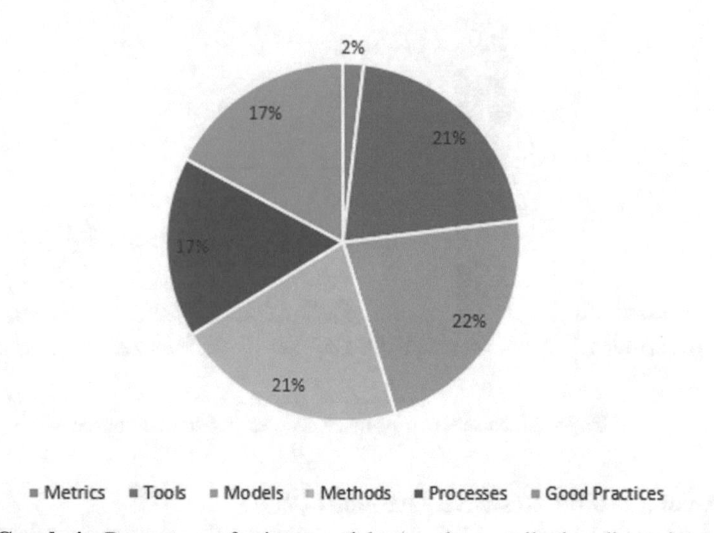

Graph 4. Percentage of primary articles per the contribution dimension

In answer to RQ2, Graph 5 shows the percentage of primary articles which address the risk management process (96%) and the remaining support and management processes (4%). Considering the total number of primary articles referring to the risk management process (96%), Graph 6 presents the distribution of such articles in relation to the main processes involved in developing a software product.

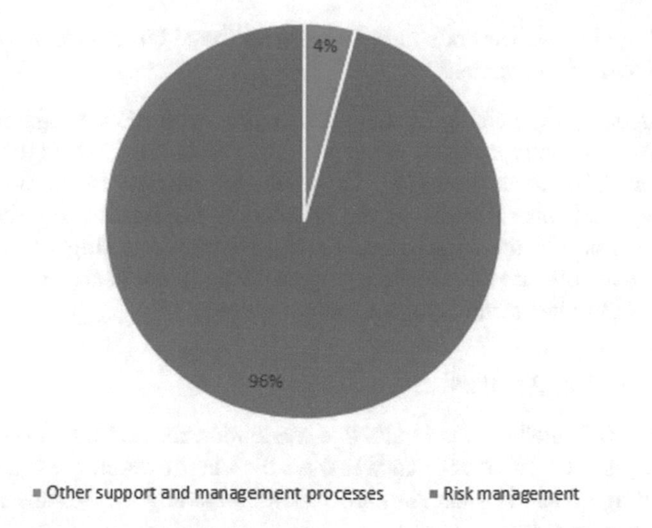

Graph 5. Percentage of primary articles per risk management processes and other support and management processes

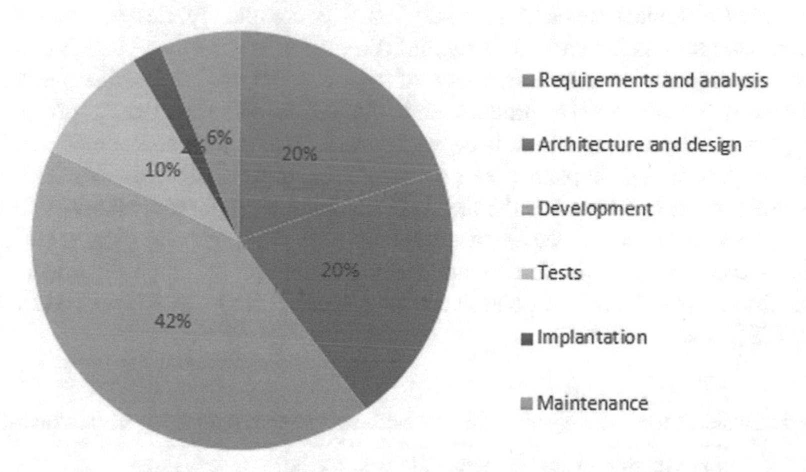

Graph 6. Percentage of studies dealing with risk management per different groups of main processes

3 Comparison of the Methodologies, Methods and Standards Dealing with Risk Management

This section presents a definition of the methodologies, methods and standards dealing with risk management which are included in the comparative analysis, together with the justification of their selection (Sect. 3.1). It further includes the guidelines for the DESMET method [12] adapted to this work (Sect. 3.2) and, finally, the results obtained are presented (Sect. 3.3).

3.1 Methodologies, Methods and Standards Considered Which Deal with Risk Management

Those methodologies, methods or standards dealing with risk management in software projects most frequently included in the primary studies are CMMI [16], PMBOK [17] and Software Risk Evaluation [18]. However, we decided to consider the Magerit methodology [16] since it is one of the pioneers in information systems risk analysis and management. On the other hand, although artificial intelligence can be used to manage risks in software development projects, we did not consider techniques in this comparison but rather methodological proposals only.

3.2 Comparative Analysis

For the DESMET method, the qualitative evaluation method based on characteristics analysis [12] is used, in order to answer the following research question: Are the risks of a computing system implantation process addressed by the existing methodologies, methods and standards?

In order to evaluate the methodologies, has been considered that the implantation process should be viewed in three dimensions. The first dimension is called "*Process*" and it includes product characteristics such as size, complexity, design characteristics, performance and quality level. The second dimension is "*Product*" and it consists of characteristics such as the complexity of the product to be installed, installation requirements for the software product, and integration with the client's infrastructure, among others. The last dimension is called "*Person*" due to the existence of peopleware and its impact in the implantation process of computing systems [7]. This three-dimensional process is based on the metric classification proposed by Pablo Vázquez in [17]. Table 4 presents the degree of compliance of the methodologies, methods and standards considered in relation to risk management for the three dimensions mentioned above. The degree of compliance is classified into three levels: Null ("N"), Partial ("P") and Total ("T").

Table 4. Evaluation of the methodologies, methods and standards dealing with risk management

Implantation phase	Methodologies, methods or standards			
	CMMI-DEV	PMBOK	S.R.E.	MAGERIT
Dimension "Process"	P	P	P	P
Dimension "Product"	N	P	P	P
Dimension "Person"	N	N	N	P

The following preliminary conclusions can be derived from the results shown in Table 4:

- In the "*Process*" dimension, all the methodologies, methods and standards considered show a partial degree of compliance.

- In relation to the "*Product*" dimension, SOFTWARE RISK EVALUATION, PMBOK and MAGERIT have a partial degree of compliance while CMMI does not address any risks in this dimension.
- Finally, as regards the "*Person*" dimension, MAGERIT shows a partial degree of compliance and the rest do not propose any risks for this dimension

The complete report of the state-of-the-art determination can be found in the Open Institutional Repository – UTN Regional Bs As [22].

4 Conclusions and Future Works

A systematic mapping of the literature (SMS) has been presented showing the methodologies, methods and standards dealing with risk management in software development projects. Then, a characteristics-based evaluation was performed following the DESMET method guidelines. The comparison was made considering a three-dimensional view of the implantation process (Process-Product-Person). We were able to identify to what extent the methodologies, methods and standards dealing with risk management in software projects support the implantation process of computing systems.

Lines of future work include: (a) the need to set risks for the implantation process of computing systems and (b) to validate the proposals in companies within the national industry.

References

1. Charette, R.: Why software fails. IEEE Spectr. **42**, 42–49 (2005). https://doi.org/10.1109/MSPEC.2005.1502528
2. Dhlamini, J., Nhamu, I., Kaihepa, A.: Intelligent risk management tools for software development, pp. 33–40 (2009). https://doi.org/10.1145/1562741.1562745
3. Johnson, J.: My life is failure: 100 things you should know to be a successful project leader. (2006)
4. Menezes Jr, J., et al.: Defining indicators for risk assessment in software. De Procedia Comput. Sci. (2013)
5. Sommerville, I.: Ingeniería de Software. Pearseon, Madrid (2006). SBN:84-7829-074-5
6. Caballero, S., Kuna, H.D.: Análisis y gestión de riesgo en proyectos software. In: XX Workshop de Investigadores en Ciencias de la Computación (WICC 2018). Universidad Nacional del Nordeste (2018). ISBN 978-987-3619-27-4
7. Panizzi, M., Bertone R., Hossian A., Proceso de Implantación de Sistemas Informáticos – Identificación de vacancias en Metodologías Usuales. In: Libro de actas de la V Conferencia Iberoamericana de Computación Aplicada, Vilamoura, Algarve, Portugal (2017). ISBN 978-989-8533-70-8
8. Kimer, T: Software engineering techniques: design for quality. In: IFIP International Federation for Information Processing, vol. 227 (2006)
9. Johnson, D.L.: Risk management and the small software projects (2009). http://www.sei.cmu.edu/iprc/sepg2006/johnson.pdf

10. Liu, D., Wang, Q., Xiao, J.: The role of software process simulation modeling in software risk management: a systematic review. In: 3rd International Symposium on Empirical Software Engineering and Measurement (Empirical Software Engineering and Measurement), pp. 302–311 (2009). https://doi.org/10.1109/esem.2009.5315982
11. Bertone, R.A., Thomas, P.J., Taquias, D., Pardo, S.: Herramienta para la Gestión de Riesgos en proyectos de software. In: XVI Congreso Argentino de Ciencias de la Computación, pp. 567–576 (2010). http://hdl.handle.net/10915/19289
12. Panizzi, M., Bertone, R., Hossian, A.: Propuesta de un Modelo de Proceso de Implantación de Sistemas Informáticos (MoProIMP). In: Libro de Actas del XXIV Congreso Argentino de Ciencias de la Computación (CACIC 2018). Universidad Nacional del Centro de la Provincia de Buenos Aires (2018). ISBN: 978-950-658-472-6
13. Marcela, G.B., José, C.-L., Mario, P.V.: Métodos de investigación en ingeniería del software. Editorial Ra-Ma (2014)
14. Kitchenham, B., Dyba, T., Jorgensen, M.: Evidence-based software engineering. In: International Conference on Software Engineering, Washington DC, USA, pp. 273–281 (2004)
15. ISO/IEC/IEEE 12207:2017: International Organization for Standardization (2017). https://www.iso.org/standard/63712.html
16. CMMI Institute: «Capability maturity model integration» (2010). https://cmmiinstitute.com/
17. Project Management Institute (2013). https://www.pmi.org/pmbok-guide-standards
18. Software Engineering Institute: «Software risk evaluation method» (1999). https://resources.sei.cmu.edu/asset_files/TechnicalReport/1999_005_001_16799.pdf
19. Kitchenham, B., Linkman, S., Law, D.T.: DESMET: a method for evaluating software engineering methods and tools, Keele University (1996)
20. Portal de administración electrónica: «MAGERIT v.3: Metodología de Análisis y Gestión de Riesgos de los Sistemas de Información» (2012). https://administracionelectronica.gob.es/pae_Home/pae_Documentacion/pae_Metodolog/pae_Magerit.html
21. Pablo, V., Marisa, P., Rodolfo, B.: Estimación del esfuerzo del proceso de implantación de software basada en el método de puntos de caso de uso. In: Libro de Actas del 6to. Congreso Nacional de Ingeniería Informática/Sistemas de Información (CoNaIISI 2018). Simposio de Ingeniería de Sistemas y de Software. Mar del Plata, Argentina (2018). ISSN: 2347-0372
22. Ortiz Felipe. Reporte Técnico Nro. 1. Trabajo de Especialidad en Ingeniería en Sistemas de Información. Establecimiento del estado del arte de la gestion de riesgos en el proceso de implantación de sistemas informtáicos. Publicado en: ria.utn.ar/ handle/123456789/1841

Semantic Representation Models of Sensor Data for Monitoring Agricultural Crops

Jorge Gomez[1(✉)], Bayron Oviedo[3], Alexander Fernandez[1], Miguel Angel Zuniga Sanchez[2], Jose Teodoro Mejía Viteri[2], and Angel Rafael Espana Leon[2]

[1] Departamento de Ingenieria de Sistemas, Universidad de Cordoba, Monteria, Colombia
jellercergomez@correo.unicordoba.edu.co
[2] Departamento de Informatica, Universidad Tecnica de Babahoyo, Babahoyo, Ecuador
[3] Universidad Tecnica Estatal de Quevedo Ecuador, Quevedo, Ecuador

Abstract. This paper shows the development semantic representation models of sensor data for monitoring agricultural crops. The purpose of this research article is explore the possible application of methodologies for ontologies in the process related to the agricultural environment and the information collected from the crop growth variables. therefore, improving interoporability becomes an important element in the treatment of large volumes of data, which will eventually help to make the right decisions and improve production in different types of crops.

Keywords: Ontology models · Internet of Things · Precision agriculture

1 Introduction

At present there is a process of constant adjustment focused on the models used for the analysis and subsequent processing of information, all this generated by the need for machines to process the data in a similar way as people do. Understanding this facilitates how to integrate the concept of IoTs and services so that physical devices are easily used as a web service and are efficiently integrated into networked applications with the required functionalities [1].

Allowing our expressions to be clear for machines has allowed new uses for ontologies, creating and relating objects to each other. Revealing the path towards the implementation of descriptive languages that make it possible for data to be formatted according to these models and different ontological processes to be applied [2]. Likewise, the constant growth of data collected by the increasingly large infrastructures for the processes of the Internet of Things (IoT), makes it necessary to employ new ways to be able to use these data efficiently and accurately.

M. Botto-Tobar et al. (Eds.): ICAETT 2019, AISC 1066, pp. 33–41, 2020.
https://doi.org/10.1007/978-3-030-32022-5_4

To promote new and better models of semantic representation of data collected by the sensors of an IoT infrastructure is an urgent need, since manually processing and analyzing these data is far from the conscious reality. The increase in data from sensors poses challenges such as making data available in a way that is easy and understandable for prospective users and their applications. It is necessary to achieve routing in order to achieve the semantic improvements with which data can be structured and organized and thus make them processable and also interoperable for machines [3].

Achieving this interoperability will open the door to the practical understanding of millions of data generated by sensor systems, impacting human daily life from different fronts, such as health (MediCare), government, infrastructure, security and many other dimensions of people's environment [4–6].

This document is divided as follows: related works, methodologies for ontologies, conceptual model, case study and conclusions.

2 Related Work

The following is a review of different points of view related to aspects that show the development of ontologies for the subsequent deployment and operation of semantic web concepts.

It is necessary to bear in mind some regulatory criteria aimed at creating a standard that serves as a reference. As well as other revisions in which it is understood how this can be applied to the infrastructures used for data collection, processing and analysis.

The aim is to improve these processes through the use of semantic representation models. These models, which will be generated from the data of the sensors, will have the purpose of making transparent the operation of Internet systems of Things (IoT) with people.

The semantic web has taken its place in different scenarios that involve the use of all those processes related to the Internet. The possibility of adding semantic and ontologically formatted metadata has allowed to develop concepts that today are regulated by the W3C. The purpose of this is to improve the use of data for a specific purpose. Achieving a clear interaction that leads to the improvement and understanding of the processes carried out in the Internet.

This has allowed the evolution and standardization of ontological languages such as the OWL 2 version. This focuses on discriminating sensors, measurements, sensor systems and how these are linked as concepts and sub concepts.

All this within a metadata system that is used to efficiently identify between intelligent agents and achieve effectiveness in the use of them [7]. Based on the use of the SSO pattern (Stimulus, Sensor, Observation), it allows to link the sensors with what they measure as well as with their results, becoming a minimum ontology of light character for the semantic measurements carried out in the web.

A practical application is shown in the implementation of the business model e3-value [8], this allows through the adoption of a consistent ontology on a business environment, to be applied at different scales in the use of IoT.

An example of this can be seen in the case of monitoring patients remotely through sensors (Tele-Healt, Tele-Care), such as glucose sensors, blood pressure, etc. Where the knowledge of the patient's condition is delivered efficiently and permanently. This is a clear way to develop a business model based on semantic representation taking into account a sustainable business form for an IoT platform [1].

The search to improve the processes by which information is obtained through the management of large volumes of information is present in the context of the management of data generated by ubiquitous systems, which from the point of view of the so-called actors are understood as Internet of Things (IoT). It is precisely this need that becomes important in the deployment of these systems based on IoT, which makes it necessary to know how to handle the information collected.

This is applied to the management of growth parameters given in crops, where there is a need to know and control them. This involves the management of a large amount of information collected throughout their processes. Hence, defining ways to develop semantic models that allow the implementation of ontologies that offer an improved management of the management of the collected data.

The generation of ontologies applied to the processing of data in agriculture after they are generated by means of IoT's environments, proposes to find ways to improve the way in which the stored information is used, allowing those involved in the agricultural processes a better deposition in each one of its phases [9]. The use of these models increases the efficiency of the consultations.

The use of middleware solutions to manage a multitude of sensors in IoT infrastructure deployments improves the way information is collected more efficiently. The use of this management through semantic context systems such as CASSARAM presents an example of how semantic representation models can be applied in an environment that seeks to achieve simple sensor management. This middleware solution proposal uses an inheritance-based comparative filtering technique based on priorities, helping the indexing of sensors faster and in different locations [10].

The constant increase in the volume of data collected makes it necessary, in effect, the development of models that allow the proper management of this information and its subsequent use. At present, important efforts are being made to achieve better models of sensor representation.

3 Methodology for Ontology Modeling

To develop an ontology requires the use of some methodology that facilitates the design process. This section describes existing methodologies in the literature, in which it is observed that different methodologies are composed of similar phases and activities. Figure 1 presents the generic life cycle for the development of ontologies.

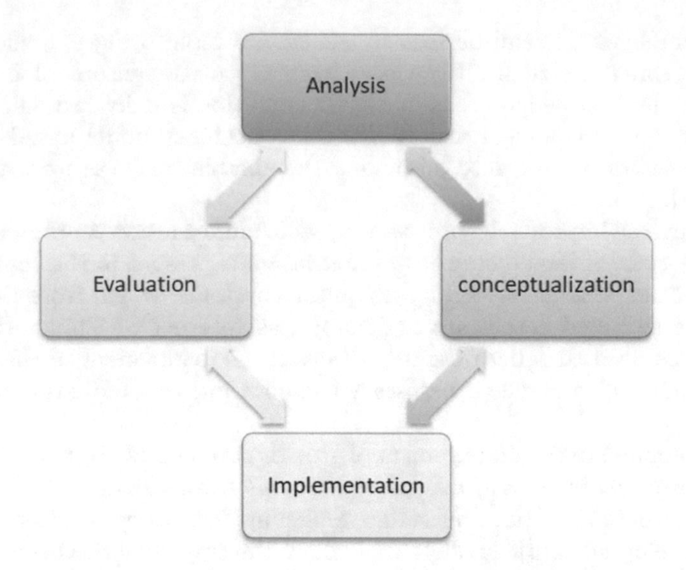

Fig. 1. Life cycle for the development of ontologies.

The general process for the development of ontology consists of four phases:

Analysis Phase. This phase contemplates the purpose of the ontology that will allow the definition of its scope, domain and reuse.

Conceptualization Phase. In this stage a conceptual model is defined that describes the ontology to be developed and at the same time must comply with the specification obtained in the previous phase. Conceptualization consists of organizing and converting the informal perception of a domain into a semi-formal specification, using intermediate representations such as diagrams or tables that can be understood by domain experts and ontology developers.

Implementation. Refers to the explicit representation in a formal language of the knowledge acquired in the previous phase. That is to say, it implies the generation of computational models according to the syntax of languages of formal representation such as RDF/S, OWL and FLogic (Frame Logic), among others.

Evaluation. It consists of making a technical judgment of the ontology, associated software environment and documentation, with respect to a frame of reference during each phase and between the phases of its life cycle. The framework can include requirements specifications, competency questions, and the real world.

Although numerous methodologies exist, the following will be described in this report: Skeletal [11], Methontology [12], On-To-Knowledge [13] and NeOn

[14]. The latter is the one adopted for the development of the ontology proposed in this work.

4 Proposed Model

This approach proposes a conceptual model of an internet system of things for monitoring environmental variables in crops. The model is characterized by eleven concepts that will be described below:

Measurements this concept represents all the measurements obtained by the sensors distributed in a crop. The purpose is to obtain information on environmental variables such as CO2, PH, solar radiation, humidity, temperature, among others. This type of concept is important to model not only because the information is extracted from the sensor data, but also a semantic significance is obtained from these measures, such as the interpretation of the search for radiation-sensitive crops.

Devices describes the hardware and software characteristics required to extract information from the monitoring system, ranging from sensors that extract information from environmental variables, actuators that execute actions such as irrigating a crop or servers responsible for hosting the extracted data, among others. In this concept, the COBRA-Device ontology was reused to extract relationships and concepts about the devices.

User includes the information of people who intervene in the system. Users have roles that allow them to have privileges in the system, such as administrator, developer, client, among others. In this concept the FOAF ontology was reused to extract the properties widely known as names, surnames, e-mail, cell phone number, web page, among others.

Event represents the events that are presented in a monitoring system, this is done through notifications. These events can be executed by the detection of some variables such as humidity level. Also it can be product of the intervals of time for the sensing of the environmental variables or anomalies in the same system.

Station represents the concepts related to monitoring stations, these can be of any nature such as weather stations, river flows and obviously the monitoring of agricultural crops.

Time describes the notion of real time for the measurement of environmental variables. That is to say, it represents the time in function to the execution of routines for the detection and data collection of the system. The time dimension is important, because it determines the execution time for the taking of a reading or activation of an actuator at the current moment or the programmed time to perform it at a future moment.

Location describes the location of monitoring stations, including sensors, actuators, users and devices in general. This information can be obtained through the devices.

Crops represents the concepts related to the variety of crops that can be monitored, also describes the characteristics of the crops among others.

In Fig. 2, you can see the general ontology with its respective concepts.

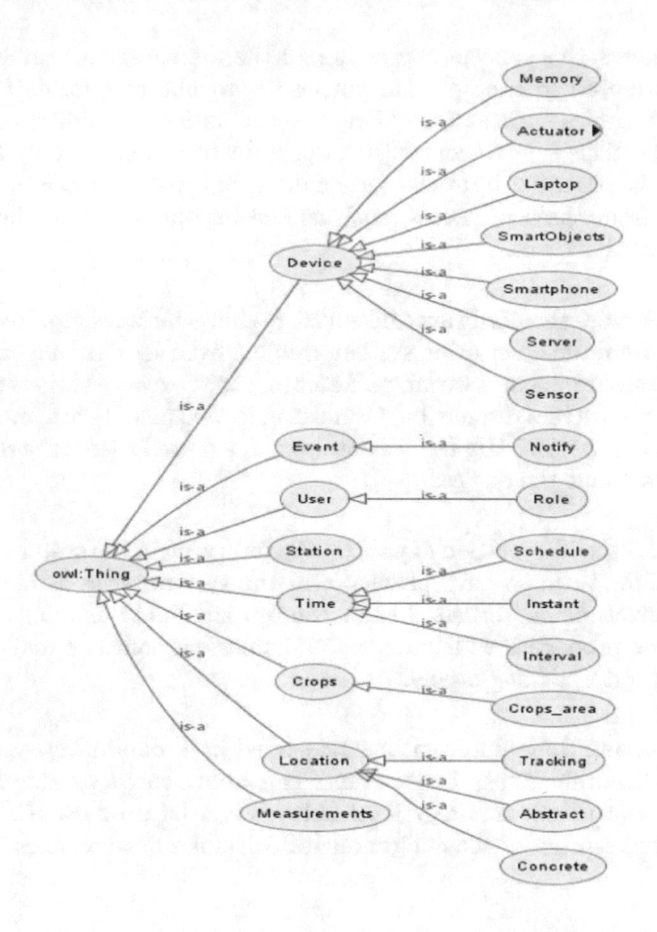

Fig. 2. General ontology for crop monitoring.

The Table 1 describes some general relationships based on the concepts described above.

Table 1. General relations of the ontology for crop monitoring.

Relation	Description	Domain and range restrictions
stationLocated	All monitoring stations that are located in one location	Domain: Station Range: Location
deviceLocated	All devices are located in one Location	Domain: Device Range: Location
stationHasDevice	All stations have a device that will allow them to monitor	Domain: Station Range: Device
Itsallows	A sensor allows one or more readings of variables	Domain: Device Range: Measurements
hasTime	Every station has a set time to get readings	Domain: Station Range: Time

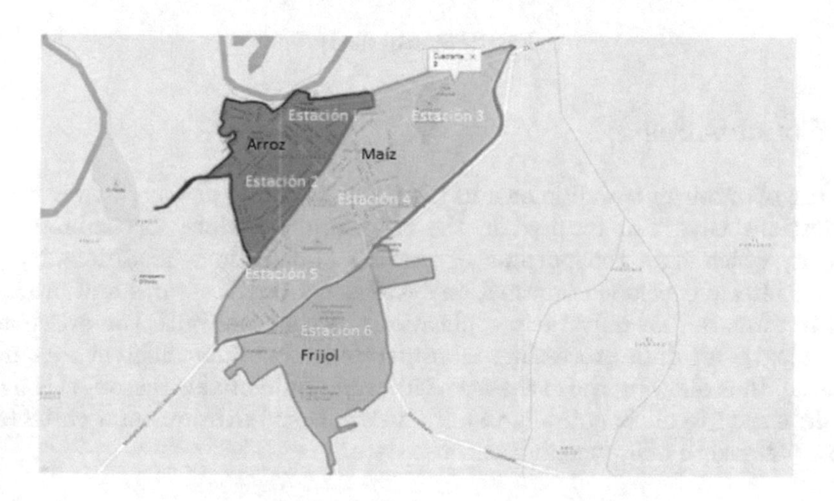

Fig. 3. Monitoring stations.

5 Case Study

To apply this model, he designed a system that allows the monitoring of environmental variables for crops at the University of Cordoba. Stations were taken from three types of crops (rice, corn and beans), as can be seen in Fig. 3.

The following consultation was established in Sparql to obtain the measurement of the environmental parameters of the three crops in all seasons, such as humidity level and UV radiation, as can be seen in Fig. 4.

```
PREFIX owl: <http://www.w3.org/2002/07/owl#>
PREFIX rdf: <http://www.w3.org/1999/02/22-rdf-syntax-ns#>
PREFIX rdfs: <http://www.w3.org/2000/01/rdf-schema#>
PREFIX IoTOntology <:http://websemantic.org/iot/station_mesaurements#>
SELECT ?schedule, measurement, ?station, ?location  WHERE {
        ?schedule
IotOntolohy:hasDateLectureBegining\""+dateBegin+"\"^^xsd:date
?schedule  IotOntolohy:hasTimeLectureBegining\""+timeBegin+"\"^^xsd:time
?schedule IotOntolohy:hasmeasuremente \""+type_measure+"\"^^xsd:string
?schedule IotOntolohy: stationLocated ?station
?schedule IotOntolohy: deviceLocated ?device

}
```

Fig. 4. Sparql query.

6 Conclusions

The use of ontology is well known in the world of semantic web, however these applications have also resulted in the description of data for heterogeneous systems, which have the purpose of sharing information. Traditionally, software architects developed information systems and architectures and processed data in formats that only their applications recognized. With the articulation of ontologies for data processing, interoperability between different systems is achieved, thus allowing the exchange of information. For this reason, this model was developed to allow not only to obtain data from environmental variables of crops, but also to offer meaning to those data.

References

1. Glova, J., Sabol, T., Vajda, V.: Business models for the Internet of Things environment. Procedia Econ. Finance **15**, 1122–1129 (2014)
2. Silega-Martínez, N., Macías-Hernández, D., Matos, Y., Febles, J.P.: Framework based on MDA and ontology for the representation and validation of components model. Revista Cubana de Ciencias Informáticas **8**(2), 85–101 (2014)
3. Ganz, F., Barnaghi, P., Carrez, F.: Automated semantic knowledge acquisition from sensor data. IEEE Syst. J. **10**(3), 1214–1225 (2016)
4. Piedra, N., Suárez, J.P.: Hacia la Interoperabilidad Semántica para el Manejo Inteligente y Sostenible de Territorios de Alta Biodiversidad usando SmartLand-LD. Revista Ibérica de Sistemas e Tecnologias de Informaão (26), 104–121 (2018)
5. Gómez, J., Oviedo, B., Zhuma, E.: Patient monitoring system based on Internet of Things. Procedia Comput. Sci. **83**, 90–97 (2016)
6. Gómez, J.E., Huete, J.F., Hernandez, V.L.: A contextualized system for supporting active learning. IEEE Trans. Learn. Technol. **9**(2), 196–202 (2016)
7. Compton, M., Barnaghi, P., Bermudez, L., GarcíA-Castro, R., Corcho, O., Cox, S., Huang, V.: The SSN ontology of the W3C semantic sensor network incubator group. Web Semant. **17**, 25–32 (2012). Science, services and agents on the World Wide Web

8. Gordijn, J.: E-Business value modelling using the e3 -value ontology, pp. 101–111 (2004)
9. Siquan, H., Wang, H., She, C., Wang, J.: CCTA 2010. Part I, IFIP AICT, vol. 344, pp. 131–137 (2011)
10. Perera, C., Zaslavsky, A., Christen, P., Compton, M., Georgakopoulos, D.: Context-aware sensor search, selection and ranking model for internet of things middleware. In: 2013 IEEE 14th International Conference on Mobile Data Management, vol. 1, pp. 314–322. IEEE, June 2013
11. Swartout, B., Patil, R., Knight, K., Russ, T.: Toward distributed use of large-scale ontologies. In: Proceedings of the Tenth Workshop on Knowledge Acquisition for Knowledge-Based Systems, pp. 138–148, November 1996
12. Fernández-López, M., Gómez-Pérez, A., Juristo, N.: Methontology: from ontological art towards ontological engineering (1997)
13. Sure, Y., Staab, S., Studer, R.: On-to-knowledge methodology (OTKM). In: Handbook on Ontologies, pp. 117–132. Springer, Berlin (2004)
14. Suárez-Figueroa, M.C., Gómez-Pérez, A., Fernández-López, M.: The NeOn methodology for ontology engineering. In: Ontology Engineering in a Networked World, pp. 9–34. Springer, Berlin (2012)
15. Gyrard, A., Serrano, M., Atemezing, G.A.: Semantic web methodologies, best practices and ontology engineering applied to Internet of Things. In: 2015 IEEE 2nd World Forum on Internet of Things (WF-IoT), pp. 412–417. IEEE, December 2015
16. Urbano-Molano, F.A.: Wireless sensor networks applied to optimization in precision agriculture for coffee crops in Colombia. J. Ciencia e Ingenieria **5**(1), 46–52 (2013)
17. Wang, W., De, S., Toenjes, R., Reetz, E., Moessner, K.: A comprehensive ontology for knowledge representation in the Internet of Things. In: 2012 IEEE 11th International Conference on Trust, Security and Privacy in Computing and Communications, pp. 1793–1798. IEEE, June 2012
18. De, S., Barnaghi, P., Bauer, M., Meissner, S.: Service modelling for the Internet of Things. In: 2011 Federated Conference on Computer Science and Information Systems (FedCSIS), pp. 949–955. IEEE, September 2011
19. Dueñas, D.E.M., Gallegos, C.G.M., Acosta, R.A.M., Fernández, R.L., Urquiza, D.E.P., Saltos, M.B.G.: Theoretical foundations of web 2.0 for teaching in higher education. Revista de Ciencias Médicas Cienfuegos **15**, 190–196 (2017)
20. Bermudez-Edo, M., Elsaleh, T., Barnaghi, P., Taylor, K.: IoT-Lite: a lightweight semantic model for the Internet of Things. In: 2016 Intl IEEE Conferences on Ubiquitous Intelligence Computing, Advanced and Trusted Computing, Scalable Computing and Communications, Cloud and Big Data Computing, Internet of People, and Smart World Congress (UIC/ATC/ScalCom/CBDCom/IoP/SmartWorld), pp. 90–97. IEEE, July 2016

Acquiring, Monitoring, and Recording Data Based on the Industrie 4.0 Standard Geared Toward the Maca Drying Process

Gianmarco Nagaro(✉), Abel Koc-Lem, Leonardo Vinces, Julio Ronceros, and Gustavo Mesones

Universidad Peruana de Ciencias Aplicadas, Av. Prolongación Primavera 2390, Santiago de Surco, Lima, Peru
{u201312024,u201210259,pcmajron}@upc.edu.pe, {leonardo.vinces,gustavo.mesones}@upc.pe

Abstract. Maca is a natural product that is characterized as a hormonal regulator and food supplement. Produced only in regions with an altitude higher than 3,800 m, which creates great opportunity for its export. For the maca to be exportable it is necessary to apply drying procedures under the condition that it conserves its nutritional properties, whose drying time varies between 12 h and 6 days, according to the current standards. That is, the collection of real-time process data to monitor and control operations is by local network, but not connected in the cloud (IoT) according to Industrie 4.0 standard. Herein, we propose a system is configured by an S7-1200 PLC within a local network and a server, with remote access with user restriction, which allows sensitive process variables, such as absorption temperature and airflow, to be controlled remotely and in real time, such that they would avoid the alteration of the properties of the maca in the drying process. In addition, data from the process can be visualized simultaneously by telemetry and curves of drying the maca from the Web to be able to apply some type of intervention if necessary. Real-time graphs were obtained in the web server and after analyzing them, it was concluded that the data sent to the server are capable of being processed in order to generate a set of results and analyze the information obtained that will help future regimes of drying.

Keywords: Industrie 4.0 · IIoT · Maca · Absorption temperature · Drying process · Web server

1 Introduction

Owing to its diverse orography and its microclimate system, Peru is rich in agricultural wealth. Maca is a natural product that strengthens human immune systems owing to its antioxidant properties [1] and its production is concentrated in the central Peruvian region at altitudes exceeding 3,800 m above sea level. In approximately 8,000 Ha of agricultural land available, there are ten thousand farmers engaged in this activity with an annual production of 50 thousand tons [2], but only 5% of it is exported as flour,

M. Botto-Tobar et al. (Eds.): ICAETT 2019, AISC 1066, pp. 42–53, 2020.
https://doi.org/10.1007/978-3-030-32022-5_5

mainly to the North America, Japan, and China. Energy bars, pills or infusions are also manufactured with maca.

Regarding its production, dry maca producers must comply with several export standards, which ensure that the product retains its important properties during its preparation process. For example, if maca dries at an inadequate temperature, its proteins and minerals may be destroyed, decreasing its nutritional properties. Therefore, maca producers seek to prepare their products in the shortest time possible and with the largest possible amount of maca deemed suitable for export.

In the research "Design, Development and Control of a Portable Laboratory for the Chili Drying Process Study" [3], a low volume drying system is observed where the data is processed by an HMI attached to the machine, these are sorted and visualized by an existing program embedded in the same drying machine. In our case, we have the ability to send this data via Web, to be processed in Azure, Anaconda Cloud or another system that allows the storage of data processing code. On the other hand, in the work "Obtencion y Modelado de Isotermas de Sorcion de Caramelos "Gummy" de Batata" [4], there is a precise way to process the data using mathematical models and get properties of the material to be dried but not it has the means to make experiments in different temperatures controlled by a system that ensures stability in the flow and temperature, in addition to having a temperature range lower than our proposed system (up to 30 °C against 55 °C).

The dryers currently used for drying small volumes of maca vary in their drying and control method and in the way data are acquired and viewed. Herein, we will consider two methods, sun-and oven-drying.

Solar dryers [5] are based on the use of solar radiation to evaporate the water in the maca, which dries the product slowly and non-aggressively. This method guarantees that it will dry at a safe temperature keeping its nutritional properties intact. This method is highly vulnerable to pests and predatory animals; however, this method is still widely used and has been improved using protection glass and safe environments to isolate the product from the environment but without considerably improving its drying time, i.e., 2–6 days.

On the contrary, oven drying [1] rapidly increases product temperature so that it loses humidity faster than with solar radiation. Drying times are reduced to just 12 h, considerably increasing efficiencies. However, this method is aggressive since there is no automatic or feedback-based control. This method does not ensure that the product is dried at a safe temperature and consequently, a significant number of production batches could not be suitable for exporting.

As these drying systems require measuring masses, temperatures, humidity, and drying control air flow, a visual procedure was developed to monitor the information obtained by the sensors in real time based on the Industrie 4.0 standard, whose action axes are shown in Fig. 1.

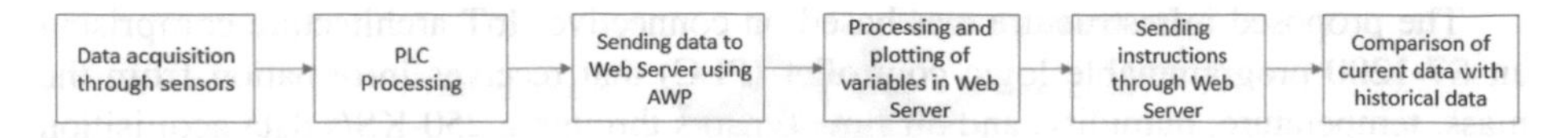

Fig. 6. Block diagram for the proposed system

2 Description of the Proposed System

2.1 Data Acquisition Through Sensors

In this stage, the measured values are received by each sensor. These values are sent to the analog input modules that capture the signal and store it in their respective memory spaces.

2.2 PLC Processing

Plant data are processed, displayed in human–machine interface (HMI) and used in proportional integral derivative (PID) controller of the actuators and the variables of interest.

2.3 Sending Data to the Web Server

The data is sent to the web server using automation web programming (AWP) commands so that the PLC may associate the variables from the HTML and Java scripts with the internal PLC variables.

2.4 Processing and Plotting of Variables in the Web Server

The variables are processed on the web server and then stored and plotted. For these purposes, JavaScript and Canvas library are used. These charts help users determine the status of the drying process and view the performance of the actuators in real time.

2.5 Sending Instructions Through Web Server

Users may send instructions to modify the drying regime in the middle of the process. Therefore, the process may be modified from any point in the cloud as long as the PLC is connected to the Internet.

2.6 Comparing of Current and Previous Data

The web server compares current parameters against parameters from previous runs, thus reporting past results to users and determines whether they are desirable for the current process.

3 Implementation

To implement this system, an S7-1200 PLC was used, which was programmed to receive the data from the analog signals. Two four-input and two 4AI/2AO analog-output models were used, with three humidity sensors, three temperature sensors, and one flow sensor. Therefore, there were seven analog PLC inputs in total. Then, the PLC was configured to recognize a folder called 'user files' and a file called 'index.htm' as the home page of the web server. This folder contains all the pages and libraries comprising web server. After this configuration, the page is programmed to receive and send data from the PLC using AWP commands, which are used by the PLC to recognize these variables as read/write only. Then, asynchronous communication is programmed using asynchronous JavaScript and XML (AJAX). This way, data may be sent and received without updating the entire page every time. Figure 7 shows that each time an event occurs, an HTML request is generated, processed on the server, and sent using JavaScript without updating the whole page but with only the value shown (See Fig. 7).

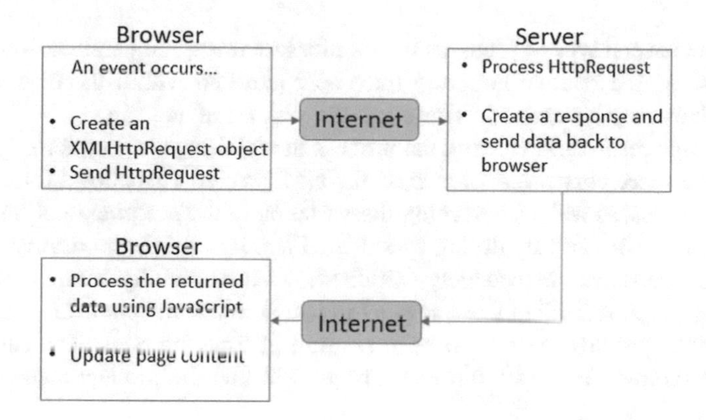

Fig. 7. Asynchronous communication through AJAX [1]

Then, values are plotted in real time using Canvas.js, which graphs a value every second and may display up to 30 min of data at a time, or the entire process at once depending on the user.

Finally, PID variables are associated to the values entered by the user within the page. This way, users may change the drying rate depending on the results from the sensors and graphics presented.

4 Maca Drying Process and Data Reading

A 304 stainless-steel dryer was designed to protect the product from pests, birds, and rodents, among others. The drying process uses a web server embedded in the S7-1200 controller. The distribution of the sensors is shown in Fig. 8.

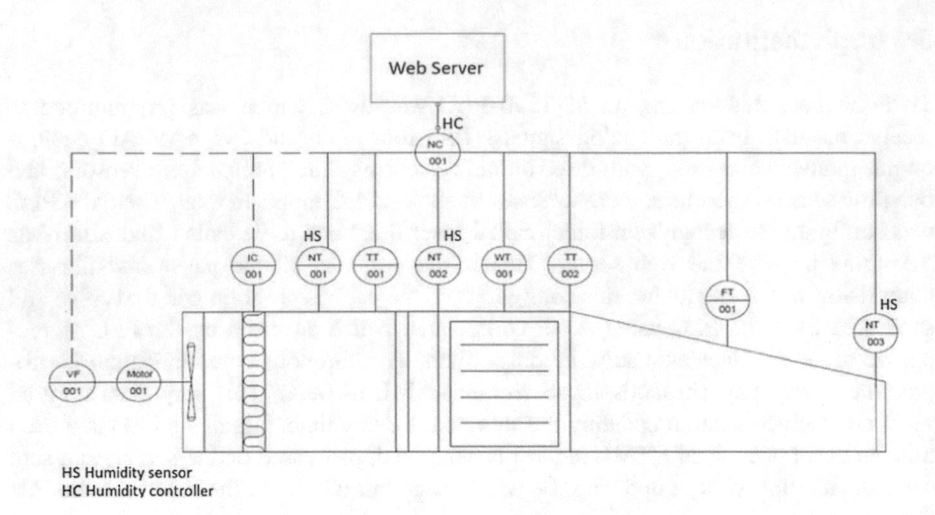

Fig. 8. Instrumentation diagram inside the maca drying plant

There are several types of sensors which measure mass, temperature, humidity, and flow. This way, the relative humidity may be controlled within the drying duct and actuator values may be updated. These are the resistance and fan.

For the final record and viewing the process in real time, the product's drying curve and its drying speed curve were plotted on the HMI screen. This allows for a study after the drying process, as well as observing the variation in the percentage of humidity that the product is losing and its drying speed. In Figs. 9 and 10, the drying and drying speed curves are shown, respectively, which shows the humidity content through time in the drying process. Each of these stages has an automatically controlled flow rate. To plot this curve, product mass must be measured in time intervals. The sampling frequency is 1 sample/s in a 150-min test. The weight that the product loses is the same

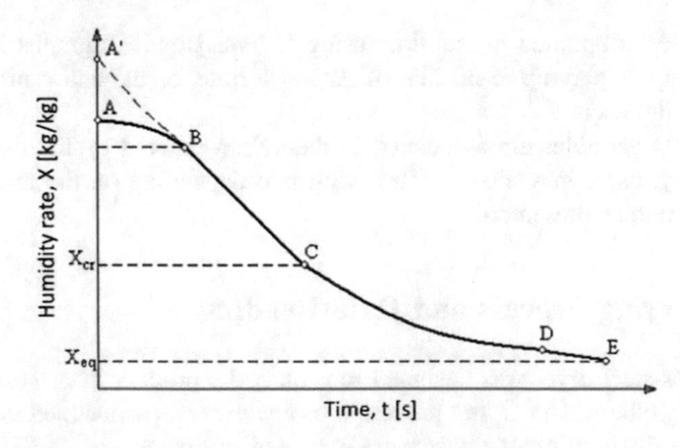

Fig. 9. Drying curve [10]

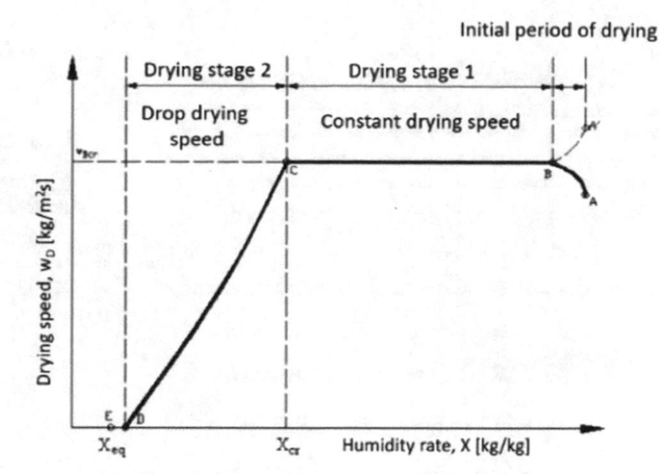

Fig. 10. Drying speed curve [10]

amount of H_2O that evaporates. defined by Eq. (1) and the product humidity percentage is defined by Eq. (2). These equations are required to plot the drying curve.

$$w_i - w_t = a_i - a_t \tag{1}$$

where w_i and w_t is the initial mass of the product and the product mass measured at any moment in time (both in Kg), respectively, a_i and a_t is the initial amount of H_2O and that of H_2O measured at any moment in time (both in kg).

$$H_t = \frac{a_t}{w_t} \times 100 \tag{2}$$

where H_t is the product's humidity percentage.

From Eq. (2), a_t may be cleared and replaced in Eq. (1). Hence, Eq. (3) may be used to determine the humidity percentage at any moment in time:

$$H_t = \frac{a_i + w_t - w_i}{w_t} \times 100 \tag{3}$$

5 Results and Information Display in the Web Server Interface

The interface may be quickly customized as per user requirements using cascading style sheets, as shown in Fig. 11.

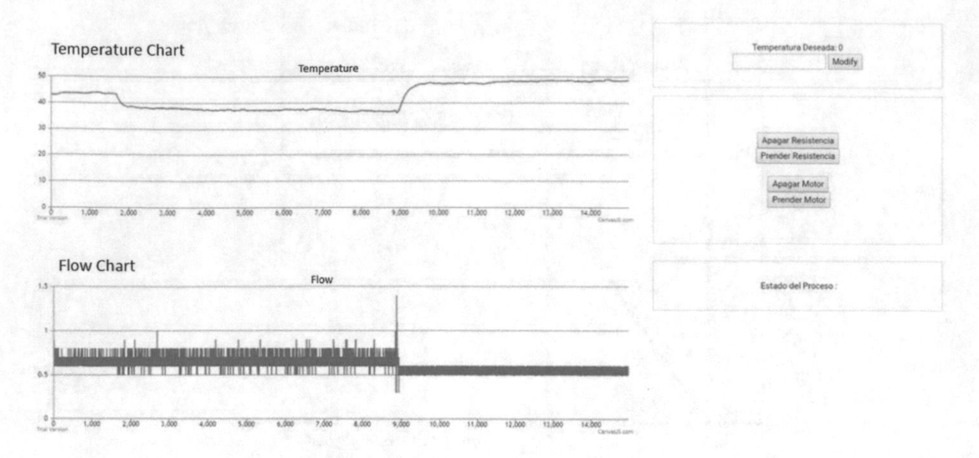

Fig. 11. Web server graphic interface

As observed from the drying process control, the drying temperature may be set and the motor or the electrical resistances that provides heat to the machine may be activated. Additionally, the temperature and air flow variables per sample unit per second are presented in real time.

Two drying tests performed at 45 °C with the respective information captured by the web server in real time are shown in Figs. 12, 13, 14 and 15, where the presented variables are current weight and humidity lost by the maca.

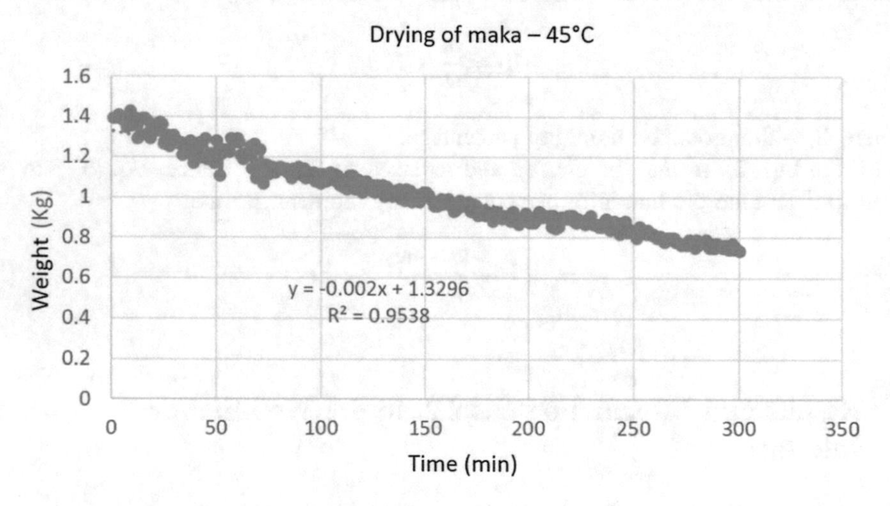

Fig. 12. First drying test.

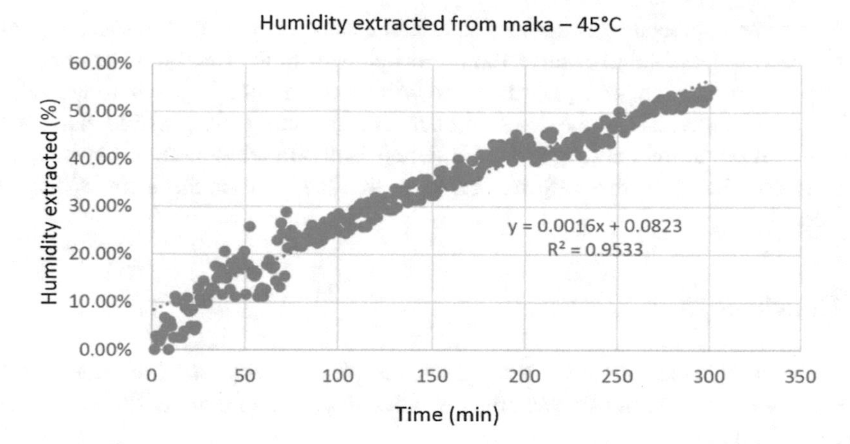

Fig. 13. First test that measures lost humidity.

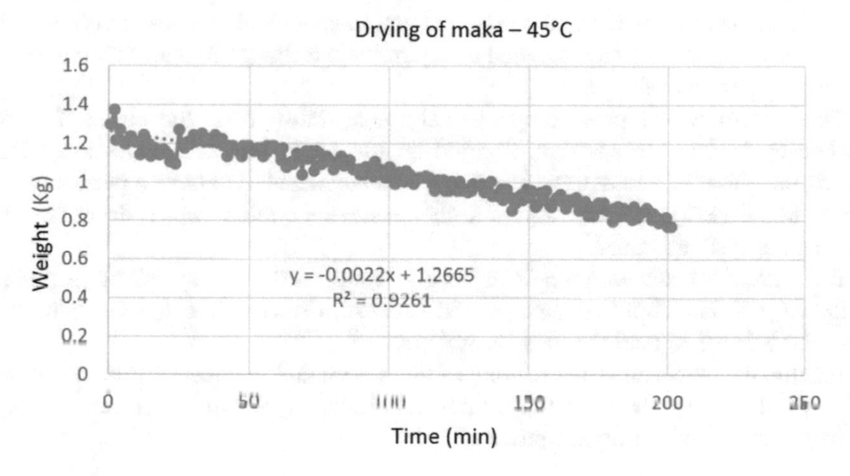

Fig. 14. Second drying test.

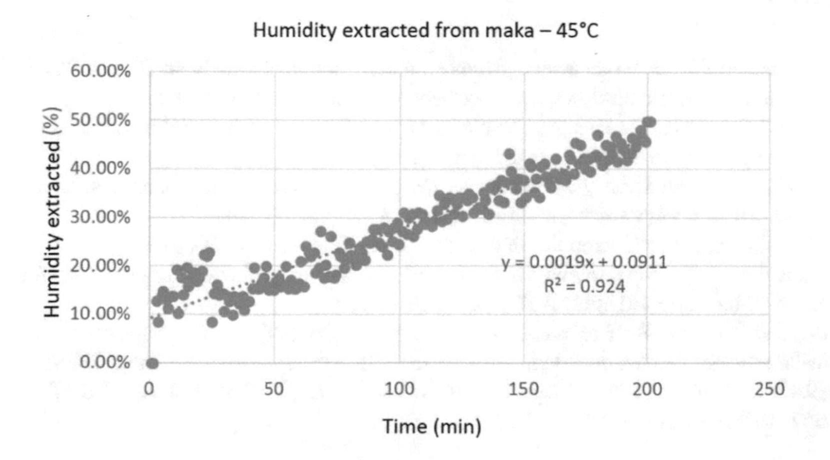

Fig. 15. Second test that measures lost humidity.

As we can appreciate, the sensors have had a sensing cycle that is short enough to produce drying graphs and observe the linear behavior in the first hours of drying, and it is expected that in longer drying times we will reach an asymptote where we will find the isotherm temperature of the dried product. This is completely possible thanks to the resolution of our system and by storing different isotherm temperatures a system could be generated that uses that information to notify the user of the state of the dried product.

6 Conclusions

The web server remained active throughout the process without losing data and was accessible from any point using Internet, which allowed us to monitor the process from other access points.

Additionally, the web server has a built-in plotter and options for changing the temperature limit of the process and for stopping and restarting it remotely.

Besides, the information was assessed to graphically obtain the maca drying curves, estimate the amount of water lost, and to determine the change in humidity reported for the product per unit of time.

The system would present operational vulnerability from the cloud if it were attacked by hackers. In addition, it could be subject to Internet connection failures, causing data loss for a long time. For that it is recommended to have a permanent data storage in the PLC and a robust system that ensures an uninterrupted drying as many times as the system is used.

If a problem occurs with the dryer infrastructure, such as a tray fall or an electrical shutdown, the data could become invalid because a reset of the weight value could cause the balance to read the data incorrectly.

Finally, the precision of the reading of the weight and the operative integrity of the trays, will depend on the mechanical vibrations in the dryer and is a current priority in the improvement of the drying process.

References

1. Ríos, A.: Metabolismo de ácidos grasos en maca durante el secado en Horno. PUCP (2018)
2. Ministerio de Agricultura www.minagri.gob.pe. Accessed 27 Feb 2019
3. Guzman-Valdivia, C.: Design, Development and Control of a Portable Laboratory for the Chili Drying Process Study. Elsevier, Mexico (2016)
4. Cerviño, V.: Obtención y Modelado de Isotermas de Sorción de Caramelos "Gummy" de Batata. III Congreso argentino de ingeniería. Argentina (2016)
5. Martínez, A.: Construcción de un Secador Solar. IPL México (2007)
6. Industrie 4.0, https://www.siemens.com/customer-magazine/en/home/industry/one-step-closer-to-industrie-4-0.html. Accessed 24 Feb 2019
7. Boyes, H., Hallac, B.: The industrial internet of things (IIoT): an analysis framework, https://reader.elsevier.com/reader/sd/pii/S0166361517307285?token=10DE60A51676883ABDA389A80D17966B2A3F056BCF1CB64A451194A2F47213074DDB294A7DDCAB1083D12B52113530CC. Accessed 11 Mar 2019

8. Anticona, O.S., Ramos, J.: Implementación de un sistema de Supervisión, Control y Adquisición de Datos a Distancia en el Secador de Bandejas para Determinar la Influencia de la Temperatura en el Tiempo de Secado de la Papa. UNT Trujillo (2005)
9. Hermann, H., Pentek, T., Otto, B.: Design Principles for Industrie 4.0 Scenarios. In: 2016 49th Hawaii International Conference on System Sciences (HICSS), Koloa (2016)
10. Dávila Nava, J.R.: Estudio experimental del efecto de la porosidad de partículas sobre el proceso de secado en un lecho fluidizado a vacío empleando aire, Puebla: Universidad de las Américas Puebla, pp. 51–73 (2004)

Design and Tests to Implement Hyperconvergence into a DataCenter: Preliminary Results

Edgar Maya-Olalla[1], Mauricio Dominguez-Limaico[1], Santiago Meneses-Narvaez[1(✉)], Paul D. Rosero-Montalvo[1,2], Sandra Narvaez-Pupiales[1], Marcelo Zambrano Vizuete[1], and Diego H. Peluffo-Ordóñez[1,3]

[1] Universidad Técnica del Norte, Ibarra, Ecuador
sjmenesesn@utn.edu.ec

[2] Instituto Tecnológico Superior 17 de Julio, Ibarra, Ecuador

[3] Universidad Yachay Tech, Urcuquí, Ecuador

Abstract. Hyperconvergence is a new technological trend that integrates and centralizes the functions of network, storage and computing in a single infrastructure, facilitating the administration, operability and scalability of a Data Center as a whole, benets that do not provide an architecture of traditional network or virtualization-specic technologies. This research based on qualitative and experimental methods suggests a model of Implementation of a HyperConvergent Architecture for the management of the Data Center of the Universidad Técnica del Norte, as a competitive and high-performance Open Source alternative for the integration of physical and virtual components. The suggested deployment model is based on the virtualization platform Proxmox VE, CEPH (Storage Software Platform), vSwitch (network scheme) and KVM (equipment virtualization). It includes a centralized domain and it provides a 99.88% availability rate making it in total harmony with functionalities requiring high availability. The results show the simplicity of the system: efficient execution of all applications, migrations of virtual machines from node to node, inactivity times between 50.3 ms and 53 ms, processing acceleration providing agility to IT operations without forgetting that its implementation and its start-up times are relatively low.

Keywords: Hyperconvergence · Proxmox · Data Center · Availability

1 Introduction

Nowadays, considering the large amount of non-integrated equipments hosted in Data Centers as well as the huge pressure to respond to IT demands associated to manual and complex processes, the versatility and power required to satisfy the most demanding requirements which are exponentially growing are not delivered. As a consequence, many companies Data Centers have become

M. Botto-Tobar et al. (Eds.): ICAETT 2019, AISC 1066, pp. 54–66, 2020.
https://doi.org/10.1007/978-3-030-32022-5_6

non-flexible environments in terms of speed, load and cost as they try to satisfy present and future requirements [1].

In the effort to tackle those challenges, Hyperconvergent systems may appear as a solid option. This new technological trend offers high integration on the physical and virtual components of an infrastructure, linking servers, storage and network to boost IT performance at a reasonable costs. As a result, applications are processed faster even by simple system. It is important to note that these infrastructure components are not only grouped together but also fully integrated, synchronized and visible for the administrator through a hypervisor [2]. Gartner predicts that up to 20% of critical applications currently implemented in the IT infrastructure of three levels known as converging will make the transition to Hyperconverging infrastructure by 2020 [3]. Considering the latter, the market and the future of this technology is promising.

It should be noted that the potential of HCI (Hyperconvergent Infrastructure) has created a lot of interest among IT professionals around the world, including in Latin America, with strong feedbacks from Brazil, Colombia as well as Chile and Argentina [5]. In 2017, Hyperconvergence gained more support not only in Data Centers, but also within the IT industry. Even though companies tend to be slow to adopt changes in their infrastructure, administrators, managers and executives know that Hyperconvergence is the future and they are already prospecting its possibilities and capacities [4].

Hyperconvergent systems take advantage of intelligence defined by software as shown in Fig. 1 enabling processing in a unique server platform thus eliminating inefficiencies, accelerating processing and providing agility to IT operations while implementation and setting times remain relatively low [5].

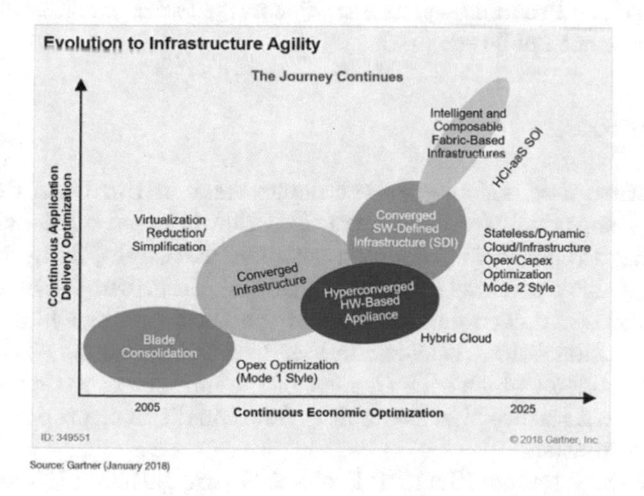

Fig. 1. Hyperconvergence evolution [5]

PROXMOX VE platform and Red Hat Hyperconverged Infrastructure appear as interesting products among open source software solutions for Hyper-convergent infrastructure. They provide high computer integration, storage and network components, backup facilitation, restoration and recovery front to failure. These services are provided through an interface or stable centralized web hypervisor. It is even proven for business solutions using CLI (Command-line Interface), log files and Linux kernel options [6,7].

The implementation of virtualization platforms in data centers is becoming more frequent, this is due to the multiple benefits it offers. Open source virtualization platforms are quite attractive for different applications, in the present document the study and analysis of Proxmox VE are done as virtualization technology.

In [24] two virtualization platforms are analyzed, Open node and Proxmox VE, where the results obtained, Proxmox has certain advantages in terms of performance; The parameters evaluated in this document are jitter, throughput and delay. On the other hand, it also highlights parameters such as management, ease of installation and configuration of Proxmox VE, which has made this option more attractive and known.

In another study [25] the performance of Proxmox VE together with Laravel as an online education platform is analyzed, Proxmox VE is used for the administration of virtual machines, which can be managed in different hardware and network configurations among themselves. performs performance tests in terms of response times to access a website that runs on the virtualization platform, reaching excellent results with 500 concurrent clients.

There are different investigations related to virtualization using Proxmox [26, 27], where they evaluate different virtualization technologies, but a virtualization technology such as Proxmox versus a traditional network architecture as such in a data center is not analyzed.

2 Methodology

The architecture used for the present deployment in the Data Center of the Universidad Tecnica del Norte in Ibarra, Ecuador, is based on the virtualization platform of the Proxmox VE, mounted with CEPH (Ceph File System) as storage system. In this case three servers are involved, described in Table 1, enabling a complete and centralized management of the Data Center while offering high availability functionalities. As shown in Fig. 2, at the beginning, the architecture is a traditional one in which each element of the infrastructure is managed individually. It doesnt feature high availability functionalities or proper management of backup information.

A total of three servers (Svr) HP Proliant G9 are gathered to create the cluster. Proxmox VE is installed on each server. Each physical server is added to the cluster forming as such a virtual architecture in which three servers are managed as if they were only one physical equipment. Table 1 details the characteristics of the servers involved.

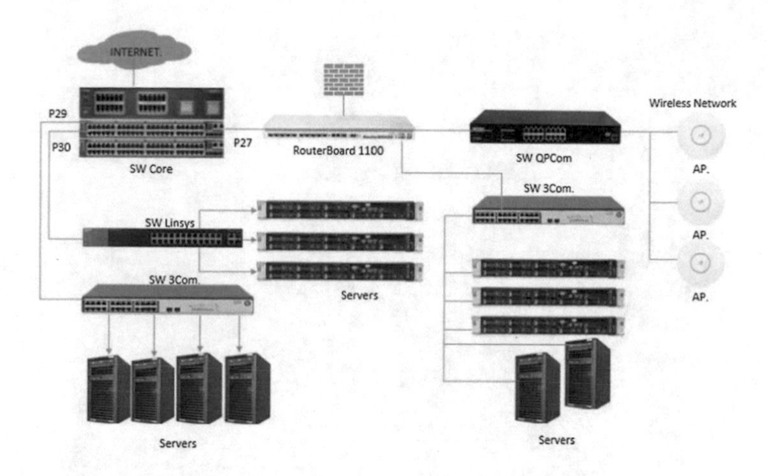

Fig. 2. Data center traditional architecture at Universidad Técnica del Norte.

Table 1. UTN server's characteristics

Characteristic	Svr1-HP	Svr2-HP	Svr3-HP
Ram	32 GB	32 GB	32 GB
HDD	2 × 600 GB	2 × 600 GB	2 × 600 GB
CPUs	16	16	16

The cluster provides a virtualized machine with the following features: 48CPUs, 96 GB of RAM, and 2.4 TB of storage, resulting from adding 600G to each node (that is 1.8T) plus 600G of storage shared by CEPH making 2.4 TB total of storage.

The shared storage system CEPH is used by this architecture. This system enables creating shared virtual disks (OSDs) which replicate information from active VMs (Virtual Machines) on the platform. One of the critical part of the CEPH storage configuration is the configuration of Ceph Pool because it is where the information replication numbers in the OSD are defined.

Figure 3 describes some terminology to understand how the CEPH storage operates [8].

- Object (obj): The object is the smallest unit of storage of information in Ceph.
- Placement Group (PG): They are a series of objects that are clustered and replicated in several OSDs.
- Pool: They are groups of several PG where the Obj Replica numbers are configured.

The following values are defined from Fig. 4: Size is set to 3 which is the number of replicas to be maintained, Min Size is assigned 2 which is the minimum

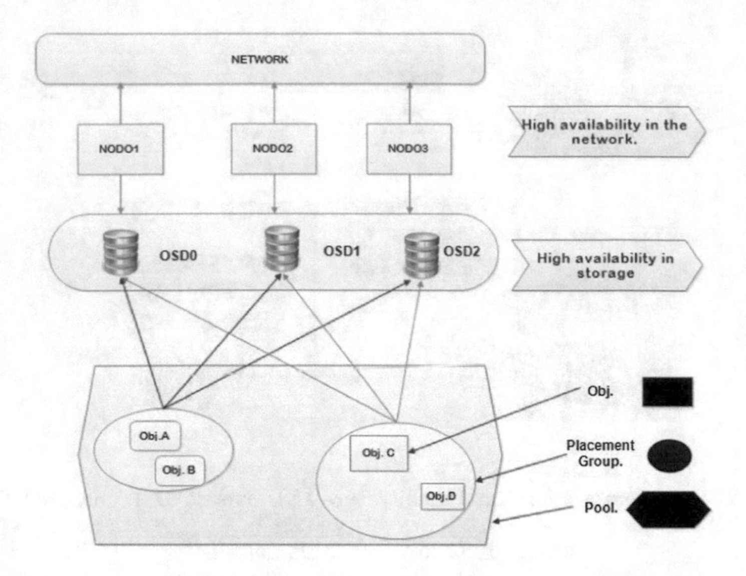

Fig. 3. Objects, PG, Pool

number of replicas, and pgnum is set to 100. According to the following Eq. 1, TPG is calculated [8].

$$TPG = \frac{\#OSD * 100}{\#REPLICAS} \tag{1}$$

where:

- $OSDNumber = 3$
- $NumberofREPLICAS = 2$
- Total Placement Groups $(TPG) = 100$

After setting up CEPH in the cluster, virtual machines high availability and hot-plug migrations functions are available. The Hyperconvergence goal is met. Figure 5 describes the implementation flow diagram [9].

The final implementation of Hyperconvergent infrastructures provides a more centralized system and easier system to operate. Figure 6 shows how a traditional architecture can be migrated in a Hyperconvergent architecture.

3 Results

Several functional tests are carried out with two virtual machines, the characteristics of the first machine (VM1) are the following: Memory = 6 GB, Processors = 2, Hard Disk = 20 GB, the characteristics of the second machine (VM2) are as follow: Memory = 6 GB, Processors = 2, Hard Disk = 2 GB, the hot-plug migration and high availability tests were carried out with a half-full disk. Both VM1 and VM2 were tested for hot-plug migration and high availability (HA).

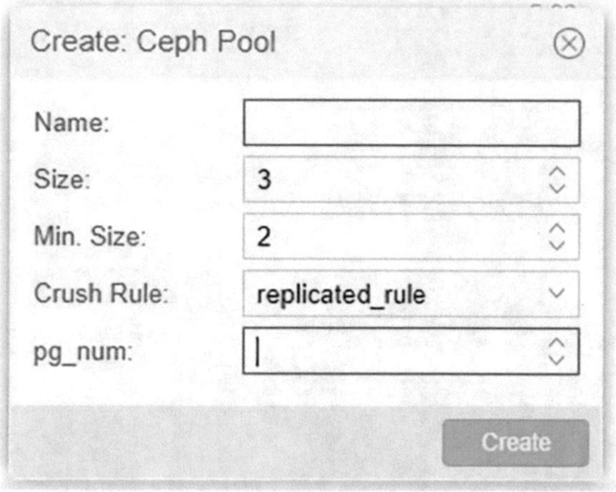

Create: Ceph Pool

Name:

Size: 3

Min. Size: 2

Crush Rule: replicated_rule

pg_num:

Create

Fig. 4. Ceph Pool Window

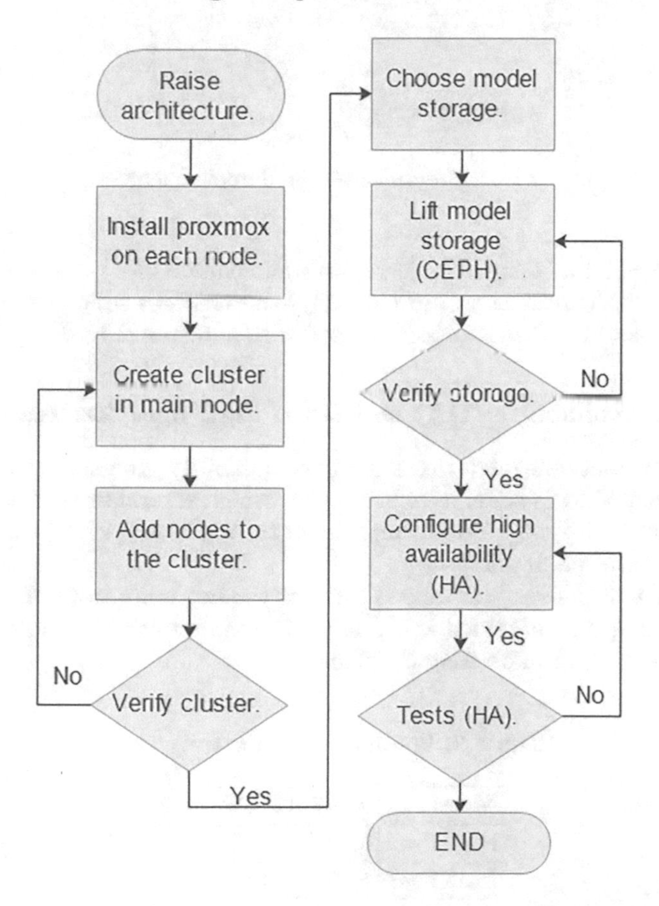

Fig. 5. Flow diagram for the deployment of the implementation

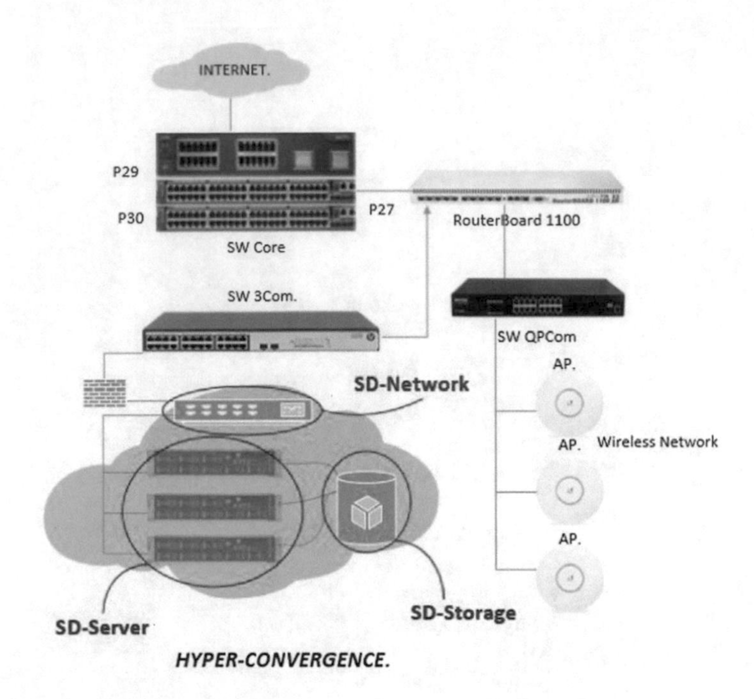

Fig. 6. Implemented final deployment.

Moreover, VM1 was tested for load tests as the Moodle platform is installed on it. Once the cluster is set up correctly, the CEPH configuration should be checked thanks to a shared disk in all nodes as shown in Fig. 6.

3.1 High Availability (HA) and Hot-Plug Migration Tests

In order to evaluate the migration and high availability process, the availability of the virtual machines is taken as reference, for which several tests of hot migration and high availability were performed in both VM1 and VM2, the results are shown in Tables 2, 3 and 4.

The result of the test tests shows that for the hot migrations of the VMs, the highest average time obtained is 50.4 s; for the same migration tests, there is an average downtime of more than 54.06 ms.

Table 2. Hot-plug migration tests TM.

Nodes	TP-VM1	TP-VM2
Pve1-Pve2	50.3 s	18 s
Pve2-Pve3	50.3 s	17 s
Pve3-Pve1	50.6 s	17 s

Fig. 7. CEPH OSD verification in each node.

Table 3 shows the average times required by the virtual machine to migrate (TM) from one node to another. For example, VM1 with its 20 GB storage migrates from node1 (Pve1) to node 2 (Pve2) in 50.3 s in average.

Table 3 shows each virtual machine average downtime (TI) while performing migration between nodes. VM1 requires an average downtime of 53 ms to perform the migration from node one to node two.

Table 3. Hot-plug migration tests TI.

Nodes	TI-VM1	TI-VM2
Pve1-Pve2	53 ms	47 ms
Pve2-Pve3	54.6 ms	52 ms
Pve3-Pve1	54.6 ms	51 ms

Based on tests carried out previously it is possible to calculate the platform availability time while running hot-plug migrations and high availability (HA) tests. The following Eq. 2 is applied to get the availability percentage [22]:

$$\%Availability = \frac{TST * DT}{TST} * 100 \tag{2}$$

Where:

- TST = Total Possible Service Time during the sample period.
- DT = Actual Drop Time recorded during the sample period [16].

Based on Eq. 2, Fig. 7 as well as Table 4 shows the percentage of availability for migrations performed during the tests. The percentage of availability is quite high for each migration with at least 99.888% service availability.

Based on the migration tests and downtime obtained, the percentage of availability of the VMs in the migration process is calculated, where it is observed

Table 4. Migration percentage of availability.

% availability	T1	T2	T3
Pve1-Pve2	99.894	99.896	99.894
Pve2-Pve3	99.892	99.888	99.894
Pve3-Pve1	99.896	99.888	99.892

that the percentage of availability in each migration is quite high since there is an average availability of service of 99,892% as shown in Fig. 8; therefore, it is considered that the implemented system successfully executes the migration and high availability processes.

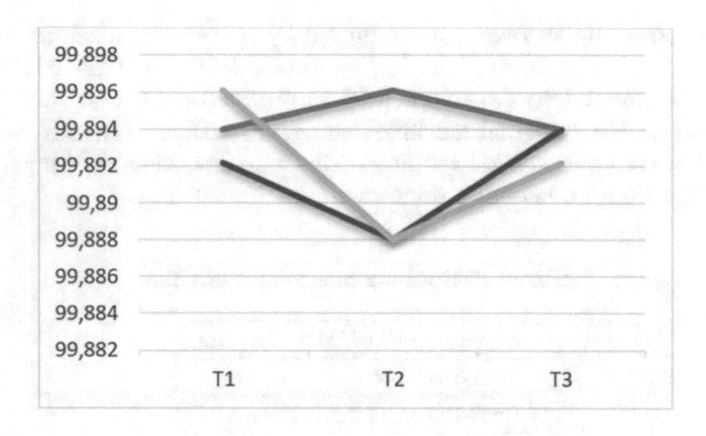

Fig. 8. Service percentage of availability while VMs migrate.

3.2 VM1 Load Tests

For the purpose of this test, the server was intentionally overloaded in order to assess its response and measure different resource parameters. These tests were carried out on both the implemented Hyperconvergent architecture vs. the traditional architecture. The test was made using the JMETER tool. Different parameters such as CPU consumption level, latency and throughput were monitored. The "mpstat 2" tool is used to measure CPU consumption level. The Linux OS is installed on the server so latency and throughput values are measured thanks to the Jmeter tool [23].

CPU Analysis. The first parameter to be evaluated to determine the performance of the system implemented with a hyper-converged architecture (HCI) versus a traditional architecture is the CPU consumption. For which by means of tests of the load to the servant with JMeter and the tool mpstat for monitoring of the consumption of CPU; according to Table 5 it is evident that in terms of performance, the HCI consumes 4% less of this resource compared to the traditional architecture.

Latency Analysis. The second parameter evaluated is the latency in the system. JMeter measures latency from just before sending the request to right after the first response has been received. Therefore, the time includes all the processing necessary to assemble the request, as well as the assembly of the first part of the response, which in general will be longer than one byte. [23], the unit of measure of the latency in milliseconds (ms). Table 5 shows values obtained related to latency, where traditional architecture is 261.36% lower than in a hyper-converged architecture with Proxmox VE, that is, the latency in HCI is reduced by more than 50%.

Table 5. Tests of overloaded servers.

Parameters	HCI	Traditional arc
CPU consumption	86%	90%
Latency	2.2 ms	5.75 ms
Throughput	3451/min	1018/min

Throughput Analysis. The third parameter to evaluate in the implemented HCI system is the throughput, this is calculated as the ratio between requests and unit of time. The time is calculated from the beginning of the first sample until the end of the last sample. This includes any interval between samples since it is supposed to represent the load on the server. The mathematical expression is: Performance = (number of requests)/(total time) [23]. For analysis purposes, the performance in the HCI implemented is 817.833% higher than in the traditional architecture as shown in Table 5.

Figures 8 and 9 show VM1 performance measured by Jmeter. Throughput is indicated in green while median and standard deviation are plotted in blue, red and violet. Figure 8 displays best results because the Hyperconvergent architecture is supported at the network level. The situation differs in Fig. 9 as it is associated to the traditional architecture [23].

The final implementation of Hyperconvergent infrastructures provides a more centralized system and easier system to operate. Figure 10 shows how a traditional architecture can be migrated in a Hyperconvergent architecture.

4 Discussion

A series of tests were carried out in order to validate the suitability of the described process and to verify the initial purpose. The Hyperconvergent system

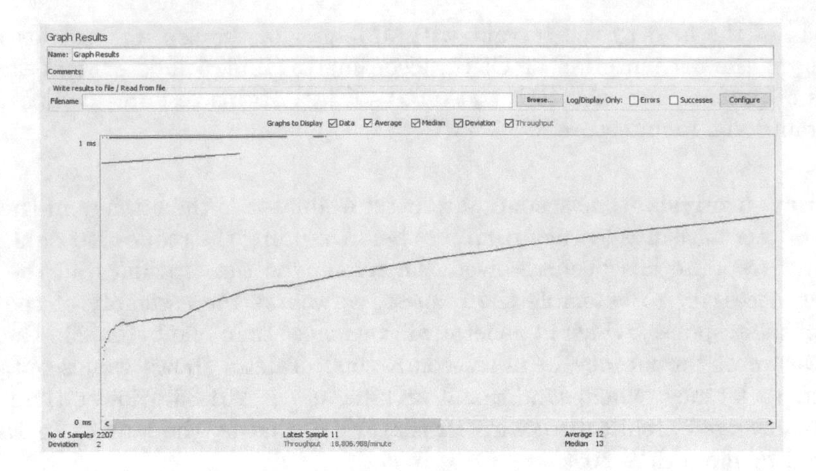

Fig. 9. HCI server tendency measures.

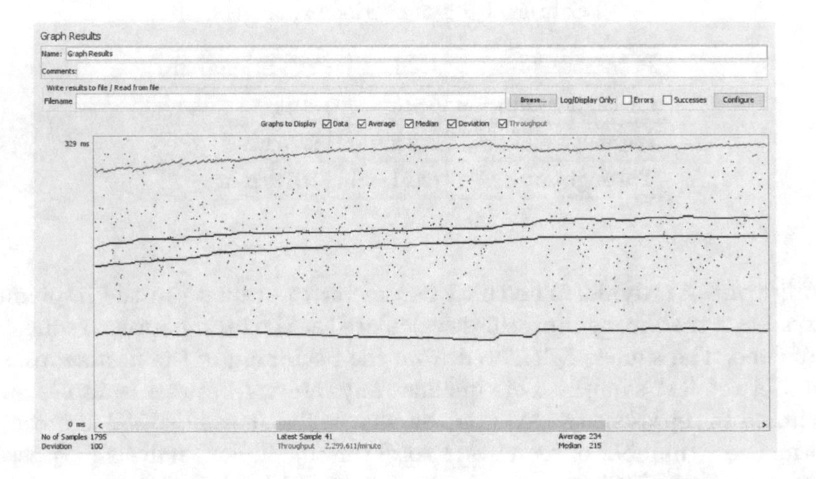

Fig. 10. Traditional architecture server tendency measures.

is named as such because it meets the following three key characteristics: it is a network, it provides storage and computation defined by software and it offers some benefits compared to a traditional system including high availability, hot-plug migration, and better server performances. The high availability and hot-plug migration tests show that VM1 maximum migration time was 50.3 s with 53 ms inactivity time. It is important to note that in case of physical equipment failure the VM migrates to another physical server and the service remains active at a 99.89% rate. Details can be seen in Tables 2, 3, and 4. The tests conducted with Jmeter are relevant to compare the performance of both architectures and to demonstrate that Hyperconvergent systems provides better performance than traditional architecture ones. Indeed, as shown in Table 5, processing time was reduced by 4%, latency and throughput in the HCI improve significantly.

5 Conclusions and Social Implications

Hyperconvergent deployment in Data Centers addresses problems such as scaling, availability and backup allowing the hosted services to be always accessible which are not the cases with traditional architectures. Hyperconvergence integrates storage, network and servers permitting plugging other technologies such as high-availability Cloud Computing at reasonable costs and allowing administrators to improve the performance of their infrastructure.

The Proxmox VE platform makes possible the creation of clusters and provides VM high-availability functionalities and hot-plug migration thanks to its CEPH storage system. This is made possible thanks to its distributed storage which generates minimum service downtime. The machines migration times are quite short but downtimes are even shorter since whenever there are services drops the downtimes are minimal.

The research presents the analysis of an HCI versus a traditional architecture, where the storage system in the HCI works with CEPH; under this storage system, it has been possible to obtain quite low migration and inactivity times, reaching 99.892% availability in the VMs.

In terms of CPU consumption, the difference between the two architectures is low, being barely 4%; On the other hand, the latency and throughput times are significant, in this sense the implementation of an HCI under Proxmox VE provides greater advantages in terms of the performance of the VMs, this without considering the different advantages in terms of management and administration for a data center.

References

1. Hewlett Packard Enterprise Development LP: Infraestructura Hiperconvergente (2018). https://www.hpe.com/es/es/integrated-systems/hyper-converged.html
2. Comstor: Qué es Hiperconvergencia? Febrero 2017. https://blogmexico.comstor.com/que-es-hiperconvergencia
3. John McArthur, J.P., Weiss, G.J., Yamada, K., Aoyama, H., Dawson, P., Chandrasekaran, A.: Magic quadrant for hyperconverged infrastructure, pp. 126, February 2018
4. Hyperconverged.org. Hyperconvergence trends in 2018. http://www.hyperconverged.org/
5. TechTarget: Qué es la hiperconvergencia y como aprovecharla (2018). https://searchdatacenter.techtarget.com/es/opinion/Riesgos-y-beneficios-de-la-infraestructura-hiperconvergente
6. PROXMOX. Open-source virtualization platform. https://www.proxmox.com
7. REVISTA BYTE TI. Red Hat Hyperconverged Infrastructure, infraestructura hiperconvergente Open Source de produccion. https://www.revistabyte.es/actualidad-byte/red-hat-hyperconverged-infrastructure

8. Cheng, S.M.: Proxmox High Availability. PACKT Publishing, Birmingham (2014)
9. Proxmox Server Solutions Gmbh: Proxmox VE Administration Guide, Viena (2018)
10. Proxmox: Install Ceph Server on Proxmox VE (2017). https://www.proxmox.com/en/training/video-tutorials/item/install-ceph-server-on-proxmox-ve
11. De Ingenieros, E.S.: Proyecto Fin de Carrera Ingeniería de Telecomunicación Diseo de un entorno de trabajo para PYMES mediante virtualización sobre Proxmox VE (2016)
12. Hernández, J.E.: Instalación, configuración y creación de Clúster en PROXMOX VE (2017). http://911-ubuntu.weebly.com/
13. Ronkainen, N.: Server virtualization, Finlandia (2003)
14. Cisco: Criterios de hiperconvergencia de última generacion. UU, EE (2016)
15. Vasquez Reyna, J.E.: Cloud Computing (2009)
16. Lastras Hernansanz, J., Requejo, J.L., Mirón García, J.D.: Arquitecturas de red para servicios en Cloud Computing, p. 71 (2009)
17. Apriliana, L., Darusala, U., Nathasia, N.D., Informatika, T., Nasional, U.: Clustering server pada cloud computing Berbasis Proxmox VE Menggunakan Metode high availability. J. Inf. Technol. Comput. Sci. **3**(1), 173–178 (2018)
18. Escobar Rodriguez, J.J.: Diseo de Infraestructura de un Data Center TIER IV de acuerdo a las especificaciones técnicas de la norma TIA-942, pp. 1–115 (2015)
19. Castillo, L.: Diseo De Infraestructura De Telecomunicaciones Para Un Data Center, p. 112 (2008)
20. Paltan, O.: Desarrollo de estandares y procedimientos para la creacion de un Data Center en la UPSE, p. 178 (2010)
21. Granda, A.: Rediseo de la red de la empresa COMWARE utilizando arquitecturas multiservicios que optimicen la administracion. Quito (2016)
22. Ritchie, G.: Introduccin a la Gestión de Disponibilidad ITIL, New Zealand (2009)
23. JMeter. https://jmeter.apache.org/usermanual/glossary.html
24. Singh, M.K.L., Quijano, W.A.T., Koneru, G.: Evaluation of network performance-type1 open source virtualization platforms. In: International Conference on Computer Communication and Informatics, ICCI 2015 (2015)
25. Chen, L., Huang, W., Sui, A., Chen, D., Sun, C.: The online education platform using Proxmox and noVNC technology based on Laravel framework. In: Proceedings - 16th IEEE/ACIS International Conference on Computer Information Science ICIS 2017, pp. 487–491 (2017)
26. Algarni, S.A., Ikbal, M.R., Alroobaea, R., Ghiduk, A.S., Nadeem, F.: Performance evaluation of Xen, KVM, and proxmox hypervisors. Int. J. Open Source Softw. Process. **9**(2), 39–54 (2018)
27. Riasetiawan, M., Ashari, A., Endrayanto, I.: Distributed replicated block device (DRDB) implementation on cluster storage data migration. In: Proceedings of 2015 International Conference on Data Software Engineering ICODSE 2015, pp. 93–97 (2016)

An Evaluation of the Relevance of Global Models of Indicators for Latin American Cities

Joe Carrión(✉), Patricio Coba, and Mario Pérez

Universidad Israel, Francisco Pizarro E4-142 Y Av. Orellana, 170522 Quito, Ecuador
{jlcarrion,ecoba,mperez}@uisrael.edu.ec

Abstract. The local governments planning is the main tool to met the citizen needs. One of the components of this tool should include the outlines to take advantage of technology. But the people responsible to plan need specific metrics to measure how they are using the technology. This paper purpose an study of the relevance of the global indicators related to Cities with the best conditions to live or visit. We compared a set of models and the data of some indicators about Latin America cities and top cities around the world.

This research have analyzed a set of models provided by academy, industry and government and using a method deductive summarize the relevant indicators for Latinoamerica. We plan evaluate global models with a set of real dataset of some cities in order to create a model of indicators relevant to Latin America cities.

Keywords: Indicators of sustainable development · Smart city · Models · City ranking

1 Introduction

The using technology as a backbone of the development projects is a goal that many cities abroad had a few years ago, now it is a reality and it can be verified when a city appears in the of ranking of the best cities to live.

According to the United Nations (UN), in its study The 2017 Revision of World Population Prospects [18], the world population by 2030 would reach 9.7 billion inhabitants and 11.2 billion inhabitants by 2050. Therefore, the intelligent distribution of resources and promoting a more efficient urban reality is fundamental, hence the beginning of the road to becoming an Intelligent City must be the main objective of an urban planning. Several groups of university, governmental and private enterprise researchers have taken the initiative to create tools to achieve a better quality of life for the inhabitants, based on a sustainable development in the consumption of resources. Although there is no standard methodology to define an intelligent city, there are several proposals to classify the level of compliance of guidelines to be a Smart City.

Supported by Universidad Tecnoógica Israel.

M. Botto-Tobar et al. (Eds.): ICAETT 2019, AISC 1066, pp. 67–78, 2020.
https://doi.org/10.1007/978-3-030-32022-5_7

For this work we have analyzed 21 proposals among models and rankings with greater scope and recognition which are shown in Table 1. We have selected 9, such proposals analyze, the Well being of citizens in relation to Economy, Citizenship, Governance, Mobility, Environment, and Quality of life, among others. Most works coincide in these matters. To select a model, the number of researchers was taken into account, the number of cities in which they are used and the ones that are most cited in research articles. From the result of comparing 9 models with 551 indicators, the three models have been selected as the most relevant and with the highest impact. The three models that are shown below:

Table 1. List of compared models. Number, Year of publication, number of indicators, cities, type (ranking or model) and author.

N	Year	Title	Ind.	Cities	Type	Author
1	-	Smart Cities Prospects by Procedia CS	-	-	Ranking	Academy
2	2011	Boyd Cohen Wheel	42	-	Model	Academy
3	2018	IESE Cities in Motion Index	84	165	Ranking	Academy
4	2007	European smart cities	74	70	Ranking	Academy
5	2015	Las Normas para las Ciudades Inteligentes	27	-	Model	Government
6	2015	Sustainable Infrastructure in Cities	42	-	Ranking	Industry
7	2016	Global Cities 2018	27	125	Ranking	Industry
8	2018	Innovation Cities Index	162	500	Ranking	Industry
9	2017	CETLA (United 4 Smart Sustainable Cities)	70	0	Model	Industry
10	2017	Safe Cities Index 2017	49	-	Ranking	Industry
11	2018	Sustainable Cities Index 2018 (Acardis)	48	-	Ranking	Industry
12	-	CDP, Carbon Disclosure Project,	-	-	Ranking	Industry
13	-	100resilientcities.org	-	-	Ranking	Industry
14	-	Green City Index 2012	-	-	Ranking	Industry
15	-	Green City Index 2010	-	-	Ranking	Industry
16	2017	Top 50 Smart City Governments	-	50	Ranking	Industry
17	2018	Smart Cities Index by Easypark	-	50	Ranking	Industry
18	2018	Global Financial Centres Index (GFCI)	103	110	Ranking	Industry
19	2018	Global Power City Index 2017 (MMF)	70	-	Ranking	No government
20	2018	ISO 37210	17	-	Model	No government
21	2018	U4SSC INDEX	91	-	Model	No government

- Universidad de Navarra (IESE): Public the Cities in Motion model (CIMI) based on four main factors: Sustainable ecosystem, Innovative activities, Equality between citizens and Territory connected. Make an annual study about the growth of cities. In its latest edition published in [6], it defines 83 indicators, including the ISO37120 standard (version 2018). The study includes 165 cities in 80 countries.
- The Center for Telecommunications Studies of Latin America (cet.la): It is led by the ASIET, Inter-American Association of Telecommunications Companies. They publish "Smart Cities: Social evaluation of Smart Cities projects", based on Markvon models that define states in specific quantities associated with a cost and effects. The states are grouped into 8 dimensions with 70 indicators in total.
- The United Nations in 2016 in coordination with the ITU, UNECE, FAO, UNESCO and other non-profit organizations established the project called U4SSC Model "United 4 Smart Sustainable Cities" [15] that has 3 dimensions Economy, Environment, and Society and Culture. The model includes 91 indicators.

Taking into account that the planet contains diverse life forms coexisting with each other and that humanity is part of it, it is mandatory to contemplate the idea that at some point the resources that the human being requires to live are exhausted and that our future generations do not have what it takes to prolong the species.

With this background, the UNDP (United Nations Development Programs) chose to create the SDGs (Sustainable Development Goals), one of the objectives of sustainable development is that of Sustainable Cities and Communities, whose principle is based on guaranteeing access to safe and affordable housing and the improvement of marginal settlements. It also includes making investments in public transport, creating green public areas and improving urban planning and management so that it is participatory and inclusive.

The UNDP Report in the UN Report Brundtland defines sustainable development as meeting the needs of the present generation without compromising the ability of future generations to meet their own needs.

The term Smart City is used worldwide, but its conceptualization is not integrated, several authors have defined smart cities as:

1. "Smart city is a place where traditional networks and services are made more efficient with the use of digital and telecommunication technologies for the benefit of its inhabitants and business" [10].
2. "A Smart City is about the City that uses a smart system, characterised by the interaction between Infrastructure, capital, behaviours and cultures, achieved through their integration" [4].
3. "The main driver for smart city birth and development is technology. Especially ICT, that permits to wire and link different actors in the urban arena and to supply digital services, by both private and public institutions" [11].
4. "The smart city (intelligent city) is the term that is at present used for denotation of communities, i.e. the residential complexes from the viewpoint of quality, performance and interactivity of urban services" [21].

   Before continuing, it is necessary to define the following terms used in this document:

– Model: Group of indicators used by ranking organisations to evaluate a city.
– Dimension: First level of grouping of indicators.
– Category: Second level of grouping of indicators. A Dimension can have several categories.
– Subcategory: Third level of grouping of indicators. A Category can have several categories.
– Indicator: Characteristic, attribute, quality or service(s) that a city has for the welfare of citizens.

2 Problem Statement

According to the cited definitions of smart cities, technology seeks to improve the quality of life of the inhabitants of a community. To achieve this result, several civil society organizations, governments, industry have proposed models of indicators that compare and classify cities at a global level (Quote UNESCO, ISO, CIMI, CETLA U4SCS). It has been analyzed that the proposed models focus on the concept of Sustainability, which is consistent with the Millennium objectives proposed by Unesco. However, a study carried out by the authors on the CIMI indicators reveals some questionable aspects, among which are considered:

1. The model has a global approach, with the risk of overlooking geographic, demographic and local cultural conditions. (Size of population, GDP, size of companies, food habits, among others)
2. Data sources for some regions are not available, so approximate data from regional sources, outdated or non-existent in the year of measurement are applied. (Number of hotels, number of digital users, photos in Panoramio, number of international congresses and others)
3. Some indicators are specific to products, services or mobility conditions that do not become a local need or do not justify their existence (number of restaurants of a certain food chain, number of airports or number of international students).

Due to the above, the indicators used in the CIMI model for certain cities result in unrealistic valuations because they do not have updated data, use approximate or incompatible data.

3 Objective

To compare global models and indicators that are used to measure sustainable cities and analyze their relevance in Latin American cities.

4 Related Work

Models of indicators or frameworks for smart cities have been proposed by various sectors of society, such as industry, academia and civil society (non-profit or professional organizations). They have been designed with a global scope and regional scope. The results of the models have been disseminated as city classifications (rankings) based on the score obtained by applying the measurement model.

At the global level, from the industry, the ISO proposes the ISO standard 37210 published in [14], from the academy the Center for Globalization and Strategy and IESE Business School publishes [6]. In the case of governments, AENOR publishes [1].

A very relevant study about Smart City corresponds to "Smart Cities Wheel", which proposes 17 aspects to consider for the sustainability of cities.

These models or rankings are based on the model of Cohen (2011) [9] 'The Smart City Wheel', has been used in several research works, this model classifies the indicators that define a smart city in 6 axes, Economy, Government, Society, Forms of Life, Mobility and the Environment.

The Table 1 shows the full list of the most relevant models we compared, most of them are referenced to themselves, they are cited for other authors as main source of information or cited by newspapers and magazines. It is showed the Year of the publication, the Title, the Number of Indicators, the Type of Author and finally the initial year of the publication.

Since the appearance of Smart Cities in 2004, a considerable number of models, studies and evaluation rankings have been developed, these are so diverse that a global standard has not been defined to establish the dimensions, categories and even less the indicators of evaluation to use. The diversity of the models and rankings is due to:

- The dimension to be included: financial, security, environmental or global;
- The scope to cover: public, private or mixed;
- The approach: technological, social or environmental;

Some models or rankings aim to measure economically strong cities or that their evaluation is oriented towards the exclusive use of technology since their basic needs have already been met. The diversity can be seen in its structure and its form of evaluation, for example the Innovation Cities Index ranking of 2THINKNOW that defines 3 dimensions, while the ISO model considers 17, another example of the difference is the ranking of Global Cities 2016 that defines 27 evaluation indicators and 2THINKNOW uses 126.

The problem raised has been addressed by [20] in his article: review of methodologies of urban indicators, among its main conclusions mentioned that: "when there is no data at municipal level extrapolation of data from the national scale is carried out. it supposes an abstraction that distorts the reality of the evaluated cities, since it is pondered with foreign cities".

Also authors of [2] proposes a maturity model to compare cities, based on the fact that some criteria may produce inconsistent results, authors consider that "establish criteria defining only based smart cities in the number of inhabitants can lead to misconceptions by not considering regional characteristics, political, social and economic of these cities", Finally, they propose a model adapted to the reality of Brazil. This proposal also is shared by [22] that mentions: "With the absence of a diagnosis, actions can become disoriented, poorly prioritized, redundant, and not deliver the expected return. In this way, the application of SMM makes it possible to verify the diagnosis by domains, thus observing in which aspect the city undergoing study stands out, as well as its imbalances".

[19] presents the article titled "Current trends in Smart City initiatives: Some stylised facts", one of the conclusions is "the number of city domains covered by smart initiatives does not seem to be correlated to the size of a city, considered in terms of population, but it is significantly correlated to the demographic density. This shows that both large and small cities exhibit some strengths and weaknesses in terms of innovation capabilities".

The Table 2 summarise a set of works about the composition of smart city models.

Table 2. List of articles reviewed related to index comparison, where authors adapt o propose a model. R means the reference and M defines if the article propose a new model of indicators

R	Summary	M
[2]	Create a Model por Barzil. Br-SCMM (Smart City Maturity Model)	Yes
[5]	Most adapted model for Moroccan smart cities	Yes
[13]	Review of different rankings of smart cities and analyze how to use them as instruments to identify strengths and weaknesses and contribute to the competitiveness of the city. Analyze the benefits and limitations of the rankings	No
[3]	Define a multi-dimensional conceptual model with 3 levels of classification. The first level with 6 categories (Living, mobility, People, Economy, Environment, Government). The model does not propose a set of indicators	Yes
[17]	It proposes a model of indicators focused on the city of San Francisco and Seoul. It uses open data sources, however it does not expose the evaluation criteria or dimensions in a model. Does not include the indicators in detail	Yes
[16]	It presents a justification of the need for smart cities, key concepts and evaluation criteria. It defines 3 general axes¿ Internet of things, Internet of people, Internet of services and Internet of data and summarizes technological tools for measurement, but does not present a set of indicators	No
[12]	It summarizes the conceptual foundations of smart cities from the point of view of the Academy, Industry and government. It presents a critique that people are not at the center of the effort, nor focus on the common good, but on innovation	No
[8]	It presents the importance of evaluating smart cities through indicators. It highlights the importance of the model of the European Union, but shows its weaknesses when evaluating small cities. Therefore, the authors propose a model based on the ISO37120 standard	No
[23]	It is a review of the literature, about smart cities and sustainability in the different scientific data bases. It is concluded that more than intelligent concepts are focused on being cities committed to sustainability	No

5 Indicator's Methodology Applied

In order for a city to be considered Intelligent, or at least be in the way of being, it is required to be evaluated. The evaluation of a city within the context of Smart Cities is very complex due to the diversity of elements that make it up, that is why many institutions, organizations and companies have proposed evaluation models or rankings, these models and rankings are organized by dimensions, categories and indicators, however, there are some that propose subcategories and levels of indicators.

We had access to information on 21 models to evaluate Smart Cities, which are shown in the Table 1.

5.1 Indicator Selection Process

The process for selecting the indicators is as follows:

- Step 1: Choice of models and initial rankings.
- Step 2: Choice of model to evaluated.
- Step 3: Select relevant cities.
- Step 4: Select the best ranked cities.
- Step 5: Analyze non-relevant indicators and unit of measure.
- Step 6: Seek the information of the selected indicators and compare results.

Indicators need to be justified for a city, in this paper a European Union study published in [7] on the criteria for the design of indicators is taken as a basis. The study mentions the following criteria that should be included or indicators: Relevance, Completeness, Availability, Measurability, Reliability, Familiarity, Non-Redundancy, and Independence.

5.2 Step 1: Choice of Models and Initial Rankings

There was access to information to a total of 23 investigations between smart city measurement models and rankings, of which 9 were selected to establish the initial model from which the dimensions and categories will be obtained. The selection was made based on the criteria of:

- Access to information, online, books and articles, if the author has published his research methodology.
- It is focused on the 17 objectives that the UN established for sustainable development.
- Global recognition the author or cited by other authors and that are as recent as possible.
- Aligned to the objective of the present research.

When applying the mentioned criteria, the 9 jobs shown in the Table 3 were obtained.

Table 3. List of initial selected models

No.	Name	Type
1	Normas para las ciudades inteligentes	Model
2	ESE Cities in Motion Index	Ranking
3	Global Cities 2018	Ranking
4	Sustainable Cities Index 2018 (Acardis)	Ranking
5	Innovation Cities Index	Ranking
6	Telos	Model
7	U4SSC United 4 Smart Sustainable Cities Index	Model
8	CETLA	Model
9	Boyd Cohen Wheel	Model

5.3 Step 2: Choice of Model to Evaluated

Once the dimensions and categories have been established, the next step is to remove AENOR because it specifically establishes norms and guidelines, and what is sought is a set of indicators. The next selection criterion is to take a model or ranking of each of the sectors that have contributed with relevant studies such as industry, academia and non-government. In the Industry sector, CETLA was chosen because it is a model that is oriented to the context of the present investigation. It should be noted that CETLA involves a set of academic institutions and companies in Latin America. The Table 4 shows the models that were finally obtained. The Cities In Motion Index of the business school of the University of Navarra defines 9 dimensions: Human Capital, Social Cohesion, Economy, Governance, Environment, Mobility and Transportation, Indicators of Urban Planning, International Reach and Technology. IESE has made five publications of the ranking of cities around the world, it is one of the most recognized worldwide. The United Nations in 2016, in coordination with the ITU, UNECE, FAO, UNESCO and several other non-profit organizations, established the initiative called U4SSC United 4 Smart Sustainable Cities model that has 3 dimensions, among which are Economy, Environment and Society and Culture. CETLA, research carried out within the Open University of Catalonia for the Center for Telecommunications Studies in Latin America, the model proposed by CETLA is oriented towards the social welfare of Latin America, despite not being a well-known model at the international level. The world was taken because it is oriented to the reality of the region, and its indicators are adjusted to the realities of small cities.

Table 4. Final model and ranking list

No	Models	Area
1	CIMI (Cities in Motion Index)	Academy
2	U4SSC (United 4 Smart Sustainable Cities Index)	No Nongovernmental
3	CETLA (Centro de Estudios de Telecomunicaciones de Amrica Latina)	Industry

The comparison of the current models allows us to focus our study on the CIMI due to its global reach as for example it includes 165 cities, of which 74 are capitals in the world; for Latin America, 20 countries with 27 cities are included.

5.4 Step 3: Select Relevant Cities

The objective of this step is to select relevant cities with reference for the city of Quito, which is located at position 140 of the ranking (CIMI-2018). This is the base line to select cities cities better qualified than Quito. From this group we chose the best classified (Buenos Aires) located at position 76 (CIMI-2018) and Montevideo located at position 100 (CIMI-2018) for being a small population city in the region and comparing country capitals.

5.5 Step 4: Select the Best Ranked Cities

With this step, the upper limit of the classification is established with the maximum scores that can be obtained. The 3 best classified cities were selected: New York in position 1, London in position 2 and Paris in position 3, data corresponding to CIMI-2018.

5.6 Step 5: Analyze Non-relevant Indicators and Unit of Measure

This step aims to identify indicators that have not been considered by other models and that their unit of measure is not relevant for a city.

For the following indcxcs we did not find matches in modeles of CETLA and US4CS: airports, Mc Donalds, number of air passengers, Apple stores, business schools, embassies, Metro length, subway stations, high-speed trains and skyscrapers.

The indicator of Table 5 were analyzed in detail by means of the exact data search to evaluate how the indicator behaves in cities of Latinamerica.

5.7 Step 6: Seek the Information of the Selected Indicators and Compare Results

We seek the data regarding four indicators from Table 5, except for Number of photos in Panoramio, because these data are not available. London City get the maximum score (5 points) for these five indicators.

These data displayed in Fig. 1 show that a subset of the indicators analyzed in recognized models have an impact on the classification of some cities that have other needs or that satisfy the demand of their inhabitants or visitors with other services. First, in the case of Quito, to compare it with the city of Montevideo in the indicators of Hotels has almost the same value. Second, the number of passengers per airport can not be compared between Quito and Montevideo with the other cities. Third, the only comparable indicator is Airports. Fourth, the Number of MacDonals is comparable among 5 cities, however in Quito it is absolutely smaller, Fifth, the number of conferences are not comparable between Quito London, New York and Paris.

Table 5. Indicator to analyse in detail

Dimension	Indicator	Source
International outreach	Number of McDonalds restaurants per city	OpenStreetMap
International outreach	Airports: Number of points where flight operations take place within a 40 km radius from the latitude and longitude defining the center of the city. It includes airports, aerodromes, airfields, and landing strips whether international, private, military or otherwise. Also included are the buildings used for processing passengers and cargo (terminals)	OpenStreetMap
International outreach	Number of passengers per airport Number of passengers per airport in thousands	OpenStreetMap
International outreach	Ranking of cities according to the number of photos taken in the city and uploaded to Panoramio (community for sharing photographs online). The top positions correspond to the cities with the most photographs	Sightsmap
International outreach	Number of international conferences and meetings that take place in a city	International Congress and Convention Association
International outreach	Hotels Number of hotels per capita	OpenStreetMap

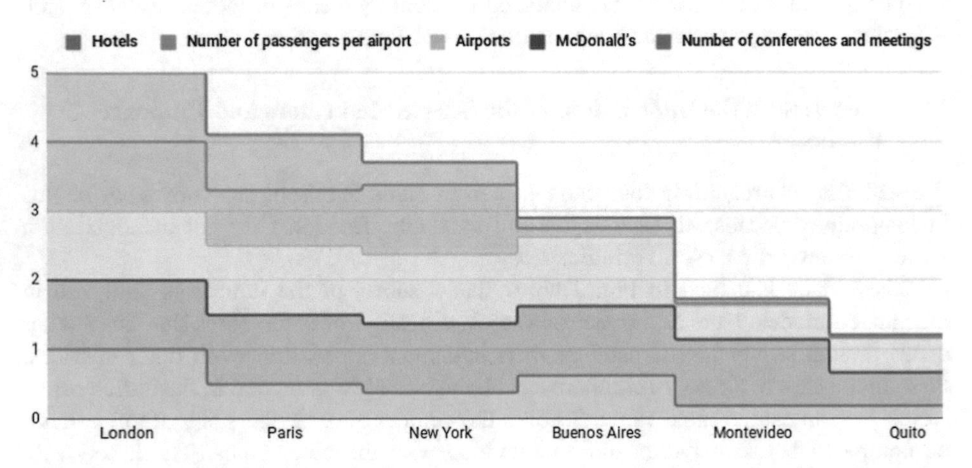

Fig. 1. Behaviour of Indicators selected about top cities and Latinoamerican selected cities

Table 6. Table of values of indicators to analysed in detail

City/Indicator	London	Paris	New York	Buenos Aires	Montevideo	Quito
Number of conferences and meetings	16	8	6	10	3	3
McDonalds	21	29	156	63	20	10
Airports	2	3	3	3	1	1
Number of passengers per airport	127248451	79185596	132000000	10340613	2071724	486141
Hotels	3821,25	3057	1259	583	232	147

6 Conclusions and Opened Lines

En el caso de los indicadores relacionados con las cadenas alimentarias, tambin es necesario analizar las necesidades locales y regionales de las ciudades latinoamericanas, ya que existen otras necesidades y demandas, por lo que no existen las mismas cadenas alimentarias en la misma proporcin. According to the problem posed, about the relevance of the indicators, the data sources not available and the existence of specific indicators, it has been identified that the International Projection indicators are not relevant to the local context of cities such as Quito or Montevideo. Indicators based on data not available as Panoramio are not relevant either, since there is no available data in this regard, so it is necessary to rethink this type of data source (Fig. 1).

In the case of Indicators related to food chains, it is also necessary to analyze the local and regional needs of Latin American cities, since there are other needs and demands that is why the same food chains do not exist in the same proportion (Table 6).

As future work, it is expected to conduct a more in-depth review of the state of the art of other global and regional models and indicators to identify their strengths and weaknesses. As an open line of research is expected to deepen studies related to sustainability to improve the quality of life of citizens and compare the indicators of smart cities. With these studies it is expected to propose a model of indicators relevant to the needs of the local population.

References

1. AENOR: Las normas para las ciudades inteligentes. Standard, Agencia Española de Normalizació, Espaa, Madrid (2015)
2. Afonso, R.A., dos Santos Brito, K., do Nascimento, C.H., Garcia, V.C., Álvaro, A.: Brazilian smart cities: using a maturity model to measure and compare inequality in cities. In: Proceedings of the 16th Annual International Conference on Digital Government Research, dg.o 2015, pp. 230–238. ACM, New York (2015). https://doi.org/10.1145/2757401.2757426, http://doi.acm.org/10.1145/2757401.2757426

3. Al-Nasrawi, S., Adams, C., El-Zaart, A.: A conceptual multidimensional model for assessing smart sustainable cities. JISTEM-J. Inf. Syst. Technol. Manag. **12**(3), 541–558 (2015)
4. Alkandari, A., Alnasheet, M., Alshekhly, I.F.T.: Smart cities: survey. J. Adv. Comput. Sci. Technol. Res. **2**, 79–90 (2012)
5. Benamrou, B., Mohamed, B., Bernoussi, A.S., Mustapha, O.: Ranking models of smart cities. In: 2016 4th IEEE International Colloquium on Information Science and Technology (CiSt). pp. 872–879. IEEE (2016)
6. Berrone, P., Ricart, J.E., Carraso, C., Ricart, R.: IESE cities in motion index 2016 (2016). Retrieved 1, 2017
7. Bosch, P., Jongeneel, S., Rovers, V., Neumann, H., Airaksinen, M., Huovila, A.: CITYkeys indicators for smart city projects and smart cities. CITYkeys report (2017)
8. Bruni, E., Panza, A., Sarto, L., Khayatian, F., ct al.: Evaluation of cities smartness by means of indicators for small and medium cities and communities: a methodology for northern italy. Sustain. Cities Soc. **34**, 193–202 (2017)
9. Cohen, B.: Smart city wheel. Retrieved from Smart & Safe City (2013). http://www.smartcircle.org/smartcity/blog/boyd-cohen-the-smart-city-wheel
10. Commission, E.: Smart cities (1999). http://web.archive.org/web/20080207010024/, http://www.808multimedia.com/winnt/kernel.htm
11. Dameri, R.P.: Searching for smart city definition: a comprehensive proposal. Int. J. Comput. Technol. **11**(5), 2544–2551 (2013)
12. Dameri, R.P.: The conceptual idea of smart city: University, industry, and government vision. In: Smart City Implementation, pp. 23–43. Springer, Heidelberg (2017)
13. Giffinger, R., Gudrun, H.: Smart cities ranking: an effective instrument for the positioning of the cities? ACE: Architect. City Environ. **4**(12), 7–26 (2010)
14. ISO: Sustainable development of communities – indicators for city services and quality of life. Standard, International Organization for Standardization, Geneva, CH, May 2014
15. Smiciklas, J., Gundula Prokop, P.S.Z.S.: Collection Methodology for key Performance Indicators for Smart Sustainable Cities. Naciones Unidas, Nueva York: (2017)
16. Khatoun, R., Zeadally, S.: Smart cities: concepts, architectures, research opportunities. Commun. ACM **59**(8), 46–57 (2016)
17. Lee, J.H., Hancock, M.G., Hu, M.C.: Towards an effective framework for building smart cities: Lessons from seoul and san francisco. Technol. Forecast. Soc. Chang. **89**, 80–99 (2014)
18. Nations, O.U.: World Population Prospects: The 2017 Revision. Naciones Unidas, Nueva York (2017)
19. Paolo, N., De Marco, A., Cagliano, A.C, Mangano, G., Scorrano, F.: Current trends in smart city initiatives: some stylised facts. Cities (2013). www.elsevier.com/locate/cities
20. Penã, S., Osvaldo, O.: Smart city: Diagnóstico de la ciudad de guayaquil (ecuador) (2018)
21. Prochazkova, D., Prochazka, J.: Safety of smart cities. Int. J. **15**(8), 6979–6985 (2014)
22. de Santana, E.D.S., de Oliveira Nunes, É., Passos, D.C., Santos, L.B.: SMM: a maturity model of smart cities based on sustainability indicators of the ISO 37122. Retrieved from Int. J. Adv. Eng. Res. Sci., 7 (2019)
23. Song, H., Srinivasan, R., Sookoor, T., Jeschke, S.: Smart Cities: Foundations, Principles, and Applications. Wiley, Hoboken (2017)

Design and Implementation of An Architecture for Electronic Billing Through Web Services and Mobile Devices

Fernando Solis Acosta[1,2](✉) and Diego Julián Pinto[2,3]

[1] Instituto Tecnológico Superior Rumiñahui, Sangolquí, Ecuador
edgar.solis@ister.edu.ec

[2] Departamento de Ciencias de la Computación, Universidad de las Fuerzas Armadas - ESPE, Sangolquí, Ecuador
efsolis@espe.edu.ec, djpinto@espe.edu.ec

[3] Informatics Engineering, Universidad Central del Ecuador, Quito, Ecuador
djpintoa@uce.edu.ec

Abstract. Electronic billing comes in response to both the needs of businesses and legal foundations defined by regulators, with an increasingly vital importance given a society that becomes more digitally based with each passing day. Electronic billing became possible in Ecuador in 2014, and in 2015, telecommunications companies and exporters began using the service, followed by public sector entities. Soon, all taxpayers are expected to apply electronic billing options. Given the incredible growth in these billing services, an application for Android mobile devices was analyzed, designed, and implemented. This application includes an architecture that integrates electronic signature processes, web services for receiving and authorizing electronic documents, with the generation, distribution, and storage of electronic vouchers over modern and user-friendly interfaces. MobileD methodology was applied because of its agility and special focus on the mobile device market, as it applies techniques and processes that are both well known and consolidated. Using test-driven development, 82 unit tests and 6 integration tests were made to correct errors, and 113 acceptance tests were implemented among 8 users to verify all of the system's functionalities and ensure that they system conformed to user's needs.

Keywords: Electronic billing · Digital certificate · Electronic signature · Android · Agile methodologies

1 Introduction

Organization's activities have evolved in relation to Information and Communications Technology (ICTs) to increase efficiency in their operations. In this context, electronic billing [8] has emerged not only as an important business opportunity, but also required by Ecuadorian regulators; namely, the SRI (Internal Revenue Service). Through this technology, a tax document is issued that replaces paper receipts and contains all information relating to the commercial transaction, payment obligations, and taxes owed.

The electronic bill is transmitted from the issuer to the recipient using communication systems that guarantee the authenticity and integrity of the document.

M. Botto-Tobar et al. (Eds.): ICAETT 2019, AISC 1066, pp. 79–89, 2020.
https://doi.org/10.1007/978-3-030-32022-5_8

Taxpayers, however, are faced with several issues that they have to overcome in this process, namely, the absence of mobile tools and appropriate infrastructure to manage accounting and billing activities.

Given that there is no system that covers all of such needs, the solution proposed herein involves a mobile application that integrates the processes of electronic signatures, generation, authorization, notification, and storage of electronic bills, using digital certificates.

2 Materials and Methods

2.1 Establishment of Interest Groups

Proposing and establishing the basis for implementation is key to developing an application, and in this specific case, the parties involved are national or foreign individuals who are performing legal economic activities, whether required to keep accounting or not, and corporations (legal entities) that have a Unique Taxpayer Registration Number (RUC) that allows them to issue and deliver SRI-authorized sales receipts for all of their transactions and submit tax statements. From among this group, micro-businesses and small businesses were chosen which, based on their sales volume, share capital, number of workers, and level of production, do not generate a large number of invoices each month.

People are required who have experience and expertise in projects relating to electronic billing and digital certificates, since they are able to verify the security of electronic invoices.

2.2 Development Tools

A series of tools were used to develop the solution, as described in Table 1, below.

Table 1. Acceptance test results

Tool	Use
Java Enterprise Edition (Java JEE)	Standards-based platform for the development of multilayer
	Corporate and web applications [4]
Apache Tomcat [11]	Java-based web application container that was created to run Servlets and JavaServer Pages (JSP) web applications
Java Server Faces (JSF)	Web development framework based on components for the Java EE platform
Primeface	Framework of user interfaces for JSF; it is lightweight and easy to use since it does not require installation or configuration
Netbeans IDE	Development environment that allows easy and fast development of Java desktop, mobile and web applications [7]
Android Studio IDE	Development environment for creating applications on all types of Android devices [6]
PostgreSQL	Object-relational database management system [3]
Hibernate ORM Object-Relational Mapping Tool for relational databases, facilitates creation, manipulation and access to data	

The use of web services such as RESTful [5] is also important for data exchange and communication between applications.

2.3 Analysis of Requirements

To create the user interfaces and the components of the application, 36 initial requirements were identified, defining their importance based on a scale. The order and priority of the tasks was defined by the product stack, which contains the total requirements of the application and served to identify the project modules, as shown in Table 2.

Table 2. Project modules

Module	Characteristic
Persistence	• Hibernate framework • Data access objects • Objects-Entities • Relational database engine • PostgreSQL remote
Users	• User registration: individuals and legal entities • Access using a username and password • Information update • User administration • Display of sent and received messages • Sending messages
Taxes	□ Display, search, creation, editing and elimination of taxes
Messages	• Display and search through sent and received messages • Reply to messages □ Write messages □ Delete messages. • Ascendant and descendant ordering of messages
Clients, products and establishments	□ Covers creation, inquiry, updating, and deletion of clients, products, and establishments
Electronic vouchers	• Creation of electronic vouchers: invoices, credit notes, and withholding vouchers in XML format using the structure in the most recent technical datasheet for electronic vouchers published by the SRI • Creation of the RIDE file (printed format of an electronic document) as a PDF • Validation of XML file structure
Repository	• Display, search, and forward issued electronic vouchers: authorized and not authorized, received, voided, and pending • Creation and display of the RIDE file • Download the authorization file in XML and the RIDE file in PDF
Electronic signature	□ Digital certificate information • Signature validation on electronic documents • Sign electronic documents using the digital signature on XML, XadES, and BES documents • Integration with SRI Web Services

2.4 Planning

Table 3 provides details on the iterations that were involved in the process and will be implemented in the future.

Table 3. Development phases

Initialization Phase	
Iteration	Description
Iteration 0	□ Scope and establish the project. □ Review and analyze initial needs.
Production phase	
Iteration	Description
Persistence Module Iteration	□ Implementation of the module. □ Debugging and testing.
Users Module Iteration Tax Module Iteration Module Messages Iteration Electronic Voucher Module Iteration □ Debug and update user histories.. Module of Customers, products and establish- □ Generation and execution of acments Iteration ceptance tests. Repository Module Iteration Electronic Signature Module Iteration	□ Implementation of the module.
Stabilization phase	
Iteration	Description
Persistence Module Iteration	□ Debugging and testing
Users Module Iteration Tax Module Iteration Messaging Module Iteration □ Refinement of user interfaces. Electronic Receipts Module Iteration □ Generation and execution of acModule of Customers, products and establish-ceptance tests. ments Iteration Repository Module Iteration Electronic Signature Module Iteration	
Phase of system tests and corrections	
System tests Iteration	Collection of acceptance tests and user interfaces

2.5 Architecture and Application Design

The mobile and web applications must be connected to the internet and interact with the SRI's electronic voucher web services. This architecture is shown in Fig. 1.

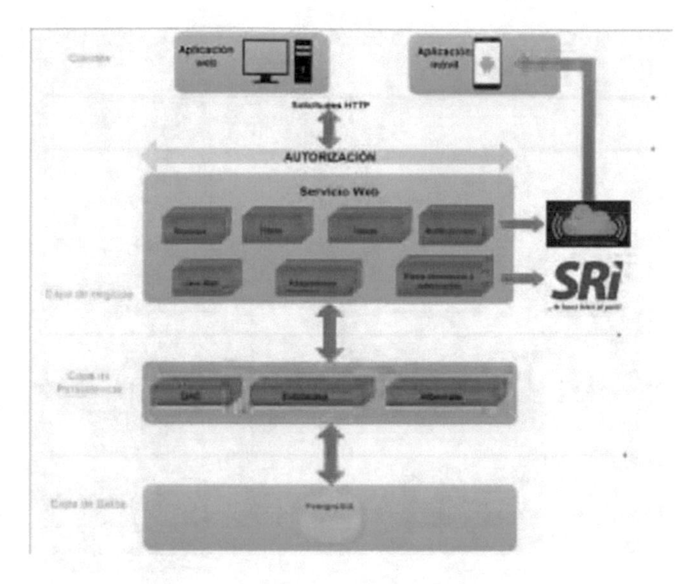

Fig. 1. Architecture

The web application navigation scheme is shown in Fig. 2. On the main screen of the mobile application, the navigation interface asks you to select a type of user (sender or receiver). Once logged in, a side menu appears, displaying the available screens.

Senders have the following main options:

- User information
- Electronic billing
- Vouchers issued
- Vouchers received
- Administration of establishments
- Customer management
- Product Management
- Messages sent and received
- Digital certificate

Receivers have the following options:

- User information
- Vouchers received
- Messages sent and received

The system has up to 6 browsing levels, including the screen to select the type of voucher, generate electronic invoice, and all the way through signing the voucher and obtaining SRI authorization.

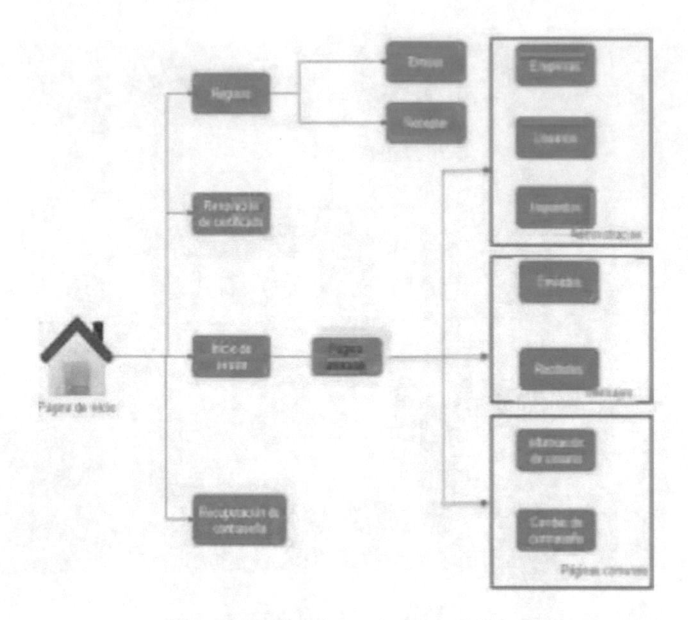

Fig. 2. Online browsing capabilities

The modeled database has 24 entities, with Table 4 explaining how the data is organized.

Table 4. Functional aspects of the database.

No. of tables	Functionality
6	Storage of electronic voucher information
7	Manage users, profiles, and companies
1	Information relevant to the digital certificate
4	Components in generating vouchers, including customers and products
2	Rates and types of tax
4	System setup

2.6 System Implementation and Testing

Technical specifications of devices compatible with the mobile application are detailed below:

- Operating System: minimum version 4.4 API 19 Kitkat.
- RAM: minimum of 512 MB, recommended 2 GB.

- Storage: minimum space of 10 MB.

The application was implemented through incremental iterative development and guided by tests (TDD) [10].

Figure 3 shows how the main website appears after validating credentials. The website facilitates business administration, digital certificate information, and the entire billing related process.

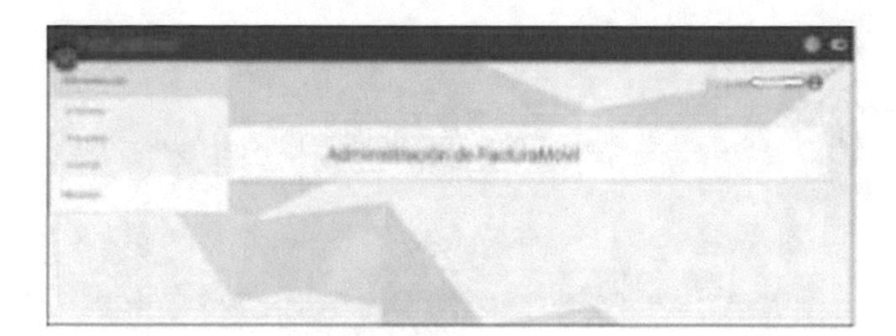

Fig. 3. Web application

After logging on to the mobile application, users have access to a drop-down menu, as shown in Fig. 4. Main functionalities include selecting the type of voucher to generate the electronic invoice, repository options, and authorized vouchers.

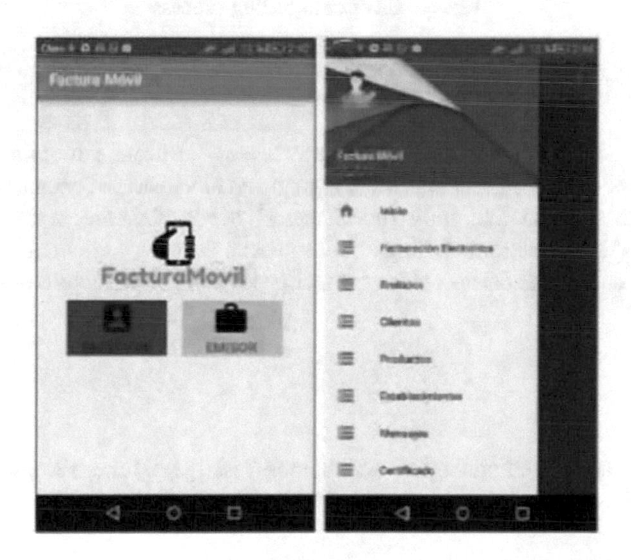

Fig. 4. Mobile app

Forty percent involves unit tests performed to ensure stability of the base code, with 50% dedicated to user interface and acceptance tests to verify that the application is actually working, with a remaining 10% for integration tests to verify that the application correctly works with other components. JUnit framework was used for the unit and integration tests, with the Espresso framework for user interface tests.

Figure 5 shows the process that begins when the application generates an XML file using the SRI format. The XML file is then sent over the web service, validating the structure, and including an electronic signature [12]. The XML is then sent to the SRI web service, notified, and stored on the server.

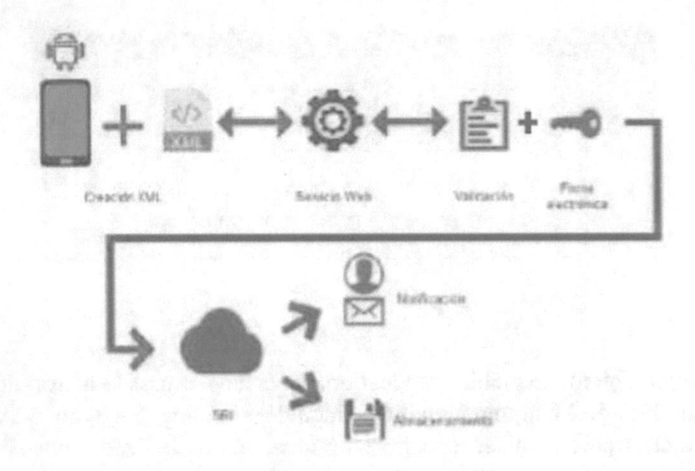

Fig. 5. Electronic billing process

3 Related Jobs

While a wide variety of projects have been developed around mobile payments, all having different characteristics and focusing on meeting various needs, there is yet no comprehensive solution that fully meets users' needs. Customers need a practical, responsive, and safe electronic billing solution that can serve them on an ongoing basis. However, the current market provides no technological infrastructure that can meet all of their needs.

4 Results

In the field of agile methodologies, acceptance tests [2, 9] are a known and recommended practice. The results are provided below in Table 5.

Table 5. Acceptance test results

Test users	Total acceptance tests	Total correct acceptance tests	Total unverified tests	Total discarded tests
8	113	904	16	4

Table 6 shows the response times for critical processes. All of the calculated times are acceptable, with the most critical process being signature and authorization, since the authorization process is performed online.

Table 6. Critical response times

URI	HTTP Action	Business operation	Average response time (ms)		
/certificate/ updateCertificate	PUT	updateCertificate	3980	POST	guardarEmpresa
/ElectronicVoucher /signatureAuthorization	POST	sendFileReprobeElectronicXML	2454		
/company/saved Company	POST	saveCompany	3566		

The minimum and maximum average response times are shown in Table 7.

Table 7. Response times

URI	Action HTTP	Business operation	Average response time (ms)
/Additionalmail/PostIdmail/ {idClient}	GET	getAdditionalAdsby Customer	6
/company/recoveryKey/ {identifierBusiness}	POST	recoverPassword	5821

The developed architecture accepts different types of content, with the most used formats being JSON and plain text.

5 Discussion

The use of an electronic signature validates the traditional physical billing processes. The electronic process, just like the traditional paper-based process, is legally valid for tax purposes, and uses an electronic signature with authorized digital certificates.

In Ecuador, electronic invoicing is a reality, and its use is growing rapidly. The expectation is that all taxpayers will eventually use electronic billing [1].

In developing the architecture, screen prototypes and browsing schemes help us to get a clearer idea of what we want to achieve.

JSON is an ideal format for exchanging information, especially on devices where resource optimization is a priority, since it is adaptable, simple, and easy to process.

6 Conclusions

To appropriately develop mobile applications, we recommend using the Mobile-D methodology, since it facilitates development, particularly in terms of analyzing requirements, product stacks, user histories, and acceptance tests, along with test-based development.

The application was developed with Android's SDK and its Material Design interface. A web application was also created using Java EE, JSF, and Primefaces to administer and register user information. To manage all of the requests and responses generated from both applications, a RESTful web service was created, primarily using JSON. For data storage and persistence, PostgreSQL was used, with Hibernate ORM framework.

The Jersey framework and the Google Gson library were two of the key components in constructing the web service, because they support the serialization and deserialization of messages in JSON format.

Management of the electronic signature digital certificate is a cornerstone of the project since as it is stored in a secure directory on the server and does not save your password, it does not directly interact with the application.

The distributed architecture that was developed provides all of the functions to manage electronic vouchers, administer and register users, connect with SRI web services, and process electronic signatures based on the Xades standard. All of this is managed over a RESTful web service having various logical layers.

References

1. Internal Revenue Service. Retrieved on 15 of 9 of 2016, from http://www.sri.gob.ec/de/10109. (2014)
2. Desai, C., Janzen, D., Savage, K.: A survey of evidence for test-driven development in academia, pp. 97–101. ACM (2008)
3. Douglas, K., Douglas, S.: PostgreSQL: a comprehensive guide to building, programming, and administering PostgresSQL databases. Developer's Library (2003)
4. Groussard, T.: Java Enterprise Edition: Development of web applications with JEE 6. eni Editions (2010)
5. Gulabani, S.: Developing RESTful Web Services with Jersey 2.0. Packt (2013)
6. Inc., G.: Android Recovered 2016 (2016) .https://www.android.com/intl/es_es/
7. Keegan, P., Champenois, L., Crawley, G., Hunt, C., Webster, C.: NetBeans (TM) IDE Field Guide: Developing Desktop, Web, Enterprise, and Mobile Applications. Prentice Hall PTR, Upper Saddle River (2006)
8. Pérez Villeda, M.: Electronic Bill. Tax Publishers, Jodhpur (2006)
9. Simoes Hoffmann, L., Guarino de Vasconcelos, L., Lamas, E., Marques da Cunha, A., Vieira Dias, L.: Applying acceptance test driven development to a problem based learning academic real-time system, pp. 3–8. IEEE (2014)

10. Vaca, P., Maldonado, C., Inchaurrondo, C., Peretti, J., Romero, M., Bueno, M., et al. Test-driven development - benefits and challenges for software development. In: XLIII Argentine Conference on Informatics and Operational Research (43JAIIO) -XV Argentine Symposium on Software Engineering, Buenos Aires (2014)
11. Vukotic, A., Goodwill, J.: Apache Tomcat 7. Springer, Heidelberg (2011)
12. WANG, X.-L.: Research on the electronic signature system with credible fusion verification and its application. In: 2013 International Conference on Computational and Information Sciences, pp. 80–83 (2013)

Migration of Monolithic Applications Towards Microservices Under the Vision of the Information Hiding Principle: A Systematic Mapping Study

Victor Velepucha(✉), Pamela Flores, and Jenny Torres

Escuela Politécnica Nacional, Ladrón de Guevera E11-253, Quito 170517, Ecuador
{victor.velepucha,pamela.flores,jenny.torres}@epn.edu.ec

Abstract. Organizations throughout time accumulate applications, which, given their old age, are generally designed in monolithic architecture. Given technological advances, enterprise has the necessity to modernize these applications, being the migration from a monolithic architecture towards microservice architecture a good alternative, however according to our research many failed attempts were found. We performed a Systematic Mapping Study in order to obtain studies to show how to migrate or modernize monolithic applications towards microservices based on some principle of Software Engineering. As a result, we found that there are different types of approaches of studies, such as: (a) solution proposals, (b) experience reports, (c) validations research and (d) opinion articles. Between the studies found, there are no studies related to migration process that is based on a Software Engineering principle, nor the Information Hiding Principle. This research indicates that there is a lack of a theoretical foundation with guidelines on how to perform a decomposition of a monolithic application towards microservices. Given this gap, we propose to migrate a monolithic application to microservices following the principles of the Information Hiding Principle.

Keywords: Monolithic · Microservice · Migrate · Migration · Modernization · Decomposition

1 Introduction

Organizations have accumulated applications, which generally follow a monolithic architecture, some developed in three layer patterns; these applications are called legacy applications [1]. Generally, these applications are deployed on physical servers, but given the technological advances of recent years these applications are now being deployed on virtual servers from On-premise to Cloud Computing. However, in this migration process there are many failed attempts [2,3].

When we talk about monolithic applications, they are defined as an application whose source code is programmed as a single block. Some of them are

M. Botto-Tobar et al. (Eds.): ICAETT 2019, AISC 1066, pp. 90–100, 2020.
https://doi.org/10.1007/978-3-030-32022-5_9

internally organized in logical layers, which follows the pattern of n-layer programming such as: (a) presentation layer, (b) business layer, (c) data access layer; nevertheless, its execution is done as a single block and the source code is on a central repository. The way to program and maintain these applications is with a team of programmers who work simultaneously taking care when they need to merge the code changed, compile and deploy. The way to scale in these applications is usually to clone the servers and enable a load balancer, with high cost to enterprises [4].

Otherwise, a microservice architecture (MSA) is composed of a set of small services, which performs a unique business responsibility, has its own data persistence repository, and uses light communication such as RPC-based API or RESTful Webservices. Creating a microservice consists of splitting an application into small cooperating components, where these components interact with each other through interfaces. These microservices bring benefits such as easy scalability, portability, maintainability, short development cycles and the opportunity to work with modern tools. Its greatest benefit is being deployed in containers in cloud computing and automatizes various phases of the software development life cycle (SDLC) through the use of DevOps tools, thereby saving organizations time and costs [1,3,5–7].

The Information Hiding Principle (IHP) was proposed by Parnas in 1972 in his article *On the Criteria to Be Used in Decomposing Systems into Modules* [8]. This principle recommends decomposing systems into quasi-decomposable modules, that is, dividing systems into modules that have the least dependence on one another, achieving quasi-decomposable systems, following guidelines: (a) Each module must be characterized by its knowledge of a design decision, (b) Hide the design decisions of other modules. The fact that each module is characterized by its knowledge of a design decision matches the definition of what a microservice is and serves the purpose of decomposing a monolithic application into microservices. By following these guidelines, we receive the benefits of: (a) Short development time, (b) Flexibility in the development of a system, (c) Comprehensibility of a system.

These days one alternative for solving the issues related with monolithic architecture is migrating from monolithic to microservices, but this process has many problems in practice, mainly because there is no software engineering behind this process. For this reason, it is necessary to know about the state of the art around the migration process between both architectures. In this sense, we have conducted a Systematic Mapping Study (SMS), following the guidelines provided by Kitchenham et al. [9]. Our motivation is to find scientific articles that show how to achieve this migration process based on a foundation of software engineering. For this study we have chosen the following research questions: RQ1: "What proposals are there for migrating monolithic applications to microservices?", RQ2: "Which of these proposals to migrate monolith applications to microservices are based on a software Information Hiding Principle?". As a result, we found that there are different types of approaches that we classify as: (a) solution proposals, (b) experience reports, (c) validations research and (d) opinion articles; that show how to accomplish this migration process. However, there are no studies based on any Software Engineering Principle to explain how to make the migration process from monolithic to microservices, only a few articles reference some principles superficially.

This article is organized as follows. In Sect. 2 we explain the detailed process of how we perform our search, specifying the method, research question, inclusion and exclusion criteria. In Sect. 3 we explain the selection process. In Sect. 4 we explain the data extraction. In Sect. 5 we discuss the results and findings and finally in Sect. 6 we present the conclusions and future work.

2 Research Method

To carry out the SMS review we will use the guidelines provided by Kitchenham et al. [9] which are: (a) Define the research question, (b) Search of relevant literature, (c) Selection of relevant articles, (d) Classification of articles, (e) Extraction and aggregation of data.

2.1 Research Questions

For this study we set to answer the following research question:

- RQ1: What proposals are there for migrating monolithic applications to microservices?
- RQ2: Which of these proposals to migrate monolith applications to microservices are based on a software Information Hiding Principle?.

2.2 Search Strategy

As a search strategy, an exploratory search is made based on the research question as shown below.

- **Search terms**
  The following criteria or search terms are defined:
  - migrate microservices
  - modernize applications
  - monolithic microservices
- **Inclusion and exclusion criteria**
  The following inclusion and exclusion criteria are defined:
  - Papers from the year 2012 and above because in this year the term of microservices appears [5]
  - Related only with the Software Engineering area
  - It belongs to the Journal Citation Reports (JCR) in the case of journals
  - It belongs to a CORE Source ranking in case of proceedings
  - Only English written papers

- **Literature resources**
  Based on the defined search terms, we perform the search of journals and proceedings in the most important scientific databases such as: Scopus, Web Of Science (WoK), IEEE Xplorer, ACM DL, Springer and Science Direct (SD).

3 Study Selection

The objective of this phase is to extract information from primary studies, reduce bias, which we define with our protocol and the way we will perform data extraction [9].

3.1 Design of Extraction Form

An exploratory review is carried out to determine the default search values of each database, finding that each one has different default search criteria. It is determined to use a search standard and the decision is made to perform the search by document title. All this selection and classification process is shown in Fig. 1:

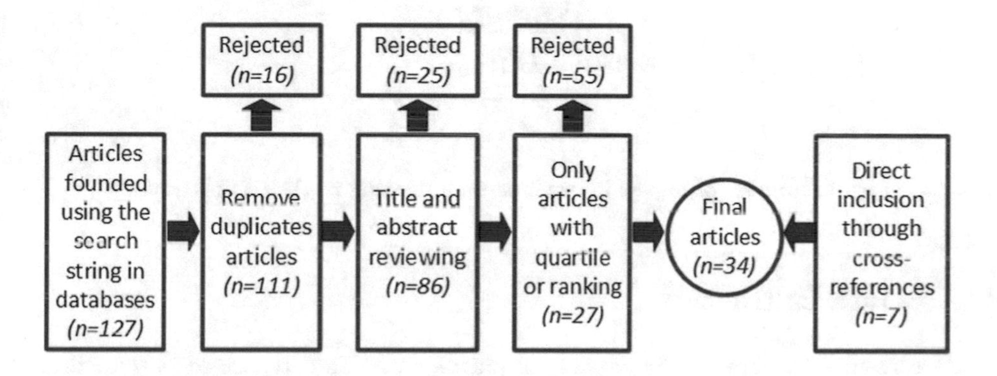

Fig. 1. Data selection process

3.2 Selection Process

On the other hand, a classification of papers by quartiles is made according to the JCR classification, where in Q1 is the top 25% of journals of a particular category, in Q2 the range of 25% to 50%, in Q3 from 50% to 75% and in Q4 the rest. This classification is shown in Fig. 2:

To show the ranking of the paper conferences, we classified according to the categories indicated by CORE, which classifies the congresses into the following categories: A +, A, B and C, being A + those that refer to excellent conferences, and C those that still do not meet minimum standards. Additionally,

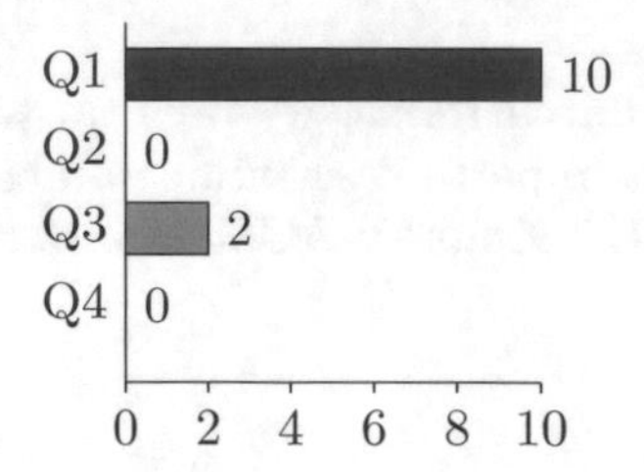

Fig. 2. Papers classified by quartiles of JCR

opportunistic searches are carried out with the purpose of finding papers that are not referring to microservice migration, finding some papers that indicate how to migrate from legacy to SOA and from legacy to Cloud Computing that are included in our research. The conference papers obtained are shown in Fig. 3.

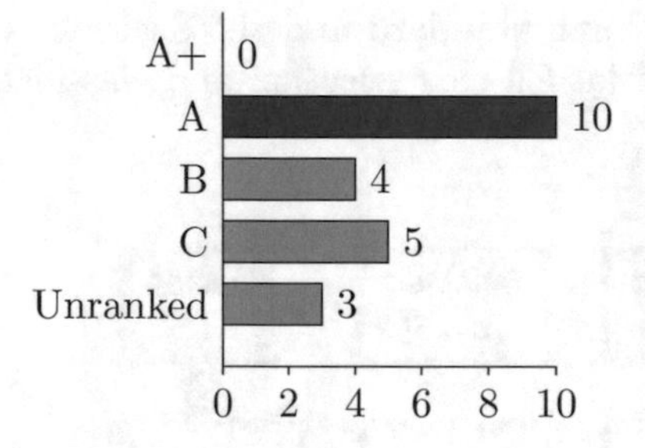

Fig. 3. Number of papers in conferences according to the CORE ranking

4 Data Extraction

As a result, we obtained 34 papers, 12 journals papers categorized in quartiles, 22 conference papers. As part of our analysis strategy, we classify the papers by types of research proposals, finding four categories shown in Fig. 4.

5 Data Synthesis

In the synthesis of data, from the data previously collected from the primary studies, we summarize them in a descriptive way [9]. It is evident that there are more Solution Proposals papers founded, because the microservice architecture is currently in the process of research, migration, testing and implementation both in industry and academia. We will discuss the most relevant proposals.

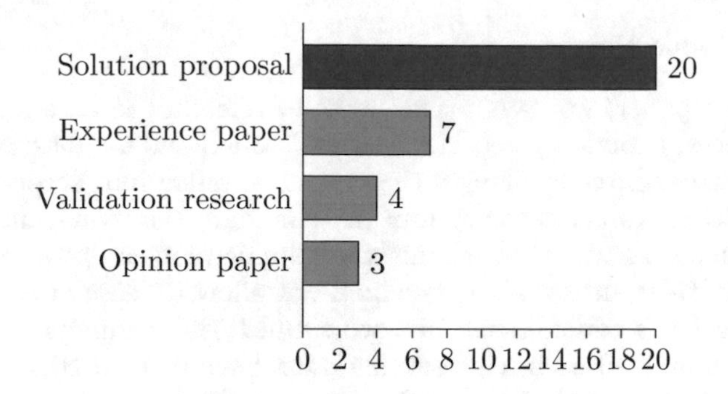

Fig. 4. Research classification

5.1 Solution Proposal

Named solution proposals, all those jobs that have generated a model, method, framework or techniques. These studies have been obtained mostly under an empirical approach. Next, the articles that are included in this category are discussed. The framework proposed by Sun et al. [10] focuses on migrating a monolith Internet of Things (IoT) application to microservices, with its proposal being scalable, expandable and maintainable, which will allow greater integration between applications and use of BigData. Its main strength is the integration of applications with the use of BigData, however, given the characteristics of IoT, its proposal cannot be generalized to other applications. In [11], the author presents a framework called SCAR, whose main objective is to evaluate monolithic applications vs microservices, basing itself on creating highly parallel applications without a server (serverless) controlled by events, emphasizing optimizations based on a cache applied to minimize response times, being useful for high performance computing. Another framework called SmartVM is shown in [12], for the implementation of microservices compatible with SLA (Service Level Agreement), to close the gap between the best practices in the industry and the workflow focused on the developed of the most SaaS providers. Taibi et al. [13], obtains a framework, after analyzing and comparing three different migration processes, adopted by the professionals that were interviewed, together with the motivations and common problems that frequently occur during migrations. Finally, Escobar et al. [14], carries out a case study around a SISINFO application and documents the migration process to microservices. His main contribution was the definition of diagrams of the source code of both the initial application and the microservices. These diagrams can be used as a reference by architects and developers to detect the architecture of the application. Of the five previous studies, it is important to remember that only two of them specify frameworks for migration.

It is interesting to note that there are studies that are based on migrating applications towards an SOA architecture, which is an architecture prior to the birth of microservices, being a great use for its study. [15–18].

5.2 Experience Paper

As Experience paper, we have placed articles referring to case studies, migration experiences, experience reports, lessons learned and demonstrations. Furda et al. [2] describes three challenges that must be taken into account to migrate legacy code to microservices they are: (a) eliminate the management of legacy code states, in such a way that the migrated application can be easily scaled, (b) implement multi-tenant capabilities which can allow to lease or sell the services of the application to different clients, with which the companies can have economic returns, and (c) each microservice must have its own BDD. Bucchiarone et al. [19] outlined the challenges when faced with modernizing the FX Core Danske Bank application towards microservices, emphasizing that monolithic applications are currently complicated to maintain and modify in an application. From their experience, the migration process should be done from the inside out, implementing only one functionality at a time and proposes to create work teams dedicated to work exclusively with microservices to allow acquired expertise in microservices and in specific business functionality. In [20], the authors show the lessons learned during the refactoring and incremental creation of microservices of a mobile commercial application and how the adoption of DevOps and its practices facilitated the migration process. Its main contribution are migration patterns that researchers can use and generate.

It is worth mentioning that although the articles in this category provide valuable results when carrying out a migration, they lack rigor at the methodological level, therefore, the results of these studies can be questioned for their generality.

5.3 Validation Research

We have called Validation Research the articles that perform validations results, some of which can lead case studies, testing, deployment and efficiency analysis.

In the literature, two articles were found [21,22] which applied case studies using an online application related to streaming of movies; Gan et al., studied the microservices bottlenecks are analyzed by having high concurrency and how this affects the design of the datacenter (hardware) and how to adjust. They apply a statistics method at the moment of execution of microservices at the level of execution latency times vs Queries per second (QPS) in order to evaluate the Quality of Services (QoS). They hope that this research can provide information on how an application behaves under specific scenarios and see the limits of how well it can work. Meanwhile, a conceptual methodology was proposed by Ahmadvand et al. [22] which indicates that there may be problems when migrating towards microservices because not all stakeholders of the Software Development Life Cycle (SDLC) of a system participate or are taken into account, specifically focused on the security and scalability requirements.

Villamizar et al. [4] evaluates the behavior of a monolith application and a microservices application deployed in cloud computing to measure its performance in scalability, for which they present a case study and show the results of the performance tests. The indicators include; costs per hour of execution, response times, evidencing that there are cost savings and better performance in a microservice application than monolithic application.

We can see that the two investigations are based on testing an application and its effect when they are migrated to microservices. This is valid since it serves as a clear example of how the use of this application can be triggered with the increase of users and shows the advantage of scaling microservices displayed in containers. Another article compares two architectures in the cloud, putting emphasis on the scalability and costs.

5.4 Opinion Paper

An Opinion Paper considers when the researchers according to their experience, give their opinion about how to migrate from a monolithic architecture to microservices. These opinions can be recommendations, challenges, issues, strategies and reflections.

Linthicum et al. [23] talks about concepts of SOA and microservices, the benefits of using containers and resources in the cloud. They indicate patterns of how to refactor an application to use containers and microservices, thereby reducing the dependency on the infrastructure, making it portable so as not to depend on a provider of a cloud. An overview is presented by Larrucea et al. [24], who indicated that by 2021 80% of cloud application development will be developed with microservices. It presents an overview of microservice technology and how to perform the migration. They indicate that the trend is to use the RPC-based API or RESTFul Webservices and use of DevOps. It emphasizes working with structured agile teams around microservices. Indicate guidelines for migrating monolith applications to microservices and benefits with the use of DevOps as automated monitoring and versioning. Killalea et al. [25] identifies some of the hidden dividends of microservices for example: disrupt trust, enable failure, etc. The developers should consider obtaining microservices, concluding that migrating to microservices should not be the rule for all companies, and that additional dividends should be considered in the migration process.

The opinions of experts is a very valuable contribution because they propose future work that can contribute to the migration, development and evolution of microservices.

After making this classification, we can extract certain findings that answer the research question. From the SMS, it is determined that only Bucchiarone et al. [19] briefly mentions the Information Hiding Principle indicating how work teams should be organized, while Killalea et al. [25] describes the Single Responsibility Principle (SRP) very superficially when creating microservices. No study is based on how to decompose a monolithic application towards microservices based on some principle of Software Engineering. The studies are oriented towards a migration towards cloud computing, specifically in the area of infrastructure and use containers an DevOps tools. The typical reasons moving towards microservice architecture are to manage complexity and scalability. A new search term "decomposition" can be included as the keyword for our future research. There is a lack of theoretical foundation that provides guidelines on how to perform a decomposition of a monolithic application towards microservices.

6 Conclusion and Future Work

The research questions formulated for this study were answered: "RQ1: What proposals are there for migrating monolithic applications to microservices?. RQ2: Which of these proposals to migrate monolith applications to microservices are based on a software Information Hiding Principle?". All the related papers reviewed in our SMS show frameworks, guidelines, challenges, validations, good practices, experiences and opinions on how to migrate to monolithic applications towards microservices under different points of view. These include; migrating to containers, using DevOps, adapting to Cloud Computing, IoT, performance in infrastructure, SLA's and QoS, whose studies show benefits such as ease in maintainability of an application, portability, scalability, ease of work distribution.

Our main finding is there are no studies that show how to decompose a monolithic application towards microservices based on a software principle. Only two papers mention briefly two principles: Single Responsibility Principle (SRP) and Information Hiding Principle. Most of the studies are oriented towards migrate monolithic applications towards cloud computing. There is a lack of a theoretical foundation that show guidelines of how can perform a decomposition of a monolithic application towards microservices.

Given this gap, our next step in the future is to present a model that shows how to carry out this process of migration from a monolithic application to microservices based on the Principle of Information Hiding, and for the validation, we will conduct a case study with a real application in a production environment.

References

1. Villamizar, M., Garcés, O., Ochoa, L., Castro, H., Salamanca, L., Verano, M., Casallas, R., Gil, S., Valencia, C., Zambrano, A., et al.: Cost comparison of running web applications in the cloud using monolithic, microservice, and aws lambda architectures. SOCA **11**(2), 233–247 (2017)
2. Furda, A., Fidge, C., Zimmermann, O., Kelly, W., Barros, A.: Migrating enterprise legacy source code to microservices: on multitenancy, statefulness, and data consistency. IEEE Softw. **35**(3), 63–72 (2018)
3. Clarke, P.M., Elger, P., O'Connor, R.V.: Technology enabled continuous software development. In: Proceedings of the International Workshop on Continuous Software Evolution and Delivery, pp 48. ACM (2016)
4. Villamizar, M., Garcés, O., Castro, H., Verano, M., Salamanca, L., Casallas, R., Gil, S.: Evaluating the monolithic and the microservice architecture pattern to deploy web applications in the cloud. In: 2015 10th Computing Colombian Conference (10CCC), pp. 583–590. IEEE (2015)

5. Fowler, M., Lewis, J.: Microservices a definition of this new architectural term (2014). https://www.martinfowler.com/articles/microservices.html
6. Thönes, J.: Microservices. IEEE Softw. **32**(1), 116 (2015)
7. Gutiérrez-Fernández, A.M., Resinas, M., Ruiz-Cortés, A.: Redefining a process engine as a microservice platform. In: International Conference on Business Process Management, pp. 252–263. Springer, Heidelberg (2016)
8. Parnas, D.L.: On the criteria to be used in decomposing systems into modules. Commun. ACM **15**(12), 1053–1058 (1972)
9. Kitchenham, B., Charters, S.: Guidelines for performing systematic literature reviews in software engineering (2007)
10. Sun, L., Li, Y., Memon, R.A.: An open IOT framework based on microservices architecture. China Commun. **14**(2), 154–162 (2017)
11. Pérez, A., Moltó, G., Caballer, M., Calatrava, A.: Serverless computing for container-based architectures. Future Gener. Comput. Syst. **83**, 50–59 (2018)
12. Zheng, T., Zheng, X., Zhang, Y., Deng, Y., Dong, E., Zhang, R., Liu, X.: SmartVM: a SLA-aware microservice deployment framework. World Wide Web **22**, 1–19 (2018)
13. Taibi, D., Lenarduzzi, V., Pahl, C.: Processes, motivations, and issues for migrating to microservices architectures: an empirical investigation. IEEE Cloud Comput. **4**(5), 22–32 (2017)
14. Escobar, D., Cárdenas, D., Amarillo, R., Castro, E., Garcés, K., Parra, C., Casallas, R.: Towards the understanding and evolution of monolithic applications as microservices. In: 2016 XLII Latin American Computing Conference (CLEI), pp. 1–11. IEEE (2016)
15. Mohagheghi, P., Sæther, T.: Software engineering challenges for migration to the service cloud paradigm: Ongoing work in the REMICS project. In: 2011 IEEE World Congress on Services, pp. 507–514. IEEE (2011)
16. Razavian, M., Lago, P.: Towards a conceptual framework for legacy to soa migration. In: ICSOC/ServiceWave 2009 Workshops on Service-Oriented Computing, pp. 445–455. Springer, Heidelberg (2010)
17. Beserra, P.V., Camara, A., Ximenes, R., Albuquerque, A.B., Mendonça, N.C.: Cloudstep: a step-by-step decision process to support legacy application migration to the cloud. In: 2012 IEEE 6th International Workshop on the Maintenance and Evolution of Service-Oriented and Cloud-Based Systems (MESOCA), pp. 7–16. IEEE (2012)
18. Malouche, H., Halima, Y.B., Ghezala, H.B.: Enterprise information system migration to the cloud: Assessment phase. In: 2016 IEEE/ACS 13th International Conference of Computer Systems and Applications (AICCSA), pp. 1–6. IEEE (2016)
19. Bucchiarone, A., Dragoni, N., Dustdar, S., Larsen, S.T., Mazzara, M.: From monolithic to microservices: an experience report from the banking domain. IEEE Softw. **35**(3), 50–55 (2018)
20. Balalaie, A., Heydarnoori, A., Jamshidi, P.: Microservices architecture enables devops: migration to a cloud-native architecture. IEEE Softw. **33**(3), 42–52 (2016)
21. Gan, Y., Delimitrou, C.: The architectural implications of cloud microservices. IEEE Comput. Archit. Lett. **17**(2), 155–158 (2018)
22. Ahmadvand, M., Ibrahim, A.: Requirements reconciliation for scalable and secure microservice (De)composition. In: 2016 IEEE 24th International Requirements Engineering Conference Workshops (REW), pp. 68–73. IEEE (2016)

23. Linthicum, D.S.: Practical use of microservices in moving workloads to the cloud. IEEE Cloud Comput. **3**(5), 6–9 (2016)
24. Larrucea, X., Santamaria, I., Colomo-Palacios, R., Ebert, C.: Microservices. IEEE Softw. **35**(3), 96–100 (2018)
25. Killalea, T.: The hidden dividends of microservices. Commun. ACM **59**(8), 42–45 (2016)

Blockchain and Its Potential Applications in Food Supply Chain Management in Ecuador

Mario Peña[1], Juan Llivisaca[2], and Lorena Siguenza-Guzman[3](✉)

[1] Research Department, Universidad de Cuenca, Cuenca, Ecuador
mario.penao@ucuenca.edu.ec
[2] Department of Applied Chemistry and Systems of Production, Faculty of Chemical Sciences, Universidad de Cuenca, Cuenca, Ecuador
juan.llivisaca@ucuenca.edu.ec
[3] Department of Computer Sciences, Faculty of Engineering, Universidad de Cuenca, Cuenca, Ecuador
lorena.siguenza@ucuenca.edu.ec

Abstract. In recent years, attention has been directed to the problem of food loss and waste. On the one hand, there is an increasing demand for higher quality fresh produce and products. On the other hand, studies suggest that at least one-third of food production is lost along its supply chain. The most crucial part is constituted by suppliers-retailers-consumers, since it presents the highest percentages of loss or waste. Several works in the literature suggest that the exchange of information is one of the most important means to reduce waste. Shared information can improve decisions regarding the quantity in supplier's orders and vendor's inventory allocation among retailers. Based on these needs, the Blockchain technology, developed in recent years to generate secure transactions on different sectors, has the characteristics of decentralization, information security, and reliability. If this technology can be applied as an underlying basis of a supply chain, it could improve exchanges of information and products among all parts of the system. In this context, the current article presents a systematic review on Blockchain in food supply chain management (FSCM) in Ecuador, its main contributions, and potential benefits. Findings indicate that Blockchain in FSCM is a recent research area whose importance is rapidly growing. However, little is known about Ecuadorian studies applying Blockchain. Additionally, no studies have reported the combination of Blockchain and Internet of Things technologies, one of the most prominent and complementary applications of Blockchain.

Keywords: Blockchain · Supply chain management · Literature review · Ecuador

1 Introduction

In the Food Supply Chain Management (FSCM) context, sustainability has become an essential issue. Globally, the average products lost throughout the food supply chain (FSC), from the production sector to the retail, is estimated to be about 35%; this means

M. Botto-Tobar et al. (Eds.): ICAETT 2019, AISC 1066, pp. 101–112, 2020.
https://doi.org/10.1007/978-3-030-32022-5_10

that one-third of the food produced is wasted [1]. It is annually equivalent to an estimated 1300 million tons [2, 3]. Likewise, a worldwide analysis of crops production and consumption found that a quarter of the crops are lost or wasted [4]. Therefore, food loss and waste has received an ever increasing attention from both the public and the private research sectors. Most of the food losses or spoilage usually occur in early stages of the FSC, such as during harvesting, transportation or storage. On the other hand, the food waste refers to foods that are ready for human consumption but are not consumed, e.g., when food or processed food is wasted in the retail, a restaurant or the consumer's home [5]. Improving the efficiency of resources in the FSC and their consumption, as well as changing the general diet is vital to guarantee the future food supply for nine billion people projected to 2050 [6]. However, some parts of the world face a surplus food and obesity, where food represents only a small part of a families budgets [7], other parts are characterized by malnutrition, food insecurity and limited natural resources [8]. Thus, it is ecologically and economically unsustainable to waste food instead of consuming it.

Minimizing food waste is therefore, an essential part of the FSC responsibility [9]. Food waste data can be used in many logistic planning systems to determine more exact placement of orders and delivery times, as well as to ensure food quality, quantity, and precise allocation. At the end of SC, food waste mainly occurs at the wholesale/retail - consumer interface [7], due to its diverse interactions. Product groups, such as bakery and fresh fruits and vegetables all contribute to food waste. Unfortunately, these products are the most important since they have a great effect on a large scale of the total business in all parts of the distribution chain. The most common reason for waste in the retail store is the expiration date [9]. The principal causes include either a food surplus or products arriving very late at the end of their shelf life [10]. The origins of these effects have been identified as the lack of an information exchange, the difficulties of forecasting and from poor management. An additional reason is inadequate planning for when delays occur in the early stages of FSC, this can damage operations at later stages [11].

Innovative organizations take advantage of an information exchange technology to intelligently manage their logistics processes, and make decisions based on accurate reliable and timely information [12]. Since the 1980s, a large number of publications on the FSM has offered several solutions to the question of how to benefit from shared information [13]. Many of these studies have focused on the exchanges of retail demand information, concluding that shared information can improve decisions regarding the number of producer orders and the inventory allocation among retailers. Demand information exchange becomes more beneficial when the demand is variable, and the product is perishable and expensive [14]. Moreover, producers can obtain a significant reduction in the inventory levels and further reduce costs when demand variability is high and correlated with time [15]. Effective communication helps to plan the crop and facilitates the demand prediction, thus avoiding one of the most common reasons for waste. Connecting FSC actors through a Blockchain distributed ledger can improve food traceability, efficiency, cost-savings and security [16]. Blockchain technology can create a transparent information exchange and generate secure transactions among relevant parts in the SC [17].

Thus, the current article presents a systematic review of Blockchain applications in the Ecuadorian FSC context, its main contributions, and potential benefits.

2 Related Work

2.1 Blockchain and Internet of Things Technologies

Blockchain technology has been developed rapidly over the recent years, capturing the attention of both experts and academics. Nowadays, more than 25 countries are investing in Blockchain, obtaining more than 2,500 patents and spending a total of 1.3 billion dollars [18]. This technology has the characteristics of decentralization, information security and reliability [19]. Blockchain, also characterized by trust, openness, collective maintenance and non-manipulation, is a type of underlying technology, based on distributed accounting. This distributed accounting book cannot be altered or falsified by cryptography. The principal area of application is the connection of cryptocurrencies in conventional banking and financial institutions. Each transaction must be authenticated through the agreement of more than half of the participants in the network [20]. This means that a participant cannot modify any data without the approval of other participants.

Nowadays, Blockchain technology has been applied beyond financial transactions, to many other transactions and applications, such as healthcare [21], public services, and government sectors [22]. For instance, Tian [23] proposes an agrifood Supply Chain (SC) and a traceability system based on Radio Frequency Identification (RFID) and Blockchain technology, guaranteeing safety and quality of the food. Blockchain is used to ensure the reliability and validity of the information shared and published.

Another state-of-the-art technology is the Internet of Things (IoT), an evolutionary paradigm that has emerged and gained wide-ranging attention in science and engineering world. Its principal objective is to solve problems without the intervention of the human workforce, creating interactions between people and machines, and among machines. Therefore, IoT is an interconnected network of devices ranging from small sensors and actuators to vehicles and other larger devices, fixed or portable, that can transfer data automatically without the need for human intervention [24]. IoT is not a unique technology, but a combination of multiple technologies that would work for overall intelligent achievement. These technologies include communication technology, information technology, electronic sensor, and actuator technology and advances in computing and analysis trends [25].

2.2 Supply Chain Management Powered by Blockchain and IoT

Several studies have tried to combine these two powerful technologies to potentiate SC management performance. For instance, according to Farooq and Obrien [26], many companies have received innovative technological support to improve their SC performance. Although Blockchain was considered a disruptive technology in the early stages [27, 28], large companies like IBM, Walmart, and Maersk have seen that Blockchain could answer the call to the problem of network traceability, products, and containers [29–31]. Another application area has been agri-food SC [32]. In this sector, Blockchain, combined with Hazard Analysis and Critical Control Points (HACCP) contributes to ensure the quality of agricultural products and to avoid health problems like the salmonella outbreak case linked to Maradol-brand papayas, or the contaminated eggs scandal in Switzerland [28, 29, 33].

According to Perboli, Musso, and Rosano [27], logistics is an essential SC part because if missed, moving raw materials from suppliers to final customers would not be possible. Due to this, logistics, in many cases, is complicated because of the level of interaction among the SC members and the level of decisions that may exist, such as, the complexity of performing logistics in urban areas with congested routes, or the synchronization of deliveries coinciding with times and other restrictions, as well as, the dispatch for different customers with diverse geographical locations. All these challenges have a common problem, communication. In this context, Blockchain becomes a valuable contribution since this new technology guarantees the transaction of information in an efficient, fast, safe and reliable manner. However, this technology presents some limitations, such as scalability and cost [28, 32]. Firstly, many Blockchain applications have been developed as prototypes in controlled labs. For instance, Hinckeldeyn and Jochen [30] present the implementation of a prototype combining IoT and Blockchain to a small container. Challenges and conclusions of this work show that existing communication protocols are insufficient for Blockchain; the role of intermediates has to be redefined; and, scalability is still insufficient for SC applications. Secondly, the financial resources required in Blockchain are substantial. The investment is necessary for: infrastructure, software, e.g., Ethereum, an open source Blockchain platform, and communication, e.g., RFID, barcodes and digital bases; which for many companies, this is out of reach.

3 Research Methodology

A literature review is a concise summary of the best available evidence that uses methods to identify and synthesize relevant studies on a particular topic [34]. The present study follows the PICO Structure [35] and the methodology proposed by Fink [36], which consists of seven main tasks: (1) selecting research questions, (2) selecting bibliographic or article databases, (3) choosing search terms, (4) applying practical screening criteria, (5) applying methodological screening criteria, (6) doing the review, and (7) synthesizing the results. Following Fink's steps, the systematic literature review starts with very specific needs for knowledge or research questions. According to Petersen et al. [37], the research questions in a systematic review aim at aggregating evidence and hence a very specific goal has to be formulated (e.g. whether or not an intervention is practically useful by industry). Therefore, the research questions addressed in this study are the following: What is the State of Art (SoA) of Blockchain in the Ecuadorian FSCM context? Is there any Ecuadorian case study applying Blockchain in FSCM? What are the main benefits of the use of Blockchain technology in FSCM? On the other hand, PICO was developed to identify keywords and formulate search strings. In this study, Population refers to Blockchain and Food Supply Chain in Ecuador. Intervention, refers to technology in Blockchain. Comparison, refers to identifying search strategies to Blockchain, and Outcomes refers to answer the research questions.

Consequently, bearing in mind the degree of relevance or specialization to the subject analyzed, as second step, a set of three search engines was selected to perform bibliographic browsing: Scopus, Scielo and Google Scholar. Scopus is the largest and most common multidisciplinary database available. Scielo is a bibliographic database especially created for Latin American publications. Finally, Google Scholar is a web search engine that indexes the full text or metadata of academic journals, books, conference papers, theses, etc. The search was operated according to the following procedure. A selection of keywords, supported by the PICO structure, was performed in order to identify terms that represent the concepts related to the topic under the study (Step 3). Based on this terms selection, relevant articles were searched by combining the following subject terms: "Blockchain", "Supply Chain", "Internet of Things", "case study", and "Ecuador". All these search terms and their combinations were searched in the full document (e.g. article title, abstract, keywords, and body), and the analysis was limited to journal articles, conference papers and theses published in English and Spanish. An overview of the criteria and its corresponding results is shown in Table 1.

Table 1. Search criteria and number of results per database

Database	Query	Number of results
Scopus	Blockchain AND Ecuador	16
Scopus	"cadena de suministro" AND "Ecuador"	18
Scielo	Blockchain AND Ecuador	0
Google Scholar	"Blockchain" AND Ecuador AND "case study" AND "supply chain management"	29
Google Scholar	"Blockchain" AND Ecuador AND "caso de estudio" AND "cadena de suministro"	1
Google Scholar	"Blockchain" AND Ecuador AND "cadena de suministro"	21
Google Scholar	"Internet de las Cosas" AND Ecuador AND "cadena de suministro"	100

When the number of documents was sufficient to conduct the analysis, the resulting literature was sorted, summarized, and discussed in order to generate a final sample consisting of 185 potentially relevant studies. Then, the full text of each document was retrieved for detailed evaluation in order to discard those that did not meet the selection criteria with the application of Blockchain in Ecuadorian SCM. Each discarded article was registered in an excluded-studies table, followed by an explanation for its separation. Next, applying practical screening criteria (Step 4), the standardized inclusion/exclusion criteria were as follows: (1) Documents written in English or Spanish were included. (2) Only the studies related to both the application of Blockchain or IoT in Ecuadorian SCM were selected. (3) Documents describing the application of Blockchain in SCM without a specific case study implemented in Ecuador were excluded. (4) Dissertations and conference papers were also included. (5) Duplicated documents obtained from Scopus, Scielo and Google Scholar were excluded. (6) Year of publication was not considered. Methodological screening criteria (Step 5) was used to determine which document found by the search string provided important contributions to the studied sector. In cases where the uncertainty of the paper relevance was found, the full-document was downloaded and main sections, such as conclusions, discussions, and results were read and analyzed. As a result, 27 primary studies were subsequently selected for doing the review (Step 6). In each study, information extracted was the following: (1) basic information (title, authors, type of publication, name of publication, country and year), and (2) information related to the study (application sector, objective, main contribution, abstract). These data were collected in a "finding matrix" for their subsequent analysis. The collection of primary studies was then analyzed quantitatively and qualitatively (Step 7).

4 Results and Discussion

After applying the screening criteria, a quantitative analysis over the final studies was performed. The 44% (12) of the selected documents are not directly related to the research questions. These documents cover themes about Crowdfunding, Industrial IoT Monitoring, Social commerce, Cryptocurrencies, Web of Things, Agriculture, Learning, Bitcoin, Security, Cloud Computing, Energy, Liquefied Petroleum Gas (LPG), Marketers and Press.

In addition, almost half (48%) of the documents deal with slightly related topics. For instance, the authors, Barrera and Valverde [38], Macías and Robles [39], and Mendoza and Sandoval [40] describe how to optimize logistics through process improvement. Martínez-Vivar et al. [41] designed a methodology to evaluate the logistics system in the transportation process, in order to reduce distribution costs and thus increase the performance of the logistics system. Sablón Cossio and Arias-Gutiérrez [42] diagnose the milk agro-alimentary chain to propose feasible development strategies. Álvarez, Aldas, and Reyes [43] apply Lean and Theory of Constraints tools to reduce delivery times and materials inventories in process. Likewise, Ruiz Carrillo and Pupo Francisco [44] proposed tools such as the critical path, program evaluation, and continuous displacement, with a focus on the theory of constraints, as the basis to develop a better activities distribution in a manufacturing system, reducing the total processing time noticeably. Rosero-Mantilla et al. [45] proposed to improve productivity in a medium-sized industry by increasing production capacities through the use of an Aggregate Production Planning model. Guerra [46] developed secure information systems using a framework based on Blockchain. The author analyzes the communication processes in an organization and proposes a secure system of transactions among suppliers, company, and customers. Córdova [47] performed a Bitcoin analysis in international commerce considering the trust and security of Blockchain, as well as, the adoption as a method of payment. Reyes et al. [48] developed an industrial software application for manufacturing planning and programming. This software contains three modules used for planning and scheduling resources, process management and a Lean manufacturing module for contingencies notification, issuance and displaying. Bermeo, Castillo, and Serrano [49] proposed the use of intelligent systems, such as Artificial Neural Networks (ANN) and genetic algorithms, together with monitoring indicators as support systems, in order to implement a management model that focuses on alignment-adaptability-agility of the SC.

The remaining papers (7%) were highly related to the research goal. The first study by Silva, Gerreiro, and Sousa [50] present a meta-model integrating Hyperledger Composer (HC) and an ontology that models business transactions and human interactions on organizations (DEMO). They considered a FSC case of study in the UK retail industry (SC for bananas from Ecuador). The model identified transaction types (e.g., Order creating, Order producing, Order Status, i.e., the current state of the orders) and the actor roles that participate in these type of transactions. The study shows how the traceability problem can be solved considering different SC roles, and the trust of the meta-model is based on HC where it is possible to consult the transactions that have been submitted. The second study by Toapanta et al. [51] present a prototype based on Blockchain technology to provide robustness against failures, attacks by third parties and mitigate information vulnerabilities of public organizations in Ecuador. The authors claim that data access through Blockchain in an Ecuadorian public organization would be an alternative to improve the security of the information. In addition, they establish that information integrity has a higher priority over confidentiality and authenticity in this kind of organization. A summary of the previous two highly related works is presented in Table 2.

Findings indicate that an acceptable number of studies on SC have been performed in Ecuador. Unfortunately, only two articles document applications of Blockchain to enhance Ecuadorian SC. No studies have been reported combining Blockchain and IoT technologies. Results show that SC is a subject highly investigated in Ecuador, as can be seen in the 13 related works. Nevertheless, the link with Blockchain is still minimal. Although the research is immature in Ecuador, potential opportunities and challenges can be found in the literature as shown below.

The literature review shows that Blockchain adoption in SC is still in early stages since, in this technological processes, dedicated software is required for applications and existing communication protocols [30]. Blockchain applications using RFID, barcode, and Arduino are increasing in different industries [52]. Despite the potential of IoT and the facilities that this technology presents (e.g., cheaper sensors), significant technological challenges are still present, such as security, connectivity, and interoperability [18].

Table 2. Summary of the closely related works.

Work	Topics	Objective	Contribution
[50]	HC, DEMO, Traceability	Develop a model between DEMO and HC, to be applied in a FSC context. It offers a possible solution to the lack of control and traceability in collaborative business processes	(1) A meta-model between DEMO and HC with defined roles (initiator and executor) (2) The development of a Blockchain solution in HC that allows a possible solution to the lack of FSC traceability
[51]	Security, Blockchain, Information	Generate a prototype in a Blockchain diagram, based on an algorithm using flowchart techniques to provide robustness against failures, third-party attacks and mitigation of information vulnerabilities	(1) Blockchain becomes a fundamental rule to provide security, robustness against failures and attacks from third parties (2) The information integrity in public organizations has higher priority over confidentiality and authenticity

The current analysis shows that Blockchain applications have two main problems: scalability and cost. The former combines security and decentralization. Bitcoin and Ethereum are designed to be decentralized and secure, but they have a few numbers of transactions (7 to 15 transactions per second) compared with Quorum and Hyperledger Fabric (3500 transactions per second). This fact compromises scalability for many industries [29]. In the latter, the cost of Blockchain includes IT team and the platform. According to Perboli et al. [27], the Blockchain implementation could comprehend an internal IT support team with a cost of about 130 K€. This fact is a problem due to the need for high investment in technological adoption [29].

Despite the many Blockchain advantages, the current investigation suggests that coordination and collaboration of all SC participants ensure adoption of Blockchain technology. According to Holma and Salo [53], the trust among partners under a win-win framework improve the Blockchain adoption. However, any large company may have to "force" its SC partners who are not always willing to use Blockchain technology [29].

Finally, the study demonstrates some benefits in the use of Blockchain technology in FSC: keeping a formal product registry, tracking possession of goods at different stages in the chain, ensuring procurement of supplies between manufacturers and vendors, tracking the identity and reputation of suppliers, and using smart contracts to automatically negotiate best prices in real time [29, 54]. All in all, Blockchain is a potential technology that can enhance the synergy among all SC components and therefore improve its holistic performance. For the development of the Blockchain adoption in FSC, and considering the expected impacts of Blockchain, the study has a valuable contribution.

5 Conclusions

Several documents including applications of Blockchain and IoT technologies have been analyzed in this systematic literature review. In the international context, much work has been developed around these technologies, including food supply chain management. By merging IoT and Blockchain technologies, synergy can be generated among SC stakeholders. However, in the Ecuadorian context, only two articles are documented implementing Blockchain in SC.

In comparison to other parts of the world, it is striking that no work has been reported using IoT and Blockchain technologies. This result may be unfavorable for industrial development; however, it can be seen as a central niche of research of valuable applications around SC, especially in FSC that, currently, present significant challenges and could worsen shortly. In the short term, Blockchain complementing with IoT and other novel technologies like quantum computing will change the way supply chains are managed. These technologies may offer coverage to the solutions needed for any SC requiring information sharing, as well as, guaranteeing its security, reliability, and transparency. The current study highlights the lack of development on these disruptive technologies in Ecuador. Although this research was not exhaustive and oriented to the application of Blockchain in Ecuadorian FSC, it serves as a comprehensive base for understanding the use of Blockchain in SC in general context as well.

Acknowledgements. This paper is supported by the Research Department of the University of Cuenca (DIUC).

References

1. Lemma, Y., Kitaw, D., Gatew, G.: Loss in perishable food supply chain: an optimization approach literature review. Int. J. Sci. Eng. Res. **5**, 302–311 (2014)
2. Gustavsson, J., Cederberg, C., Sonesson, U., et al.: Global food losses and food waste. Food Agric Organ UN, pp. 1–38 (2011). https://doi.org/10.1098/rstb.2010.0126
3. Kim, W.-G., Nam-Chol, O., Pak, H.-S.: Waste Composition for Solid Waste Management and Its Characteristic Analysis, a Case Study. Landscape Archit. Reg. Plan. **2**, 72–77 (2017). https://doi.org/10.11648/j.larp.20170203.11
4. Kummu, M., de Moel, H., Porkka, M., et al.: Lost food, wasted resources: global food supply chain losses and their impacts on freshwater, cropland, and fertiliser use. Sci. Total Environ. **438**, 477–489 (2012). https://doi.org/10.1016/j.scitotenv.2012.08.092
5. Normann, A., de Hooge, I.E., Bossle, M.B., et al.: Key characteristics and success factors of supply chain initiatives tackling consumer-related food waste – a multiple case study. J. Clean. Prod. **155**, 33–45 (2016). https://doi.org/10.1016/j.jclepro.2016.11.173
6. Foley, J.A., Ramankutty, N., Sheehan, J., et al.: Solutions for a cultivated planet. Nature **478**, 337–342 (2011). https://doi.org/10.1038/nature10452
7. Stuart, T.: Waste: Uncovering the Global Food Scandal. Penguin, London (2009)
8. McGuire, S.: FAO, IFAD, and WFP. The state of food insecurity in the world 2015: meeting the 2015 international hunger targets: taking stock of uneven progress. Rome: FAO, 2015. Adv Nutr. **6**, 623–624 (2015). https://doi.org/10.3945/an.115.009936

9. Stenmarck, Å., Werge, M., Hanssen, O.J., et al.: Initiatives on prevention of food waste in the retail and wholesale trades. Nordic Council of Ministers (2011)
10. Mena, C., Adenso-Diaz, B., Yurt, O.: The causes of food waste in the supplier-retailer interface: evidences from the UK and Spain. Resour. Conserv. Recycl. **55**, 648–658 (2011). https://doi.org/10.1016/j.resconrec.2010.09.006
11. Parfitt, J., Barthel, M., MacNaughton, S.: Food waste within food supply chains: quantification and potential for change to 2050. Philos. Trans. R. Soc. B Biol. Sci. **365**, 3065–3081 (2010). https://doi.org/10.1098/rstb.2010.0126
12. Ellram, L.M., la Londe, B.J., Margaret Weber, M.: Retail logistics. Int. J. Phys. Distrib. Logist. Manag. **29**, 477–494 (1999). https://doi.org/10.1108/09600039910291993
13. Caro, M.P., Ali, M.S., Vecchio, M., Giaffreda, R.: Blockchain based traceability in agri food supply chain management: a practical implementation. In: 2018 IoT Vert Top Summit Agric - Tuscany IOT Tuscany 2018, pp. 1–4 (2018). https://doi.org/10.1109/IOT-TUSCANY.2018.8373021
14. Ferguson, M., Ketzenberg, M.E.: Information sharing to improve retail product freshness of perishables. Prod. Oper. Manag. **15**, 57–73 (2006)
15. Lee, H.L., So, K.C., Tang, C.S.: The value of information sharing in a two-level supply chain. Manag. Sci. **46**, 626–643 (2000). https://doi.org/10.1287/mnsc.46.5.626.12047
16. Kim, M., Hilton, B., Burks, Z., Reyes, J.: Integrating blockchain, smart contract-tokens, and IoT to design a food traceability solution. In: 2018 IEEE 9th Annual Information Technology, Electronics and Mobile Communication Conference IEMCON, pp. 335–340 (2018). https://doi.org/10.1109/IEMCON.2018.8615007
17. Tse, D., Zhang, B., Yang, Y., et al.: Blockchain application in food supply information security. In: IEEE International Conference on Industrial Engineering and Engineering Management 2017, pp. 1357–1361 (2018). https://doi.org/10.1109/IEEM.2017.8290114
18. Roriz, R., Pereira, J.L.: IoT Applications Using Blockchain and Smart Contracts. **2**, 426–434 (2019). https://doi.org/10.1007/978-3-030-02351-5
19. Li, J., Wang, X.: Research on the application of blockchain in the traceability system of agricultural products. In: In: DEStech Transaction on Social Science Education and Human Science, pp. 2637–2640 (2018). https://doi.org/10.12783/dtssehs/ichss2017/19584
20. Crosby, M., Nachiappan, P.P., et al.: Blockchain technology: beyond bitcoin. Appl. Innov. Rev. (2016). https://doi.org/10.1515/9783110488951
21. Tama, B.A., Kweka, B.J., Park, Y., Rhee, K.H.: A critical review of blockchain and its current applications. In: Proceeding 2017 International Conference Electronic Engineering Computer Science Sustain Cultural Heritage Smart Environment Better Future, ICECOS 2017, pp. 109–113 (2017). https://doi.org/10.1109/ICECOS.2017.8167115
22. Christidis, K., Devetsikiotis, M.: Blockchains and smart contracts for the internet of things. IEEE Access **4**, 2292–2303 (2016). https://doi.org/10.1109/ACCESS.2016.2566339
23. Tian, F.: An agri-food supply chain traceability system for china based on RFID & blockchain technology (2016). https://doi.org/10.1109/ICSSSM.2016.7538424
24. Dorri, A., Kanhere, S.S., Jurdak, R.: Blockchain in internet of things: challenges and solutions (2016). https://doi.org/10.1145/2976749.2976756
25. Kumar, N.M., Mallick, P.K.: Blockchain technology for security issues and challenges in IoT. Proc. Comput. Sci. **132**, 1815–1823 (2018). https://doi.org/10.1016/j.procs.2018.05.140
26. Farooq, S., Obrien, C.: A technology selection framework for integrating manufacturing within a supply chain. Int. J. Prod. Res. **50,** 2987–3010 (2012). https://doi.org/10.1080/00207543.2011.588265
27. Perboli, G., Musso, S., Rosano, M.: Blockchain in logistics and supply chain: a lean approach for designing real-world use cases. IEEE Access **6**, 62018–62028 (2018). https://doi.org/10.1109/ACCESS.2018.2875782

28. Saberi, S., Kouhizadeh, M., Sarkis, J., Shen, L.: Blockchain technology and its relationships to sustainable supply chain management. Int. J. Prod. Res. **57,** 1–19 (2018). https://doi.org/10.1080/00207543.2018.1533261
29. Bowen, T., Jiaqi, Y., Si, C., Xingchen, L.: The Impact of Blockchain on Food Supply Chain: The Case of Walmart, vol. 11373, pp. 149–158. Springer, Switzerland. https://doi.org/10.1007/978-3-030-05764-0
30. Hinckeldeyn, J., Jochen, K.: (Short paper) developing a smart storage container for a blockchain-based supply chain application. In: Proceedings of 2018 Crypto Valley Conference Blockchain Technology CVCBT 2018, pp. 97–100 (2018). https://doi.org/10.1109/CVCBT.2018.00017
31. Min, H.: Blockchain technology for enhancing supply chain resilience. Bus. Horiz. **62**, 35–45 (2019). https://doi.org/10.1016/j.bushor.2018.08.012
32. Queiroz, M.M., Fosso Wamba, S.: Blockchain adoption challenges in supply chain: an empirical investigation of the main drivers in India and the USA. Int. J. Inf. Manag. **46**, 70–82 (2019). https://doi.org/10.1016/j.ijinfomgt.2018.11.021
33. Malik, S., Kanhere, S.S., Jurdak, R.: ProductChain: scalable blockchain framework to support provenance in supply chains. In: 2018 IEEE 17th International Symposium *on* Network Computing and Applications, NCA 2018, pp. 1–10 (2018). https://doi.org/10.1109/NCA.2018.8548322
34. Dyba, T., Digsoyr, T., Hanssen, G.: Applying systematic reviews to diverse study types: an experience report. In: Proceedings 1st International Symposium on Empirical Software Engineering and Measurement, pp. 126–135 (2007). https://doi.org/10.1109/ESEM.2007.59
35. Kitchenham, B., Charters, S.: Guidelines for performing systematic literature reviews in software engineering technical report. Software Engineering Group EBSE Technical Report Keele University, Department of Computer Science University of Durham 2 (2007)
36. Fink, A.: Conducting Research Literature Reviews: from the Internet to Paper, 4th edn. SAGE, Thousand Oaks (2014)
37. Petersen, K., Vakkalanka, S., Kuzniarz, L.: Guidelines for conducting systematic mapping studies in software engineering: An update. Inf. Softw. Technol. **64**, 1–18 (2015). https://doi.org/10.1016/j.infsof.2015.03.007
38. Barrera, J., Valverde, F.: Influencia de las TIC's en la Logística Integral de las PYMES para la Internacionalización de Bienes y Servicios (2018)
39. Macías, D., Robles, G.: Análisis del proceso logístico y su incidencia en exportaciones en Isotanques en la empresa J. B. Logistic S.A (2018)
40. Mendoza, J., Sandoval, A.: Análisis del proceso logístico de importación de insumos médicos en la empresa VEIMPEX S.A (2018)
41. Martínez-Vivar, R., Sánchez-Rodríguez, A., Pérez-Campdesuñer, R., García-Vidal, G.: Contribution to the logistic evaluation system in the transportation process in santo domingo, Ecuador. J. Ind. Eng. Manag. **11**, 72–86 (2018). https://doi.org/10.3926/jiem.2422
42. Sablón Cossio, N., Cárdenas Uribe, M.B., Pérez Quintana, M.L., Bravo Sánchez, L., Manjarrez Fuentes, N.: Milk agro- alimentary chain sustainable development strategy in the conditions of the Ecuadorian Amazon region. In: Proceedings of the International Conference on Industrial Engineering and Operations Management, pp. 943–951. IEOM Society International, Bandung (2018)
43. Álvarez, K., Aldas, D., Reyes, J.: Towards lean manufacturing from theory of constraints: a case study in footwear industry. In: 2017 International Conference Industrial Engineering, Management Science and Application, ICIMSA 2017 (2017). https://doi.org/10.1109/ICIMSA.2017.7985615

44. Ruiz Carrillo, J.A., Pupo Francisco, J.M.: Manufacturing system improvement in shrimp processing: analysis case MARECUADOR S. A. exporter. Revista ESPACIOS **38**, 17 (2017)
45. Rosero-Mantilla, C., Sánchez-Sailema, M., Sánchez-Rosero, C., Galleguillos-Pozo, R.: Aggregate production planning, casestudy in a medium-sized industry of the rubber production line in ecuador. In: IOP Conference Series Mater Science Engineering, vol. 212 (2017). https://doi.org/10.1088/1757-899X/212/1/012018
46. Guerra, M.: Como potenciar el modelo de sistemas de información (SCM) cadena de suministros en la empresa Schréder Ecuador S.A. (2018)
47. Córdova, J.: Incidencia del Bitcoin en el Comercio Internacional en el Periodo 2014–2016 (2018)
48. Reyes, J.: Faps system: a prototype for lean manufacturing scheduling in footwear. Presented at the XII Jornadas Iberoamericanas de Ingeniería de Software e Ingeniería del Conocimiento y Congreso Ecuatoriano en Ingeniería de Software, pp. 13–25 (2017)
49. Bermeo, M.J.P., Castillo, C.H.P., Serrano, V.S.: Neural networks and genetic algorithms applied for implementing the management model "Triple A" in a supply chain. case: collection centers of raw milk in the Azuay Province. In: MATEC Web Conference, vol. 68, pp. 06008–06008 (2016). https://doi.org/10.1051/matecconf/20166806008
50. Silva, D., Guerreiro, S., Sousa, P.: Decentralized enforcement of business process control using blockchain. In: Aveiro, D., Guizzardi, G., Guerreiro, S., Guédria, W. (eds.) Advances in Enterprise Engineering XII, pp. 69–87. Springer, Cham (2019). https://doi.org/10.1007/978-3-030-06097-8_5
51. Toapanta, M., Mero, J., Huilcapi, D., Tandazo, M., Orizaga, A., Mafla, E.: A blockchain approach to mitigate information security in a public organization for Ecuador. Presented at the IOP Conference Series: Materials Science and Engineering (2018). https://doi.org/10.1088/1757-899X/423/1/012164
52. Kshetri, N.: Can blockchain strengthen the internet of things? IT Prof. **19**, 68–72 (2017). https://doi.org/10.1109/MITP.2017.3051335
53. Holma, H., Salo, J.: Global Logistics: New Directions in Supply Chain Management: Improving Management of Supply Chains by Information Technology, Water, G.,. London (2014)
54. Treiblmaier, H.: The impact of the blockchain on the supply chain: a theory-based research framework and a call for action. Supply Chain Manag. **23**, 545–559 (2018). https://doi.org/10.1108/SCM-01-2018-0029

Agricultural Information Management: A Case Study in Corn Crops in Ecuador

Fernando Sichiqui[1], Jaime Gustavo Huilca[1], Andrea García-Cedeño[1(✉)], Juan Carlos Guillermo[1], David Rivas[2], Roger Clotet[3,4], and Monica Huerta[1]

[1] GITEL, Telecommunications and Telematics Research Group, Universidad Politécnica Salesiana, Cuenca, Ecuador
{pschiqui,jhuilca}@est.ups.edu.ec,
{agarciac,jguillermo,mhuerta}@ups.edu.ec

[2] Universidad de las Fuerzas Armadas ESPE, Sangolquí, Ecuador
drrivas@espe.edu.ec

[3] Networks and Applied Telematics Group, Universidad Simón Bolívar, Caracas, Venezuela
roger.clotet@campusviu.es

[4] Universidad Internacional de Valencia, Valencia, Spain

Abstract. This document, proposes the development of a multiplatform application called Platano-Maiz deployed on a Umbler server, this application allows farmers community, through the Internet to: monitor, administer and manage information of climate and soil variables that influence the growth of corn crop, which is located in Guapán, Cañar - Ecuador. The main objective of developing this application is to provide farmers with a secure tool to support them in making optimal and timely decisions based on the information graphically presented using statistical graphs. In addition, through the application, farmers can monitor the entire growth process of their crop, improve production, optimize resources (water, fertilizers, time, money, among others) and allow them to view and receive notifications via email. The web application is developed with free software tools and libraries, allowing the community of developers to extend its development (data analysis, data prediction, etc.), the graphical interface is user-friendly and intuitive for any user and the monitored data are completely open and accessible.

Keywords: Agricultural · Information management · Corn

1 Introduction

The agricultural production of corn, as well as a wide variety of crops, is the main food and economic contribution of various nations. These countries are dependent on the primary sector of production for the generation of jobs and raw materials, which in turn gives rise to the development of other activities; therefore, the relevance of agriculture demands proper management of soil, water and fertilizers to ensure a productive harvest without affecting the environment

M. Botto-Tobar et al. (Eds.): ICAETT 2019, AISC 1066, pp. 113–124, 2020.
https://doi.org/10.1007/978-3-030-32022-5_11

[22]. In the case of Ecuador, the presence of a wide variety of soils, climates and geographical features gives rise to a wealth of crops, a positive factor for agricultural practice, which represents the general income of rural communities belonging to 37% of the population [4,15].

Globally, maize is considered as a staple food due to its contribution of nutrients, which in addition to its benefits for human consumption, corresponds to the main ingredient for livestock feed, as well as in the development of industrial products [23]. Consequently, the cultivation of this cereal is of utmost importance, especially in Latin America, a region that in 2017 produced 259 million tons of maize according to the Food and Agriculture Organization of the United Nations (FAO) [2]. In Ecuador, maize maintains economic and historical relevance, represents a product worked for centuries that adapts to different ecosystems: from the tropics to the Andean [11]. The study area of this article are the Andean of Ecuador.

In Ecuador, the importance of maize crops and the benefits they provide have motivated different government entities to implement strategies in favor of the grain production. However, public policies are geared towards the industrial agricultural sector, generating a challenge for small and medium-sized producers [15,18]. According to the III National Agricultural Census, more than 3 million inhabitants of rural areas of Ecuador practice family farming, and 62% of all agricultural production units correspond to subsistence family farmers characterized by conditions of poverty and very low or null economic income from the sale of their crops [9]. Based on this problem, the scientific community has developed different innovative systems that allow to increase the results in sowing and harvesting processes and, hence, provide solutions for small and medium-sized farmers inclusion in economic and technological contexts. Among the various existing technologies, the trend of Wireless Sensor Networks (WSN) for monitoring soil, plant and environmental data stands out as a mean for increase productivity [1].

In order to improve development and production of maize plantations, it is proposed to design a web interface for the management of agricultural information provided by a WSN. The objective is to contribute to the farmer's decision making through the correct administration and visualization of data: humidity and temperature of the environment, humidity and temperature of the soil, luminosity and radiation. The user will be able to follow the development of the crop through a user-friendly virtual environment, which doesn't require technical knowledge for its utilization. It consists of an economically accessible solution based on free software. The information will be vital for optimal management of resources and prevention of pests and diseases.

The beneficiaries of this project are part of the community of small farmers of the Guapán parish in the city of Azogues in Cañar province of Ecuador. The inhabitants of this sector are dedicated to the cultivation of maize, where climatic variations, such as changes in temperature and humidity, have hindered the development of the plant which causes a decrease in quality and production [21], causing economic and social losses in terms of food.

The present document is structured in different headings: in Sect. 2, the background in reference to maize production in the study area is detailed, as well as the main characteristics of the product; Sect. 3 describes the development methodology; the results of the final design of the interface are presented in Sect. 4; and finally, in Sect. 5 the conclusions are established.

2 Related Work

Different works implemented in the agriculture sector have been analyzed, focused on WSN and graphic platforms to visualize the monitored information.

The project by Jara et al. concerns a monitoring system for environmental parameters in order to optimize the use of water in agriculture, where a network of sensors made up of ZigBee technology is used to send the information. This data will be shown in an interface designed in LabVIEW and over an Android application [16]. Similarly, Erazo et al. implement a WSN using ZigBee and WiFi protocols to monitor the climatic variables of a greenhouse [13].

In Valencia-Spain (2017) the network of low energy mobile sensors PLATEM PA was implemented in hydroponic agriculture. The purpose of the system is to acquire environmental data to monitor the plant condition and make appropriate decisions for proper development. The gateway is in charge of collecting the data and sends operations to the actuators, it is connected to Internet through a 3G mobile network and stores the information in its multimedia platform [6]. Likewise, the authors of [1] present a proposal based on the same technology, however it is intended for coffee crops monitoring in Santa Isabel, Ecuador.

Another project analyzed is a fuzzy system of irrigation applied to the growth of Habanero pepper in Yucatan, Mexico. The authors analyzed the crop coefficients, soil type, wind speed, solar radiation, temperature, relative humidity and rainfall level [7].

Guillermo et al. propose an architecture based on Internet of Things, composed by a WSN for agricultural monitoring of cacao crops in Ecuador. This system is focused on increasing productivity, preventing pests and optimizing resources. It stands out for being low cost through the use of open source software and hardware [15].

On the other hand, in [12] a WSN application analysis is presented to monitor and optimize flower crops, which is one of Ecuador's main export products. The acquired data are displayed through a graphical interface designed in LabVIEW, this allows the visualization of alerts when the variables exceed the adequate ranges. Also, in a second paper they analyzed energy consumption of proposed system [14].

In project [5] a WSN is designed for greenhouse crops, using as protocol 6LoWPAN and RPL. It measures humidity, temperature, light and water content, information that is stored on a LAMP server and then displayed on a dynamic website.

3 Maize Production in Cañar province

Maize, also known as Zea-Mays, is one of Ecuador's most significant crops. It is a cereal that adapts to different ecological and soil conditions and can be classified into two groups: soft maize and hard maize. The first category is grown for personal consumption and for domestic trade, which is planted in the Ecuadorian highlands. The hard maize is cultivated in the zone of the coast and is characterized for being of yellow color, this one serves for the agroindustry [22]. The advantage of this agricultural product is that it can be used as food at any stage of its growth [24].

Cañar belongs to one of the provinces of the Ecuadorian sierra where diversified family agriculture is most practiced. The main crops, which cover 58.20% With regard to the production of the latter, it constitutes 23.64% and is destined for self-consumption [3,18]. Within Cañar province (see Fig. 1), a zone whose predominant crop is corn has been chosen, therefore, the study area used corresponds to the parish Guapán [8].

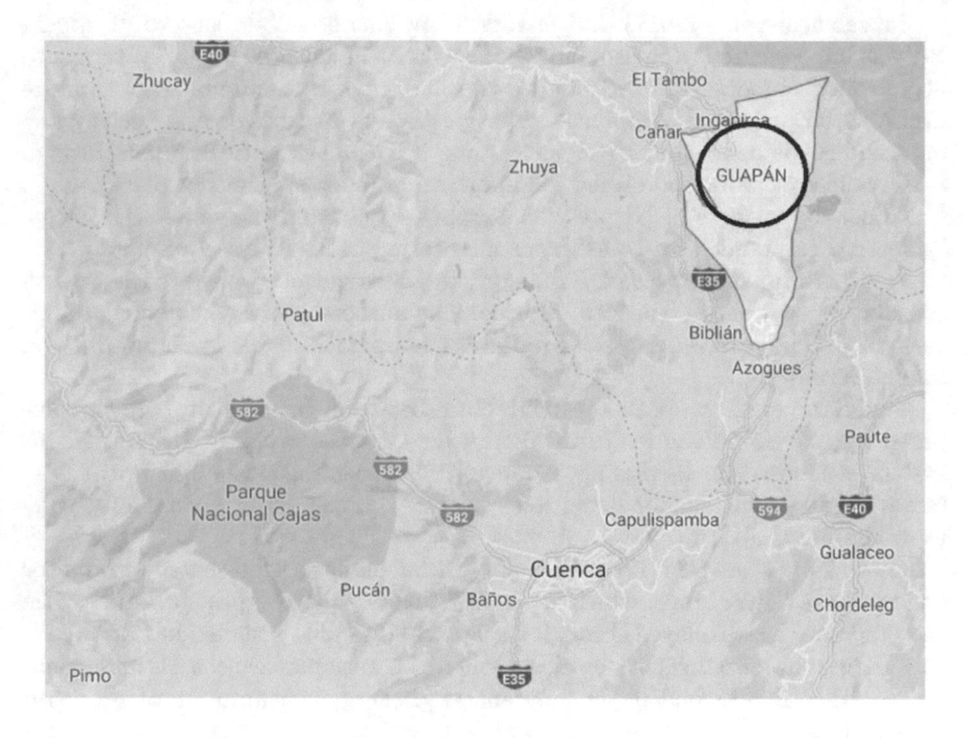

Fig. 1. Study area localization

3.1 Botanical Characteristics of Maize

The corn plant develops easily with appropriate climatic conditions, it can reach a height of between 2.70 to 3 m. Among the main characteristics is its root system, stem and leaves. In the root system there are three types of temporary, permanent and anchoring roots. The temporary ones appear when the grain sprouts and later the permanent roots appear they can arrive at a depth of 2 m, that is to say that the radicular system this formed by roots that are united that arrive at reaching a radius of 1.3 m, the development of the roots this related to the humidity of the ground and stops its growth if there is excess or lack of water. The stem is of robust cylindrical form with knots of 8 to 38, this structured by the epidermis that is in charge of the protection, the wall that has some channels where the elaborated sap is transferred that goes from the roots up to the leaves. The medulla, which is a mass in the centre of the stem, is where proteins and water are stored. The leaves can measure up to 1 m and each corn plant can have between 8 to 30 leaves, this formed by the pod is wrapped around the stem, the limb which has a central vein and parallel veins, and the ligule has the function of protecting against water and dust that does not enter the stem [19].

3.2 Climate and Soil Requirements for Maize Crops

The development of the corn plant depends on environmental conditions such as temperature, humidity, and solar radiation; if there are droughts and high temperatures, early ripening may occur. The soils that are destined for sowing must have a texture capable of retaining humidity, capable of storing water and necessary nutrients, this is important because the roots grow to a depth of 60 cm demanding a drained and aerated soil [4].

The preparation of the soil is very important, they are carried out plowed and crossed to approximately 30 cm or 40 cm of depth obtaining a better drainage, retaining water and at the same time, cleaning grasses and weeds. In the sowing the best seeds are selected so that they can resist to the diseases and plagues. It is recommended to sow, preferably by hand, when the soil has a temperature of 12°C [17]. The main climatic and soil conditions suitable for the development of maize correspond to:

- Precipitations from 600 to 1200 mm during the entire development cycle.
- Light from 1000 to 1500 h during the development cycle.
- Temperature between 10–20°C and maximum 30–32°C
- The type of soil should be neither too sandy nor too compact.

3.3 Maize Diseases

Maize in tropical ecosystems is attacked by many pests, this causes production losses as hot and humid climates favor the development of pests and diseases [20]. The diseases that most affect this type of crop are: Coffee stain, tropical rust and asphalt stain:

- The brown spot is caused in areas where there is abundant rainfall and high temperatures, which directly affects the leaves, pods and stems. The first symptoms are when small spots appear in the part of the central rib in a circular and brown, this disease can cause the stem to rot [10].

- Tropical rust is a fungus that appears in hot and humid tropical climates, sprouts in the form of sores in a circular or oval, with a yellow and white color where this disease produces a hole, until causing the fall of the leaf [10].
- The asphalt stain complex occurs in humid areas where bright black spots appear, then develop necrotic areas in the tissue which if the stains cause complete burning of the foliage [10].

4 Methodology

WSN is composed of two nodes, each one is responsible for acquiring climatic and soil variables that influence the development of the corn crop as: humidity of the environment, soil moisture, air temperature, relative temperature, Luminosity and Radiation.

Through Python script each package of sensor node, that contend id and read values from sensors, is processed to store the values in the Raspberry memory.

Average values from WSN lectures are stored in a MySQL database. Through WAMP server this values can be requested using GET call from a Lynx browser.

The stored data can be analyzed through a statistical graphs, created using Highcharts image manager based on JavaScript. The data are classified by date and time so that through a graphical form the last data received from the nodes can be shown, also graphs show maximum, average and minimum values for each hour in the day. The user can also select aggregation of data by week or month.

The system components are the WSN, the server, the visualization interface and the graphic manager, the relation between them are show in Fig. 2.

The graphic interface for the system was designed and implemented with free software, friendly and entertaining for the users. The interface is divided into four sections: Banner, Content Box, Menu and End of Page Label as observed in Fig. 3.

Below is detailed each of the sections of the graphical interface:

1. Banner: In this section, we show a logo that represents the "Plátano-Maíz" project, we also find the administrator icon where we will be redirected to allow to create new users or change log in user.
2. Content Box: It shows the contents of the different options of the menu: graph the data of the climatic variables obtained in real time, and also the graph of a day, week and month.
3. Menu: Permits to select between: Analysis and visualization options of each node, allows users to select the type of graph that they want to observe either the data of a day, week and month; Display the data that is stored in database; Contact the people who perform the project and its location; and show project information.
4. End of Page Label: Project name and year. When clicking, the page will be reloaded again.

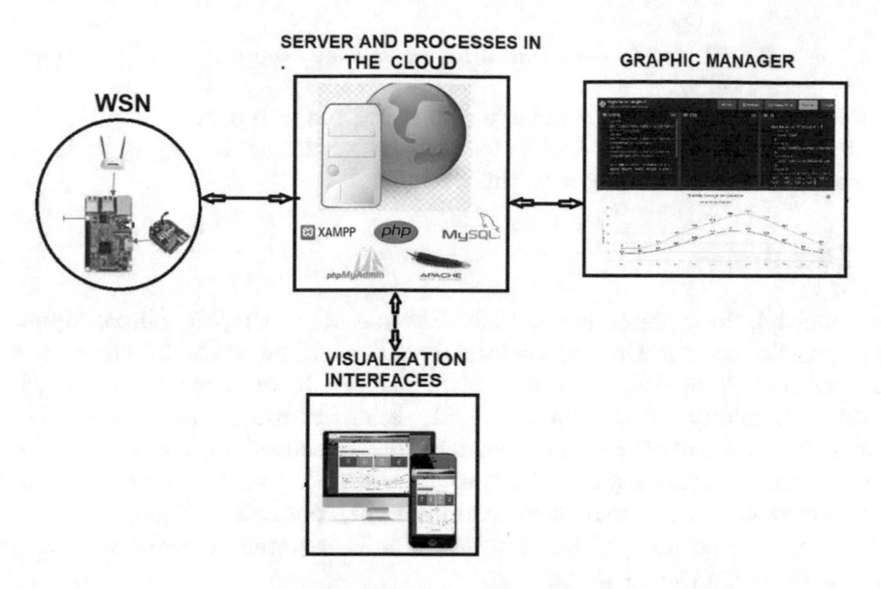

Fig. 2. System components

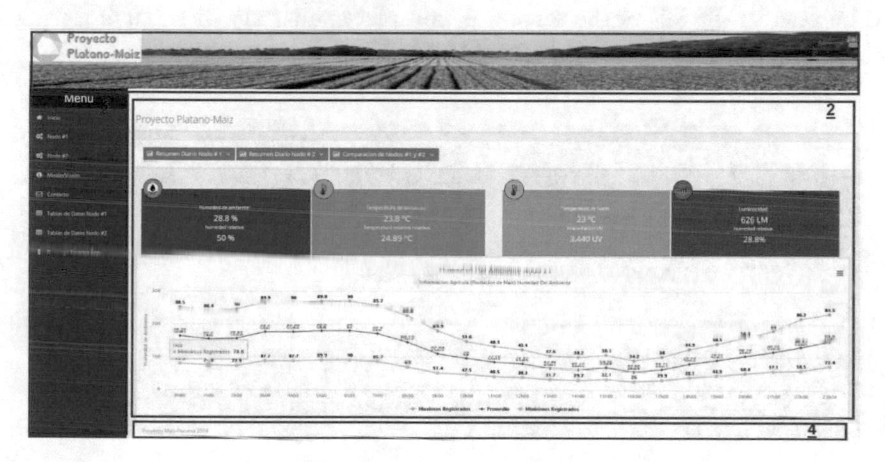

Fig. 3. Graphic user interface

4.1 Programming Languages

- HTML5 is used to creat webpages to be visalized in a browser.
- CSS 3.0 is used for managing the styles of the website making it more user friendly.
- Bootstrap 4.0 tookid is used to manage the sensitivity of browsers in terms of screen sizes by adjusting their content to different devices. Java Script is used to reduce the load of server-level processes such as validations, handling of strings, handling of dates.

- JQuery 3.7 library is used handling of special events in jointly as part of JavaScript code.
- Python 3 is used to communicate nodes and the web page.
- JSON is used as lightweight data-interchange format to paging node information and transmitting it to the web page.

5 Results

The multiplatform interface of the Platano-Maiz Project allows dynamic and graphic visualization of historical and realtime state of climatic and soil variables provided by the WSN. The WSN contains 2 nodes which measure humidity of the environment, temperature, relative temperature, luminosity and radiation. The visualization, download and query results of WSN data measurements, is for public use and can be accessed through http://proyectoplatano-maiz-com-ec.umbler.net/?open. For changes of visualization or management of the databases, it is required to enter with a user and password previously authorized.

In the graphic interface of Fig. 4, the main menu with its respective options can be seen to the left of the screen. A complete summary of the monitoring in real time of the variables acquired by the WSN can be visualized.

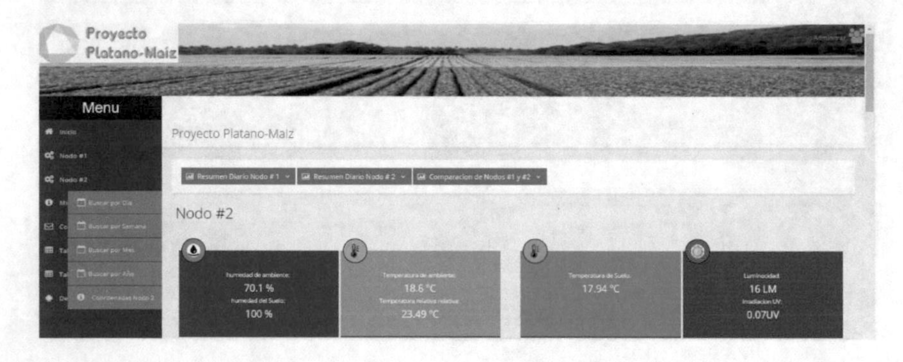

Fig. 4. Graphic interface (start menu)

As seen in Fig. 5 and Fig. 6, in the main menu of the application the options Node 1 and 2 can be found, with which a sub-menu is opened with different alternatives to visualize the maximums, minimums and averages of the data obtained during the monitoring of the climatic and soil variables. The visualization is displayed by graphs and statistical information can be selected daily, weekly, monthly and annually.

The system allows observing the behavior of the different variables that influence the growth and production of the corn crop, in the event that the values exceed the pre-established maximum and minimum ranges. The system automatically sends or notifies the farmer with a message, email or phone call, so

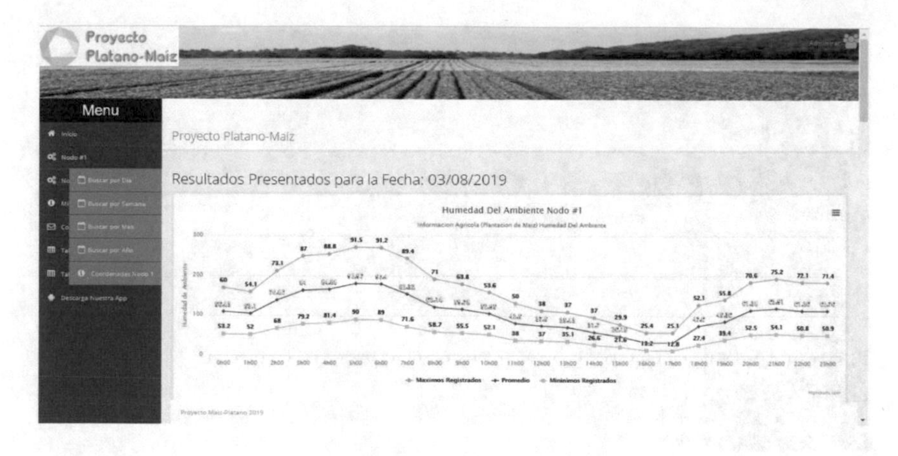

Fig. 5. Data display per day

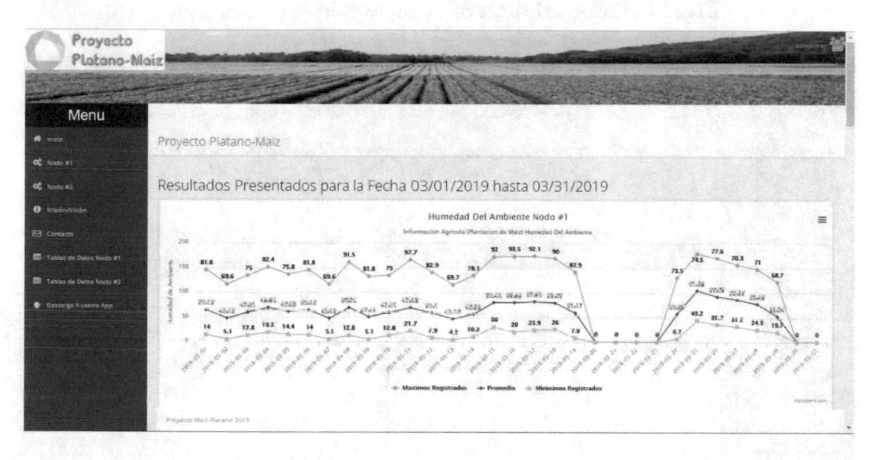

Fig. 6. Data display by month

he can make decisions immediately. The application allows knowing the mission and vision of the project, for the benefit of small farmers. Additionally, you have the option to visualize a table with the historical data of the climate and soil variables obtained from the WSN as shown in Fig. 7.

Finally, the application has the option to manage the web page, for this, it is necessary to create a user (name, surname, email, user's name) and a password. Once the user is registered, it is possible to enter the system to manage, manage, store information and add more nodes, among others. According to the assigned role, the user can add, modify and eliminate certain options of the application as shown in the Fig. 8.

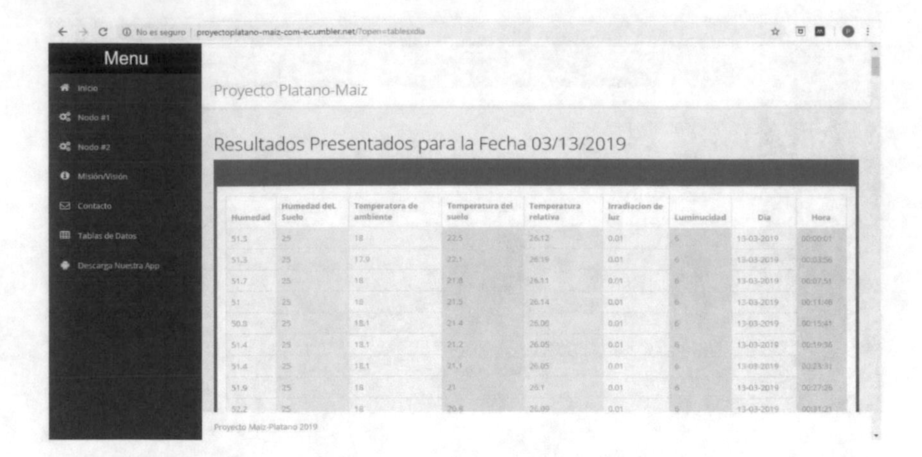

Fig. 7. Historical data of climate and soil variables

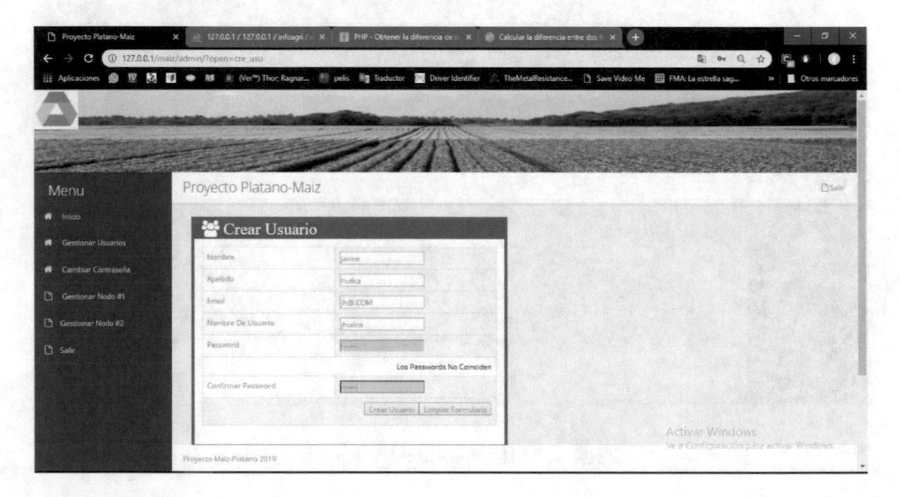

Fig. 8. Option to manage the web page

6 Conclusions

The Platano-Maiz multiplatform application was of great help, because with the data obtained during the monitoring stage, farmers were able to make a concise evaluation, follow up and adequate control of the corn crop from any geographical location through the internet, they received adequate notifications of changes in climate and soil variables, optimized time and resources, minimized the efforts involved in the production of the crop, contributes to the right decision making by the farmer, specialists, researchers, and so on. It also allows to control pests and diseases, critical periods, inferences about data and so on.

The system can be implemented anywhere with maize crops that need to measure climate and soil variables, thus bridging the current gap between technology and small farmers. The application's graphical interface is user-friendly and intuitive, and is developed with open source hardware and software tools.

The information or historical data from the monitoring carried out is completely accessible to any person either for later analysis or for use in any other system or application.

According to the analysis of related works, in most of the provinces of Ecuador, there is currently a very large digital gap in the use of technology in agriculture, mainly due to the following factors: lack of knowledge of the benefits of the application of technology in the fields by farmers, this causes farmers to lose interest, and lack of economic support and training by government institutions in charge of agricultural development in Ecuador.

For future work, the monitoring of agricultural variables through the "Banana-Maize" project is contemplated in periods from the sowing stage to the harvest -approximately 6 months-. The objective is to obtain a database of reference to the behavior of several crop cycles for their respective analysis according to the final agricultural production. The acquired results will form the basis for determining the optimal parameters for the efficient development and management of the sowing.

It will be implemented a module for the analysis of monitored data using Big Data techniques, in order to make weather and soil predictions using Machine Learning techniques. It is also planned to implement web service, database and front end files in the institutional server.

References

1. Abad, J., Farez, J., Chasi, P., Guillermo, J.C., García-Cedeño, A., Clotet, R., Huerta, M.: Coffee crops variables monitoring: a case of study in Ecuadorian Andes. In: International Conference of ICT for Adapting Agriculture to Climate Change, pp. 202–217. Springer, Heidelberg (2018)
2. Organización de las Naciones Unidas para la Alimentación y la Agricultura, O.: Producción de cereales de américa latina y el caribe en 2017 superó en 20 por ciento el nivel de 2016 (2017). http://www.fao.org/americas/noticias/ver/es/c/1072212/
3. Amaya, M., Aguirre, O., Iñaguazo, C., Vallejo, R., Espinel, R.: PERFIL Agropecuario Provincial del Ecuador 2008 (2008). http://www.ecuadorencifras.gob.ec/documentos/web-inec/Bibliotecas/Estudios/Estudios_Economicos/Perfil_Agropecuario_2008.pdf
4. Basantes Morales, E.R.: Manejo de cultivos andinos del ecuador (2015)
5. Cama Pinto, A., Montoya Gil, F., Gómez López, J., García Cruz, A., Manzano Agugliaro, F.: Sistema inalámbrico de monitorización para cultivos en invernadero (2014)
6. Cambra, C., Sendra, S., Jimenez, J., Lloret, J.: Red de sensores inalámbricos de bajo consumo energético en agricultura hidropónica. XIII Jornadas de Ingeniería telemática (JITEL 2017). Libro de actas, pp. 55–62 (2018)
7. Ceballos, M.R., Gorricho, J.L., Palma Gamboa, O., Huerta, M.K., Rivas, D., Erazo Rodas, M.: Fuzzy system of irrigation applied to the growth of Habanero pepper (capsicum Chinense Jacq.) under protected conditions in Yucatan, Mexico. Int. J. Distrib. Sens. Netw. **11**(6), 123543 (2015)

8. Clavijo, W.: Estudio de Impacto Ambiental y Plan de Manejo Ambiental para el Coprocesamiento de Desechos Peligrosos en la Planta Industrial Guapán de la Empresa Unión Cementera Nacional. Technical report, WCR Consultoría Ambiental (2017)
9. Dávalos, D.: Caracterización de la agricultura familiar campesina en el Ecuador (2015)
10. De León, C.: Enfermedades del maíz: una guía para su identificación en el campo (1984)
11. Crop Science Ecuador: El maíz, un alimento fundamental de identidad (2017). https://www.cropscience.bayer.ec/Noticias/Noticias/2017/Septiembre/Maiz-alimento-ancestral.aspx
12. Erazo, M., Rivas, D., Pérez, M., Galarza, O., Bautista, V., Huerta, M., Rojo, J.L.: Design and implementation of a wireless sensor network for rose greenhouses monitoring. In: 2015 6th International Conference on Automation, Robotics and Applications (ICARA), pp. 256–261. IEEE (2015)
13. Erazo-Rodas, M., Sandoval-Moreno, M., Muñoz-Romero, S., Huerta, M., Rivas-Lalaleo, D., Naranjo, C., Rojo-Álvarez, J.: Multiparametric monitoring in equatorian tomato greenhouses (i): wireless sensor network benchmarking. Sensors **18**(8), 2555 (2018)
14. Erazo-Rodas, M., Sandoval-Moreno, M., Muñoz-Romero, S., Huerta, M., Rivas-Lalaleo, D., Rojo-Álvarez, J.: Multiparametric monitoring in equatorian tomato greenhouses (ii): energy consumption dynamics. Sensors **18**(8), 2556 (2018)
15. Guillermo, J.C., García-Cedeño, A., Rivas-Lalaleo, D., Huerta, M., Clotet, R.: IOT architecture based on wireless sensor network applied to agricultural monitoring: A case of study of cacao crops in ecuador. In: International Conference of ICT for Adapting Agriculture to Climate Change, pp. 42–57. Springer (2018)
16. Jara, M., Arauco, L., Abad, P., Paucar, R.: Implementación de un sistema de monitoreo de parametros ambientales y del suelo desde dispositivos fijos y móviles usando la tecnología zigbee para el uso eficiente del recurso hídrico en la agricultura (2014)
17. López Marcillo, B.R.: Análisis de la producción de maíz en la provincia de Manabí y su aporte al desarrollo local. Periodo 2012-2017. B.S. thesis, Universidad de Guayaquil. Facultad de Ciencias Económicas (2018)
18. Martínez Valle, L.: La Agricultura Familiar en El Ecuador. Serie Documentos de Trabajo N 147, No. 28, Rimisp (2013)
19. Ortas, L.: Cultivo del maíz: fisiología y aspectos generales (2008)
20. Paliwal, R.L.: El maíz en los trópicos: mejoramiento y producción, No. 28, Food & Agriculture Org (2001)
21. Ríos Marín, M., Sánchez Padilla, L., Gómez Gómez, J.: Sistema de monitoreo agrícola mediante redes inalámbricas de sensores para el monitoreo de variables ambientales-sismoagro. Ingeniería al Día **2**(2), 4–22 (2016)
22. Sánchez Ortega, I., Pérez-Urria Carril, E.: Maíz i (zea mays). REDUCA Biol. **7**(2), 151–171 (2014)
23. Shiferaw, B., Prasanna, B.M., Hellin, J., Bänziger, M.: Crops that feed the world 6. past successes and future challenges to the role played by maize in global food security. Food Secur. **3**(3), 307 (2011)
24. Tovar, C.D.G., Colonia, B.S.O.: Producción y procesamiento del maíz en colombia. Revista Guillermo de Ockham **11**(1), 97–110 (2013)

Pillars for Big Data and Military Health Care: State of the Art

Diana Martinez-Mosquera[1](✉), Sergio Luján-Mora[1], Luis H. Montoya L.[2], Rolando P. Reyes Ch.[3], and Manolo Paredes Calderón[3]

[1] Departamento de Lenguajes y Sistemas Informáticos, Universidad de Alicante, Alicante, Spain
sdmml@alu.ua.es, sergio.lujan@ua.es
[2] Departamento de Ciencias de la Ingeniería, Universidad Israel, Quito, Ecuador
lmontoya@uisrael.edu.ec
[3] Universidad de las Fuerzas Armadas ESPE, Sangolquí, Ecuador
{rpreyes1,mparedes}@espe.edu.ec

Abstract. Big Data is a buzzword used to describe the processing of high volumes of data. Some types of health data are considered as Big Data due to the huge amount of data originated in this sector. Researchers have consolidated their efforts to present new tools and platforms for Big Data in health care, especially with the exponential growth observed on remote sensors. Although no specific studies have been presented at the military health context, the collected experience from several reviews proves the need for applying Big Data techniques to ensure efficient military operations. In this paper, we present the attained results from state of the art studies about Big Data and health case reviews published during the 2014 to 2018 timeframe. As a result, 17 relevant studies were found from several scientific digital libraries; the main proposed approaches and methodologies that are able to be included into the military health care domain were summarized into acquisition, storage, processing, management, security, and normative pillars. The results reveal the need for further studies regarding the military health care using Big Data approaches in order to improve the military life. It is important to mention that militaries are constantly exposed to health risks and this is the main reason for monitoring their health status.

Keywords: Big Data · Health care · Military · Review · State of the art

1 Introduction

Big Data is currently one of the most popular buzzwords in the area of information systems, used to describe massive volumes of data [1]. Health data can be considered as Big Data due to the explosive growth of generated data provided by medical devices for health monitoring and tracking, thus expanding the available patient information, his behaviors, and his health status such as heart rate, blood pressure, and sleep patterns, among others [2].

M. Botto-Tobar et al. (Eds.): ICAETT 2019, AISC 1066, pp. 125–135, 2020.
https://doi.org/10.1007/978-3-030-32022-5_12

It is widely known that militaries face danger at every turn and are constantly exposing their health to sometimes unknown dangers; thus, a study [3] presented that 62% of federal information technology (IT) professionals in United States believe that including Big Data will improve the health care in the military life. Furthermore, physical preparation of the militaries is usually outdoors and they are generally assigned to isolated places and exposed to adverse environments; therefore, the use of remote monitoring, considered as a technique to deal with the health care, is suitable to monitor the military health status and perform prompt clinical decisions assisted by Big Data analytics platforms.

Considering the presented issues, this work consolidates all the reviews for Big Data and health care performed during the 2014 to 2018 timeframe, since after reviewing the state of the art studies a specific study targeted to military health care is missing. The difference between Big Data analytic of civilian health and military health are the conditions, because as it is known, during the military physical training or missions, the personnel suffer overload of physical work for a long time.

Our efforts were focused in review studies, since they consolidate the scientific research in a period, with the aim of extracting the suitable approaches and methodologies proposed at the health care and Big Data areas that could be applied to the military health care domain. The raised research questions in our work are: What researches about Big Data and military health care have been published? What approaches and methodologies regarding Big Data and health care can be applied to the military domain?

As result, 17 review studies were deemed relevant and after a deep analysis of the collected scientific evidence from these studies, we propose six pillars for the use of Big Data in a military context: acquisition, storage, processing, management, security, and normative.

The remainder of this paper is organized as follows. Section 2 presents the method used to perform this study explaining in detail every phase followed by the review: objectives and justification, research questions, strategy and how the studies were collected. Section 3 summarizes the main findings from every relevant review study. Section 4 presents the answers to the research questions and based on the results we propose the six fundamental pillars for Big Data in military health care. Finally, conclusions and future works are presented in Sect. 5.

2 Method

The state of the art study will be based on the guidelines proposed by Kitchenham [4], the main phases are: planning, conducting, and reporting the study. During the next sections, each phase will be further elaborated.

2.1 Planning the State of the Art Study

The main goals of this phase are explained in detail below:

Identification of the Need
We have searched for state of the art studies related to Big Data and military health care but found no studies, thus, in this article we propose to consolidate reviews about Big Data and health care in order to know the researchers' perspective in this field and how they can be applied to the military health care domain.

Objectives and Justification
This state-of-the-art study includes the relevant issues of the reviews in Big Data and health care area in order to identify the existing proposals and methodologies and how a Military Scientific and Technology Research Center can apply them. The main objective is conducting the research in order to satisfy the needs from military community regarding the health of their effectives.

Research Questions
This study raises the following research questions:

RQ1: What researches about Big Data and military health care have been published?

Rationale: Our interest is to consolidate all research efforts related to Big Data and military health care with the aim of employing identified approaches and methodologies in a Military Scientific and Technology Research Center.

RQ2: What approaches and methodologies regarding Big Data and military health care can be applied?

Rationale: Our goal is to map the relevant findings according to the military needs.

Strategy
Our strategy to conduct the compilation of studies comprises three steps:

1. Searching for review studies about Big Data and health care from four renowned scientific digital libraries:
   - ACM Digital Library.
   - Google Scholar
   - IEEE Xplore.
   - Scopus.
2. Filtering the studies with the following inclusion criteria:
   - Only studies written in English.
   - Only secondary studies.
   - Only studies published in conferences or journals.
   - No cover letters, editorials, reports, essays, nor book reviews were considered.
   - Only studies from 2014 to 2018 to consolidate the latest information concerning the topic of interest.

   The used search string was "Big Data" AND "health" AND ("review" OR "state of the art" OR "mapping" OR "survey") in the abstract, title, or keywords fields.
3. Selecting the relevant studies after performing a quick review of the abstract, title and keywords fields.
4. Reviewing the full content of every study to answer the research questions raised during the previous stage.

2.2 Conducting the State of the Art Study

In Table 1, we summarize the results of the collection stage. We can highlight the following findings:

- Google Scholar was the highest-ranking library from all considered sources with 18 works collected, followed by Scopus with 15 works, IEEE Xplore with 3 works and ACM Digital Library with only 1 study.
- After performing a review of the abstract, title and keywords fields, the resulting selected studies were reduced to 17: 1 from ACM Digital Library, 7 from Google Scholar, 3 from IEEE Xplore, and 6 from Scopus.

Table 1. Results of collection stage.

Year/source	ACM Digital Library	Google Scholar	IEEE Xplore	Scopus
2014	0	7	1	0
2015	0	6	1	0
2016	1	4	1	4
2017	0	1	0	8
2018	0	0	0	3
Total by source	1	18	3	15

3 Results

In [2], the authors present an overview of computational methods used in the analysis of massive amounts of medical data. They are focused into computational approaches and present three main topics: challenges for Big Data in health, processing pipeline of health care informatics, and future directions.

They mention the main approaches for Big Data in health care are related to: personalized care, clinical operation, public health, genomic analytics, fraud detection, and device remote monitoring. They define device remote monitoring as a way to capture and analyze continuous health care data in vast amounts from medical devices with the purpose of monitoring medical conditions and preventing afflictions. They also emphasize that device remote monitoring is an opportunity to be explored for Big Data and health care. Here, we found the first approach that can be applied in military health, the use of device remote monitoring due to the fact that military physical preparation is mostly performed outdoors. Militaries are generally assigned to work at isolated places where they may face adverse environments; thus, remote monitoring can be used to monitor in real-time the military health status and perform early clinical decision assisted by Big Data Analytics platforms.

In this study, the authors also mention the mobile health as a scenario for Big Data in health informatics, where the use of mobile technology offers a great platform for health care. That can provide patient's information, behavior, health status and can monitor the patient's status including heart rate, blood pressure level, sleep patterns. It is considered like a useful approach in the militar domain for the effective management of this data through a Big Data solution.

In [5] the authors explain the advances in Internet of Things (IoT) for health care technologies and consider Big Data as a fundamental part to deal with the massive amount of information produced by health devices. Although this study does not present the use of Big Data in detail, we consider that the IoT approaches presented in this paper can be employed in military proposals, for instance, IoT health care network, topology, architecture, platform, services, and applications.

Another study [6] presents Big Data examples used in genomic medicine, health of the population, and clinical issues. The authors consider the management of Big Data information in the health care domain as an important challenge due to the explosive growth of health data. A research related to Big Data and cancer study is presented as an important opportunity to solve the IT personal issues regarding the management of the large amount of data from each patient. The study provides interesting facts such as, 63% of Federal IT professional are aware that Big Data technologies will help the population health and 62% indicate Big Data will improve the military life care. With the later, we highlight the importance of the Big Data approach in the military health domain.

In [7] the researchers present Big Data concepts and methodologies for health care. They mention that Big Data can be used in early detection and rapid response to disease outbreaks, for instance: disease diagnose and treatment support in the clinical decisions, clinical data, patient behavior and support information. The authors also present health care applications using Big Data tools, from which we consider important for military health: early diagnosis of diseases, use of health data, population health, images processing and signals processing. The paper is mainly concentrated into chronic diseases; however, there are important ideas for our approach. As an important statistic, they present that health care data has now grown to the range of petabytes.

In study [3], general concepts of Big Data and health care are presented, as important findings they mention that in the last 10 years the stored data has increased; thereby, there is a great wealth of data to be analyzed. Current computational analytical techniques are not enough to deal with health care, specifically in the genome area, one of the mostly studied fields. This study comprises a literature review from 2005; thus, it could be an important source for studying key items to our topic of interest.

Article [8] presents the main health care functions: clinical decision support, disease surveillance and population health management. These three functions are also suitable for military health care; therefore, this study can provide a guidance to know the platforms, tools, methodologies in Big Data applied in health care. As important findings, the researchers present data from the United States where the health care system achieved 150 exabytes in 2011 and they forecast that zettabytes and yottabytes will be reachable soon.

In [9] the study discusses the existing activities and future opportunities related to health; for instance: medical and health informatics, translational bio-informatics, sensor informatics, imaging informatics. It also proves that health care information is Big Data since it complies with at least six features: value, volume, velocity, variety, veracity, and variability. Concerning to the subject matter of our study, we deemed important the sensor informatics information covered in this paper and the authors consider that new technologies have allowed to sense health data and that there are nowadays different types of sensors used in health care, such as wearable, implantable and environmental. These sensor types' areas are useful for military health care and, according to the provided findings in the paper, smartphones are highly used for this proposal. An important presented conclusion is contemplating the personal data as the new oil of the 21st century due to the fast growing of smartphones, sensors, networks, and software applications.

In [10] there is a compilation of articles in clinical medicine for Big Data. In summary, a diversity of clinical data is presented, methods to deal with Big Data, challenges, and constraints. Once health data from the army has been attained, we deduce this study can help on how to deal with clinical Big Data.

In [11] the authors present the associated implications of using Big Data in the health care novelty. The relevant findings applied to our matter of subject are related to the usage of remote sensors for health care. The study mentions the sensor main uses; for instance, automatic signal tracking for behaviors, risk factors, trends. Some of the variables to follow up can be food consumption, diet data, glucose monitor, physical activity, sleeping hours, etc.

Study [12] presents a literature review from 2000 to 2015 about Big Data applications in biomedical research and health care. The review examines and discusses Big Data application in four disciplines: bioinformatics, clinical informatics, imaging informatics and public health. The authors refer to the generated data in health care as Electronic Health Records (EHR) and consider the data provided by medical sensors as unstructured. Although the article does not present sensor information in detail, the presented concepts are useful to our study.

In [13] the challenges and opportunities in the Big Data area applied to health care are reviewed. Specially, challenges related to data structure, security, data standardization, storage, and transmission, as a result of the limited skills in this field. The main opportunities and advantages are the care quality improvement, early detection of diseases, population health management, data quality, structure, and affordability. These opportunities can also be achievable in the military health care.

In [14] the authors present the definitions, sources, applications, techniques and technologies for Big Data in health care domain. Their findings detect an absence of consensus in the Big Data operational definition and the need for transforming raw data in useful information. In the health care area, the study demonstrates that health data is used by hospitals, clinics, governments, laboratories, pharmacies, medical journals, etc.; the often-used technique is Natural Language Processing (NLP) and the most used platform is Apache Hadoop. Related to the type of data, the authors classified health data as unstructured and semi-structured.

In [15], the article presents the impact of Big Data in health care, identifying the stakeholders, sources of Big Data, nature of analytics and analytical techniques, used tools, etc. An important contribution related to the military health care is the stream and real-time data processing tools that can be used; for instance, Apache Storm, SQL Stream Blaze, and Apache Kafka.

In [16], the authors present a review of mechanisms to assure Big Data in health; among them, encryption algorithms, architectures, data validation, MapReduce variants using the Open Secure Socket Layer (SSL), privacy techniques, protocols, database models, etc. We consider data security as an important and fundamental subject in military health care; therefore, this study will aid in this intent.

In [17], the authors refer to Big Data sources and techniques in health sector and the most used techniques in chronic diseases prediction. As relevant findings, the study presents summary data mining techniques to predict chronic diseases, Big Data platforms, databases used for health in Big Data. This study could be used as basis for the analysis of Big Data in order to detect dangerous conditions in the army.

In [18] challenges of Big Data in health care are discussed, in summary they highlight the data quality, the analysis, the user skills, the privacy, the economic, lack of standards and the facility. In the military health care it is important to consider these points of view as an early knowledge of the challenges in any implementation.

4 Discussion

In this section, after an in-depth review of every relevant study presented in Sect. 3 Results, we will answer the raised research questions. The following is a more detailed explanation of these feedbacks.

RQ1: What researches about Big Data and military health care have been published?

Feedback: There is a large amount of studies about Big Data and health care and, although we did not find researches specific to military health care, there are important ideas and useful concepts in the literature review for the military health care domain.

RQ2: What approaches and methodologies regarding Big Data and military health care can be applied?

Feedback: Based on the results from the presented review studies, in Fig. 1 we summarize the main findings about approaches and methodologies for Big Data and military health care. Our proposal is mapped onto six pillars: acquisition, storage, processing, management, security, and normative.

The remote devices used for monitoring the military's health status compose the acquisition pillar. According to the state-of the art study, the IoT solutions are mainly used for health care, for instance: IoT health care network, topology, architecture, platform, services, and applications. The following is also used in large scale: sensors and smartphones. The main objective of the wearable technologies are oriented to monitor clinical and health applications [19].

As the physical preparation of the military is mostly outdoors and they are generally assigned to work in isolated places, we consider suitable the usage of remote devices while aligned with Big Data for health trends. We propose the use of non-obtrusive sensors for monitoring vital signs as pulse, respiratory rate, body temperature, blood pressure and blood oxygen. Some examples of the possible locations of the sensors are presented in the Fig. 2.

After the data acquisition, the raw data like images and signals must be stored as EHR in a storage platform. Thus, the storage pillar must be able to store unstructured and semi-structured data so that the processing pillar, by means of NLP processing converts the raw data into useful information. NLP refers to any type of computer

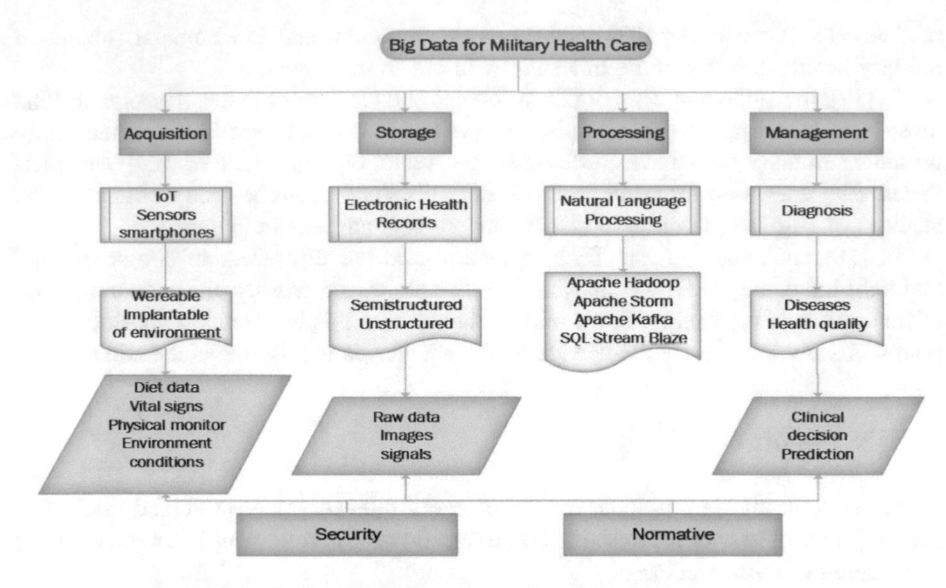

Fig. 1. Pillars for Big Data and military health care.

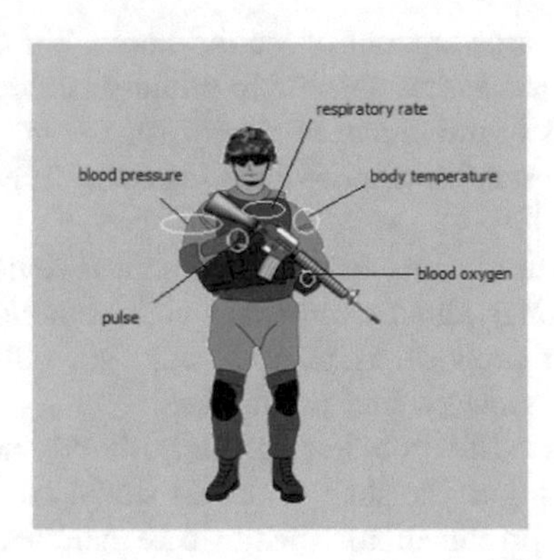

Fig. 2. Location of non-obtrusive sensors for monitoring vital signs.

manipulation of natural language, which means the language that is commonly used in everyday human communication [20].

In the state-of-the-art review, the studies propose some platforms like Apache Hadoop, Storm, Kafka, and SQL Stream Blaze for data processing. With this information in the management pillar, health personal can quickly provide health diagnosis for military personnel; detect diseases, health quality, among other activities. Moreover, depending on the used techniques for analyzing the information, a clinical predictive decision can be done and thus avoid potential complications.

Apache Hadoop is a processing framework with storage and computational capabilities in analytical tasks [21]. Apache Storm is a real time processing tool developed for a huge amount of data [22]. Apache Kafka is an open source platform that provides a real-time publish-subscribe solution for big data volumes [23]. SQL Stream Blaze is a platform for streaming analytics based on Apache Kafka. Hadoop is currently the mostly used tool but it only processes the data in batch mode, when real-time data processing is now a common requirement for data analysis [22].

The Fig. 3 presents a possible implementation of a system for monitoring health care in militaries.

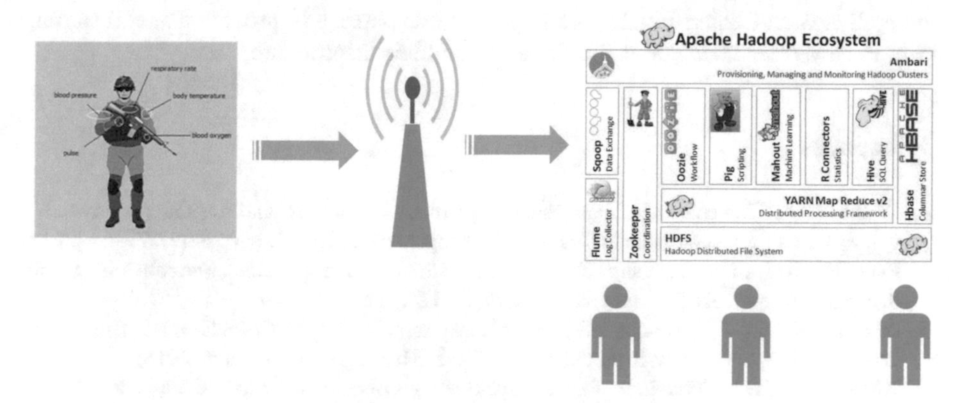

Fig. 3. System for monitoring health care in militaries

The security pillar is implicit into the previous four pillars, since privacy, consistency, accuracy, trustworthiness, and availability must be ensured from the moment the data is collected until the entire cycle is completed. Furthermore, the normative pillar ensures compliance to local and international standards in order to attain quality in the process.

Our proposed pillars were based on the fact that military health data deserve a different treatment since they keep being exposed to several health risks. As we mentioned before, 62% of federal IT professionals in the United States agree that including Big Data will improve health care in the military life. Therefore, we consider suitable to monitor military health status through Big Data systems.

5 Conclusions and Future Work

In this paper, we have presented a state-of-the-art of reviews about Big Data and health care. From the studies collected from digital libraries regarding the period from 2014 to 2018, we identified 17 relevant studies after the inclusion and selection stages. Every relevant research was deeply studied and analyzed so to respond the two raised research questions. Although none of the studies directly relates to our topic of interest, we found general approaches and methodologies relevant to the military health care domain. We consider this apparent lack of scientific articles about the topic to be the result of confidential commitment that military institutions require in order to safeguard the public and their own security.

As a result, we proposed six pillars: acquisition, storage, processing, management, security, and normative. Then we explained how to apply them in the military environment.

This study determines the need for further research in the military health care context and presents an approach for implementing Big Data based on proposals from the research community in the health care area. As future work, we hope to present case studies of application of each one of these pillars in a Military Scientific and Technology Research Center. Finally, we hope that this paper will provide a useful starting point to new researches about Big Data and military health care.

References

1. Waller, M.A., Fawcett, S.E.: Data science, predictive analytics, and big data: a revolution that will transform supply chain design and management. J. Bus. Logist. **34**(2), 77–84 (2013)
2. Fang, R., Pouyanfar, S., Yang, Y., Cheng, S.: Computational health informatics in the big data age a survey. ACM Comput. Surv. **49**(1), 12.1–12.36 (2016)
3. de la Torre Díez, I., Cosgaya, H.M., Garcia-Zapirain, B., López-Coronado, M.: Big data in health: a literature review from the year 2005. J. Med. Syst. **40**(9), 209 (2016)
4. Kitchenham, B.: Procedure for undertaking systematic reviews. Computer Science Department, Keele University and National ICT Australia Ltd., Australia (2004)
5. Islam, S.R., Kwak, D., Kabir, M.H., Hossain, M., Kwak, K.S.: The internet of things for health care: a comprehensive survey. IEEE Access **3**, 678–708 (2015)
6. Onyejekwe, E.R.: Big data in health informatics architecture. In: IEEE/ACM International Conference on Advances in Social Networks Analysis and Mining, pp. 728–736 (2014)
7. Thara, D.K., Premasudha, B.G., Ravi, R.V., Suma, R.: Impact of big data in healthcare: a survey. In: 2nd International Conference on Contemporary Computing and Informatics, pp. 729–735 (2016)
8. Raghupathi, W., Raghupathi, V.: Big data analytics in healthcare: promise and potential. Health Inf. Sci. Syst. **2**(1), 3 (2014)
9. Andreu-Perez, J., Poon, C.C., Merrifield, R.D., Wong, S.T., Yang, G.Z.: Big data for health. IEEE J. Biomed. Health Inform. **19**(4), 1193–1208 (2015)
10. Wang, W., Krishnan, E.: Big data and clinicians: a review on the state of the science. JMIR Med. Inform. **2**(1), 1–16 (2014)
11. Hansen, M.M., Miron-Shatz, T., Lau, A.Y.S., Paton, C.: Big data in science and healthcare: a review of recent literature and perspectives. Yearb. Med. Inform. **9**(1), 1–11 (2014)
12. Luo, J., Wu, M., Gopukumar, D., Zhao, Y.: Big data application in biomedical research and health care: a literature review. Biomed. Inform. Insights **8**, 1–10 (2016)
13. Kruse, C.S., Goswamy, R., Raval, Y., Marawi, S.: Challenges and opportunities of big data in health care: a systematic review. JMIR Med. Inform. **4**(4), 1–14 (2016)
14. Mehta, N., Panditb, A.: Concurrence of big data analytics and healthcare: a systematic review. Int. J. Med. Inform. **114**, 57–65 (2018)
15. Palanisamy, V., Thirunavukarasu, R.: Implications of big data analytics in developing healthcare frameworks – a review. J. King Saud Univ.-Comput. Inf. Sci., 1–11 (2017)
16. Hamrioui, S., de la Torre Díez, I., Garcia-Zapirain, B., Saleem, K., Rodrigues, J.J.: A systematic review of security mechanisms for big data in health and new alternatives for hospitals. Wirel. Commun. Mob. Comput. **2017**, 1–7 (2017)

17. Alonso, S.G., de la Torre Díez, I., Rodrigues, J.J., Hamrioui, S., López-Coronado, M.: A systematic review of techniques and sources of big data in the healthcare sector. J. Med. Syst. **41**(11), 183 (2017)
18. Stylianou, A., Talias, M.A.: Big data in healthcare: a discussion on the big challenges. Health Technol. **7**(1), 97–107 (2017)
19. Cedillo, P., Sanchez, C., Campos, K., Bermeo, A.: A systematic literature review on devices and systems for ambient assisted living: solutions and trends from different user perspectives. In: International Conference on eDemocracy & eGovernment (2018)
20. Bird, S., Klein, E., Loper, E.: Natural Language Processing with Python. O'Reilly Media Inc, Sebastopol (2009)
21. Sarnovsky, M., Butka, P., Paulina, J.: Social-media data analysis using tessera framework in the hadoop cluster environment. In: 37th International Conference on Information Systems Architecture and Technology, vol. 2, pp. 239–251 (2017)
22. Iqbal, M.H., Soomro, T.R.: Big data analysis: apache storm perspective. Int. J. Comput. Trends Technol. **19**, 9–14 (2015)
23. Garg, N.: Apache Kafka. Packt Publishing Ltd., Birmingham (2013)

Temporal Analysis of 911 Emergency Calls Through Time Series Modeling

Pablo Robles[1], Andrés Tello[1], Lizandro Solano-Quinde[2](✉), and Miguel Zúñiga-Prieto[1]

[1] Department of Computer Science, University of Cuenca, Cuenca, Ecuador
{pablo.robles,andres.tello,miguel.zunigap}@ucuenca.edu.ec

[2] Department of Electrical, Electronic and Telecommunications Engineering, University of Cuenca, Cuenca, Ecuador
lizandro.solano@ucuenca.edu.ec

Abstract. We present two techniques for modeling time series of emergency events using data from 911 emergency calls in the city of Cuenca-Ecuador. We study state-of-the-art methods for time series analysis and assess the benefits and drawbacks of each one of them. In this paper, we develop an emergency model using a large dataset corresponding to the period January 1st 2015 through December 31st 2016 and test a Gaussian Process and an ARIMA model for temporal prediction purposes. We assess the performance of our approaches experimentally, comparing the standard residual error (SRE) and the execution time of both models. In addition, we include climate and holidays data as explanatory variables of the regressions aiming to improve the prediction. The results show that ARIMA model is the most suitable one for forecasting emergency events even without the support of additional variables.

Keywords: 911 calls · Emergency calls · Temporal models · GP · ARIMA

1 Introduction

Predicting emergency events is a relatively new area of research with important applications for decision makers in areas of civil security and urban planning. In addition to planning and resource allocation, it increases citizens security and overall urban well-being. Several approaches for modeling the information of emergency events use temporal, spatial, and demographics data to create predictive models [2,3,18,20]. The objective is to identify optimal solutions for timely and cost-effective assignment of resources, creation and maintenance of infrastructure, and contingency planning.

The work of Chandrasekar et al. [3] classifies crime events based on the time and location of occurrence of the event using data from the city of San Francisco. The authors suggest that, instead of classifying crime events into very specific classes, grouping events into larger categories allows to find patterns in

M. Botto-Tobar et al. (Eds.): ICAETT 2019, AISC 1066, pp. 136–145, 2020.
https://doi.org/10.1007/978-3-030-32022-5_13

the data. Bappee et al. [2] use machine learning to find the relationship between criminal activity and geographical regions. They use spatial information from open street map data, and crime hotspots based on density clustering algorithms as features to train classifiers of different types of crime. Other approaches focus on predicting traffic incidences. The work of Yuan et al. [20] uses a dataset of vehicle crashes from 2006 to 2013, in the state of Iowa, USA, as well as various weather attributes with one hour of granularity to create prediction models. An interesting approach to detect traffic accidents is shown in [18]. This work use machine learning algorithms trained with location information, obtained from smart-phones located at vehicles not involved in the incident, to detect traffic accidents.

In this paper, we introduce mathematical models to describe the distribution of emergency events in time using ECU-911 data. ECU-911 [1] is the largest system for monitoring emergencies, which coordinates the aid for these incidents through call-centers and dispatchers. The tasks of ECU-911 agents facilitate the deployment and support from various institutions such as, National Police, Public Health Ministry, Social Security Office, Army, Fire Department, National Secretary of Risk Management, and Transit Commission. However, to this day there is still the need to optimize the work of ECU-911. This optimization could improve the pre-assignment and deployment of resources, according to the level of incidences, the type of event, the geo-spatial, and the historical data. We focus on the case of the city of Cuenca and consider three types of events: *Civil Security, Transit*, and *Hospital and Health Services*. We propose several models and compare the performance of these models and investigate other variables (climate, holidays) for improving the prediction of the number of daily events.

This article is organized as follows. Section 2 describes previous work in this field. In Sect. 3 we introduce our proposed models based on geo-spatial data. Section 4 discusses the experiments to evaluate our models and the obtained results. Finally, Sect. 5 presents the discussion of our results and conclusions.

2 Related Work

Predicting emergency events, as registered through 911 calls in various cities in the world, is a complex and relatively new problem. Such predictions are important for decision making because they would help to optimize costs and response-times, as well as resource allocations and for develop of new infrastructure. City planning and maintaining the overall security of a city can benefit from research in this area. Many researchers, for instance [4–6, 11, 15, 17, 19] have shown the importance of this problem and have provided valuable insights on how to model emergencies.

Some studies model the occurrence of emergency events as time series of counts for different practical applications, e.g., criminal incidents, traffic accidents, insurance claims, among others. For instance, The work of Lee and Lee [12] presents a model for criminal incidents using count time series. The authors propose a causality test for crimes and temperatures in Chicago, based on previous studies that shows the influence of an environmental factor on criminal behavior. They examine the relationship between those variables using conditional mean equation of Poisson INGARCH models, and construct the test using the least squares estimator (LSE). They assessed their method using the Chicago's crime data set and demonstrated that the sexual crimes have a causal relationship with temperature.

Ihueze and Onwurah, in [10], analyzed road traffic crashes in Anambra State, Nigeria, and created prediction models using auto-regressive integrated moving average (ARIMA) and auto-regressive integrated moving average with explanatory variables (ARIMAX) techniques. They computed the prediction accuracy using lower Bayesian information criterion, mean absolute percentage error, root mean square error; and higher coefficient of determination (R-Squared). This metric was used to compare the performance of their models. In their experiments, ARIMAX model outperformed the ARIMA (1, 1, 1) model that they used as the basis model for comparison.

On the other hand, other works show that modeling crime with Gaussian processes can produce high accuracy. The work of Flaxman [8], proposes a model based on Gaussian Processes, and short-term spatiotemporal forecasting of crime, up to 12 weeks into the future. The dataset used for this work is the crime data of the city of Chicago. The data are aggregated crime counts by type of crime, neighborhood and week of the year. Although the performance of this model is comparable and, in some cases, better than other approaches, a noticeable limitation is its computational burden.

Another example of crime prediction using Gaussian processes is the work of Flaxman et al., [7]. They propose a model for spatiotemporal forecasting of crime events using Gaussian processes with auto-regressive smoothing kernels in a regularized supervised learning framework. The results of their approach outperformed the baseline KDE estimates. Their model was applied in the context of the Real-Time Crime Forecasting Challenge proposed by the National Institute of Justice of the United States. The metrics used to assess the model were the prediction accuracy index (PAI), and the prediction efficiency index (PEI). The contest included 20 categories, 4 crime types with 5 different forecasting windows, the model proposed in this work won a total of 9 categories. The best PEI results in all crimes (1 week, 1 month, 3 months), burglary (1 week, 2 weeks), street (2 weeks), and vehicle (1 week). They also won in PAI entries for burglary (1 week and 2 weeks). The results showed that their approach is generalizable because it outperformed other models for different crime types and across different timescales.

3 Methods

The aim of this work is to predict the number of daily events in the city of Cuenca based on historical spatiotemporal data of three types of events: *Civil Security, Transit*, and *Hospital and Health Services*. We achieve this by using two time series models to describe the distribution of events in time. Specifically, the

Gaussian Process and the Auto-regressive - ARIMA models were used for this purpose. The dataset of reported emergencies of the ECU-911 from January 1st 2015 to December 31st 2016 was used to create and validate our models.

For this purpose, we create the time-series data by grouping the events per day and considering the daily count as the variable to predict. This led to the sequence $T_0 = \{t_1, \ldots, t_n\}$ that contains the number of events per day.

3.1 Gaussian Process

This model aims to create a description in time of a sequence of events. The main objectives are: (i) to investigate whether there are Gaussian relations in the sequence of events, and (ii) to verify whether there is a global function of the data that could determine the temporal dependencies.

In the time series, created as described above, we consider the domain of a Gaussian process as $x_1, \ldots, x_n = 1, \ldots, n$, and the function to model as $y(x_i) = t_i$. Thus, the Gaussian process is defined as $y(x_i) = \mu + z(x_i)$, where μ is the expected value, with dimension n, $z(x_i)$ is the Gaussian process with expected value 0 and covariance defined by $var(z(x_i)) = \sigma^2; covar(z(x_i), z(x_j)) = \sigma^2 R_{ij} \quad \forall i, j$, and $\mu = (\mathbf{I}_n^T R^{-1} \mathbf{I}_n)^{-1} (\mathbf{I}_n^T R^{-1} T_0^T)$. $R_{ij} = \exp\left(-\theta (x_i - x_j)^2\right)$ is the correlation function of the exponential type. σ^2 is the variance of the process. $\mathbf{I}_n$ is a vector of all 1 of dimension n and θ is the parameter of the correlation function. The values μ, σ^2, θ are obtained from the count data T_0.

3.2 Auto-regressive Model - ARIMA

Time series are useful to model the number of events, observed over time, per time period [13]. There are several phenomena that can be modeled as count time series, e.g., number of daily hospital admissions, the number of defective items, the number of stock market transactions, among others. For the purpose of this work, count data is defined as the number of events per time interval [16]. Hence, the observations include only non-negative integer values, ranging from zero to some value. When data includes a large number of distinct values, some researchers model counts as if they were continues data [9].

Our objective is to verify the suitability of auto-regressive models for modeling emergency events as compared alternative models. In this context, we consider the model *auto-regressive integrated moving-average* (ARIMA): $t_j = -(\Delta^d t_j - t_j) + \phi_0 + \sum_{i=1}^{p} \phi_i \Delta^d t_{j-i} - \sum_{i=1}^{q} \theta_i \varepsilon_{j-i} + \varepsilon_j$. This model consists of a regressive model summarized with three parameters: p, d, q. These are positive integer numbers that model three components of the regression: the auto-regression itself, the integration, and the moving-average. The value of d defines the multiplicity of the unit-root of the model, p is the number of terms of the regression, and q is the number of terms of the moving-average, which can be considered as a linear function. Furthermore, $\Delta^d t_j = t_j - t_{j-1}$. The remaining coefficients $\phi_i, \theta_i, \epsilon_j$ are coefficients and constants of the regression.

For the verification purposes, the time series, created as explained above, was used as the signal to model.

4 Experiments

In this Section, the results obtained by the two models and a comparison of the performance between them are presented.

Gaussian Process. To evaluate the Gaussian process we considered the data of *transit and mobility* events and organized them sequentially as a count of events per day (The time series T_0 described aerlier). In our implementation we used the library GPfit of the R-language [14]. Since GPfit accepts values of the independent variable in the range $[0, 1]$, we transform T_0 from the integer range $[1, 731]$ to $[0, 1]$.

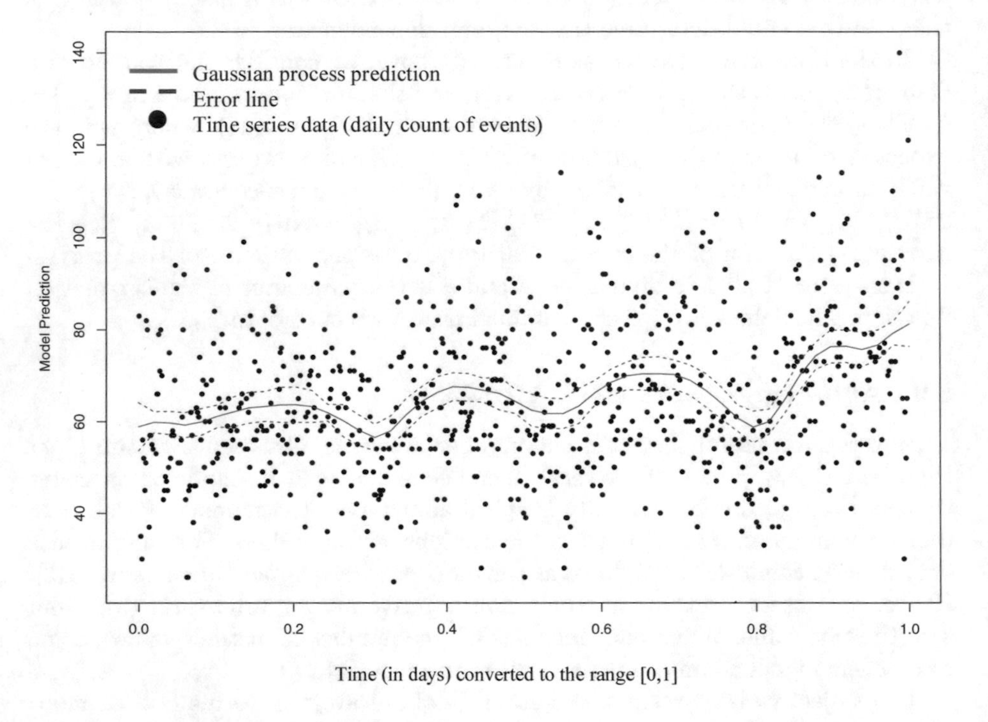

Fig. 1. Gaussian process model - prediction of transit and movility events

Figure 1 shows the distribution of events in time as approximated with a Gaussian process. Black dots represent the count data. The blue line shows the prediction as due to the Gaussian process, while the red line shows the error of the prediction with respect to the real values. As it can be seen in the Figure, the Gaussian process describes a function that attempts to represent the moving average of the events in the neighborhood of the events. If we consider the distance of the error to the prediction we can notice an error that is approximately constant (visually 5–10 events).

ARIMA. In order to verify the quality of the predictions of the ARIMA model, we created models at various time scales (daily, weekly, monthly, tri-monthly, and yearly). In this document we report the results for the daily temporal scale. We perform the quantitative analysis of the performance by computing the error and comparing the results against those of the gaussian process. The objective of this analysis is to determine which model offers the best predictive performance of the data of our interest. Because the auto-regressive analysis requires historical information to perform the predictions, we considered the first two months of data as the training set, using temporal windows of different sizes for prediction purposes. Figure 2 shows the result of the daily predictions for daily windows. As in the previous models we considered the residual standard error of the forecast which shows that ARIMA performs favorably with respect to the alternative models. The advantage of ARIMA is that the seasonal information obtained from other models (e.g., STL) can be incorporated in the predictions.

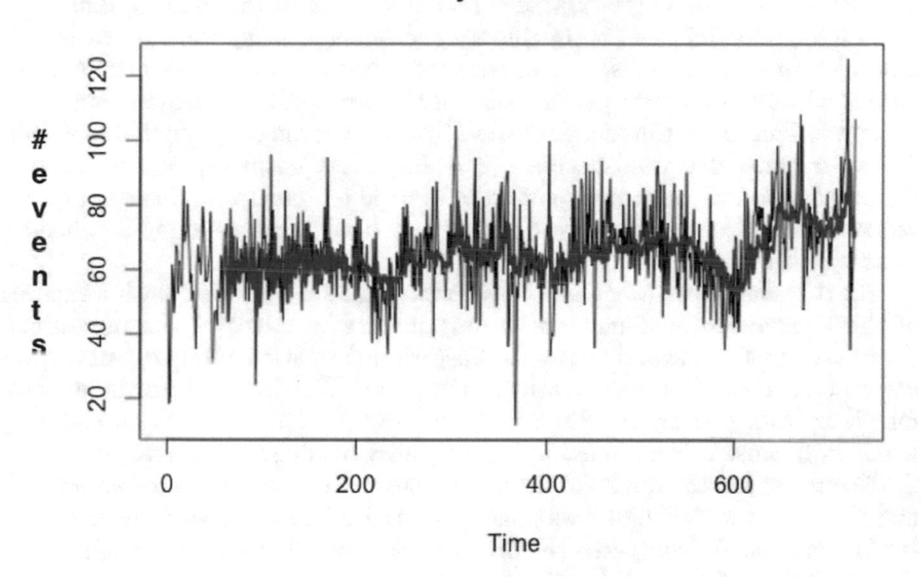

Fig. 2. ARIMA model - prediction of transit and mobility events

Comparison of Performance of Predictive Models. We consider the standard residual error - SRE (or the residuals' mean squared error) $\sqrt{\frac{\sum_{i=1}^{n}(t_i - t'_i)^2}{n-1}}$ where t_i and t'_i are the ith value of the original series and its prediction, respectively, and n is the number of entries in the series.

Table 1 shows the SRE of the models discussed in this document. These results indicate that ARIMA is the most adequate model due to both prediction performance and computational cost. Thus, ARIMA is the most suitable

Table 1. Prediction error comparison

Model	Gaussian process	ARIMA
SRE	15.60	11.33
Time	2 h 25 min	1 s

model for modeling emergency events for ECU-911. Another important qualitative assessment is that the Gaussian process model does not provide a direct description of the seasonal patterns in the data. This could only be discovered by introducing latent models that could extract the seasonality. Hence, combine other models such as Seasonal Trend Decomposition STL with ARIMA could improve the results.

Climate and Holiday Information. In this set of experiments we aimed to verify the effectiveness of additional variables to improve the prediction of the number of events or to give statistical stability. Due to the existing difficulties when integrating information of climate and holidays using the various models due to computational costs, variable dependencies, etc, we consider the ARIMA model, which has the best performance in the previous set of experiments.

In addition to the time series created by counting the daily number of events, T_0, we considered two sequences $C_0 = \{c_1, \ldots, c_n\}$ and $U_0 = \{u_1, \ldots, u_n\}$, that correspond to the sequence of daily average temperature (climate) and the sequence of indicators of the type of day $t_i \in \{0, 1\}$ whether a day is a holiday or not, respectively.

For this analysis we used the ARIMA model and incorporated the information of the temperature and holidays as explanatory variables of the regressions. As in our previous experiments, the historic information used as training set correspond to the first two months of the series. We also perform the analysis for other sizes of temporal windows in the test set but we focus on the daily windows because it is more useful from a policy-making perspective.

Figure 3 shows the results of the analysis using the daily average temperature and the holiday status information using an ARIMA model with daily predictive (test) windows. We computed the SRE of this prediction and compared it with the predictions of an ARIMA without explanatory variables.

Table 2. Prediction error comparison

Model	ARIMA	ARIMA + CLIMATE + HOLIDAYS
SRE	11.33	12.32
Time	1 s	1 s

Table 2 shows the results of our analysis. This shows that the use of additional explanatory variables does not improve the predictions in terms of error

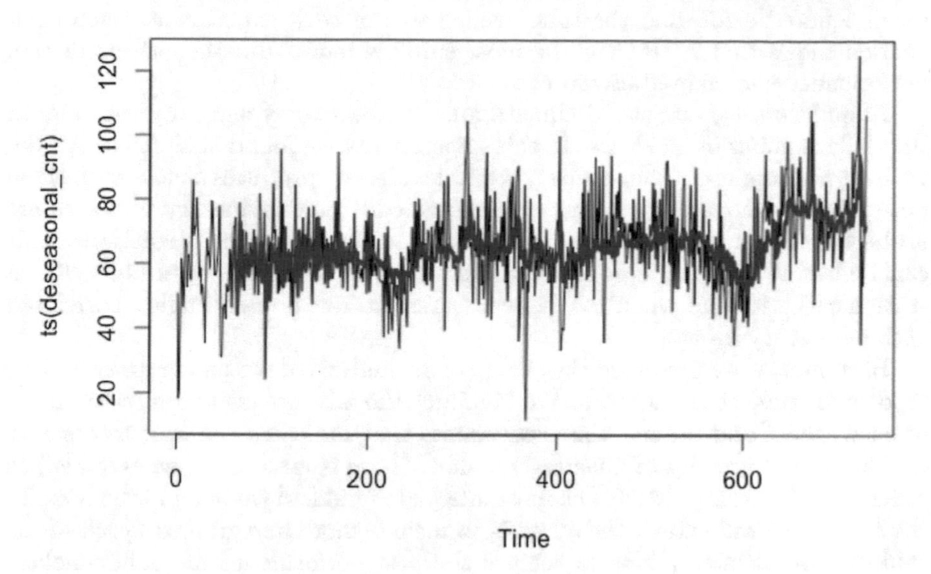

Fig. 3. ARIMA model - prediction of transit and mobility events

reduction. However, this may be due to a possible over-fitting of the original ARIMA model. Thus, adding the explanatory variables may add stability to the predictions compared to the case of a simple auto-regressive model. Also, it is important to notice that the fact that ARIMA is a good model in comparison to the alternatives may be due to the periodicity in the behaviour of this type of events for this specific city.

5 Conclusions

In this document we describe the experiments to model emergency events, as time series, using data from Ecu-911 emergency calls. First, we studied two models to verify their suitability and select the best performing model. Second, we added explanatory variables to the selected model in order to assess whether they contribute to the improvement of the results.

The performance comparison and selection was performed using two models: Gaussian Process and Auto-regressive Integrated Moving Average (ARIMA). While the main drawback of the Gaussian process is the computational cost, its main advantage is the stability, as the error remains similar in the time domain. In addition, the Gaussian process does not provide a direct way to study the seasonal patterns within the time series. This difficulty can be overcome by models such as ARIMA. ARIMA gives the flexibility that it can be combined with other models to improve the results (e.g., seasonal information obtained from the "Seasonal and Trend decomposition using Loess - STL" model could be added). To assess the performance of both models we compare the residuals' mean squared error, and the time needed to run each experiment. Such comparison shows that ARIMA is the most suitable model due to both prediction performance and computational cost.

In addition, we integrated climate and holidays as explanatory variables of the daily number of incidents. In this experiments we found that ARIMA even without the support of the auxiliary variables already produces a close to optimal performance among the two models we studied. The explanatory variables are useful to provide stability to the models and to generalize the predictions. This can be useful in the case of cities where the count of events does not define a stationary series and when the temperatures are not already highly correlated with emergency events.

In summary, we presented the results of an analysis of the most representative models for time series analysis. We highlight the advantages and disadvantages of each model and present the experiments that indicates the best alternative for the case of the city of Cuenca-Ecuador. Those results could be extended to other cities in order to verify the most adequate model in these new scenarios. In the future, we will extend the analysis to include more fine grained detail of the spatiotemporal interactions to achieve the best performance for policy-makers and 911-service managers.

Acknowledgements. This article is part of the project "Análisis predictivo de la ocurrencia de eventos de emergencia en la provincia del Azuay", winner of the "XV Concurso Universitario de Proyectos de Investigación" funded by the Dirección de Investigación de la Universidad de Cuenca. The authors also thank the Servicio Integrado de Seguridad ECU911 - Zona 6 for their collaboration and data provided.

References

1. Ecu 911 website. http://www.ecu911.gob.ec
2. Bappee, F.K., Júnior, A.S., Matwin, S.: Predicting crime using spatial features. CoRR abs/1803.04474 (2018). http://arxiv.org/abs/1803.04474
3. Chandrasekar, A., Raj, A.S., Kumar, P.: Crime prediction and classification in San Francisco city
4. Chirigati, F., Doraiswamy, H., Damoulas, T., Freire, J.: Data polygamy: the many-many relationships among urban spatio-temporal data sets. In: Proceedings of the 2016 International Conference on Management of Data, SIGMOD 2016, pp. 1011–1025. ACM, New York(2016). https://doi.org/10.1145/2882903.2915245
5. Chohlas-Wood, A., Merali, A., Reed, W.R., Damoulas, T.: Mining 911 calls in New York City: temporal patterns, detection, and forecasting. In: AAAI Workshop: AI for Cities (2015)
6. Cramer, D., Brown, A.A., Hu, G.: Predicting 911 calls using spatial analysis, pp. 15–26. Springer, Heidelberg (2012)
7. Flaxman, S., Chirico, M., Pereira, P., Loeffler, C.: Scalable high-resolution forecasting of sparse spatiotemporal events with kernel methods: a winning solution to the NIJ "real-time crime forecasting challenge". arXiv preprint arXiv:1801.02858 (2018)

8. Flaxman, S.R.: A general approach to prediction and forecasting crime rates with Gaussian processes. Carnegie Mellon University, Heinz College Second Paper, Pittsburg (2014)
9. Hilbe, J.M.: Modeling count data. In: International Encyclopedia of Statistical Science, pp. 836–839. Springer (2011)
10. Ihueze, C.C., Onwurah, U.O.: Road traffic accidents prediction modelling: an analysis of Anambra State, Nigeria. Accid. Anal. Prev. **112**, 21–29 (2018)
11. Kim, S.Y., Maciejewski, R., Malik, A., Jang, Y., Ebert, D.S., Isenberg, T.: Bristle maps: a multivariate abstraction technique for geovisualization. IEEE Trans. Vis. Comput. Graph. **19**(9), 1438–1454 (2013). https://doi.org/10.1109/TVCG.2013.66
12. Lee, Y., Lee, S.: On causality test for time series of counts based on poisson ingarch models with application to crime and temperature data. Commun. Stat.-Simul. Comput. 1–11 (2018)
13. Liboschik, T., Fokianos, K., Fried, R.: tscount: an R package for analysis of count time series following generalized linear models. Universitätsbibliothek Dortmund (2015)
14. MacDonald, B., Ranjan, P., Chipman, H.: GPfit: an R package for fitting a Gaussian process model to deterministic simulator outputs. J. Stat. Softw. **64**(12), 1–23 (2015). http://www.jstatsoft.org/v64/i12/
15. Malik, A., Maciejewski, R., Maule, B., Ebert, D.S.: A visual analytics process for maritime resource allocation and risk assessment. In: 2011 IEEE Conference on Visual Analytics Science and Technology (VAST), pp. 221–230 (2011)
16. Plan, E.L.: Modeling and simulation of count data. CPT: Pharmacomet. Syst. Pharmacol. **3**(8), 1–12 (2014)
17. Razip, A.M., Malik, A., Afzal, S., Potrawski, M., Maciejewski, R., Jang, Y., Elmqvist, N., Ebert, D.S.: A mobile visual analytics approach for law enforcement situation awareness. In: 2014 IEEE Pacific Visualization Symposium, pp. 169–176 (2014)
18. Thomas, R.W., Vidal, J.M.: Toward detecting accidents with already available passive traffic information. In: 2017 IEEE 7th Annual Computing and Communication Workshop and Conference (CCWC), pp. 1–4, January 2017. https://doi.org/10.1109/CCWC.2017.7868428
19. Towers, S., Chen, S., Malik, A., Ebert, D.: Factors influencing temporal patterns in crime in a large American city; a predictive analytics perspective. SSRN (2016)
20. Yuan, Z., Zhou, X., Yang, T., Tamerius, J., Mantilla, R.: Predicting traffic accidents through heterogeneous urban data: a case study. In: UrbComp 2017 (2017)

Towards an Ontology for Supporting Dynamic Reconfiguration of IoT Applications

Noemi E. Sari(✉), Marlon Ulloa, Lizandro Solano-Quinde, and Miguel Zuñiga-Prieto(✉)

Computer Science Department, Universidad de Cuenca, Cuenca, Ecuador
{elizabeth.sari21, lizandro.solano, miguel.zunigap}@ucuenca.edu.ec, marlon.ulloa.amaya@gmail.com

Abstract. In the last years, several ontologies have been developed in the Internet of Things (IoT) domain to specify intelligent environments. Where, environment descriptions include both the software services, that expose the functionalities of devices; and resources, i.e. software components that control physical sensors and actuators, hosted on devices that sense or change the state of objects in the environment. This work proposes an ontological semantic network that supports the development of IoT applications that establish, at runtime, the binding with device services that allow them to fulfill their functionalities. This is done by enhancing device service descriptions with properties that allow inferring relationships among the attributes of the environment that an application is designed to control and the environment properties that device services are able to control. The applicability of the proposed model is shown through the population of the ontological network where an application establishes at runtime the binding among its functionalities and device services.

Keywords: Ontology · Web services · Internet of Things

1 Introduction

In the last years, technology has had great advances and impulses that are present in the day to day of people. One of these advances is Internet of Things (IoT) that by extending the Internet to the physical world, facilitates the interaction with physical entities (objects) in the environment; easing the way how human beings carry out their daily activities. For instance, it is feasible to find automated environments that turn on/off a fan according to observed values of temperature, identify the products that are needed in the fridge and place an order via the Internet. All of this is thanks to the communication and interconnection between components, as well as their interaction with the environment. However, the key problem that arises in most cases is the semantic description of IoT is not adopted or understood as expected [2]; therefore, it is necessary to apply semantic technologies that solve this problem.

Semantic technologies like ontologies allow modeling knowledge, improving the understanding of the semantics in any domain or study area. Currently, there are several ontologies designed to gather information from IoT domains. Among them are: SSN

M. Botto-Tobar et al. (Eds.): ICAETT 2019, AISC 1066, pp. 146–154, 2020.
https://doi.org/10.1007/978-3-030-32022-5_14

(Semantic Sensor Network) [1], which support the description of sensors; or SAN (Semantic Actuator Network) [6], which support the description of actions of actuators. There are other ontologies that extend the already mentioned, such as: IoT lite [2], which is an instantiation of the SSN ontology to describe key concepts of IoT that allow interoperability as well as the discovery of sensory data in heterogeneous platforms; Sensor Observation Sampling Actuator Ontology - SOSA [3], which extends SSN and SAN, has as general purpose the formal specification for the modeling of interactions between the objects involved in observation. However, IoT applications are designed to create smart environments in various domains for providing functionalities to IoT Devices, existing a lack of proposals that support the description of properties of the environment that an IoT Device service is able to control as well as attributes of real world objects that could be observed or changed in the environment.

This paper proposes an ontological network DS-IoT (Device Services for Internet of Things) which support the description of IoT Devices within an environment and their functionality exposed as services. For this, the SOSA and IoT-Lite ontologies were integrated by reusing them: from the first one the concepts and relationships related to objects involved in observation of sensors and actions of actuators where reused; and from the second one concepts for the discovery of data in heterogeneous platforms where reused.

This article is structured as follows: Sect. 2 describes the related works on IoT ontologies. Section 3 describes the modeling of the DS-IoT ontology. Section 4 proposes a case study to validate the ontology. And finally in Sect. 5 describe this work conclusions and future works.

2 Related Work

In recent years, several researches have been developed due to the need to improve the processes, techniques and methodologies in IoT domains. There are ontologies (i.e., SOSA, IoT-Lite) that propose solutions for different types of resources such as sensors and actuators; describing either observable attributes of objects in the environment or actuations of these objects through operations.

For example, the SOSA ontology, which extends SSN, provides concepts to represent entities (objects), their properties (attributes) and operations involved in their observation; as well as sampling and performance. On the other hand, IoT-Lite, an ontology for representing IoT concepts, facilitates the interoperability and discovery of sensor data in heterogeneous platforms through a light semantics. However, although a relationship can be established between software services that access resources hosted on IoT Devices, these ontologies do not take into account concepts that allow describing the properties that a service could either observe or change in the environment (i.e., the lighting attribute in a sensor).

In [7], the authors propose an ontology called SmartFrameNet, based on SSN, to provide information on sensors in the agriculture domain (i.e., humidity sensors, light sensors) in order to support recommendation systems that increase productivity based on the analysis of sensors data. However, the concepts considered in this ontology are closely related to agriculture, which makes it difficult to reuse them in other domains.

On the other hand, in [8] an ontology is proposed, which is aimed for description of context information in a home domain; providing knowledge about the interconnected devices. In addition to its simplicity, it is also possible to use it in domains that are similar to the one for which it was designed. However, the sensors must be connected to a specific device, in this case, to a computer, without having the opportunity to analyze the information from other devices. In [9], the authors propose a strategy based on an ontology and IoT for the rehabilitation of patients who have suffered accidents or illnesses. This ontology, support the description of information related to patients (i.e., age, weight, limitations), devices required for rehabilitation, devices for control of vital signs (i.e., devices to measure the pressure, cardiac frequency meters), and information related to doctors (i.e., location, agenda). Additionally, a middleware receives data from each of the devices that describe patients' status and, by using inferences, provides recommendations of what should performed on patients for recovery, as well as, alerts the doctor when patients are in a critical condition. However, its functionality is limited when it comes to other areas, since its concepts only describe detailed information of devices in the medical domain.

3 Device Services for Internet of Things Ontology (DS-IoT)

According to [2], an entity (object or thing) in the IoT domain could be a television, refrigerator, lamp, a person with a heart monitor implant, or any other natural or artificial object that has an Internet connection and the ability to transfer data through a network. To be part of the digital word an *Object* requires of *Devices*, hardware components, that manage its interactions; where *Devices* are attached to an *Object* or are part of its environment, being able to monitor it. From a digital point of view, *Devices* host software components (*Resources*) that provide information on the *Object* or enable controlling *Devices*. *Resources* are representations in the digital world of sensors and actuators; sensors that measure attribute values of *Objects* in the physical world and actuators that change the physical state of *Objects*. Finally, *Services* (i.e., Web services), which provide well-defined interfaces, expose the functionality of *Devices* by accessing their hosted *Resources*.

This work proposes an ontological solution to describe, manage and control the resources of an IoT environment (sensors or actuators) according to the information provided by an ontological network (DS-IoT). DS-IoT not only describe *Devices* and their *Services* functionalities; but also environment attributes or properties that could be monitored/changed by the *Resources* that Services use or control.

DS-IoT is made of two perspectives. The first perspective focuses on describing semantically physical IoT Devices together with the attributes and actions that are able to detect or change. DS-IoT takes a perspective centered on an environment and revolves around observations made by sensors related to objects in order to obtain attributes and operations.

The second perspective focuses on the digital point of view of IoT Devices, through semantically describing software services that facilitate the development of new applications that integrate software components and objects within an environment.

Next the two perspectives of the ontological network are detailed, which were developed using the protégé tool.

3.1 DS-IoT of Object Observation Perspective

In the IoT domain, semantic models allow the detailed description of concepts used to represent relationships between IoT systems. This section defines the main classes, class properties and data properties related to the concepts that underlie the IoT domain. As part of the semantic description of the objects of an IoT environment, the proposed ontological network allows defining the Object's *Location* (its longitude, latitude, altitude) within an environment, as well as *Attributes* (i.e., color, state) and *Operations* carried out on the object to change its status (i.e., turn off, turn on, regulate).

Figure 1 shows a diagram in the form of a graph of the ontological network seen from the perspective of the *Object*, presenting the main classes and properties of the *Object* model. An Object has certain characteristics, which includes (*Object hasOperation Operation*), (*Object hasAttribute Attribute*), (*Object hasLocation Location*) and (*Object inside Environment*). A particular object can have one or all of the characteristics; in addition, resources such as Sensors can be either embedded inside the object or external to it, observing its behavior (*Object observedBy Sensor*). In order to describe the performed actions on an object the ontological network proposes the relation *Actuator madeActuation Object*. DS-IoT reuses SOSA concepts such as *Observation, Actuation, FeatureOfInterest, Sensor* and *Actuator* which help in describing the object's behavior and features.

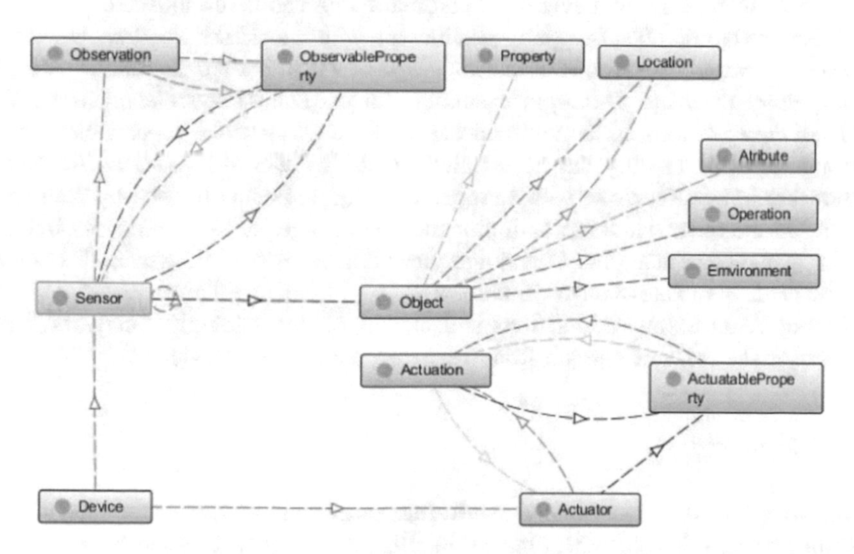

Fig. 1. Extract of classes from the perspective of the object

DS-IoT reuses classes and properties from SOSA and IoT-Lite, allowing the modeling of observations made by the sensors and actions executed by actuators, in conjunction with the modeling objects, obtaining information on attributes, properties, location among others. Having all this information in a single ontology helps in monitoring and managing objects within an IoT environment. DS-IoT facilitate the building of applications that by using inferences are able to find objects that have similar operations or are related to each other; allowing to change dynamically bindings when the environment change.

3.2 DS-IoT Perspective from Service

This section defines the main concepts, interactions and properties that describe the digital point of view of IoT Devices corresponding to software services that access resources (sensors and actuators) hosted on devices, and facilitate the building of applications that manage intelligent environments and facilitate user interactions. For instance, an application hosted in a mobile device could manage interactions among objects, environment and users; integrating, providing context and reasoning about data (i.e., obtained by sensor resources) related to heterogeneous physical environment objects, and producing changes in the state (i.e., initiated by actuator resources) of physical environment objects.

The ontological network uses the classes Device, Resource and Service to describe the generic services that the application will support and the concepts "Observable *Property*" and "Actuable *Property*" to describe the properties that you want to read or change. In turn, each entity (which is also an intelligent object) describes the services it offers. Having information about what the application needs for its operation, as well as the services offered by the devices, an automatic link can be established.

A Service has an object property relationship with the classes *Request*, *MethodGet*, *MethodPut*, *Response*, *DetailConnection*, *Input*, which have data relation as *URL*, *Token*, *Port*, *Protocol*, *Description* among others. The *MetodoGet* class uses the *callToMethod* relationship to communicate with the *ObservableProperty* class, which is responsible for reading the observations made by a sensor. *MetodoGet* has the relationship of sentResponse with Response to send the data obtained by reading the sensor. On the other hand, the option of making changes to the objects is offered by sending parameters using the MetodoPut class that is related to ActuationProperty so that the actuators make changes in the sensors.

In Fig. 2, an extract of relations and classes of the ontological network can be observed in the form of a graph from the perspective of the service.

4 Case Study

To illustrate how the objectives set out in this paper were met, the following scenario was implemented: In the home, office or intelligent environment of the future, not only home appliances will be connected to the Internet remotely; but also, different objects in the home will be able to identify the needs of home inhabitants, turning on lights, fans, coffee machines, radio, television, or initiating other actions. Objects have either external or embedded sensors/actuators that allow them to observe and change the environment, those sensors or actuators monitor or change objects' attributes

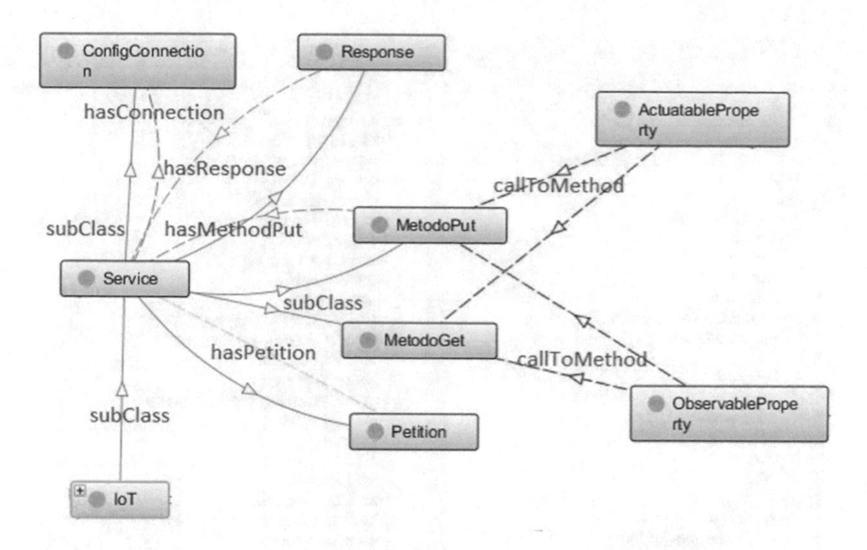

Fig. 2. Extract of classes from the perspective of the service

respectively (i.e., the light intensity of the lamp, the speed with which the fan rotates, the temperature of the water in the coffee machine, the volume level for the radio or television), and its own operations (i.e., turn on or off lights/lamp/TV/radio, set the speed of the fan, disable the alarm, get the location of objects within a home). The DS-IoT ontology shown in Fig. 3 was developed by using the Protégé ontology editor.

Once the data was loaded in the ontological network, the information was exposed and managed through services consumed by an application developed in the Java language using the Apache Jena library and perform SPARQL queries over the ontology to obtain the services that complete application functionalities. The developed application not only helps to consume each one of the operations of the detected objects, but also to find the device services whose *ObservableProperty* or *ActuableProperty* is compatible with the objects' attributes to be monitored or changed; allowing to bind at run-time application functionalities with services exposed by devices. For example, in order to shows the current temperature of the room (attribute Temperature belonging to the Object room) where the user is located, the application queries the ontological model looking for *Device Services* whose *ObservableProperty* is *Temperature*, then binds the corresponding *Device Service* with the application functionality showing the current room temperature. It is important to remember that *Device Services* access to sensor *Resources* to hosted in Devices to obtain measurements.

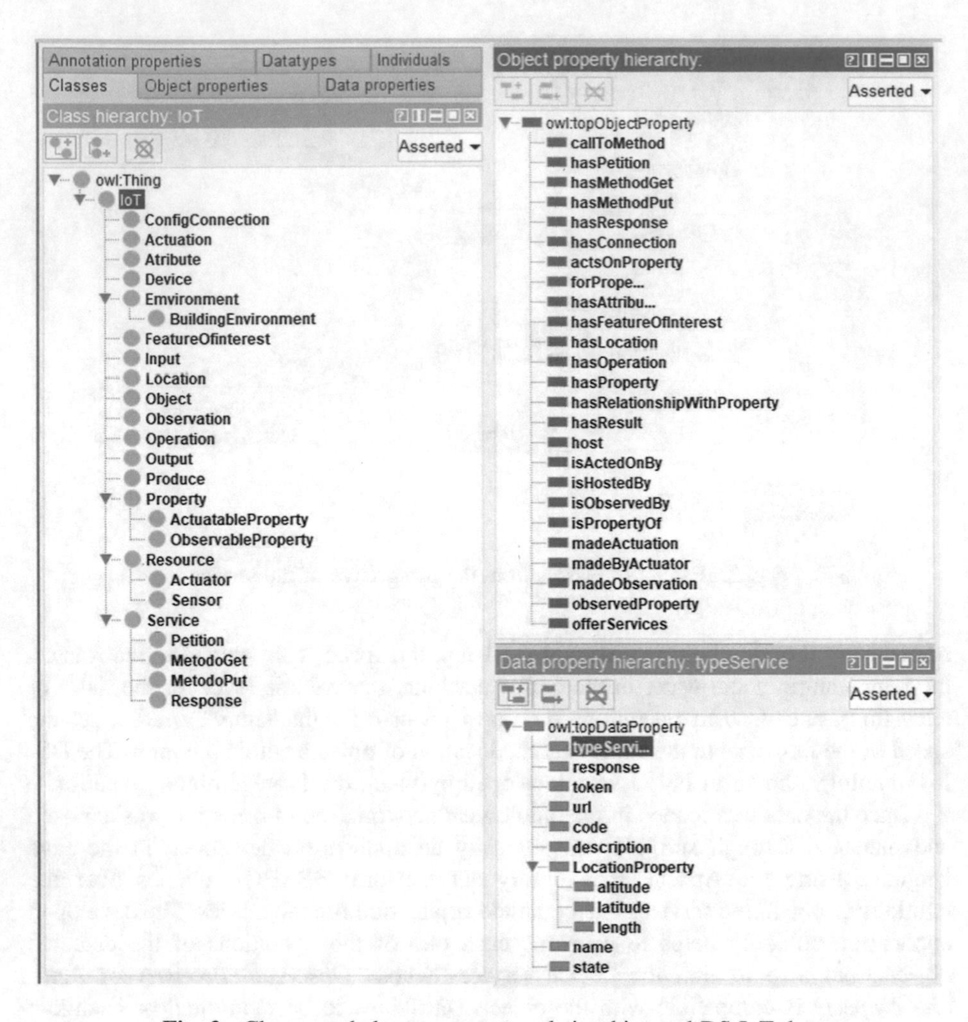

Fig. 3. Classes, subclasses, property relationships and DS-IoT data

Figure 4 illustrates the instance of the ontology in the form of a graph of the proposed scenario.

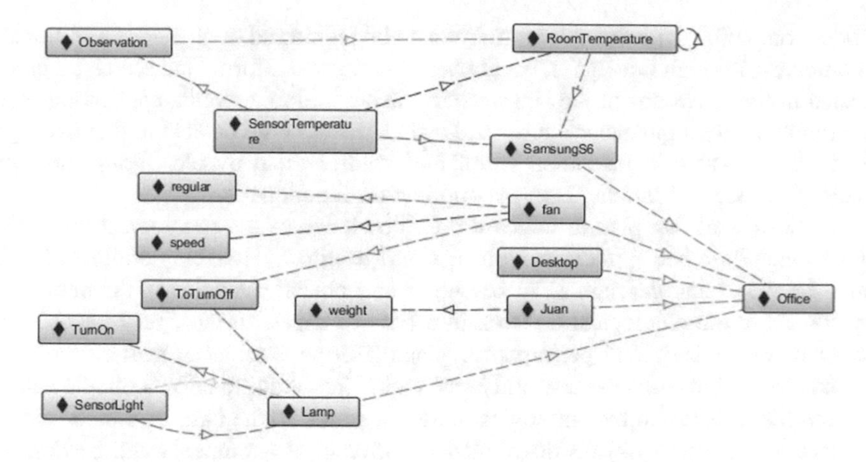

Fig. 4. Instance of the scenario ontology raised

Figure 5 shows the interface that allows the user to manipulate the environment with all detected objects and visualize the different services of each of the objects.

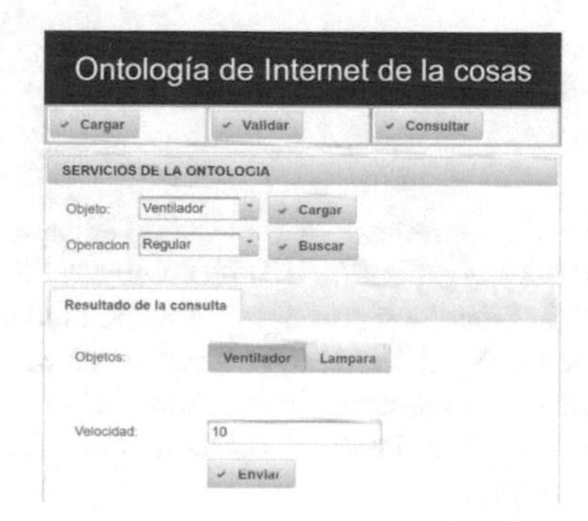

Fig. 5. Application to interact with services

5 Conclusions and Future Work

Internet of things is nowadays a trend that helps people in several activities of his daily life, where semantic technologies are used to interpret data and to obtain knowledge of IoT environments. In this work we proposed the DS-IoT ontological network, designed for IoT environments, which facilitates the description of the attributes of physical environment objects that could be monitored or changed, as well as environment properties that software services are able to measure o change by accessing to resources (sensors and actuators) hosted on devices. The proposed ontological network eases the run-time binding between application functionalities and device services that support the interaction with IoT environments. This binding is achieved by finding device services compatible with the object attributes to be monitored or changed. Additionally, this ontology, through inferences, should help to improve information search on objects detected in the environment. For instance, by using a mobile phone application which main functionality implements a universal switch, users could be able to find intelligent objects that are able to be turned on/off and control them by identifying the corresponding service and binding it to the application functionality.

As future work we plan to increase the knowledge of the ontological network in order to include other application domains and to allow describing additional information by increasing the number of resources and objects available in the ontological network. Store the ontological network in a base of triplets (ontological base), which offers better efficiency and performance when making inferences in the ontological network, which increases the flexibility and scalability. Additionally, a middleware for communication between the ontological network stored in the base of triplets with the resources and physical objects distributed in different environments will be designed and implemented. We also plan to develop a graphical interface that will allow users to describe the characteristics of devices they want to control in the environment, where characteristics will be taken from the knowledge stored in the ontology. Finally, we plan to validate empirically the proposed ontological network.

Acknowledgment. This research is supported by the DIUC_XIV_2016_038 project, University of Cuenca.

References

1. Compton, M., et al.: The SSN ontology of the W3C semantic sensor network incubator group. Web Semant. Sci. Serv. Agents World Wide Web **17**, 25–32 (2012)
2. Bermudez-Edo, M., et al.: IoT-Lite: a lightweight semantic model for the Internet of Things. In: 2016 Intl IEEE Conferences on Ubiquitous Intelligence & Computing, Advanced and Trusted Computing, Scalable Computing and Communications, Cloud and Big Data Computing, Internet of People, and Smart World Congress (UIC/ATC/ScalCom/CBDCom/IoP/SmartWorld). IEEE (2016)
3. Janowicz, K., et al.: SOSA: A lightweight ontology for sensors, observations, samples, and actuators. J. Web Semant. **56**, 1–10 (2018)
4. Bonino, D., Corno, F.: Dogont-ontology modeling for intelligent domotic environments. In: International Semantic Web Conference. Springer, Heidelberg (2008)
5. Wang, W., et al.: A comprehensive ontology for knowledge representation in the internet of things. In: 2012 IEEE 11th International Conference on Trust, Security and Privacy in Computing and Communications. IEEE (2012)
6. https://lov.linkeddata.es/dataset/lov/vocabs/SAN
7. Jayaraman, P., et al.: Internet of things platform for smart farming: experiences and lessons learnt. Sensors **16**(11), 1884 (2016)
8. Sezer, O.B., Can, S.Z., Dogdu, E.: Development of a smart home ontology and the implementation of a semantic sensor network simulator: an Internet of Things approach. In: 2015 International Conference on Collaboration Technologies and Systems (CTS). IEEE (2015)
9. Fan, Y.J., et al.: IoT-based smart rehabilitation system. IEEE Trans. Ind. Inform. **10**(2), 1568–1577 (2014)

Automation of the Municipal Inspection Process in Ecuador Applying Mobile-D for Android

Cathy Guevara-Vega[1(✉)], Jaime Hernández-Rojas[1], Miguel Botto-Tobar[2,3], Iván García-Santillán[1], Andrea Basantes Andrade[4], and Antonio Quiña-Mera[1]

[1] Facultad de Ingeniería en Ciencias Aplicadas, Network Science Research Group e-CIER, Universidad Técnica del Norte, Av. 17 de Julio, Ibarra, Ecuador
{cguevara, jghernandez, idgarcia, aquina}@utn.du.ec

[2] Eindhoven University of Technology, Eindhoven, The Netherlands
m.a.botto.tobar@tue.nl

[3] Universidad de Guayaquil, Guayaquil, Ecuador

[4] Facultad de Educación Ciencia y Tecnología, Network Science Research Group e-CIER, Universidad Técnica del Norte, Av. 17 de Julio, Ibarra, Ecuador
avbasantes@utn.edu.ec

Abstract. A municipality is an administrative entity that refers to a city. Ecuadorian municipalities regularly perform manual inspection processes in stores to calculate municipal charges. However, they have identified that these manual processes are time-consuming and have high costs. Also, the human mistakes existence in obtaining information from stores and the difficulty to manage the inspection processes themselves are problematic. In this context, it raises the technological needed of having real-time information from processes carried out for the municipality. Thus, the goal of this study is to automate the inspection process in economic activities at Ibarra – Ecuador municipality through the mobile app development using Mobile-D methodology. The MVC software architecture (Model, View, Controller) and the ScriptCase, Android Studio, PostgreSQL and REST web services tools were applied. The mobile app quality of use was evaluated by using ISO/IEC 25022 standard, and USE questionnaire (Usability, satisfaction, ease of use). And moreover, it was considered a Wilcoxon statistical analysis to verify the improvement of the automated process compared to manual. From results, it was observed that process automatization improved the information obtained at Ibarra municipality, and 84,32% was achieved with respect to the quality of use by considering the mobile app users.

Keywords: Android · ISO/IEC 25022 · USE questionnaire · Mobile-D · MVC

1 Introduction

Mobile devices are intensifying worldwide and have proven to be an indispensable part of daily life (Deloitte 2018). The penetration of smartphone users in Latin America is 39.1%, and it is estimated that in 2018 it would reach 43.2%. As the world grows, and

M. Botto-Tobar et al. (Eds.): ICAETT 2019, AISC 1066, pp. 155–166, 2020.
https://doi.org/10.1007/978-3-030-32022-5_15

the digital offer migrates from the web (www) to the mobile, the trend of smartphone use continues to grow (GSMA-LA 2018).

In Ecuador, there are currently 14.8 million mobile connections, which 90.8% of the Ecuadorian population has access to 3G and 4G technology, 68% depicts to unique subscribers, and 57% has a smartphone, and it is estimated by 2025 about 73% of Ecuador will adopt a smartphone (GSMA-LA 2018). For this, it is necessary to have new generation networks, and thus have quality telecommunications services (MINTEL 2018). Currently, companies look at the needs of investing in a significant way to meet the demand of users that generating the smartphones. In this context, it is necessary that public and private companies update their services using mobile devices to provide better access to information, they generate growth and employment (GSMA-LA 2018).

The municipalities are institutions that enjoy political, administrative and financial autonomy. The main purpose of decentralization is to strengthen local development and bring authority closer to society (Benalcázar 2013, p. 33). However, most municipalities in Ecuador have difficulties in effective tax management, through which the necessary economic resources are captured to fulfill their purposes and objectives in favor of their communities. Under the article 548 of the COOTAD (2010), to exert a commercial, industrial or financial activity, an annual patent must be obtained, prior registration in the registry that will keep each municipality for these purposes. The patent is the necessary municipal permit for the daily exercise of an activity (Farfán 2017). As we mentioned earlier, the patent is essential within the tax management. For the issuance of this tax, it is necessary to have up-to-date and timely information on the economic activity carried out by a person or taxpayer.

Vinueza-Morales et al. (2015) stated "the organizational processes are carried out manually without the support of technology, this creates the need to present a proposal that promotes a correct process automation structure"; thus, it arises to adopt new technologies such as mobile applications for entering real-time information. Technologies play a crucial role in automation since providing support for identifying, eliminate, and prevent the spread of errors, duplicates, or inconsistencies as well as reduce costs and saving times (Gascó and Ramón-Rodriguez 2007); that is to say, automation processes are linked to efficacy and efficiency and would allow municipalities to achieve the best quality in their offered services and increase its competitivity.

In this sense, this study proposes the process automation through a mobile application in order to verify the efficacy and efficiency improvements in terms of quality of use; thus, our research question is: How to improve the inspection process of economic activity at Ibarra municipality, and how to obtain real-time information?

Next, the research methodology used in this study is presented.

2 Research Methodology

The research methodology was carried out in three stages: (i) research design (research type, sample, development tools, methodology selected, software evaluation model definition); (ii) software development (web and mobile application architecture); and (iii) evaluation, it is presented in Sect. 3.

2.1 Research Design

Through a quantitative and qualitative approach, including surveys and observations in order to measure user satisfaction. We started by surveying the eight people who work at the Department of Revenue at the Ibarra municipality (www.ibarra.gob.ec).

The software tools (Table 1) used to evaluate the quality of use are based on ISO/IEC 25022 standard, and USE questionnaire (Usability, satisfaction, ease of use).

Table 1. Software tools.

Tool	Description
Android Studio	A multiplatform framework for the free integrated development environment (IDE) based on IntelliJ IDEA, for Android applications (Molina, Sandoval and Toledo 2012)
ScriptCase	A PHP code generator, generates web systems and applications, creates quick and safe customized reports (Scriptcase 2019)
PostgreSQL	A database manager-relational, open source. It uses a client/server and multi-process model to guarantee system stability (Postgresql 2019)
JSON web services	It is used to represent structured data, ideal for the exchange of information. It is used to display or send information in web applications (Mozilla 2018)

The tools mentioned above are relevant to this study because they are used in the software developed as detailed later.

We used MOBILE-D as the methodology to the development of mobile applications and which is focuses on agile development (Rahimian and Ramsin 2008), including short development cycles and directed for small teams like in this study. It is based on three recognized and consolidated methodologies, where the development practices (XP, extreme Programming), the scalability of Crystal methodologies methods and the coverage of RUP life cycles were taken Abrahamsson et al. (2004). The design is based on a five-phase process presented in Amaya-Balaguera et al. (2013) as follows:

- **Exploration:** The planning and basic concepts of the project are carried out. The project scope is carried out, and the functionalities are established. In this phase, special attention is paid to the customers participation.
- **Initialization:** In this phase, the project is configured; all necessary resources are prepared and verified; the technological and communication. The technical environment of the project is established.
- **Production:** The programming of the three days is carried out (Initial planning, work day and release day), interactively until all the necessary functionalities of all the modules are carried out. The system tests are also carried out to verify the correct functioning of the development.
- **Stabilization:** In this phase, the integration of all the modules is carried out where the full functioning of the system is verified. Besides, the success and the quality of the implementation of the project are ensured.

- **System tests:** A finished product is delivered as determined by the customer; different tests are carried out with the requirements provided by the client and corrections and repair of possible errors are made.

Quality Metrics in Use of the Product Based on ISO/IEC 25022. The quality in use model analyzes the characteristics of the interactions in different interest groups with the product (ISO/IEC 25022 2016). This model is the newest where the metrics used to measure the quality in use of the product are the effectiveness, efficiency, satisfaction, risk freedom and software context coverage (Rodríguez Carrillo, Jiménez Builes and Paternò 2015). Table 2 lists the quality metrics in use of the software product, as well as its subcategories, the assigned weight (depending on the importance for this study), and the guidelines questions based on each metric (ISO/IEC 25022 2016).

Table 2. Metrics defined in the ISO/IEC 25022.

Category	Subcategory	Weight category	Weight sub category	Questions guidance	Application method
Efficacy	Tasks completed	40%	17%	How many tasks assigned to the inspection in the economic activity could perform them without assistance?	Observation
	Objectives achieved		13%	Does the field inspection system meet the needs of the company?	Survey
	Task errors		10%	How many tasks assigned to the inspection in the economic activity had at least one error in the execution?	Observation
Efficiency	Tasks time	30%	10%	How long does it take to complete an economic activity inspection task?	Survey
	Time efficiency		15%	How efficient is an "inexperienced" user compared to an expert user?	Survey
	Profitability		5%	Do you consider that you have improved your productivity with the implementation of the inspection system in the economic activity?	Survey
Satisfaction	Utility	20%	10%	Are you satisfied with the use of the inspection's application in the economic activity?	Survey
	Confidence		6%	Did you file at least one complaint during the use of the application due to system failure?	Survey
	Comfort		4%	Do you consider that the inspections system in the economic activity field present ease and little effort in its use?	Survey

(*continued*)

Table 2. (*continued*)

Category	Subcategory	Weight category	Weight sub category	Questions guidance	Application method
Risk freedom	Reduction of environmental risks	5%	5%	How much do you agree that the use of the application decreases the environmental impact?	Survey
Context coverage	Integrity	5%	3%	In what proportion of the intended use contexts of the product can it be used with the ease of acceptable use?	Survey
	Flexibility		2%	To what extent can the product be used in additional contexts of use?	Survey

Satisfaction Metrics Based on Statistical Analysis Using the Wilcoxon Test. Statistical analysis was performed by using the Wilcoxon test in order to demonstrate that the software contributes significantly to improve the inspection process in the field of economic activity. A survey was applied before and after the software implementation. We used an adapted four-variable questionnaire (Belzunegui, Brunet and Pastor 2012) of USE (Usefulness, Satisfaction, and Ease of use - Usability, satisfaction, ease of use). The items are evaluated using a five-point Likert-scale (Likert 1932). Table 3 shows the selected questions and the measured variables (Alvites-Huamaní 2016).

Table 3. Selected questions and measurement variables adapted from the USE questionnaire.

Questions (after)	Variables
Do you consider that the economic activity system helps you save time and resources?	Utility
Do you consider that the economic activity system helps you to take an organized control of the information?	
Do you think that the mobile field inspection application is easy to use?	Ease of use
Do you think the inspection system is easy to learn how to use?	Ease of learning
Does the inspections system in the economic activity field meet your needs?	Satisfaction

Software Evaluation Model Definition. To evaluate the quality in use of the implementation of the inspection's application in the economic activity field. Two procedures were carried out: (i) Using quality metrics in use of the product based in the ISO/IEC 25010 (ISO 25000 2018); to measure efficiency, effectiveness, satisfaction, risk freedom and context coverage; and (ii) through statistical analysis based on the Wilcoxon test and boxplot diagram.

To compute the results, a five-point Likert-scale was used (Table 4). The items are measured with the same intensity that is desired to measure the attitude, and it is the respondent who gives a score. It consists of a survey with alternatives response where they might express their agreement or disagreement degree. The scores obtained concerning all the statements are added altogether to show whether the overall assessment is favorable or unfavorable.

Table 4. Five-point Likert-scale (Likert 1932).

Scale	Strongly agree	Somewhat agree	Neither agree nor disagree	Somewhat disagree	Strongly disagree
Value	5	4	3	2	1

2.2 Software Development

As per MOBILE-D methodology includes five stages as follows:

Exploration Phase. Stakeholders are defined, in this case, they are the institutions that have the Integrated Municipal Management System (SIGET), that is, those that need to optimize and keep track of the inspections in the economic activity. Next, the requirements of the system were developed, which are the following:

- **The administration module.** It is a web application that has the following features: Users creation for the use of the mobile application, types of inspectors' creation, inspection types creation, inspection types parameterization, registration of new inspections, inspection report, report of pending inspections, report of inspections carried out, inspection reports by state.
- **Field inspection module.** It is a mobile application, which is responsible for surveying economic activity. It contains a geolocation option to allow users identifying the initial route to the inspection destination or place (Quiña-Mera et al. 2019).

Initialization Phase. The development environment is defined, as well as the development tools aforementioned in Table 1. Besides, the system architecture is defined, comprising the architecture of the web application and the mobile application which are explained as follows:

a. **Web Application Architecture.** The Web Application Architecture Integrated in the Municipal Management System (SIGET) is based on layered architecture; which has three main components: web application, application server, and database server.
b. **Architecture Mobile application.** It consists of four components: mobile application, web service, application server, and database server (Sainz 2016). Once the tools were configured, an initial analysis of the functional requirements (Guevara-Vega et al. 2019a, b) was carried out using a use case. The initial iterations planning was carried out, and the following stages:

   - **Initialization:** Establishments of the Inspector type requirement.
   - **Production:** Implementation of the requirement definition of Inspector type, Refinement of interfaces, and Generation of acceptance tests.
   - **Stabilization:** Refactoring of the Inspector type definition requirement, Refinements of interfaces, and Execution of acceptance tests.
   - **Tests:** Evaluation of the tests and analysis of results.

c. **Production phase.** The development of the system is performed based on the Iterations that were raised in the initialization stage. The first task was the design of the relational database used to store the information.
d. **Stabilization phase.** The integration of all the modules is carried out, the correct functioning of the system is verified. The success and quality of the implementation are assured, and finally, the documentation generation is carried out such as a user manual, and a technical manual.
e. **Testing and system repair phase.** A unit testing plan is made according to the requirements. Once completed, corrections and repair of possible errors are made. In this way, it is finalized, and a functional system is available according to the customers' requirements.

Next, the results obtained in this study are presented.

3 Results

We analyze the results obtained in the evaluation of the inspection system for the economic activity field (SIGET) from two perspectives: (i) based on ISO/IEC 25022; and (ii) according to Wilcoxon's statistical analysis.

A Quality Analysis in Use According to the ISO/IEC 25022
Table 5 (column 7 in specific) shows the results obtained after applying the data collection methods to the users of the inspections system in the economic activity, to determine the acceptance of the developed system. The sum of the weighted scores in the evaluation of the inspection system in the field was x = 0.8432, that is, it is in the interval $0.8 < x \leq 1$ according to the scale established in Table 5, which values are between [0 and 1] being 1 the best value.

Table 5. Metrics defined in the ISO/IEC 25022.

Category	Subcategory	Weight category	Weight subcategory	Measure	Result	Result achieved	Expected result
Efficacy	Tasks completed	40%	17%	0,70	11,90%	31,60%	40%
	Objectives achieved		13%	0,90	11,70%		
	Tasks error		10%	0,80	8,00%		
Efficiency	Tasks time	30%	10%	0,93	9,25%	26,63%	30
	Time efficiency		15%	0,88	13,13%		
	Profitability		5%	0,85	4,25%		
Satisfaction	Profit	20%	10%	0,93	9,25%	17,25%	20
	Confidence		6%	0,75	4,50%		
	Comfort		4%	0,88	3,50%		
Risk freedom	Reduction of environmental risks	5%	5%	1,00	5%	5,00%	5
Coverage context	Context integrity	5%	3%	0,77	2,31%	3,85%	5
	Flexibility		2%	0,77	1,54%		
Total		100%	100%			84,32%	100%
Total/100		1	1			0,8432	1

It is observed that when developing the software, the users agree with the implementation, that is, "Strongly agree" fulfilling the quality characteristics in use, according to the ISO/IEC evaluation model: 2016 25022.

Satisfaction Analysis According to Statistical Analysis. To determine whether users agree with the automated inspection process, the sub-characteristics and satisfaction metrics to be evaluated were selected using a questionnaire of four questions (Table 3) adapted from the USE questionnaire (Alvites-Huamaní 2016). The test used the Likert-scale (5 levels) and was applied to the same sample of 8 users before and after the development of the inspection system in the economic activity.

In the statistical analysis, the types of variables and the sample used was considered. In this case, the measured variables are of the ordinal type (Likert-scale) (Likert 1932), and the sample is related (the same group of individuals), thus the statistical test used was that of Wilcoxon (Juma et al. 2019). This test consists of a non-parametric test applied to two related samples and tries to debate whether the results emitted by both samples are the same or different. Specifically, the null hypothesis (H_0) will show that the starting distributions of the populations from which the samples have been obtained are the same, as opposed to the alternative hypothesis (H_1) that there is a difference between both distributions. In this study, the two hypotheses were established as follows:

H0: There are no significant differences with the automated inspection process.
H1: There are significant differences with the automated inspection process.

Statistical analysis was performed using IBM SPSS Statistics software, version 24-2018. The decision rule in the hypothesis test is the following:

If *p_value* > 0.05 then, H_0 is accepted, otherwise H_1 is accepted. Table 6 shows the results for all tests.

Table 6. Wilcoxon test.

Statistics test	Utility (after) Utility (before)	Use (after) Use (before)	Learning (after) Learning (before)	Satisfaction (after) Satisfaction (before)
Z	−2,585b	−2,714b	−2,236b	−2,714signedb
Sig. Asymptotic (bilateral)	,010	,007	,025	,007

Additionally, the Wilcoxon rank test was used to compare two related measurements and determine if the difference between them is due to chance or not. Table 7 shows the results of the test applied to the satisfaction questionnaire.

Table 7. Wilcoxon test ranges.

Ranges	Utility (after) (before)			Use (after) (before)			Learning (after) (before)			Satisfaction (after) (before)		
	N	Avg	Sum	N	Avg	Sum	N	Avg	Sum	N	Avg	Sum
Negative ranges	0	0	0	0	8	0	0	0	0	0	0	0
Positive ranges	8	4,5	36	8	4,5	36	5	3	15	8	4,5	36
Draws	0	0		0	36		3	0		0	0	
Total	8			8			8			8		

It is observed that for each indicator (variable before and after), the positive ranges (Average range), prevail on the negative ranges and draw, which indicates that there is a favorable contribution thanks to the development and implementation of the inspection system in the economic activity field.

Figure 1 shows the trend of the four variables measured (utility, ease of use, ease of learning and satisfaction) before and after the development of the inspection system for the economic activity field. The four variables measured had an increase of at least one level in the Likert-scale showing a significant impact on user satisfaction when using the inspections system.

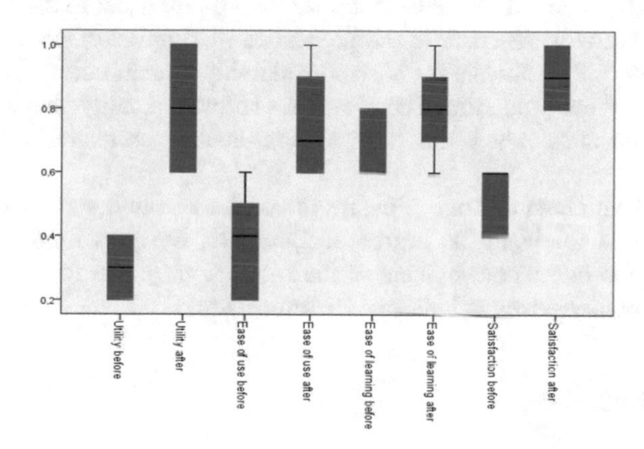

Fig. 1. Boxplot diagram of the four medium variables before and after the development and implementation of the inspections system of the economic activity field.

4 Discussion

This study agreed with Vinueza-Morales et al. (2015) in need to automate process in public and private business in order to optimize tasks, improve the customer service, etc.; and moreover, it was possible to go further by implementing a tool to measure its impact; its result (84,32%) shows a high rate of acceptance. Besides, it was verified that automation by using mobile technologies in public institutions offered the likelihood to obtain real-time information efficiently; with minor differences with Jácome et al. (2015) since they measured, analyzed and verified the improvement in the automation process.

In software development, it is necessary to adopt agile processes and practices such as Mobile-D used in this study. According to Abrahamsson (2005) suggested that using agile methods make systems more productive, easier to maintain, better-tested code, even the cost of change is lower than in the development of traditional systems.

Agile methodologies present good practices and methods to achieve a quality product when developing small projects. Amaya (2013) stated that the agile methods are an excellent alternative to guide software development projects of size reduced, as is the case of applications for mobile devices, thanks to the great ease of adaptation they possess (Alsabi and Dahanayake 2016); however, these need to be adapted to their particular characteristics to obtain quality products.

To obtain a quality software as well as to consider that the requirements analysis phase is essential to prevent setbacks in the final deployment of the software and the negative effects on the productivity of the company (Guevara-Vega et al. 2019a, b). Follow a software methodology meticulously is not everything; it is also essential to use measurement standards such as ISO/IEC 25000 (Roa, Morales and Guitiérrez 2015) where it is clear that when using a software quality standard. Better quality software is obtained that meets the specific needs of the client, where this measurement is made in a real environment, that is, where the system is running. It agrees with (Mora and Jamaica 2016) where it indicates that after taking into account the basic principles of usability and how these relate to the properties of simplicity, since, together, they make the quality of the software increase considerably in comparison to a product that does not include these properties. Therefore, the software quality should be a relevant aspect for companies, where the use of low-quality software, cause economic repercussions.

It is worth mentioning that one of the results of this research was the degree work of one of the authors to obtain the degree of Computer Systems Engineer (Hernández 2018). Finally, the future development of the inspection system for economic activity field on iOS mobile devices is indicated as future work.

5 Conclusions

This work presents the mobile application development by using software tools such as Android Studio, ScriptCase, JSON web services, PostgreSQL and Mobile-D software methodology; in order to contribute to the automation of the inspection process in the economic activity at Ibarra. The beta application was tested for 30 days, where it was evaluated by using quality metrics in use, defined in the software quality model of the ISO/IEC: 25010 and 25022 where it can be shown that customer satisfaction results in 84.32% concerning the implementation of the system. Also, nonparametric analysis was performed using the Wilcoxon tests and boxplot diagrams. Through this study, it was determined that the process automation with cutting-edge technologies contributes to the improvement of services in public or private institutions, allowing users of having real-time information.

As future work, it is proposed to improve the software through information assurance using a others DBMS that allows synchronizing the inspection information in the first connection to the internet.

References

Abrahamsson, P., Hanhineva, A., Hulkko, H., Ihme, T., Jäälinoja, J., Korkala, M., Koskela, J., Kyllönen, P., Salo, O.: Mobile-D: an agile approach for mobile application development (2004). https://doi.org/10.1145/1028664.1028736

Abrahamsson, P.: Keynote: mobile software development – the business opportunity of today. In: Proceedings of the International Conference on Software Development, Reykjavik, Iceland, pp. 20–23 (2005)

Alsabi, E., Dahanayake, A.: Modelado inteligente para métodos ligeros de desarrollo de aplicaciones móviles. In: Ivanović, M., et al. (eds.) Nuevas tendencias en bases de datos y sistemas de información, ADBIS 2016. Comunicaciones en Informática y Ciencias de la Información, vol. 637 (2016)

Alvites-Huamaní, P.: Usabilidad: páginas web, entornos y educación virtual. Hamutay **3**(1), 71–79 (2016). https://doi.org/10.21503/hamu.v3i1.1002

Amaya-Balaguera, Y.: Metodologías ágiles y desarrollo de aplicaciones móviles. In: COMTEL, pp. 184–190 (2013). http://repositorio.uigv.edu.pe/bitstream/handle/20.500.11818/832/Memoria-COMTEL-2013-177-184.pdf?sequence=1&isAllowed=y

Belzunegui, E., Brunet, I., Pastor, G.: Multi-level qualitative analysis design: a practical application for the analysis of interviews. Rev. de Metodología de Cienc. Sociales 15–44 (2012). https://doi.org/10.5944/empiria.24.2012.841

Benalcázar, G.J.: La autonomía financiera municipal en el Ecuador, con especial referencia a los ingresos tributarios y a la potestad tributaria de los municipios. J. IUS 27–40 (2013). http://www.scielo.org.mx/pdf/rius/v7n32/v7n32a3.pdf

COOTAD (2010). https://www.habitatyvivienda.gob.ec/wp content/uploads/downloads/2015/06/Codigo-Organico-de-Organizacion-Territorial-cootad.pdf

Deloitte (2018). https://www2.deloitte.com/do/es/pages/technology-media-and-telecommunications/articles/pr-global-mobile-consumer-trends.html

Farfán, D.: (2017). http://dspace.ucuenca.edu.ec/bitstream/123456789/27574/3/Trabajo%20de%20Titulaci%C3%B3n.pdf

Gascó, M., Ramón-Rodriguez, J.: Gobierno electrónico y transformación de la administración pública municipal: El caso del Ayuntamiento de Barcelona. In: VIII AECPA Conferencia sobre Ciencia Política y Administración Pública, Valencia, España (2007). https://doi.org/10.13140/RG.2.1.4406.2803

GSMA-LA (2018). https://www.gsmaintelligence.com, https://www.gsmaintelligence.com/research/?file=4880883454cefe7a3cf9b9a2d6183ead&download

Guevara-Vega, C., Basantes, A., Guerrero, J., Quiña-Mera, A.: Software estimation: benchmarking between COCOMO II and SCOPE. In: Communications in Computer and Information Science, vol. 895, pp. 176–190. Springer (2019a). https://doi.org/10.1007/978-3-030-05532-5_13

Guevara-Vega, C.P., Guzmán-Chamorro, E.D., Guevara-Vega, V.A., Andrade, A.V.B., Quiña-Mera, J.A.: Functional requirement management automation and the impact on software projects: case study in Ecuador. In: Advances in Intelligent Systems and Computing, vol. 918, pp. 317–324. Springer (2019b). https://doi.org/10.1007/978-3-030-11890-7_31

Hernández, J.: Desarrollo del módulo de inspecciones de campo para actividad económica integrado al sistema de gestión tributaria (SIGET) aplicando dispositivos móviles android y metodología ágil mobile-D para la empresa bypros sistemas incorporados Cia. Ltda. (2018). http://repositorio.utn.edu.ec/handle/123456789/8616

ISO/IEC 25022 (2016). https://www.iso.org/standard/35746.html

ISO 25000 (2018). http://iso25000.com/index.php/normas-iso-25000

Jácome, L., Gonzalez, A. (2015). http://dspace.unl.edu.ec/. Obtenido de http://dspace.unl.edu.ec/jspui/handle/123456789/11471

Juma, A., Rodríguez, J., Naranjo, M., Caraguay, J., Quiña, A., García-Santillán, I.: Integration and evaluation of social networks in virtual learning environments: a case study, pp. 245–258. Springer (2019). https://doi.org/10.1007/978-3-030-05532-5_18

Likert, R.: A technique for the measurement of attitudes. Archives of Psychology (1932)

MINTEL (2018). https://www.telecomunicaciones.gob.ec/wp-content/uploads/2018/07/Libro-Blanco-de-la-Sociedad-del-Informaci%C3%B3n-y-del-Conocimento.pdf

Molina, J., Sandoval, J., Toledo, A.: (2012). http://repositorio.utp.edu.co/dspace/bitstream/handle/11059/2687/0053M722.pdf?sequence=1&isAllowed=y

Mora, J., Jamaica, K.: (2016). http://repository.udistrital.edu.co/bitstream/11349/…/MoraHernándezJuanGabriel2016.pdf

Mozilla (2018). https://developer.mozilla.org/es/docs/Learn/JavaScript/Objects/JSON

Postgresql (2019). https://www.postgresql.org/

Quiña-Mera, J.A., Saransig-Perugachi, E.R., Trejo-España, D.J., Naranjo-Toro, M.E., Guevara-Vega, C.P.: Automation of the barter exchange management in Ecuador applying Google V3 API for geolocation. In: Advances in Intelligent Systems and Computing (2019). https://doi.org/10.1007/978-3-030-11890-7_21

Rahimian, V., Ramsin, R.: Designing an agile methodology for mobile software development: a hybrid method engineering approach. In: 2008 Second International Conference on Research Challenges in Information Science, Marrakech, pp. 337–342 (2008). https://doi.org/10.1109/RCIS.2008.4632123

Roa, P., Morales, C., Gutiérrez, P.: Norma ISO/IEC 25000, pp. 27–33 (2015). http://revistas.udistrital.edu.co

Rodríguez Carrillo, P.J., Jiménez Builes, J.A., Paternò, F.: Monitoreo de la actividad cerebral para la evaluación de la satisfacción. Eng. J. 16–22 (2015). https://dx.doi.org/10.16924/riua.v0i42.750

Sainz, I.: (2016). http://oa.upm.es, http://oa.upm.es/44639/1/TFM_IVAN_RODRIGUEZ_SAINZ_MAZA.pdf

Scriptcase (2019). https://www.scriptcase.net/, https://www.scriptcase.net/funcinalidades/

Vinueza-Morales, M., Rodas-Silva, J., Chacón-Luna, A.: Strategic planning through ICT in rural autonomous governments canton Milagro. Rev. Cienc. UNEMI **8**(14), 40–49 (2015)

VYTEDU-CW: Difficult Words as a Barrier in the Reading Comprehension of University Students

Jenny Ortiz Zambrano[1(✉)], Arturo MontejoRáez[2], Katty Nancy Lino Castillo[1], Otto Rodrigo Gonzalez Mendoza[1], and Belkis Chiquinquirá Cañizales Perdomo[1]

[1] University of Guayaquil, Cdla. Salvador Allende, Av. Delta and Av. Kennedy, Guayaquil, Ecuador
{jenny.ortizz,katty.linoc,otto.gonzalezm, belkis.canizalesp}@ug.edu.ec

[2] University of Jaén, Campus Las Lagunillas s/n., 23072 Jaén, Spain
amontejo@ujaen.es

Abstract. We present VYTEDU-CW (Educational Videos and Texts - Complex Word) as a corpus in Spanish made up of university educational texts containing difficult words. In its construction process, it worked with students of the different careers that make up the University of Guayaquil (Ecuador); the students were selected according to the level of studies with respect to the content of the texts of the corpus, this with the purpose that the text is not very easy for students who are in higher grades. The work consisted in the recognition and labeling of the complex words contained in the different texts that make up the VYTEDU corpus. It should be noted that one of the difficulties in reading comprehension is complex words, which become a barrier especially for those with special cognitive abilities. This new corpus had its origin in the VYTEDU corpus, whose demonstration was made at the congress of the SEPLN (Murcia-Spain, 2017) and on which several experiments were carried out. Currently, VYTEDU-CW is being used to continue researching the Automatic Simplification of Spanish Texts in the field of Natural Language Processing. A sample of students was taken from the different faculties and from the results the corpus statistics were obtained. The objective of this work is to determine that difficult words in educational texts become one of the main problems in the reading comprehension in the day to day of university students.

Keywords: Difficult words · University texts · Barrier · Reading comprehension

1 Introduction

1.1 The Information Society

The comprehension of a text is not an easy task for all, there are those who have a difficult linguistic understanding. Within the information society, people should be able

M. Botto-Tobar et al. (Eds.): ICAETT 2019, AISC 1066, pp. 167–176, 2020.
https://doi.org/10.1007/978-3-030-32022-5_16

to access all available information easily and simply, so improving access to written language is a topic of growing interest [1].

Many of the texts are written in a complex way, using a sophisticated and specialized vocabulary such as information from areas such as administrative, government, health, news texts or popular magazines [2] among others, which should be accessible to all members of the society especially for that large and heterogeneous target group [1] which causes a barrier in the compressibility to many readers [3] who find it difficult to understand how the children [4] and also especially those people who possess ASD [5].

According to the research carried out, it has been shown that sentences whose structures are complex such as those that are long or passively written, as well as the use of infrequent words, become difficult for many people to understand [3].

In recent years, the University of Guayaquil has strengthened scientific development through the proposal of research projects as a way to search for new technologies that contribute to the learning of students in their training process. A doctoral project is currently underway in conjunction with the University of Jaén in the field of Natural Language Processing, oriented towards the simplification of texts whose origin is based on the transcriptions of educational videos made within the university classrooms.

The objective of this work is to deepen the knowledge of the linguistic comprehension difficulties of the students. In order to study this phenomenon, a new corpus containing labeled complex words based on the texts of the transcriptions of academic videos of the VYTEDU corpus has been built.

This work constitutes a third advance in the development of the research of a doctoral project in the area of the Simplification of Texts that is currently being carried out for the University of Guayaquil, being the first the construction of the VYTEDU corpus [7] and the second one advance the application of the complexity metrics in the texts of the transcriptions of the corpora videos (VYTEDU) to determine the lexical complexity contained in the texts [8].

The rest of the work is organized as follows:

Section 2 presents the related works that have been taken as reference for the development of this Project.

Section 3, the corpus labeled VYTEDU-CW is presented, and its development is explained, how the complete words were tagged, and the construction of its data library.

Section 4, the statistics that were obtained from the generated results of the corpus are exposed.

Section 5, the conclusions are exposed.

Section 6, presents work proposals and future experiments that will be developed in the later stages of the project.

Finally, thanks to the team of researchers who have contributed to the research.

1.2 Origin of the VYTEDU Corpus

VYTEDU[1] was born in 2017 as the origin of the VYTEDU corpus, a first stage in the proposed research project due to the scarcity of resources for the Spanish language and particularly in the educational field to advance research in the area of Simplification of texts. There are other corpus used to investigate the complexity of the text, as is the case of the AMI corpus created by Carleta, 2007 on video conferences [6]; the corpus de Rosas et al., 2013, which constitutes a set of data about the analysis of feelings in Spanish on a general topic, where 105 online videos of Youtube social media were selected [7], but they reduce the resources in Spanish that serve of support for education [7].

VYTEDU contains the transcriptions of educational videos made to a group of teachers in the classrooms of the University of Guayaquil about the different topics that correspond to different subjects of the academic programs that are offered. The creation of this corpus is an essential source of data for carrying out research in the systems of simplification of texts in the educational area whose presentation was carried out by means of a demonstration at the congress of the SEPLN[2] 2017 [7] and later several experimentations were carried out from corpus [8].

1.3 The Simplification of Texts

It is a subject highly investigated [4]. It is the process of automatically producing a simplified text [9], it is a branch of the PLN[3] that maximizes the understanding of written texts, its purpose being to simplify texts by using natural language processing tools to make information more accessible [10], providing readers with a better compressibility of the information made by reducing the complexity of the texts [11] and improving their readability [12] since it will contain simple sentences expressed in a common vocabulary [9] providing them with this forms easier access to information, ensuring readers' comprehension of the various environments [12], thus reducing the efforts and costs associated with human simplification [13].

2 Related Work

Many projects in simplification of texts that have been implemented are based on the first step in the identification of difficult words or sentences that exist in the content of the text. A first step to carry out projects of simplification of texts is the classification of sentences as easy enough or too difficult, so his research focused on the identification of difficult words as a first task and then proceed to the lexical substitution of words [14].

[1] VYTEDU - Videos y Transcripciones en el ámbito educativo.

[2] SEPLN - Spanish Society for Natural Language Processing.

[3] PLN - Natural Language Processing.

Another example of this is the PSET[4] project directed to the creation of accessible texts for aphasic readers. As part of the simplification process, difficult words were identified and then the lexical simplification process was carried out, which consisted of replacing the difficult words with others that were easier to understand, thus reducing the lexical complexity of the text [14]. Consisted of a simplifier of texts for English in which both lexical and syntactic simplification was implemented. In the lexical simplification phase, the detection of complex words in which they were identified in the text was carried out through the frequency of values and the use of psycholinguistic resources [15].

An automatic text simplification system for German presented a focus-based on simplification rules supported by linguistic motivation guidelines to transform standard German into simplified German. The process of simplification of text was based on the identification of difficult words, then they were replaced by their respective definition from the dictionary of simplified definitions in German called Hurraki, in this way a more readable text was obtained [16].

A pioneering work on Text Simplification was carried out in the Czech Republic. For the identification of difficult words, the author marked what he considered to be complex, then he commissioned three scorers to add or delete some complex words and then the three scorers checked the scores they made. All words marked as complex by at least one annotator were used in the set of evaluations.

Finally, the generation of substitution was tested using these words, it was first evaluated separately for the sources based on dictionaries and for the sources based on inlays, and then jointly [17].

3 VYTEDU-CW

It is important to note that VYTEDU–CW[5] was born from the VYTEDU corpus and represents an important advance in the proposed research project for the University of Guayaquil in the area of Text Simplification. VYTEDU-CW contains the transcripts of the educational videos of VYTEDU but the particularity of this corpus lies in the labeling of the complex words that appear in the content of the texts that comprise it.

3.1 Construction Process of the VITEDU-CW Corpus

About the EIL Application

A software was built where the texts of the VYTEDU corpus were loaded, which was named EIL[6] (See Fig. 1) and on which the texts of the VYTEDU corpus were loaded and then proceeded to perform the annotation process with the students belonging to the different careers of the University of Guayaquill.

[4] PSET – Practical Simplification of English Text.

[5] VYTEDU–CW - Videos y Transcripciones en el ámbito educativo- Complex Word.

[6] EIL – Natural Language Environment.

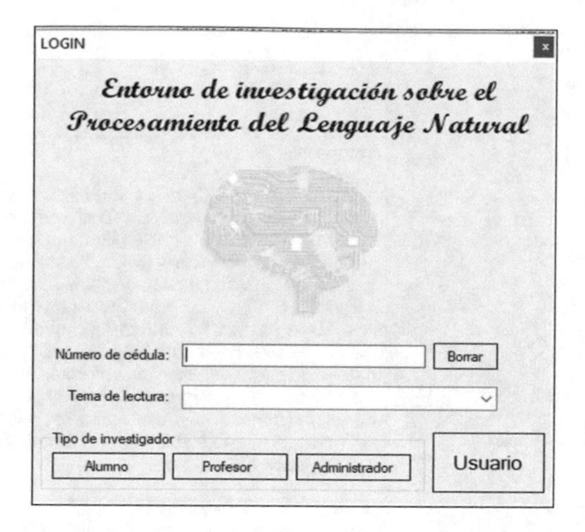

Fig. 1. EIL software - research environment on Natural Language Processing.

Labeling of Complex Words

Students were selected according to the level of studies regarding the content of the texts of the corpus, this with the purpose that the text is not very easy for students who are in higher grades. One of the data that the students had to enter the system was their personal identification and then choose the career to which they belonged and then press on the student button (see Fig. 2).

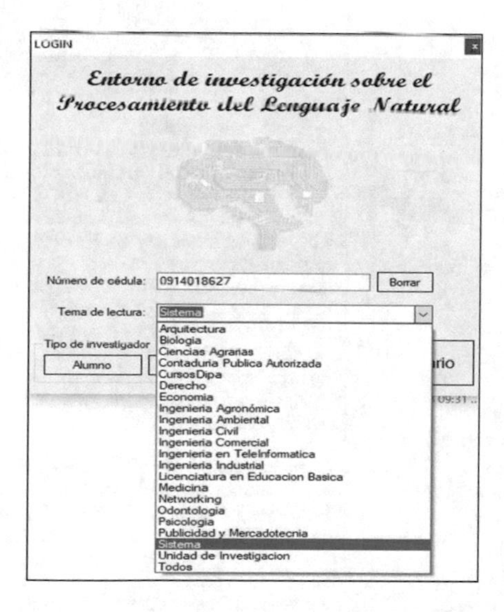

Fig. 2. Careers of the University of Guayaquil.

Showing up next the Figs. 3 and 4 which present some examples of the labeling of difficult words.

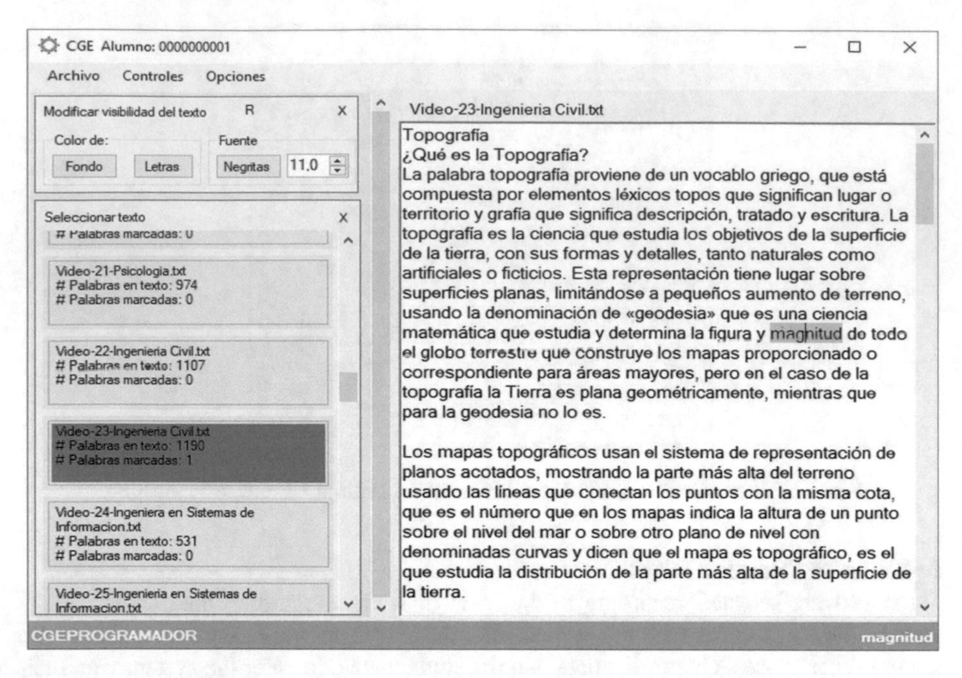

Fig. 3. Labeling of difficult words identified in the text.

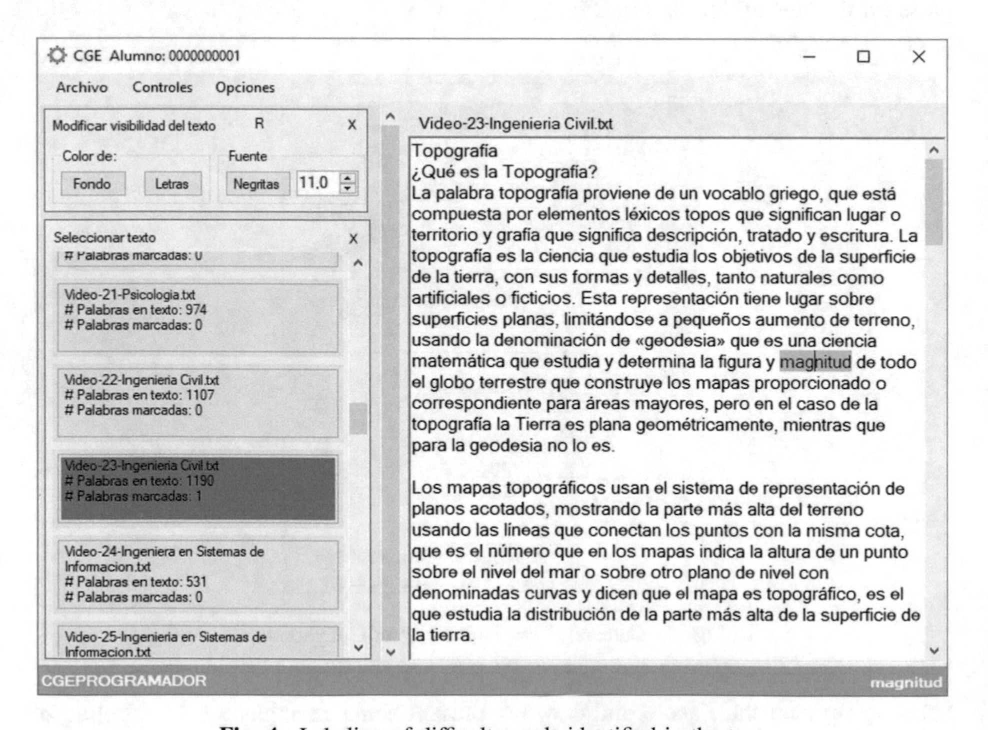

Fig. 4. Labeling of difficult words identified in the text.

Subsequently, the application loads the texts related to the career to which the student belongs, who must begin to choose the text and start the reading. During the process of reading the material, the student proceeds to identify the word or words that are difficult to understand and then proceed to label it as a difficult word, the same as the system will show shaded. This operation is repeated throughout the text with all words equal to the one chosen by the student (see Figs. 5 and 6).

Fig. 5. Students labeling difficult words in texts.

Fig. 6. Students labeling difficult words in texts.

In this way the word that was identified as difficult is stored, the type of user who accessed the system (since it could be a teacher or a student), the type of word that has been marked (since it can be a difficult word or specialized word, where the specialized words are specific to the subject and as such should be learned without alternatives), the user who tagged it, the name of the file of the text that was read, the initial position of the word in the text (it is a number), the length of the word, and the date and time when the text was read (see Fig. 7).

	A	B	C	D	E	F	G	H	I
1	cnc	Dificil	706487014	Alumno	Video-52-Arquitectura.txt	6485	3	11	20/02/2019 11:53
2	foélica	Dificil	706487014	Alumno	Video-52-Arquitectura.txt	8712	7	17	20/02/2019 11:53
3	jalaba	Dificil	915272801	Alumno	Video-54-Arquitectura.txt	654	6	2	20/02/2019 12:10
4	seprvive	Dificil	915272801	Alumno	Video-54-Arquitectura.txt	1818	8	4	20/02/2019 12:10
5	ciprés	Dificil	915272801	Alumno	Video-54-Arquitectura.txt	1848	6	4	20/02/2019 12:10
6	pinito	Dificil	915272801	Alumno	Video-54-Arquitectura.txt	1878	6	4	20/02/2019 12:10
7	hepatización	Dificil	915272801	Alumno	Video-54-Arquitectura.txt	3479	12	6	20/02/2019 12:10
8	atañe	Dificil	951875822	Alumno	Video-15-Contaduria Publica Autorizada.txt	294	5	4	20/02/2019 16:42
9	psicométrico	Dificil	951875822	Alumno	Video-15-Contaduria Publica Autorizada.txt	2918	12	22	20/02/2019 16:42
10	preliminar	Dificil	951875822	Alumno	Video-15-Contaduria Publica Autorizada.txt	4538	10	28	20/02/2019 16:42
11	fungirá	Dificil	951875822	Alumno	Video-15-Contaduria Publica Autorizada.txt	6656	7	38	20/02/2019 16:42
12	costes	Dificil	931692842	Alumno	Video-18-Contaduria Publica Autorizada.txt	825	6	7	20/02/2019 16:52
13	priori	Dificil	931692842	Alumno	Video-18-Contaduria Publica Autorizada.txt	1669	6	12	20/02/2019 16:52
14	racionalment	Dificil	931692842	Alumno	Video-18-Contaduria Publica Autorizada.txt	2138	13	15	20/02/2019 16:52
15	coste	Dificil	931692842	Alumno	Video-18-Contaduria Publica Autorizada.txt	3895	5	24	20/02/2019 16:52
16	umbrales	Dificil	931692842	Alumno	Video-18-Contaduria Publica Autorizada.txt	4286	8	25	20/02/2019 16:52
17	rubro	Dificil	944172162	Alumno	Video-45-Contaduria Publica Autorizada.txt	1082	5	4	20/02/2019 16:56
18	operativas	Dificil	944172162	Alumno	Video-45-Contaduria Publica Autorizada.txt	2597	10	8	20/02/2019 16:56
19	acciones	Dificil	944172162	Alumno	Video-45-Contaduria Publica Autorizada.txt	11449	8	32	20/02/2019 16:56
20	dividendos	Dificil	944172162	Alumno	Video-45-Contaduria Publica Autorizada.txt	11491	10	32	20/02/2019 16:56
21	abordaje	Dificil	940516677	Alumno	Video-35-Contaduria Publica Autorizada.txt	3721	8	11	20/02/2019 17:12
22	heterónoma	Dificil	915463483	Alumno	Video-21-Psicologia.txt	3387	10	13	21/02/2019 10:17
23	dislexia	Dificil	915463483	Alumno	Video-21-Psicologia.txt	5388	8	27	21/02/2019 10:17
24	fitologia	Dificil	924849128	Alumno	Video-11-Ciencias Agrarias.txt	39	9	1	21/02/2019 10:46
25	phyton	Dificil	924849128	Alumno	Video-11-Ciencias Agrarias.txt	213	8	3	21/02/2019 10:46

Fig. 7. The data library of the corpus VYTEDU-CW.

4 Corpus Statistics VYTEDU-CW

Very important data was obtained in the progress of this investigation regarding the identification of difficult words that contain educational texts. We can appreciate the following relevant data (see Fig. 8).

- The number of words in all the texts contained in the corpus is 68.301.
- The total number of users who have tagged words is 422 students.
- 573 students were the ones who accessed the application.
- There is a total of 661 different words that were marked throughout the corpus.
- 242 is the number of users who have entered the application once.
- 9175 is the total number of words in the whole corpus without repeating.

The data Corpus vs marks mean in terms of percentages that 2.38% is the percentage of words tagged from the total number of words in the corpus, while the% of terms without repetitions is the percentage of words that do not repeat with respect to the total number of words. words in the corpus.

Todos los registros	
# de terminos:	1628
# de usuarios:	422
# de entradas al corpus:	573
Registros sin repeticion	
# de terminos:	661
# de usuarios:	242
Corpus	
# de palabras:	68301
# de palabras sin repeticion:	9175
Corpus vs marcas	
% terminos:	2,38%
% terminos Sin repeticion:	7,20%

Fig. 8. The statistics of the corpus VYTEDU-CW.

5 Conclusions

This project is relevant because it constitutes an experimentation based on research works carried out in the field of Text Simplification as pro-posed by [14–17].

The research closely followed the work done by [17]. The application presented allowed the students to identify and label the difficult words that contained the texts of the corpus concerned demonstrating that the texts contain complex words that hinder reading comprehension.

A statistical study of the results achieved by VYTEDU-CW has also been carried out, which allows us to show that the achievement proposed in this third stage of the research project on Text Simplification in the proposed educational field for the University of Guayaquil has been achieved.

6 Proposals for Future Work/Experimentation

The result of this research project is very important because the proposal of many works carried out by different authors was able to identify the difficult words and then carry out the lexical simplification that consisted in replacing the difficult words by simpler ones.

The following work that we propose to present to the scientific community will consist in the construction of a software that allows complex words to be replaced by a word with a higher frequency of use, with this a product that is much easier to understand will be obtained as a product. This way derive the barriers of reading comprehension in university students which will constitute a significant advance in the research project in simplification of texts whose objective in the future will be to contribute to learning in higher education.

Acknowledgement. I want to express my thanks to the PhD. Arturo Montejo-Ráez from the University of Jaén who with his valuable contribution directs the tutelage of the doctoral project, also to the authorities of the different faculties of the University of Guayaquil who have shown interest and contribute in a very collaborative way so that the development of this Project achieve its objective. I would also like to thank Christian González Espinoza, student of the Software Engineering degree at the University of Guayaquil, who contributed to the development of the EIL application.

References

1. Dell'Orletta, F., Montemagni, S., Venturi, G.: Read-IT: assessing readability of italian texts with a view to text simplification. In: Proceedings of the Second Workshop on Speech and Language Processing for Assistive Technologies, pp. 73–83 (2011)
2. Štajner, S., Glavaš, G.: Leveraging event-based semantics for automated text simplifition. Expert Syst. Appl. **82**, 383–395 (2017)
3. Torunoglu-Selamet, D., Pamay, T., Eryigit, G.: Simplification of Turkish sentences. In: The First International Conference on Turkic Computational Linguistics, pp. 55–59 (2016)
4. Štajner, S., Evans, R., Orasan, C., Mitkov, R.: What can readability measures really tell us about text complexity. In: Proceedings of Workshop on Natural Language Processing for Improving Textual Accessibility, pp. 14–22 (2012)
5. Carletta, J.: Unleashing the killer corpus: experiences in creating the multi-everything AMI Meeting Corpus. Lang. Resour. Eval. **41**(2), 181–190 (2007)
6. Rosas, V.P., Mihalcea, R., Morency, L.P.: Multimodal sentiment analysis of Spanish online videos. IEEE Intell. Syst. **28**(3), 38–45 (2013)
7. Ortiz Zambrano, J.A., Montejo Ráez, A.: VYTEDU: un corpus de vídeos y sus transcripciones para investigación en el ámbito educativo (2017)
8. Ortiz Zambrano, J.A., Varela Tapia, E.A.: Reading comprehension in university texts: the metrics of lexical complexity in corpus analysis in Spanish. In: International Conference on Computer and Communication Engineering, pp. 111–123 (2018)
9. Aranzabe, M.J., de Ilarraza, A.D., Gonzalez-Dios, I.: First approach to automatic text simplification in basque. In: Proceedings of the Natural Language Processing for Improving Textual Accessibility (NLP4ITA) Workshop, pp. 1–8 (2012)
10. Siddharthan, A.: Syntactic simplification and text cohesion. Res. Lang. Comput. **4**(1), 77–109 (2006)
11. Al-Subaihin, A.A., Al-Khalifa, H.S.: Al-Baseet: a proposed simplification authoring tool for the arabic language. In: 2011 International Conference on Communications and Information Technology (ICCIT), pp. 121–125. IEEE (2011)
12. Bott, S., Saggion, H.: Spanish text simplification: An exploratory study. Procesamiento del lenguaje Nat. **47**, 87–95 (2011)
13. Carroll, J., Minnen, G., Canning, Y., Devlin, S., Tait, J.: Practical simplification of English newspaper text to assist aphasic readers. In: Proceedings of the AAAI-1998 Workshop on Integrating Artificial Intelligence and Assistive Technology, pp. 7–10 (1998)
14. De Belder, J., Moens, M.F.: Text simplification for children. In: Proceedings of the SIGIR Workshop on Accessible Search Systems, pp. 19–26. ACM (2010)
15. Ferrés, D., Marimon, M., Saggion, H.: YATS: yet another text simplifier. In: International Conference on Applications of Natural Language to Information Systems, pp. 335–342. Springer, Cham (2016)
16. Suter, J., Ebling, S., Volk, M.: Rule-based automatic text simplification for German. In: Proceedings of the 13th Conference on Natural Language Processing (KONVENS 2016), pp. 279–287, Bochumer Linguistische Arbeitsberichte (BLA), Bochum (2016)
17. Burešová, K.: Text simplification in Czech (2017)

Mapping of the Transportation System of the City of Aguascalientes Using GTFS Data for the Generation of Intelligent Transportation Based on the Smart Cities Paradigm

Raul Alejandro Velasquez Ortiz[1](✉), Francisco Javier Álvarez Rodríguez[1], Miguel Vargas Martin[2], and Julio Cesar Ponce Gallegos[1]

[1] Departamento de Ciencias de la Computación, Universidad Autónoma de Aguascalientes, Av. Universidad, Ciudad Universitaria, 20131 Aguascalientes, Mexico
raul.velasquez.sc@gmail.com, {fjalvar,jcponce}@correo.uaa.mx

[2] Ontario Tech University, 2000 Simcoe Street North, Oshawa, ON L1G 0C5, Canada
miguel.martin@uoit.ca

Abstract. Some city services across the world have been able to make use of technology for a more seamless delivery and massive use. As cities grow in population and complexity, new problems and challenges emerge. In particular, the growth of the city of Aguascalientes has seen an increasing demand for more and better city services over time, complicating the way in which public transportation services are provided, not only in the city but in the metropolitan area as surrounding municipalities become practically part of it. Public transportation does not have accurate and updated information that allows for technology to help in the efficient delivery of the service, causing a number of problems such as the inability to pinpointing the location of busses at any given time, resulting in challenges to providing a better service for users. This paper is a first attempt at solving the problem through the means of the GTFS (General Transit Feed Specification) format defined in [1]. We believe that our paper can help the development of software solutions by means of e.g., smartphone apps that can help the city offer better public transportation using a Smart City paradigm [2].

Keywords: GTFS · GTFS-RT · Smart mobility · Mapping · Smart city

1 Introduction

Across several cities around the world, public transportation has been of vital importance in providing a service to society for better mobility over long distances and support for those who do not have a vehicle of their own. However, cities such as

M. Botto-Tobar et al. (Eds.): ICAETT 2019, AISC 1066, pp. 177–185, 2020.
https://doi.org/10.1007/978-3-030-32022-5_17

Portland in the United States suffered from little information on the management and location of public transportation, so in 2002, under the premise of locating this information through the Internet suggested by the computer manager of geographic information systems of the city mentioned above, Bibiana McHugh, and in collaboration with Google, developed a tool capable of managing traffic data. Google adapted it and called it GTFS (General Transit Feed Specification) which, from that moment, began to be implemented through various cities around the world as indicated in [3].

The large number of API's has facilitated the use and manipulation of GTFS generating a large increase in possibilities for the creation of applications for tracking public transportation, whether buses, trains, etc., without the need to have a large amount of knowledge to make the programming of these tools.

The GTFS format is generated from a collection of 13 tables that allow the recording of information about bus units, being the most important tables agency, routes, trips, calendar, stop times and stops respectively. Each one of the tables requires attributes, for example, in the stops table the stop ID, stop name, stop latitude and stop longitude are required. These attributes allow programmers to identify the location of stops on Google Maps and tag them by ID and name and simplify the work of developing transportation planning applications as indicated in [1].

We will define the problems that the city of Aguascalientes suffers in the area of public transportation, as well as the definition and use of the GTFS model that will be used for the mapping and subsequent update in Google Maps.

As the city of Aguascalientes does not have a stable and well-defined mobility model, it lacks a quality service in public transport, which has repercussions on the acceptance of users and consumers. The GTFS model to be implemented to transform public transport into an intelligent system will therefore be defined.

The city has a transportation system that has GPS technology but is not used for tracking units to inform users. It has an application that gives static information about the waiting times of a bus, but these times are variable because there is no control of departure and arrivals of the trips made by buses, generating that users do not have the real information of the journeys and therefore do not know the exact time of the trip of the unit.

Likewise, the schedules established in the public transport units are not carried out, causing problems of collection time at bus stops and therefore, waiting times ranging from 30 to 45 min of waiting per unit.

The purpose of this article is to define and build a Smart City in Aguascalientes starting with its public transport, transforming it into intelligent transport, which although the city's transit system is semi-formal or informal due to the scarce relationship that transport units have with technology, is a more challenging work but possible to achieve. Section 2 will cover the state of the art and will focus on how in various cities with the same situation as the city of Aguascalientes and in even more complicated situations, it has been possible to generate applications from the inclusion of the GTFS to its transport database. It will also analyze the work done in Mexico City, being the first city in Mexico to implement GTFS to its transport systems. Section 3, methodology, it will indicate the different methods and techniques, as well as standards for the resolution of the problem detailed above. Section 4 will cover the results obtained and it will be possible to visualize the mapping by means of Google

Maps from the GFTS files. To conclude, Sect. 5 will give the relevant conclusions about the work, and the approach that will be taken for the future work, focused mainly on the construction of the application based on the GTFS files and the progressive transformation of the city of Aguascalientes to a Smart City.

2 State of the Art

In [4], the authors mention the importance of the GTFS format and its use with the various transportation agencies or concessionaires as an option to offer interoperability is their transportation system. The use of the model established by GTFS is simple to understand and, therefore, achievable in a short time. The ease of implementation, as well as the use of map technology such as Google Maps for the integration of GTFS tables, has facilitated the development of different databases around the world to record open data on public transport information in general, providing information on arrival time, location of routes, and optimal routes depending on the location of the user, facilitating the mobility of users.

Once GTFS was implemented in Portland, USA, a register of different transport agencies from different countries of the world was started to obtain a more intelligent and optimal transport. Countries such as Beijing, Paris, Bogota, Cleveland, to mention a few, transformed their transportation systems to the same format, radically changing the use of these as shown in [3]. This study succeeded in innovating, since previously it was very complicated to achieve a system that could plan public transportation trips through applications, and it was not until the creation of the standard already mentioned that a significant change was achieved. However, being a methodology of recent creation, at the time lacked tools for the development of applications.

Similarly, in 2016, Mexico City decided to implement the GTFS format to control and facilitate the use of public transport of its different transport companies. A work team was integrated in which academics, governments and society were dedicated to record the information they requested since the concessionaires had nothing digitized and the information was not very precise.

Based on the technology and the use of crowdsourcing methodology, they developed an application called Mapaton, with which the bus stops of the users would be registered, in a massive way, with the purpose of generating a reliable database to obtain this information. After a week, the information was collected and could be recorded in such a way that they were able to make various applications for the management of public transport, whether buses, trains, or metro, to name a few. However, no proposal was made for the analysis of the data, because many users recorded erroneous locations, repeated or non-existent, generated a large amount of time in the analysis and validation of real data, which left the project with difficulties for implementation.

Mexico City when performing the Mapaton project achieved the registration of GFTS data and due to the size of the metropolis, this did not achieve the expected impact, due to the lack and little literature contained on the same subject. The project was completed, but more documentation is required to support it, as it could not be completed in its entirety, making the lack of information on GTFS implementation and its use noticeable. Its work would have been greatly simplified, using applications in charge of facilitating the verification of information capture, which in this sense, was carried out by users. Despite this, the proposal was interesting and managed to make a change that the other states of the Mexican Republic still do not make in their public transport.

Currently there are 1069 registered concessionaires that have implemented the GTFS model and 584 locations around the world that use it to manage their public transport, being from this, Mexico City, the only city within the Mexican Republic, which integrates and uses its public transport services from the GTFS format [5], generating at the same time intelligent mobility in their means of transport, which in turn enables the city as a Smart City as indicated in [6].

All the cities that have managed to implement transport services and transformed them have generated an intelligent mobility based on technology for the optimization of this one, in such a way that, as it was mentioned, they have managed to establish an intelligent city, particularly in the area of transport, which causes great changes that speed up and improve the quality of the services.

3 Methodology

This section will define the techniques and standards that will be used to solve the problem that the city of Aguascalientes suffers in the area of public transport, as well as the definition and use of the GTFS model that will be used for mapping and subsequent updating in Google Maps. As the city of Aguascalientes does not have a stable and well-defined mobility model, it lacks a quality service in public transport, which has repercussions on the acceptance of users and consumers. The GTFS model to be implemented to transform public transport into an intelligent system will therefore be defined.

The steps to consider are the implementation of the GTFS standard that is essential for the generation of schedules and position buses in order to generate a feed that can be used in the future in the development of applications, either own or third-party with free distribution. Also, the validation part, which is supported, is required to be displayed on Google Maps, which is a fundamental part of the solution.

Similarly, it requires the use of GTFS-RT that will define and execute GTFS data already implemented on maps with functions in real time, and this is a crucial point because it is necessary to achieve the objective of problem solving.

3.1 GTFS Static Model

The general model of GTFS contemplates 13 files or tables that conform the total of the format for its use, being 6 files necessary for its correct operation and 7 optional GTFS files. The official files of the GTFS format are: agency.txt, stops.txt, routes.txt, trips.txt, stop_times.txt, calendar.txt, calendar_dates.txt, fare_attributes.txt, fare_rules.txt, shapes.txt, frequencies.txt, transfers.txt y feed_into.txt. As shown in, the GTFS model required is the one that will be used to begin with the development of the mapping of the city of Aguascalientes (See Fig. 1), as indicated in [7].

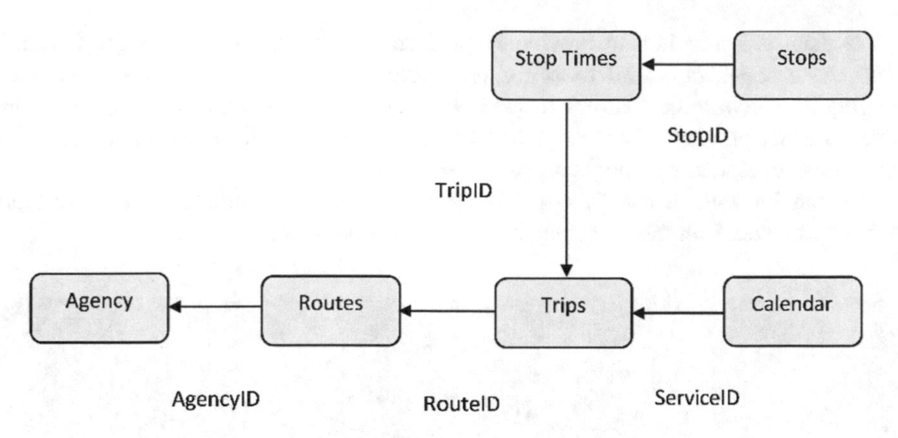

Fig. 1. Basic structure of the GTFS table model.

The consumption of the GTFS files are necessary for the real time validation of the route movements as well as the establishment of the stops. The city of Aguascalientes for the moment reflects the minimum use of GTFS files because it requires a digitization of services as soon as possible and that is available to everyone, in this case, being published to Google maps as it establishes [9].

The defined structure suggests how the different files communicate with each other, by means of established IDs, this to reflect the proper communication between files and at the time of performing the relevant validation.

3.2 GTFS RT

An extension of the GTFS format is the specification called GTFS-RT (General Transit Feed Specification Real Time), which once implemented the GTFS files to Google Maps servers or mobile application, can be processed to obtain information in real time, converting the GTFS static files already implemented into dynamic files that provide detailed information on the location of buses, service alerts in case of any congestion and vehicle positions which provides information on transport units.

The format for the interaction of data with GTFS files established is made from protocol buffers, and define the data within the file gtfs-realtime.proto, in order to generate the code and perform the transmission of data [9].

4 Results

4.1 Displaying GTFS Data in QGIS

In this section the GTFS files will be reflected in specialized geographic software, to show an image before the update that will be obtained once the state of Aguascalientes is placed and updated with the GTFS files previously generated.

The visualization of GTFS data requires a partnership with the mobility agency of the city and in this case, in order to perform a management of validation and correction of the data, it is required to previously perform tests with some software. Certainly there are a series of useful tools for this purpose, we used the QGIS software, a geographic software that allows to preview the GTFS files using Google Maps and have an exact preview of the stops and routes that will be displayed once the GTFS database is uploaded to Google Maps servers.

As can be seen in Fig. 2, once the Stops.csv file is completed, these are automatically captured on the map, obtaining its exact location.

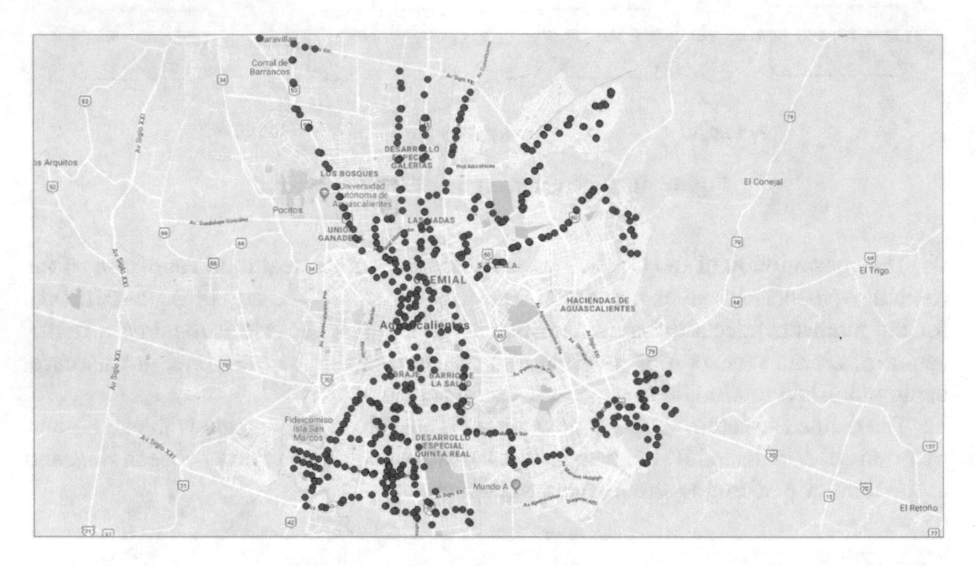

Fig. 2. City mapping using GTFS format.

The points indicated in the map show the stops that have been registered in the GTFS files, specifically stops.txt, these in turn, will be placed with the coordinates already registered in the Google maps map. The tool that provides this support is QGIS and tells us in real time how it will be displayed once uploaded to Google Maps.

QGIS is a tool with many geographic capabilities, which is useful because you can view, edit and visualize the information of the stops in this case, because before uploading the information to Google, it is necessary a filtering and analysis of GTFS data [8].

The advantage of using the standard GTFS is clear, to have a more solid control of the transport units, being specific, the urban transport. Currently Mexico City as well as the city of Leon, Guanajuato, are the only cities to implement the methodology established by Google, for the creation of intelligent mobility. Most states do not yet have the implementation of GTFS and much less GTFS-RT, because the issue is unknown, there are bus units with GPS technology for tracking units but is still scarce due to the lack of information they can provide and is not accurate information such as the standard mentioned. The city of Aguascalientes, Mexico contemplates to be the third city of the republic in generating intelligent mobility when giving these results.

4.2 OneBusAway as a Development Tool

Once the main GTFS files have been completed, further development of the application is considered. For this step there is a lot of information and open source tools for the manufacture of applications. There are a lot of products for the consumption of GTFS. The applications can be generated through transit agencies or through third parties or external companies, as indicated in [8].

The amount of applications of free distribution or open source are very varied, from the use of Google as manager of the GTFS files, to applications such as Moovit, Challo App, Citymapper, Transit App, OneBusAway, to mention a few. The ease of use of these free applications implies a production of applications in a very short period if you had to develop from scratch as mentioned in [3].

OneBusAway (OBA) is a free distribution project which since 2010 has managed to generate impact due to the great simplicity of modification in addition to having a large community that supports users of various mobility agencies that require to develop their own without the need to start a project from scratch. Currently, OBA's development team extends not only to the government but also to academia with the support of the universities of Washington, South Florida, City College of New York, just to mention a few, as well as different mobility agencies that contribute to the development of the application and its growth, with a variety of them between the United States and Canada [11].

Although OBA is a nonprofit organization and which has its application available on various platforms, a few cities have chosen to customize the OBA system to suit their needs, generating new brands totally independent of the original application, finally offering more application options.

Therefore, it is decided to make the decision to generate an application, which in order to speed up the process, gives us the freedom to create an independent brand and with the ease of using the GTFS and GTFS-RT files [10].

4.3 Discussion of Results

Preliminary results indicate that although the data capture of the mobility agency, focused only on one means of urban transport, in this case, buses, has been agile thanks to the fact that the information is digitalized. Making a comparison with the Mapatón project in Mexico City you can see that although they managed to capture a large amount of information, due to the amount of population that resides in that city, were only GTFS static data, validated to be displayed on Google Maps [12], which our advantage is that not only remains in the register of the standard GTFS, but goes beyond, the implementation of GTFS-RT and more than that, to a practical use, an application that manages this data and helps to communicate the population with the means of transport, since in the way it was done in Mexico City, and although it is a significant advance due to the size of the city, maintaining only static data, does not offer a long-term solution, in addition to the constant maintenance that must be offered because a change of that nature must be gradual, by taking the various types of transport one by one, managing and monitoring them. Currently the project is inactive and the application of Mapaton is out of service which is a retrocession in the progress made towards the transformation of an intelligent transport.

The Mapatón advantage is undoubtedly the support of many people for the collaboration of GTFS data capture, but without proper monitoring, the project is abandoned and can no longer be funded. The Mapatón advantage is undoubtedly the support of many people for collaboration in capturing GTFS data, but without proper monitoring, the project is abandoned and can no longer be funded.

The advantages of this project, goes beyond simple capture, goes towards a real practical use, ending in an app, which can be configured and edited for city purposes, that's why OneBusAway was decided as the basis of the application. Both projects have defenses, but it is necessary to use the GTFS data in a real time application like in this project.

5 Conclusions and Direction for Future Research

The greatest contribution is that the beginning of the development of an application is given using free distribution tools, in order to implement the GTFS standard to a public transport agency without the need to have any cost and with the possibility of generating comparatives in the transit data of both the application and Google Maps, as it serves as a reference for both the calculation of time as well as a second option for the user in the use of consumer applications transit data.

In this article is contemplated the mapping of routes and stops of the bus concessioned by the city of Aguascalientes using the GTFS standard to achieve it. It is possible to obtain an approach of the magnitude of the problem and it can be observed how it is possible to have a management not only of the reflected bus stations but a possible adaptation of the GTFS generated towards some application.

The work provided for the future is the conclusion of the GTFS base files, as well as the development of an application based on the GTFS generated, thus the city of Aguascalientes would be the third city, apart from Mexico City and the city of Leon that implements the geolocation of public transport by means of applications based on Google Transit technology and with the use of the standard already mentioned, and thus transforming the city to a Smart City.

Acknowledgements. We thank anonymous reviewers for their useful feedback, as well as CONACYT and the Universidad Autónoma de Aguascalientes for funding this research. The first author was a visiting scholar at Ontario Tech University during most of the elaboration of this paper. We thank the Coordinación General de Movilidad de Aguascalientes (CMOV) for sharing information about the city's transportation system.

References

1. Zhang, T., Chen, M., Lawson, C.: General transit feed specification data visualization. In: 2014 22nd International Conference on Geoinformatics, 25–27 June 2014, pp. 1–6 (2014). https://doi.org/10.1109/geoinformatics.2014.6950839

2. Strasser, M., Albayrak, S.: A pattern based feasibility study of cloud computing for smart mobility solutions. In: 2016 8th International Workshop on Resilient Networks Design and Modeling (RNDM), 13–15 September 2016, pp. 295–301 (2016). https://doi.org/10.1109/rndm.2016.7608301
3. Krambeck, H.: Introduction to the General Transit Feed Specification (GTFS) and Informal Transit System Mapping (Self-Paced). World Bank Group (2019). https://olc.worldbank.org/content/introduction-general-transit-feed-specification-gtfs-and-informal-transit-system-mapping. Accessed 27 Dec 2018
4. Friedrich, M., Leurent, F., Jackiva, I., Fini, V., Raveau, S.: From transit systems to models: purpose of modelling. In: Gentile, G., Noekel, K. (eds.) Modelling Public Transport Passenger Flows in the Era of Intelligent Transport Systems: COST Action TU1004 (TransITS), pp. 131–234. Springer, Cham (2016)
5. OpenMobilityData: OpenMobilityData (2019). https://transitfeeds.com/feeds. Accessed 24 Mar 2019
6. Faria, R., Brito, L., Baras, K., Silva, J.: Smart mobility: a survey. In: 2017 International Conference on Internet of Things for the Global Community (IoTGC), 10–13 July 2017, pp. 1–8 (2017)
7. Braga, M., Santos, M.Y., Moreira, A.: Integrating public transportation data: creation and editing of GTFS data. In: Rocha, Á., Correia, A., Tan, F., Stroetmann, K. (eds.) New Perspectives in Information Systems and Technologies, Volume 2, pp. 53–62. Springer, Cham (2014)
8. Girish, M.: Better bus implementing bus tracking app with GTFS data. In: Mai-yah, G. (ed.) p. 137 (2019). https://www.amazon.com/BetterBus-Implementing-Tracking-GTFS-dataebook/dp/B07QDCWHTP/ref=sr_1_1?keywords=betterbus&qid=1556889659&s=gateway&sr=8-1
9. Google: Programa de partners de Google Transit (2016). http://maps.google.com/help/maps/mapcontent/transit/
10. Ferris, B., Watkins, K., Borning, A.: OneBusAway: results from providing real-time arrival information for public transit. Paper presented at the Proceedings of the SIGCHI Conference on Human Factors in Computing Systems, Atlanta, Georgia, USA (2010)
11. OneBusAway: The Open Source Platform for Real Time Transit Info (2010). https://onebusaway.org/
12. Téllez, R.: A Case from Mexico City: Laboratorio para la Ciudad's Mapatón CDMX (2016). http://legiblepolicy.info/a-case-from-mexico-city-laboratorio-para-la-ciudads-mapaton-cdmx/

A ConvNet-Based Approach Applied to the Gesticulation Control of a Social Robot

Edisson Arias[1], Patricio Encalada[1(✉)], Franklin Tigre[1], Cesar Granizo[1], Carlos Gordon[1], and Marcelo V. Garcia[1,2]

[1] Universidad Tecnica de Ambato, UTA, 180103 Ambato, Ecuador
{earias4645,pg.encalada,fg.tigre,cesar_granizo, cd.gordon,mv.garcia}@uta.edu.ec

[2] University of Basque Country, UPV/EHU, 48013 Bilbao, Spain
mgarcia294@ehu.eus

Abstract. This document presents the enforcement of a facial gesture recognition system through applying a Convolutional Neural Network (CNN) algorithm for gesticulation of an interactive social robot with humanoid appearance, which was designed in order to accomplish the thematic proposed. Furthermore, it is incorporated into it a hearing communication system for Human-Robot interaction throughout the use of visemes, by coordinating the robots mouth movement with the processed audio of the text converted to the robot's voice (text to speech). The precision achieved by the CNN incorporated in the social-interactive robot is 61%, while the synchronization system between the robot's mouth and the robot's audio-voice differs from 0.1 s. In this way, it is pretended to endow mechanisms social robots for a naturally interaction with people, thus facilitating the appliance of them in the fields of childrens teaching-learning, medical therapies and as entertainment means.

Keywords: Deep Learning · Human-Robot interaction · Social robots · Neural networks · Visemes

1 Introduction

Constantly advancement of technology over the last decades, into automating applications, robotics and artificial intelligence, have encouraged the research sector to be developed of new knowledge and methodologies to be implemented in the design, make and control of robots. Social robotics, part of robotics services is focused on providing robots with the skills that human beings present for their raid into society, thus facilitating Human-Robot interaction (HRI) [1,2].

At the present time, there is a variety of contributions that have allowed the evolution of social robotics, among these contributions we have the interactive robot developed in [3], whose aim is the treatment of people with autism spectrum disorder (TEA). Ribo is a social robot with a specialized design to make

M. Botto-Tobar et al. (Eds.): ICAETT 2019, AISC 1066, pp. 186–195, 2020.
https://doi.org/10.1007/978-3-030-32022-5_18

it look more like an artificial social being instead of just like a mechanical robot [4]. Ribo had a great reception by its facial design and its way of speaking. The humanoid robot head developed by Lapusan in [5] is used in studies of social interaction, which reproduces the movement of the human head and implements computer vision such a sensory system.

Over the course of several years, the facial expressions recognition such a skill of social robots has been the theme of continues researches [6], and it still remains a major challenge in real-time applications [7]. The first step of great importance in the detection of gestures is demarcated by the extraction of characteristics into the faces recognition, for instance, the histogram of oriented gradients (HOG) attached to a linear vector support machine (SVM) and trained to perform the classification of the expression or gesture [8], or the Viola-Jones notation algorithm used to detect objects in real time, nevertheless, its main application is the detection of faces due to its high accuracy rate that presents in front of others [9].

Another segment implicit in developing social robots is the embodiment of algorithms that allow communication through simultaneous coordination between the robot's mouth movement and the audio generated like voice of the robot. In the synchronization of speech with the mechanism of mouth displacement there is an acoustic unit called viseme, which is the visual representation of the generated sound and the exact position of the mouth [10–20]. This allows the articulated movements of the robot to be similar to the movements generated by the human mouth.

In this regard, the first step of the work is the design and make of a mechanism that simulates the movement of eyes, eyebrows, eyelids and lips of an interactive humanoid-looking social robot under the humanoid robot structure [11]. Secondly, the gesticulation and speech control systems of the robot are developed. Finally, the robot and algorithms of control are evaluated.

This document relies on 6 sections, including the introduction. Section 2 deals about the methodology used, Sect. 3 is over the robot design, as well as the hardware stage implemented. Section 4 covers the development of social robot controllers software, finally, Sects. 5 and 6 show the results obtained and conclusions of the work carried out.

2 Methodology

The methodology used in the development of this study is composed of following points: First a background analysis on interactive social robots, design, construction and control mechanisms. Later, authors of this research explain the construction of social robot through a previous selection the components to be used by the robot. In addition, the conditioning of the electronic and mechanical elements of the robot. Next, researches develop a gesticulation control system based on an algorithm of convolutional neural network and speech control system of the robot through the use of visemes. Finally, evaluation of the joint performance of the designed social robot and control systems.

3 Proposed Hardware Architecture

The control of gesticulation and speech of a social interactive robot with human appearance is a complex task that eventually requires a series of steps necessary for its correct development, in this section the design and hardware construction of the prototype with humanoid appearance is thoroughly explained, able to correctly perform voice synchronization and gesticulation assignments

3.1 Development of Social Robot

Design starts from the idea of bringing about a robotic structure that has approximately the appearance of a man, the structure that fulfills this aim is the InMoov robot. Additionally, some mechanisms were designed in Computer-Aided Design (CAD) Software and incorporated into itself to show the movements of eyes, eyebrows, eyelids and lips. These are:

- **Eyebrows movement mechanism:** it is composed of 2 degrees of freedom to move the eyebrows up-down.
- **Eyes movement mechanism:** Formed by 4 degrees of freedom that allow to generate the displacement raise, lower, right and left of the eyes of the robot, also provides the movement of eyelids.
- **Lips movement mechanism:** It is formed by 6 degrees of freedom to simulate the different muscles of the human mouth. Figure 1 shows the final design of the social robot.

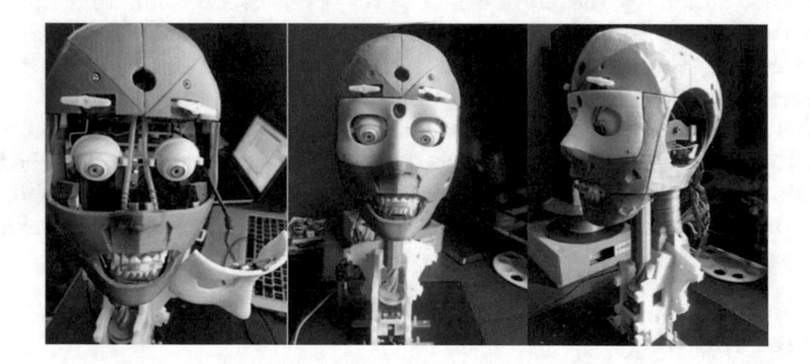

Fig. 1. Final design of the social robot.

Electrical and control system is composed by the SSC-32 servomotor control board, device that allows to control the 15 freedom degrees of the robot, which is powered with 5 V and 2 A. Furthermore, the computer takes the camera data and executes the control algorithms, coding above Python language and through the use of the serial communication protocol the position values (PWM) are sent to the controller card for moving each servomotor to the desired position. In Fig. 2 it is shown the electrical diagram that controls the actuators of the robot.

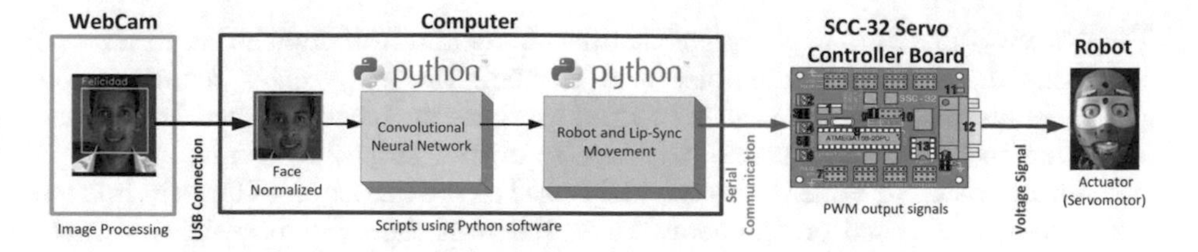

Fig. 2. Robot's electrical system.

4 Proposed Software Architecture

The facial recognition of the robot is based on Deep Learning algorithms which are used in artificial intelligence applications. Programming is done using Python since it owns several libraries specialized in the development of CNN as: Spyder and Keras. The acquisition and processing of images is done using OpenCV.

4.1 Gesticulation Control System

Gesticulation controller incorporates a convolutional neuronal network algorithm used in deep learning applications [12], to identify the facial gestures that the human being performs through the acquisition of images in real time [13], registering them and that the robot can imitate them [14]. In Fig. 3, it can see the structure of the gesticulation controller.

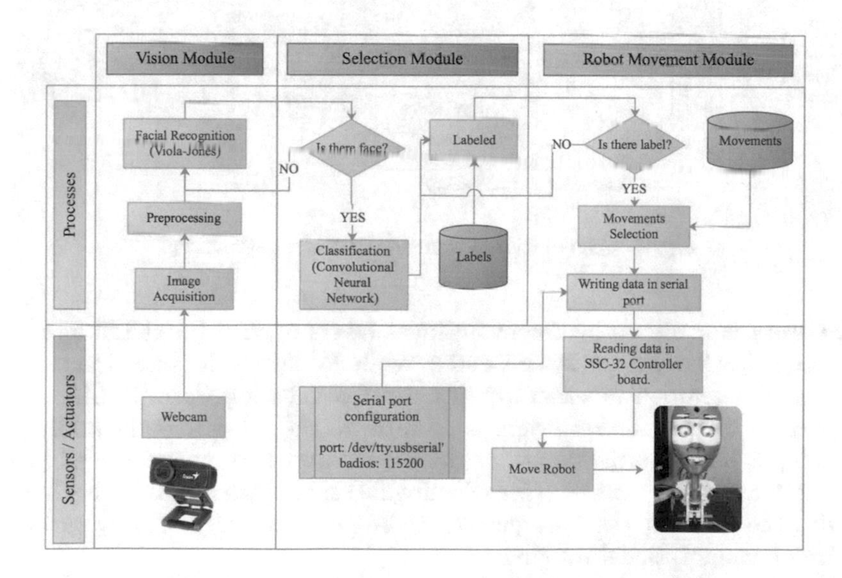

Fig. 3. Gesticulation controller structure.

In the following subsections the construction and operation process of each module that take part in the robot's gesticulation controller is described.

Vision Module: It is in charge of standardizing and unifying the camera data; scale the image to a suitable size and through the gray scales segmentation generates an image, which is sent to the face detector based on the Viola-Jones algorithm to obtain the image of the face as output [15,16].

The algorithm of Viola-Jones is widely used in the detection of faces in images and videos, it is based on the comparison between the light intensities of rectangular regions of the images called Haar characteristics, which are calculated applying an integral image.

Selection Module: It is composed of a convolutional neuronal network model [17], for which it was started by the preparation of data, then the neural network model was created and finally the model was trained and validated. Using the methodology developed in [18], the face database of the FERC-2013 is used, which was provided on the Kaggle facial expression contest [17], it is composed in 2 sets intended for training and testing of the model.

Once the data is prepared, the convolutional neuronal network model is created. The model used in this document is the one proposed in [18], which is arranged by the set of layers as indicated in Fig. 4.

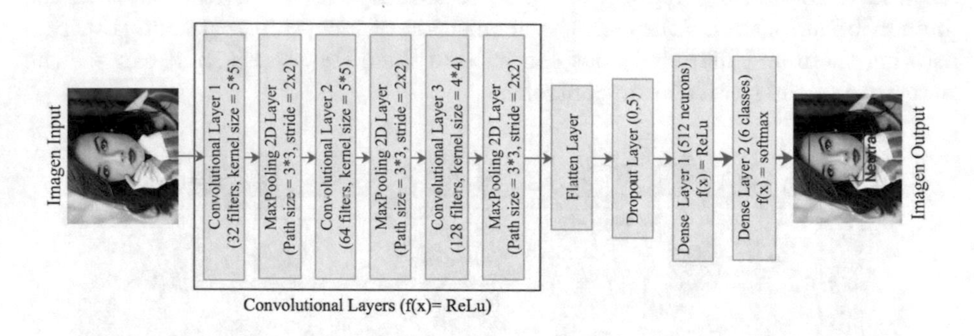

Fig. 4. Convolutional neuronal network structure.

Network is made up by 3 convolutional layers of 32, 64 and 128 filters, the step size is 5×5 pixels for layer 1 and 2, while for layer 3 it is 4×4 pixels, each layer has a activation function type Linear Rectification Unit (ReLU) [19,20], the purpose of convolution layers is to capture the most relevant information of the input image as the details of the faces. After each convolutional layer is encountered a reduction layer (MaxPooling2D) that takes like stride 3×3 pixels, allowing the reduce of the data quantity to be processed, significantly optimizing the use of computational resources.

Afterwards, it is placed the flattening layer is linked in order to yield a one-dimensional array that is then connected to the dropout layer, and finally it is joined the dense layers, one of them has 512 neurons while the other counts with 6, due to the number of classes that are desired as output, the activation functions for these layers are ReLU and Softmax [19].

The outline shows in Fig. 5 the elaboration of the neural network model by using the Keras library upon the Python language, which facilitates the creation of neural network models.

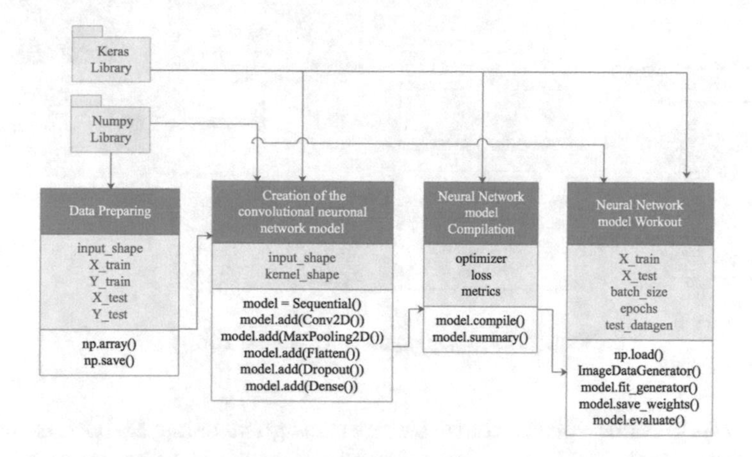

Fig. 5. Functions of CNN development.

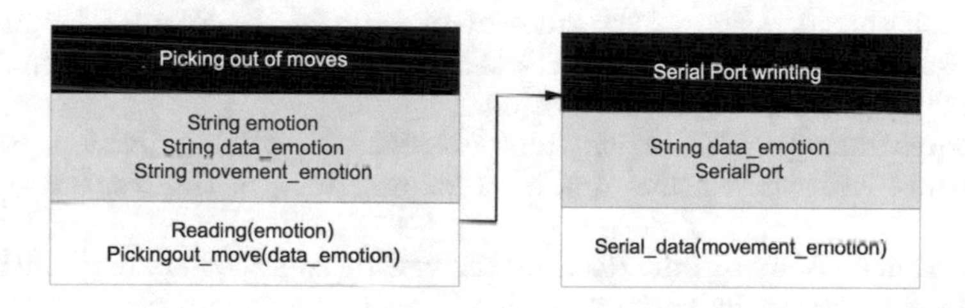

Fig. 6. Functions of the robot movement module.

Robot Movement Module: It takes the selection module prediction and compared it with the previously emotions configured data, then movements of each DOF is established to generate the detected gesture, these functions are visualized in Fig. 6

4.2 Speech Control System

Lip-sync of an interactive social robot represents a plausible characteristic of the way in which human beings communicate through speech. It is illustrated in Fig. 7 the processes involved lip-sync system.

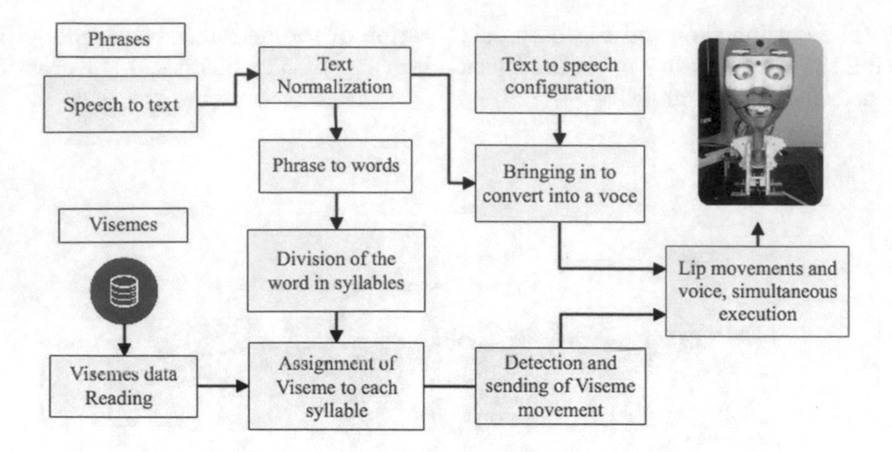

Fig. 7. Lip-sync system of social robot.

1. **Text-to-speech:** speech converter was designed using MAC OS by its following features: (i) Multiple languages for voices. (ii) Voice speed adjustable. (iii) Easy implementation of libraries.
2. **Text normalization:** remove symbols, spaces, and other characters different from the 27 letters of the Spanish alphabet of the text entered through filters.
3. **Syllables number and syllabic division:** The syllabic division was made with the consideration of the rules of division of the Spanish language [8], therefore, each chain of characters that leaves the process of normalization enters the process of syllabic division.
4. **Visemes Assignment:** Each value obtained in syllabic division process is compared with the visemes database for assigning of the respective viseme code.
5. **Servomotor Movement:** Receives the viseme codes, compares it with established, and writes the position of this in each servomotor through the serial port.

5 Discussion and Results

The robot keeps a neutral state where the movements of the robotic face remain static in front of any input image that does not reflect human face traits. When detecting a person presence, the gesticulation system is stimulated and the facial gestures present in the individual are identified.

The results achieved during the experiment carried out are the gesticulation control system operation of the interactive social robot are visualized in Table 1 first row, which shows the facial gesture of the user detected and labeled respectively, as well as the gesture by the robot. After several tests, the visemes raised for the mouth movement of the robot were reduced to a total of 7 due to their kinship and repetition of patterns in the visemes codes database. See Table 1 second row.

Table 1. Robot gesture operation results and Visemas for lip-sync.

Gesture Label	User Gesture	Font Robot Gesture
Anger		
Viseme Name	Defined Viseme	Arising Viseme
Viseme A (A)		

The robot audio signal is divided into 6 portions, in which the visemes forming the segment are established. The audio synchronization results and movement of the robot can be seen in Fig. 8.

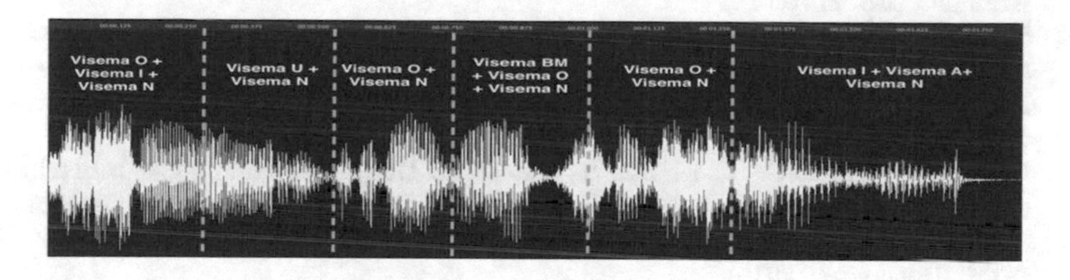

Fig. 8. Audio signal output.

6 Conclusions and Ongoing Work

In this article, a convolutional neuronal network was incorporated into an interactive social robot designed, which is based on the humanoid robot so that upon receiving an image of the facial gestures of a person, being able to imitate said gesture, thus contributing to the imitation field of facial expressions of a human being in real-time. The system can recognize and imitate human facial gestures correctly and accurately with a precision rate of 61.1%.

In addition to the neural network mainstreaming, a thorough investigation was also carried out in the labial communication field so that the robot can communicate aurally. The lip-sync system allows the robot to simulate the movements of the human mouth. The synchronization time variation between the audio signal generated by the conversion of a text string into audio, and the sequential positioning of the servomotors that respect the viseme code is minimal value, therefore it is considered an efficient sync.

Future work will improve the accuracy of the recognition rate, and give more comparison results among different methods to further verify the performance of our system.

Acknowledgment. This work was financed in part by Universidad Tecnica de Ambato (UTA) and their Research and Development Department (DIDE) under project 1919-CU-P-2017.

References

1. Barnes, J., FakhrHosseini, M., Jeon, M., Park, C.-H., Howard, A.: The influence of robot design on acceptance of social robots. In: 2017 14th International Conference on Ubiquitous Robots and Ambient Intelligence (URAI), pp. 51–55. IEEE, Jeju (2017)
2. Mead, R., Mataric, M.J.: Autonomous human-robot proxemics: a robot-centered approach. In: The Eleventh ACM/IEEE International Conference on Human Robot Interaction, pp. 573–573. IEEE Press (2016)
3. Benavides, J.: Diseo y construccin de un robot interactivo para el tratamiento de personas con el trastorno del espectro autista (TEA). Universidad de las Fuerzas Armadas (ESPE) (2016)
4. Sojib, N., Islam, S., Rupok, M.H., Hasan, S., Amin, M.R., Iqbal, M.Z.: Design and development of the social humanoid robot named Ribo. In: 2017 IEEE Region 10 Humanitarian Technology Conference (R10-HTC), pp. 314–317. IEEE, Dhaka (2017)
5. Lapusan, C., Rad, C.-R., Besoiu, S., Plesa, A.: Design of a humanoid robot head for studying human-robot interaction. In: 2015 7th International Conference on Electronics, Computers and Artificial Intelligence (ECAI), pp. WR-15–WR-18. IEEE, Bucharest (2015)
6. Chen, L., Zhou, M., Su, W., Wu, M., She, J., Hirota, K.: Softmax regression based deep sparse autoencoder network for facial emotion recognition in human-robot interaction. Inf. Sci. **428**, 49–61 (2018)
7. Faria, D.R., Vieira, M., Faria, F.C.C., Premebida, C.: Affective facial expressions recognition for human-robot interaction. In: 2017 26th IEEE International Symposium on Robot and Human Interactive Communication (RO-MAN), pp. 805–810. IEEE, Lisbon (2017)
8. Chen, J., Chen, Z., Chi, Z., Fu, H.: Facial expression recognition based on facial components detection and hog features. In: International Workshops on Electrical and Computer Engineering Subfields, Istanbul, Turkey, pp. 884–888 (2014)
9. Soni, L.N., Datar, A., Datar, S.: Implementation of Viola-Jones Algorithm based approach for human face detection. Int. J. Curr. Eng. Technol. **7**, 1819–1823 (2017)
10. Fernndez, R., Montes, H. (eds.): RoboCity16 Open Conference on Future Trends in Robotics. Consejo Superior de Investigaciones Cientificas, Madrid (2016)
11. Cheng, H., Ji, G.: Design and implementation of a low cost 3D printed humanoid robotic platform. In: 2016 IEEE International Conference on Cyber Technology in Automation, Control, and Intelligent Systems (CYBER), pp. 86–91. IEEE, Chengdu (2016)
12. LeCun, Y., Bengio, Y., Hinton, G.: Deep learning. Nature **521**, 436 (2015)
13. Kumar, P., Happy, S.L., Routray, A.: A real-time robust facial expression recognition system using HOG features. In: 2016 International Conference on Computing, Analytics and Security Trends (CAST), pp. 289–293. IEEE, Pune (2016)

14. Meghdari, A., Shouraki, S.B., Siamy, A., Shariati, A.: The real-time facial imitation by a social humanoid robot. In: 2016 4th International Conference on Robotics and Mechatronics (ICROM), pp. 524–529. IEEE, Tehran (2016)
15. Fernandez, M.C.D., Gob, K.J.E., Leonidas, A.R.M., Ravara, R.J.J., Bandala, A.A., Dadios, E.P.: Simultaneous face detection and recognition using Viola-Jones Algorithm and Artificial Neural Networks for identity verification. In: 2014 IEEE Region 10 Symposium, pp. 672–676. IEEE, Kuala Lumpur (2014)
16. Wang, Y.-Q.: An analysis of the Viola-Jones face detection algorithm. Image Process. On Line **4**, 128–148 (2014)
17. Sang, D.V., Van Dat, N., Thuan, D.P.: Facial expression recognition using deep convolutional neural networks. In: 2017 9th International Conference on Knowledge and Systems Engineering (KSE), pp. 130–135. IEEE, Hue (2017)
18. Ashwin, T.S., Jose, J., Raghu, G., Reddy, G.R.M.: An E-learning system with multifacial emotion recognition using supervised machine learning. In: 2015 IEEE Seventh International Conference on Technology for Education (T4E), pp. 23–26. IEEE, Warangal (2015)
19. Vu, T.H., Nguyen, L., Guo, T., Monga, V.: Deep network for simultaneous decomposition and classification in UWB-SAR imagery. In: 2018 IEEE Radar Conference (RadarConf 2018), pp. 0553–0558. IEEE, Oklahoma City (2018)
20. Khan, S., Rahmani, H., Shah, S.A.A., Bennamoun, M.: A Guide to Convolutional Neural Networks for Computer Vision. Synthesis Lectures on Computer Vision, vol. 8, pp. 1–207 (2018)

An Approach of a Control System for Autonomous Driving Based on Artificial Vision Techniques and NAO Robot

Carlos Carranco[1(✉)], Patricio Encalada[4], Javier Gavilanes[2], Gabriel Delgado[3], and Marcelo V. Garcia[4,5]

[1] Universidad Politecnica Salesiana, UPS, 170146 Quito, Ecuador
ccarranco@ups.edu.ec
[2] Escuela Politecnica del Chimborazo, ESPOCH, 60155 Riobamba, Ecuador
javier.gavilanes@espoch.edu.ec
[3] Universidad del Azuay, UDA, 010107 Cuenca, Ecuador
gabrieldelgado@uazuay.edu.ec
[4] Universidad Tecnica de Ambato, UTA, 180103 Ambato, Ecuador
{pg.encalada,mv.garcia}@uta.edu.ec
[5] University of Basque Country, UPV/EHU, 48013 Bilbao, Spain
mgarcia294@ehu.eus

Abstract. The joint application of robotics and artificial vision for driving a car, it has been a very important study in recent years, since a small loss of concentration can cause the vehicle to deviate from its trajectory and move to the other lane or get off the road. The new applications for autonomous driving of a vehicle provide serenity in the different situations that the driver usually carries out in the routine journey or retention in a driving track. The present scientific article presents an NAO robot software architecture for autonomous driving of an electric car, this approach implements a system to control robot joints, trajectory correction based on people and track detection allowing successfully autonomous navigation.

Keywords: NAO robot navigation · Autonomous navigation algorithm · Artificial vision · Track detection · Open CV techniques

1 Introduction

A small loss of concentration can cause the vehicle to deviate from its trajectory and move to the other lane or get off the road. More than a third of truck accidents or serious incidents on the roads are related to accidental lane departure, especially at the end of a long and tiring day of work, drivers are more likely to be distracted by external factors. Road accidents are attributed to driver errors, drowsiness, fatigue or distractions seem to be the main causes [4].

M. Botto-Tobar et al. (Eds.): ICAETT 2019, AISC 1066, pp. 196–206, 2020.
https://doi.org/10.1007/978-3-030-32022-5_19

The joint application of robotics and artificial vision for the driving of an electric car by the humanoid bipedal robot NAO, is a reduced adaptation of an autonomous vehicle, such as the prototypes of Google, Tesla that are conducted autonomously, loaded with sensors: cameras, ultrasounds, high precision GPS and expensive laser localization instruments known as lidar. These devices help the vehicle create a composite image of the world around it in order to drive safely [2].

Some systems consist of sensors for the detection of lane lines, an electronic control unit that processes the information received by the sensor or sensors, calculating at all times the position and trajectory of the vehicle with respect to the lines delimiting the lane, and an interface of user that consists of the system start-up switch and the user warning systems (acoustic, optical and/or by vibrations of the steering wheel or the seat).

The detection technologies used are three at this time. The simplest and lowest cost is that which uses infrared sensors capable of detecting or "reading" the lines of the road, while the remaining two use artificial vision and laser scanning of the environment [8].

This research is focused on the implementation of the control of the movements of the humanoid robot NAO, which should be carried out based on the information obtained from captured images and duly processed in real time, following the path to follow and avoiding collisions with people, thus allowing to control the direction and displacement of an electric car to scale so that in the future this application can be made in an electric vehicle or combustion aimed at the human being, supported by the different technological applications for safe navigation in uncontrolled environments.

This document relies on 6 sections, including the introduction. Section 2 explains the architecture of NAO Robot, Sect. 3 explains about the study case used in this research. Section 4 covers the development of proposed solution using artificial vision for autonomous car navigation, finally, Sects. 5 and 6 show the results obtained and conclusions of the work carried out.

2 NAO Robot

The goal of this section is to explain the most important theoretical foundations for the development of research, which explains the characteristics of the NAO Robot, programming frameworks and computer vision systems that allow the robot to execute movements with the aim of autonomous management.

Bipedal robotics is an area of research with great growth in recent years; Although the displacement by wheels is more efficient and allows greater speed, the robots with legs are more versatile and can move in uneven terrains. In particular, bipeds are essentially suitable for handling in our environment, because they have characteristics similar to humans, so without having to modify our homes and workplaces, these robots can perform tasks for us, being particularly interesting applications in jobs that put people's health or lives at risk [3].

The humanoid robot NAO measures 58 cm, is interactive and fully programmable, with 25 degrees of freedom. It has a network of sensors that includes two cameras, four microphones, nine touch sensors, sonars and pressure sensors, as well as a speech synthesizer and two high-fidelity speakers.

The NAO robot has an integrated computer on board, with an Intel Atom 1.6 GHz processor, runs a Linux kernel and Aldebaran middleware called NAOqi. Available with two high definition cameras, connected to an FPGA allowing the simultaneous reception of images with better speed and performance, which allows a better image processing, even with low illumination (See Fig. 1).

The NAOqi Framework, is the operating system that runs internally in the NAO robot, allows to create new functions for the robot being, fast, safe and multi-platform. NAOqi allows a homogeneous communication between the different modules that the NAO robot allows to work (motor, audio, video).

The ALMotion Module oversees the general movement of the NAO, manipulating the motors that act on the articulations of the robot. When the movement of a certain joint is required, the call is made through ALMotion, which works in cycles taking 20 ms. [5]

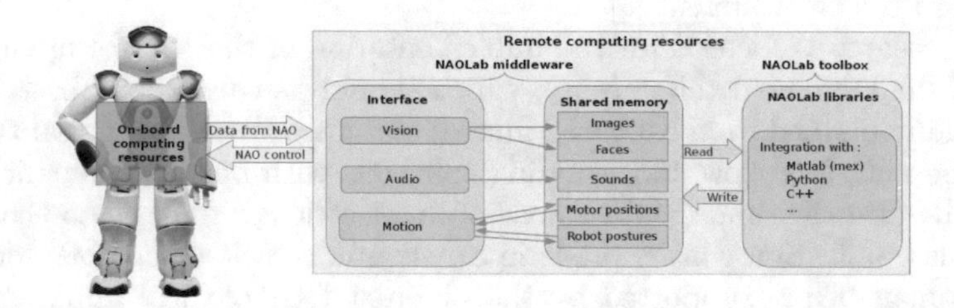

Fig. 1. Nao robot architecture.

The Nao robot has two video cameras located on the front of the head, it uses the upper camera which will acquire the image of the environment where the robot will navigate. With the use of the Open CV libraries to perform the processing of the images captured with a camera. First step to use Nao robot is the calibration of the camera. It allows obtaining the intrinsic and extrinsic parameters (internal and external) of the camera. For the use of positioning algorithms it is essential to have these parameters used in obtaining images [6].

The intrinsic parameters define the internal geometry and the optics of the camera. This parameters are (i) Main point is the intersection point between the plane of the image and the optical axis (line perpendicular to the plane of the image that passes through the center of camera).(ii) Focal distance is the length of a camera is the distance between it and the main point. (iii) Optical center is the point where the camera is located. By default, it is considered in the center of coordinates.

Extrinsic parameters defines the relationship (translation and rotation) between an absolute coordinate system (external to the camera) and another linked to the camera. They include six parameters: three for translation (Tx, Ty, Tz) and three for the angles rotated on each of the axes (α, β, γ) [1].

The goal of performing a camera calibration is the reduction of tangential and radial distortion. To calibrate the camera of the NAO robot, it is done with 6×4 chessboard boxes, images obtained with the upper camera of the robot and with the resolution that you want to work, to carry out this procedure you must have at least 20 images in different positions of the board.

3 Study Case

As previously explained, the case study is the autonomous driving of an electric car made by the NAO Robot. This robot will make movements on the steering wheel and the pedal; The way to obtain the movements and relate them to the programming of the vision system is detailed. The process is the following the robot humanoid NAO must be in the seated position on a mobile platform, such that he can reach with his right foot the drive pedal. The position of the arms and hands on the steering wheel, to be able to perform the drive in straight, left or right.

The use of the vision system through the robot's camera allows the recognition of the environment, to make the appropriate decision, performing the search process of lines that delimit the track being a very important factor to keep the vehicle in a straight line or correct the Deviation on the steering wheel and in turn to the car at the time of travel. The study proposed case is shown in Fig. 2.

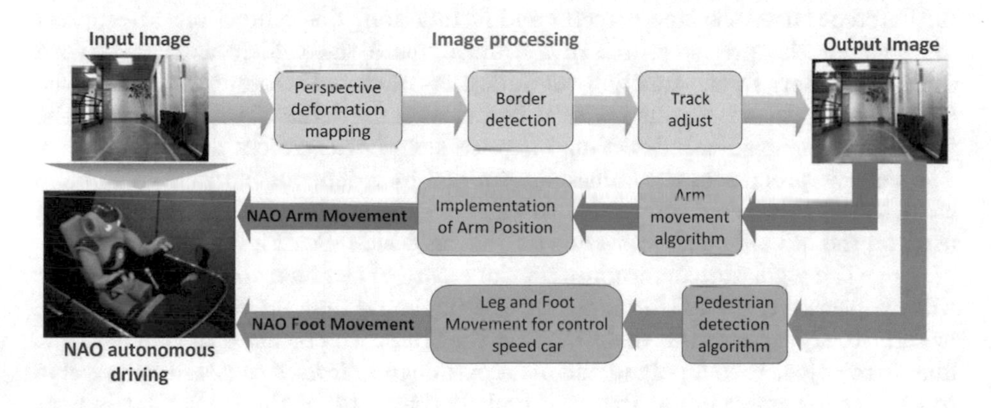

Fig. 2. Study case proposed.

4 Proposed Solution Implementation

This section presents an architecture for artificial vision which aims (i) the visualization of the track based on horizontal vision and (ii) the identification of people in front of the vehicle based on the horizontal and vertical vision of the NAO robot. This architecture can be seen in Fig. 3.

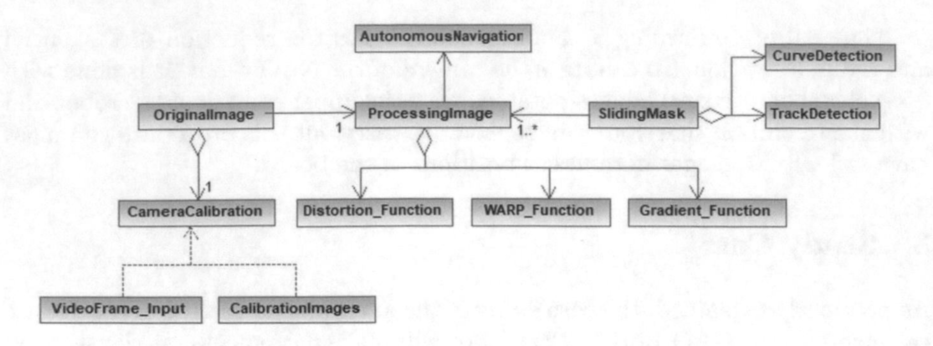

Fig. 3. Artificial vision proposed architecture.

The aim of this algorithm programmed in Open CV is (i) compute the camera calibration matrix and distortion coefficients given a set of chessboard images (see Fig. 4(a)). (ii) Use the Function Gaussian Blur to reduces the noise in the image, a process similar to convolve the image. (iii) Use of function unidistort in order to apply a distortion correction to raw images (see Fig. 4(b)). (iv) Use color transforms, gradients, etc., to create a thresholded binary image (see Fig. 4(c)). (v) Next, the algorithm apply a perspective transform to rectify binary image ("birds-eye view").

(vi) Later, this algorithm detect lane pixels and fit to find the lane boundary. In this step is used the lane_detector and fit function, These functions are applied together for the precise search of the lines where the vehicle will travel with the NAO robot. In the function the image is processed without distortion, the Gaussian function is applied with a kernel already defined, to then apply the function of the gradient: direction, magnitude and orientation, described above. The color is processed, the values determined in independent tests of the most ideal range, the colors are combined since to the study case the yellow line is used on the left side and white line on the right side. See Fig. 4(d) [7].

(vii) The algorithm determines the curvature of the lane and vehicle position with respect to center. Determine the steering angle that the robot corrects, left or right to try to keep the vehicle within the track. To calculate the curve of the line, is to adjust it to a polynomial of second degree, from this it will be possible to extract the information that is useful. For the line of the track that is near the vertical, a line can be adjusted using the formula $f(y) = Ay^2 + By + C$ where A, B and C are the coefficients. Since A gives us the data of the curve of the line, B indicates the direction that the line points and C the position of the line, based on how far it is from the left of the image ($y = 0$) [10]. See Fig. 4(e).

(viii) Finally, Warp the detected lane boundaries back onto the original image. Deformation of the image, as if you were looking from above (bird's eye). Transforming the points of the image obtained by the NAO robot camera in different image points. It is considered that the camera is in a fixed position, and the relative position of the track are always the same, it must be taken into account that the camera is located in the head of the NAO robot. The OpenCV

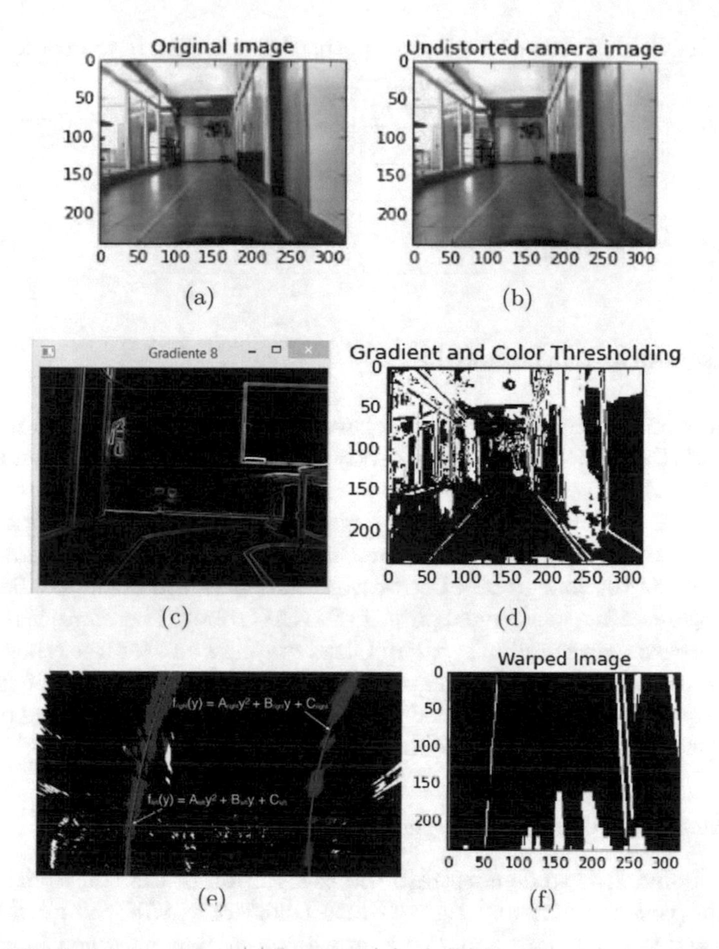

Fig. 4. Artificial vision process. (a) Image original. (b) Image using undistort function. (c) Image with gradient process. (d) Image using lane_detector function. (e) Detection of lines to determine the curve. (f) Image using warp function.

warp function needs four points: origin (src) and destination (dst), as can be seen in Table 1, defined in the function code and applied with the NAO robot camera for the place where the detection of the lines is required [2,9].

Based on the image that is captured by the camera of the NAO robot and applying the values of Table 1. The region of interest is defined, the final result can be seen in Fig. 4(f). Output visual display of the lane boundaries and numerical estimation of lane curvature and vehicle position. For the detection of people, the integrated HOG + Linear SVM detector is applied in the OpenCV library, which allows detecting people both in images and in video sequence.

Table 1. Points of origin and destination detection of the track.

Source (src) (x, y)	Target (dst) (x, y)
[0, 218]	[0, 0]
[320, 218]	[320, 0]
[320, 242]	[320, 288]
[0, 242]	[0, 288]

5 Discussion and Results

When carrying out the investigation regarding image processing and mechanical driving with the NAO robot, the respective tests are carried out when executing the algorithm.

With the use of programming in Python, making modules that link to obtain the results before the various situations that arise in the displacement with the car and process the images in a better way, the operation of the artificial vision part is shown and the mechanical part of the NAO robot. Therefore, it is analyzed with the program executed in Python, Choregraphe and Monitor (these last two are softwares that are used for the programming and visualization of robot data respectively). Next, the tests carried out, with the robot both in image processing and in movements of the NAO robot.

5.1 Tests with Camera Images of the NAO Robot

As a first point, it is to demonstrate the recognition of the track with the help of the left (yellow color) and right (white color) lines with values determined in tests carried out before applying them directly. It is observed in Fig. 5, the sequence of the frames for the left and right lines, thus painting the defined region of interest and adjusting as the image is changing in its trajectory, it can be visualized that we have the image of origin, that after having carried out the image processes the warp is shown and ending with the objective of indicating the defined area.

5.2 Detection of People with the NAO Robot Camera in the Displacement

The detection of lines as well as the detection of people see Fig. 6, are carried out with the camera of the NAO robot, whose purpose is to identify, to take some action on the pedal. In this test, after having performed the respective action on the NAO robot when detecting people in front of the car, the robot stops accelerating with the right foot and at the same time begins to say a message "Be careful", in the case that the person is not detected, the robot sends a voice message "Accelerates".

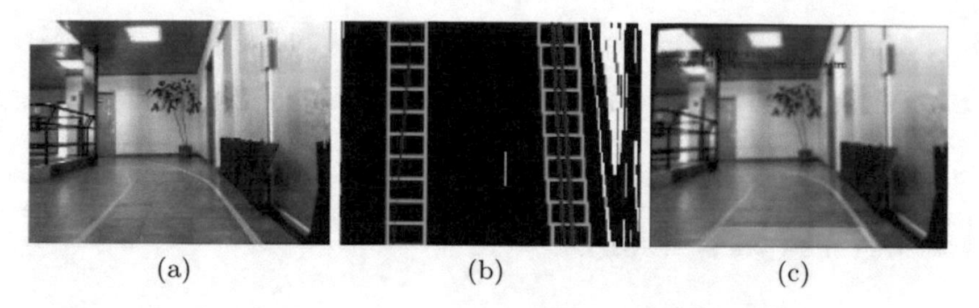

Fig. 5. Driving track recognition. (a) Image original. (b) Image using wrap function. (c) Image processed.

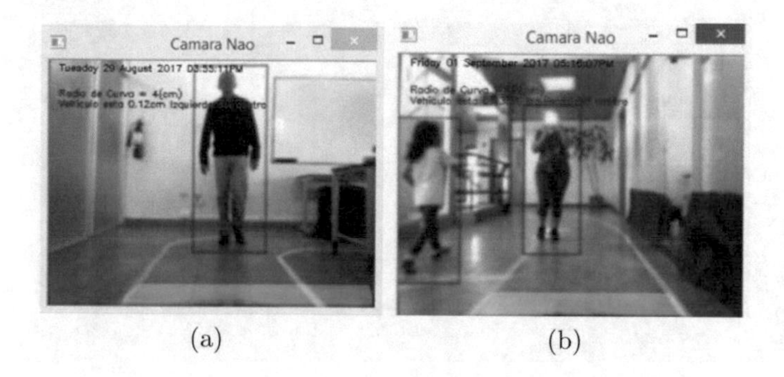

Fig. 6. People detection. (a) Center people detection. (b) Various people detection.

5.3 Right Foot Movement Acceleration and Deceleration of the Car

In Fig. 7, the movement can be observed with reference to the time that the right foot performed on the pedal to accelerate or decelerate, this operation is related to the actions that were previously mentioned with respect to, that if the robot has to move with the vehicle or at the moment that a person is detected in front of the car.

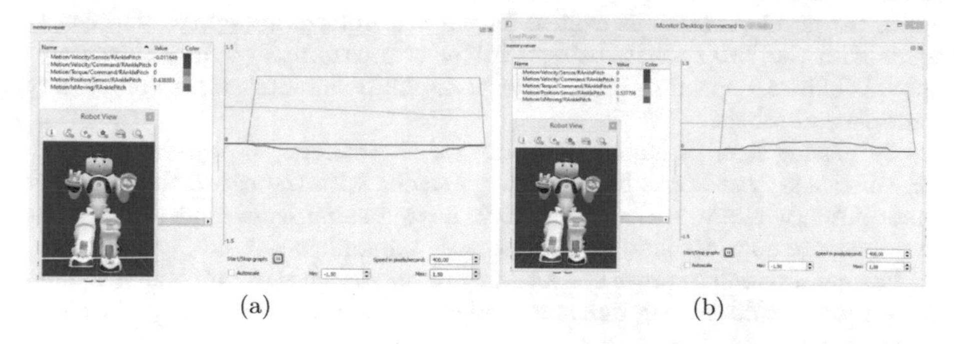

Fig. 7. Right foot movement. (a) Deceleration position right foot. (b) Acceleration position right foot.

5.4 Movement of Arms in the Displacement of the Car

At the moment the vehicle is moving on the track, there will be instants in which the NAO robot must correct the trajectory of the car, trying to keep it inside the track. Taking into account the value assigned to the center thereof, it is related to the programming of the line detection, which can be visualized at the moment of executing the Python script. See Fig. 8.

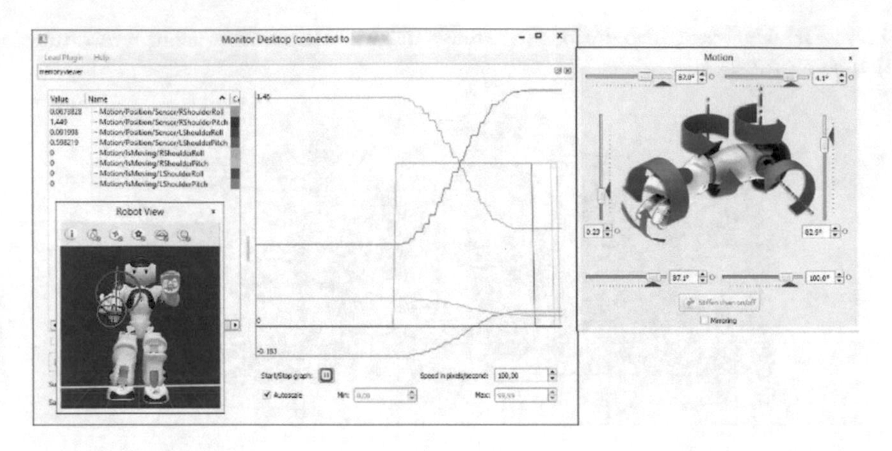

Fig. 8. Full turn to the right by the arms (right arm angles).

6 Conclusions and Ongoing Work

In this project, the implementation of a humanoid robot NAO in the driving of an electric car with artificial vision techniques has been verified. Using the different applications with which the humanoid robot NAO is handled, such as: the image processing, as well as the control of movements of the articulations in the controls of the car.

In the development it is evident that a very interesting application can be made with the NAO robot, being a driver of a car, mainly, for children with special abilities, depending on the level of disability: intellectual, motor, sensory, multiple and others.

By placing it on a mobile platform, for the transfer of objects over fairly long distances, you would have an energy saving with the use of the humanoid robot NAO, where it is located to work, as well as museum guide, in airports, hospitals, are quite striking to interact with human beings.

Future work will improve the accuracy of the recognition rate, and give more comparison results among different methods to further verify the performance of our system.

Acknowledgment. This work was financed in part by Universidad Tecnica de Ambato (UTA) and their Research and Development Department (DIDE) under project 1919-CU-P-2017.

References

1. Aguilera-Castro, D., Neira-Carcamo, M., Aguilera-Carrasco, C., Vera-Quiroga, L.: Stairs recognition using stereo vision-based algorithm in NAO robot. In: 2017 CHILEAN Conference on Electrical, Electronics Engineering, Information and Communication Technologies (CHILECON), pp. 1–6. IEEE, Pucon, October 2017. https://doi.org/10.1109/CHILECON.2017.8229674. http://ieeexplore.ieee.org/document/8229674/
2. Deng, G., Wu, Y.: Double lane line edge detection method based on constraint conditions Hough transform. In: 2018 17th International Symposium on Distributed Computing and Applications for Business Engineering and Science (DCABES), pp. 107–110. IEEE, Wuxi, October 2018. https://doi.org/10.1109/DCABES.2018.00037. https://ieeexplore.ieee.org/document/8572535/
3. Dong, E., Wang, D., Chen, C., Tong, J.: Realization of biped robot gait planning based on NAO robot development platform. In: 2016 IEEE International Conference on Mechatronics and Automation. pp. 1073–1077. IEEE, Harbin, August 2016. https://doi.org/10.1109/ICMA.2016.7558711. http://ieeexplore.ieee.org/document/7558711/
4. Dudek, W., Banachowicz, K., Szynkiewicz, W., Winiarski, T.: Distributed NAO robot navigation system in the hazard detection application. In: 2016 21st International Conference on Methods and Models in Automation and Robotics (MMAR), pp. 942–947. IEEE, Miedzyzdroje, August 2016. https://doi.org/10.1109/MMAR.2016.7575264. http://ieeexplore.ieee.org/document/7575264/
5. Korkmaz, S.A., Akcicek, A., Binol, H., Korkmaz, M.F.: Recognition of the stomach cancer images with probabilistic HOG feature vector histograms by using HOG features. In: 2017 IEEE 15th International Symposium on Intelligent Systems and Informatics (SISY), pp. 000339–000342. IEEE, Subotica, September 2017. https://doi.org/10.1109/SISY.2017.8080578. http://ieeexplore.ieee.org/document/8080578/
6. Korkmaz, S.A., Binol, H., Akcicek, A., Korkmaz, M.F.: A expert system for stomach cancer images with artificial neural network by using HOG features and linear discriminant analysis: HOG_lda_ann. In: 2017 IEEE 15th International Symposium on Intelligent Systems and Informatics (SISY), pp. 000327–000332. IEEE, Serbia, September 2017. https://doi.org/10.1109/SISY.2017.8080576. http://ieeexplore.ieee.org/document/8080576/
7. Lim, K.H., Seng, K.P., Ang, L.M., Chin, S.W.: Lane detection and Kalman-based linear-parabolic lane tracking. In: 2009 International Conference on Intelligent Human-Machine Systems and Cybernetics, pp. 351–354. IEEE, Hangzhou (2009). https://doi.org/10.1109/IHMSC.2009.211. http://ieeexplore.ieee.org/document/5335970/
8. Munoz, J.M., Avalos, J., Ramos, O.E.: Image-driven drawing system by a NAO robot. In: 2017 Electronic Congress (E-CON UNI), pp. 1–4. IEEE, Lima, November 2017. https://doi.org/10.1109/ECON.2017.8247303. http://ieeexplore.ieee.org/document/8247303/

9. Truong, Q.B., Lee, B.R., Heo, N.G., Yum, Y.J., Kim, J.G.: Lane boundaries detection algorithm using vector lane concept. In: 2008 10th International Conference on Control, Automation, Robotics and Vision. pp. 2319–2325. IEEE, Hanoi, December 2008. https://doi.org/10.1109/ICARCV.2008.4795895. http://ieeexplore.ieee.org/document/4795895/
10. Weixing, L., Haijun, S., Feng, P., Qi, G., Bin, Q.: A fast pedestrian detection via modified HOG feature. In: 2015 34th Chinese Control Conference (CCC), pp. 3870–3873. IEEE, Hangzhou, July 2015. https://doi.org/10.1109/ChiCC.2015.7260236. http://ieeexplore.ieee.org/document/7260236/

The Augmented Reality in the Teaching-Learning Process of Children from 3 to 5 Years Old

Narcisa María Crespo Torres(✉),
Ana del Rocío Fernández Torres, María Isabel Gonzales Valero,
Nelly Karina Esparza Cruz, and Joffre León-Acurio

Universidad Técnica de Babahoyo, Babahoyo, Ecuador
{ncrespo,aferandez,mgonzalez,
nesparza,jvleon}@utb.edu.ec

Abstract. Augmented reality (AR) is a technology that allows the perception of the real world. It provides the opportunity for the human being to interact with the objects already created in an environment of a supposed reality. The aim of the RA is to study the digital processing of images, the creation of patterns and the recognition of markers in 3D modeling designs. Through an analysis in a sample of children aged between 3 to 5 years, the advantages, disadvantages that the RA has over them in their cognitive advances with the experience of the objects are identified. Observation, experimental and inductive methods were used, as a result of the project, it was found that children love learning in a more practical, playful and positive way. The use of AR serves to motivate them and encourage them to learn in a meaningful, creative and animated way so that they can grasp and adopt learning as part of their environment. The augmented reality, therefore, offers the possibility of improving the processes of education in children with or without learning problems in the city of Babahoyo.

Keywords: Vision by the computer · Digital image processing · Markers · Virtual reality

1 Introduction

At present, communication technology is immersed in all areas such as architecture, education, medicine, among others. At the national level, companies or institutions use systems that allow improving the processes that are developed in their daily practice. So these make it possible to look in detail at what has been done and reflect critically on what is being achieved, something that generally, for many different reasons, does not occur.

Globalization has been very much in the technological aspect, ICT is necessary for everyday life, Ortega (2005) understands ICT applied to education as a great source of information that motivates and awakens the interest of students, due to its strong playful connotations. These tools provide interesting environments to investigate, experiment and, therefore, where to learn (Quse et al. 2011; Morilla 2012).

M. Botto-Tobar et al. (Eds.): ICAETT 2019, AISC 1066, pp. 207–218, 2020.
https://doi.org/10.1007/978-3-030-32022-5_20

One of the most useful applications of augmented reality is in education. Programs and applications with this technology have been in use for a long time in the classroom and the results are more than satisfactory in the classrooms since it allows:

- Establish a natural contact of children with new technologies, whose use is widespread.
- It teaches them that ICT also has a formative content, not just playful.
- It favors attention when dealing with applications that bet by funny teaching and that looks for to surprise to big and small.
- Improves interactivity and participation in the classroom thanks to the use of a learning system that focuses on entertainment.

According to Pons and Roquet-Jalmar (2008), cognitive development can be considered as a process by which people from birth can think, have knowledge and understand.

Due to the technological revolution it is a fact that these new technologies have facilitated our lives, which has caused an acceleration of this (Cabero and Barroso 2016).

The interactions we establish with others until the way we approach and generate knowledge have changed significantly due to the influence of Information and Communication Technologies (Román et al. 2002).

Over the years, due to the emergence of technological advances, the inclusion of ICT in classrooms of all levels and of all educational stages is considered a great opportunity (Cárdaba and Palomero 2011).

Paradoxically, the school continues to be a space where a greater technological inclusion is still necessary because there are reluctance or even technophobias that keep students away from the undoubted improvements that ICT brings to the classroom.

That is why the education system must be modern, innovative and inclusive in the 21st century (Cabero et al. 2016). Therefore, the school must be a real digital literacy space, where the teacher must be the responsible figure to design environments that favor learning and take advantage of technology as a mediator in the construction of knowledge and social interaction (Calderero et al. 2014).

According to Cabero (2015), the effort must be invested in building innovative teaching models that make the most of the technologies that currently exist.

There are two models: Learning from Technology and Learning from Technology. From the Learning of Technology model, ICTs are used as an effective means of transmitting information. Teachers see this as an efficient means of transmitting instructional content to students (Montes and Solanlly 2006).

One of the technologies that in recent times is emerging and with real possibilities of impacting the educational field is the so-called "Augmented Reality" (Augmented Reality) "(RA), technology that," according to different reports Horizon and the EduTrend Report of the Technological Observatory of Monterrey, will have a strong penetration in the educational instances in a horizon of three to five years (Cabero et al. 2016, page 63).

Although it is true that since the beginning of the 90 s the term Augmented Reality (AR) was investigated and approached through a great variety of fields, such as medicine, aeronautics, robotics or tourism, but it has it has been in recent years when it has been reaching a higher peak, especially as regards its incorporation into training (Bower et al. 2014).

Augmented reality (RA) consists in using a set of technological devices that add virtual information to physical information, in order to create a new reality, but where both real and virtual information play a significant role (Cabero et al. 2016). "Augmented reality is a subclass of mixed reality that globally can be conceptualized as a new space by association between a virtual environment and a real one where the person can perform sensorimotor behaviors as well as cognitive activities" (Cabero and García 2016, p. 19).

2 Applications with Augmented Reality

2.1 Layar y Aurasma

They allow managing contents according to the subject and specific needs of the teacher, who can create both the image that will activate the augmented reality (trigger) and the content to be observed in the form of videos, web pages, animations or 3D models (augmented reality).

In this case, we are talking about an augmented reality application for children, but also for teachers, parents who want to create their own images of augmented reality, both for educational and recreational purposes. With Aurasma it is possible to do it from your mobile phone with Android and IOS operating system. You only need to download them, for free, and start experimenting with the educational advantages of augmented reality.

2.2 Aumentaty Author

It is a company focused on the development of Motors, Applications and Projects in Augmented Reality (Fig. 1).

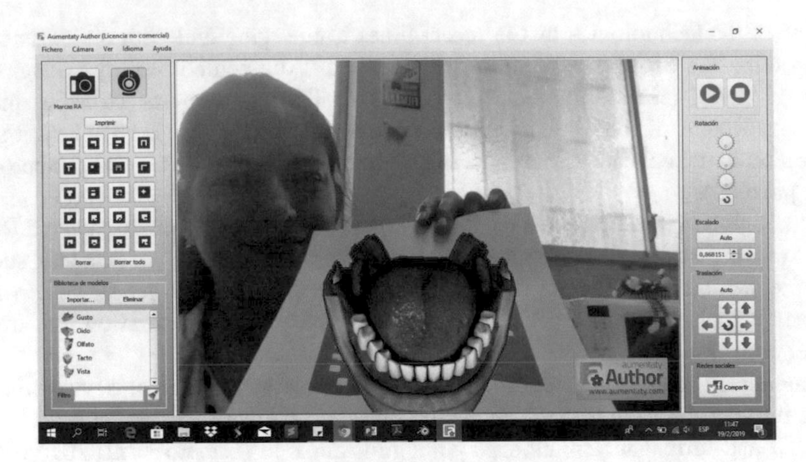

Fig. 1. Aumentary author application

2.3 Elements 4D, Quiver and Chromville

They are applications that offer already created contents, to teach some subject or to encourage the creativity of the students. In this case, you must download the templates from the application's website (which will work as triggers) and then observe them through the application on your mobile or tablet (Fig. 2).

Fig. 2. Quiver application

2.4 Ciberchase 3D Builder

A fun application of augmented reality for children that allows the little ones to become familiar with geometric shapes and simple mathematical concepts thanks to the downloading of the contents that appear on their website.

2.5 Aug that!

It is a perfect augmented reality application to entertain classes and provide additional content to lessons of any subject. Once you have generated the contents, you only have to ask your students to pass their mobile devices through the cards to begin interactive learning.

For years, Google Translate incorporates the augmented reality option so you do not have to write. Simply scan or focus your mobile to a phrase or poster in another language to activate the automatic translation service. A fun way for little ones to gain vocabulary in other languages.

In children, the play is the primary tool from their development, creating enthusiasm in their learning. The human being is a social being from birth. Throughout his life, he is part of different groups and constantly interacts with other people. Socialization "is the process by which each human being becomes an active and full member of the society of which he is a part (Riesco Díaz 2017).

The theory of constructivism articulates based on the mechanisms by which people who learn, knowledge is internalized. Jean Piaget was a Swiss psychologist of the early twentieth century, he argues that through processes of accommodation and assimilation the child is able to build knowledge from their experiences, that is why this research is researching learning through augmented reality in children, (Ares 2009) in illustration #1 you can show the factors of knowledge (Fig. 3).

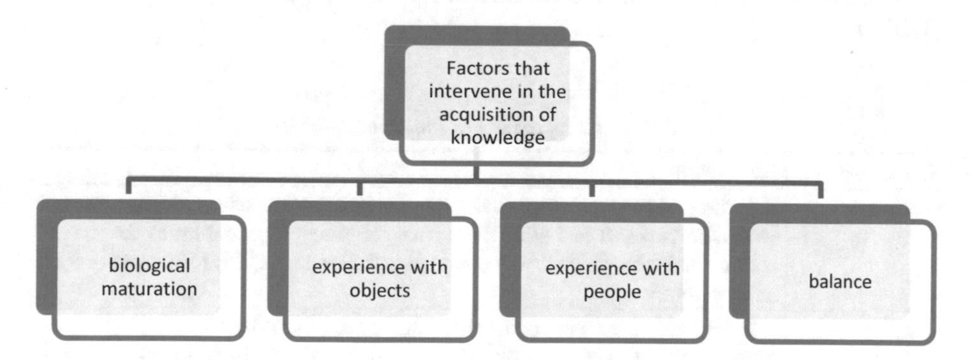

Fig. 3. Acquisition of knowledge, according to Piaget. (Source: Muñoz and López 2011)

2.6 ICT in the Teaching and Learning Process of Children

The use of ICT (Information and Communication Technologies) in learning environments has enriched the processes and allowed the evolution of the educational model.

Initially, the model was centered on the Educator and this was responsible for getting and providing the necessary information while the students adopted the knowledge using technology as a simple vehicle for transmitting information, in this model the most important theories were Behaviorism and Objectivism (Dodge 2012) (Fig. 4).

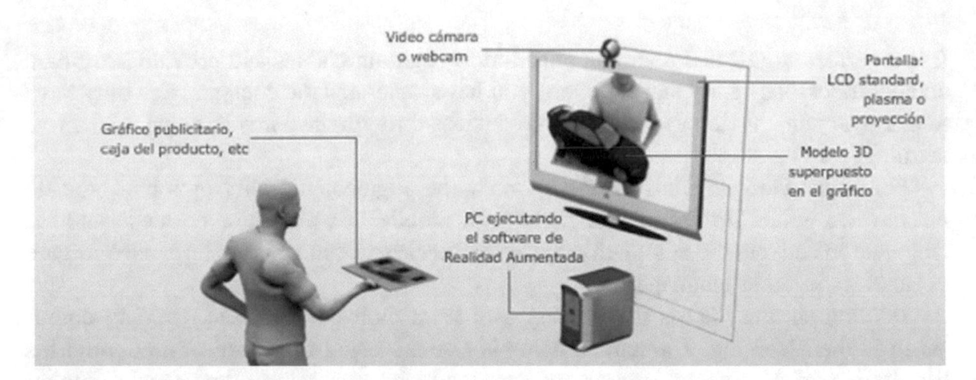

Fig. 4. Architecture of an augmented reality system (Source: http://jportiz.wordpress.com/peinuevos-territorios/coferencias/realidad-aumentada)

Augmented Reality is the term used to define a direct or indirect view of a physical environment of the real world, whose elements are combined with virtual elements for the creation of mixed reality in real time. It consists of a set of devices that add virtual information to the existing physical information, that is, add a virtual synthetic part to the real one. This is the main difference with the virtual reality, since it does not replace the physical reality, but it superimposes the computer data to the real world (Dodge 2012), in the illustration #2 the architecture of the augmented reality can be observed (Table 1).

Table 1. Components of augmented reality

Components	Description
Software	It is the computer program that takes the information received from the real world, processes it and adds the virtual information according to the instructions with which it was programmed. Example Quiver, Increase author, among others
Display screen	It is the hardware component where the result of the Augmented Reality will be seen, that is, where the objects with all the added virtual information will be appreciated
Camera	It is the hardware component that captures the real environment and feeds the system with this information
Markers	These are specific points of the real environment that are identified by the software as the points where the virtual information should be included

3 Materials and Methods

The population is made up of children from 3 to 5 of the schools of the City of Babahoyo. The gathering of information was carried out through interviews, questionnaires and work meetings, which allowed obtaining the necessary information to propose a solution to the problem.

Once the data and information were obtained, the analysis and interpretation were carried out using quantitative and qualitative metrics. The use of augmented reality is still experimental and there are many challenges for its massive implementation in the classroom. But his tuning in with the student of today, his way of learning and consuming information is undeniable.

Therefore, more and more research is addressing the pedagogical use of augmented reality and its impact on students and their significant learning.

Computers and other aspects of Information and Communication Technologies (ICT) allow children and young people to carry out a wide variety of activities and experiences that can promote learning. But many of these transactions do not take place in traditional schools. In fact, many cannot be considered "educational" according to our conventional understanding of the term. For most of us, the debate about learning is intimately related to formal education systems (how schools should be organized, managed and carried out). However, any interest in the role of ICT in children's learning requires recognition that many children are involved in ICT-related activities in their homes and with their friends. This recognition forces us to recognize a broader 'ecology' of education in which schools, homes, recess, the library and the museum all play their part (Sefton-Green 2006, p. 2).

RA has been applied experimentally in both school and business environments. Although not as much as the classical methods of education and training during the last two decades. In addition to that, now that the technologies that facilitate the RA are much more powerful than ever and compact enough to offer RA experiences not only for corporate environments but also for academic sites through personal computers and mobile devices. Different educational approaches to RA technology are more feasible. Also, wireless mobile devices, such as smartphones, tablet PCs, and other electronic innovations, are facilitating the entry of RA into the mobile space. Where applications offer great promise, especially in education and training. It is very likely that the RA can make educational environments more productive, enjoyable and interactive than ever before.

The RA not only has the ability to encourage a student to engage in a variety of interactive ways that were not possible before, but can also provide each individual with a unique discovery path with rich 3D content. Improves augmented reality student learning? A proposal of museum experience increased environments and three-dimensional models generated by computer (Kangdon 2012, p. 14 and p. 19) (Figs. 5 and 6).

Fig. 5. Augmented reality application

Fig. 6. Augmented reality application

4 Result: Analysis and Interpretation

According to observations in children who are between 3 years in the schools analyzed: motor control and reaction to physical objects are 44% good. The development of verbal skills with objects is 47% excellent. Magic ideas of cause and effects with objects as are very small children are 63% bad (Figs. 7, 8 and 9).

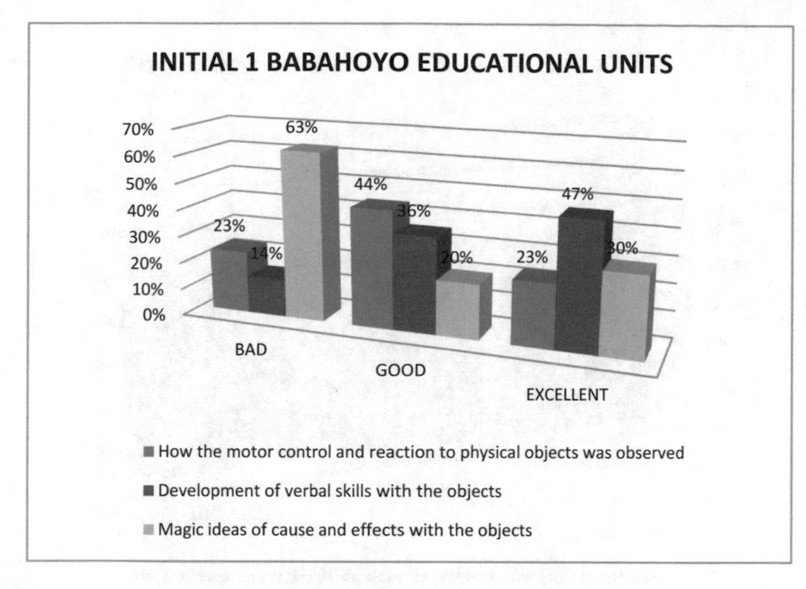

Fig. 7. Initial 1 Babahoyo Educational Units

According to our observations before the three approaches that expressed the motor control and reaction to physical objects in the children of initial two is 44%, excellent and in development of verbal skills with objects 37% good and magic ideas 60% bad.

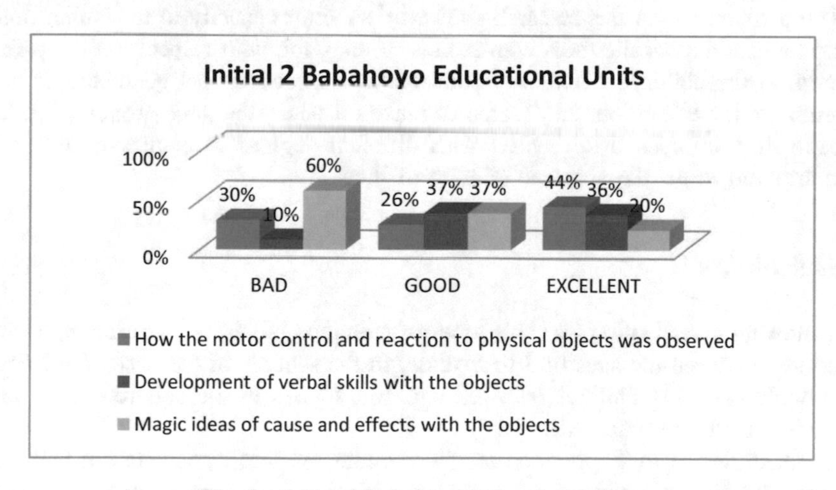

Fig. 8. Initial 2 Babahoyo Educational Units

According to our observations before the three approaches that expressed the motor control and reaction to physical objects in the children of first of basic is 65%, excellent and in development of verbal skills with objects 38% excellent and magic ideas 40% good.

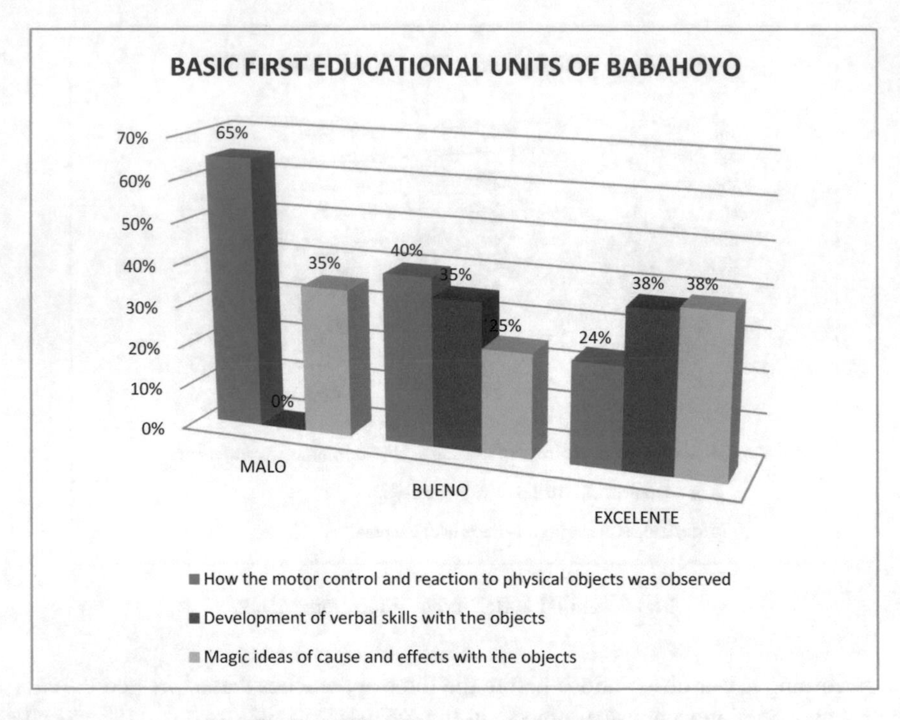

Fig. 9. Basic First Educational Units of Babahoyo

The participants for this research showed a superior capacity to remember images when they could associate them with scenes of daily life with respect to their ages. In addition, in the children a noticeable emotional contraction was generated from the beginning of the exhibition, they feared to make mistakes, the three groups of children corresponding between 3 to 5 years with different degrees of cognitive knowledge, made an effort to get the approval of each of them.

5 Discussion

The following investigation was able to work with the Quiver tools, managing to work in children between the ages of 3 to 5 years, in the schools of the City of Babahoyo. The sample is of 174 children, showed patterns, figures to the children and a lot of satisfaction could be observed.

The audiovisual tools provided by the Augmented Reality generate a high emotional impact and are decisive for remembering ideas and concepts in children from 3 to 5 years old. The maximum attention and concentration occurs during the pattern recognition experience, a fact that was not only observed during the course of the same, but also corroborated by the participants in the common goal.

A lower level of cognitive maturation or a lower use of ICT tools does not prevent the assumption of Augmented Reality technology. All the children in the experience concluded the experience as designed. There are no difficulties to interpret the juxtaposition between reality and the images generated by a computer in real time or to direct the camera of the mobile to the images to re-produce the associated content.

The name is also important, specialists say that students do not locate the material by name, but by the image, working in this way with the three channels of communication: kinésico, auditory and visual (María Reina Zarate Nava 2013).

6 Conclusions

- The augmented reality has the potential to favor the teaching and learning processes in children, integrating multimedia tools, from a connectivist perspective of network learning.
- Education should be added ICT, as an important tool today, countering the digital divide in education.
- Motivation in school children, creating by participating in them as they are interactive, dynamic activities, where the teacher can experiment and manipulate different situations.
- Educational education must understand the innovative process through the incorporation of new trends and emerging technologies, counting on the training and interest of the teaching staff in the realization of the corresponding adaptations.

References

Ares, C.M.: El uso de las TICS en el aula de eduación infantil. Standard Copyright License, México (2009)

Dodge, J.C.M., De León, I.S.: Uso de Realidad Aumentada para Enseñanza de Conceptos Básicos de Física. Ingeniare **12**, 11–26 (2012)

Nava, M.R.Z., González, C.F.. Marcadores para la Realidad Aumentada para fines educativos. Computación e Informática 12 (2013)

Riesco Díaz, B.: (2017). https://docplayer.es. https://docplayer.es/63554663-Entornos-de-realidad-aumentada-en-educacion-infantil-3-6-anos-augmented-reality-environments-in-early-childhood-education-3-6-years.html

Kangdon, L.: Augmented reality in education and training. TechTrends **56**(2), 13–21 (2012)

Sefton-Green, J.: Report 7: Literature Review in Informal Learning with Technology Outside School. FutureLab, Bristol (2006)

Muñoz, V., López, I.: Manual de psicología del desarrollo aplicada a la educación. Ediciones Pirámides, España (2011)

Bower, M., et al.: Augmented reality in education – cases, places and potentials. Educ. Media Int. **51**(1), 1–15 (2014)

Cabero, J.: Aplicación a las nuevas tecnologías al ámbito socioeducativo. IC Editorial, Málaga (2015)

Cabero, J., Barroso, J.: Redes sociales y Tecnologías de la Información y la Comunicación en Educación: aprendizaje colaborativo, diferencias de género, edad y preferencias. RED. Revista de Educación a Distancia **51** (2016). Artic.1

Cabero, J., García, F. (coords).: Realidad aumentada: tecnología para la formación. Sintesis, Madrid (2016)

Cárdaba, A., Palomero, M.: Servicios socioculturales y a la comunidad. Expresión y comunicación. Santillana, Madrid (2011)
Carcajona, P., Galindo, B.: Servicios socioculturales y a la comunidad. Desarrollo cognitivo y motor. Santillana, Madrid (2010)
Montes, J.A., Solanlly, A.: Apropiación de las tecnologías de la información y comunicación en cursos universitarios. En acta colombiana de psicología **9**(2), 87–100 (2006). Universidad Javeriana, Colombia. http://www.redalyc.org/articulo.oa?id=79890209
Pons, E., Roquet-Jalmar, D.: Desarrollo cognitivo y motor. Altamar, Barcelona (2008)
Román, P., Cabero, J., Barroso, J.M., Bermejo, B., Morales, J.A., Ballesteros, C., López, E., Martínez, A., Oliva, E.: Diseño de entornos telemáticos de formación Blanca Riesco Díaz Trabajo de Fin de Grado 2016/2017 108 para alumnos que cursan asignaturas de nuevas tecnologías. Revista de Enseñanza Universitaria **19**, 173–184 (2002)

Proposal for a Support Tool for the Study of Corneal Biomechanics and Its Influence in the Human Eye

María Isabel Cordero[1(✉)], Roberto Coronel[1], Eduardo Pinos-Vélez[2], William Ipanque[3], and Carlos Luis Chacón[4]

[1] GIIATA, Research Group on Artificial Intelligence and Assistive Technologies, Universidad Politécnica Salesiana, Cuenca, Ecuador
{mcorderom, rcoronelb}@est.ups.ecu.ec

[2] GIIATA, Research Group on Artificial Intelligence and Assistive Technologies and GIIB, Research Group on Biomedical Engineering at Universidad Politécnica Salesiana, Cuenca, Ecuador
epinos@ups.edu.ec

[3] PhD of Engineering Computer Science and Control, Universidad de Piura, Piura, Peru
william.ipanaque@udep.pe

[4] Board Member of the Ecuadorian Society of Glaucoma, Clinical Santa Lucía, Quito, Ecuador
clchacon@clinicasantalucia.com.ec

Abstract. In search of more effective methods in the diagnosis vision and reduce the risk of post-operative ectasia, this article considers the importance of the biomechanical properties of the cornea and its close relationship with the development of corneal and possible conditions diseases in the human eye. For that reason, part of the literature is the review and analysis of the methods raised to remove accurate information from medical equipment, especially in the equipment Oculus Corvis and ORA from the Clinic "Santa Lucia" in the city of Quito. The corneal indices derived from these instruments can measure biomechanical parameters, but that is not accurate because of features such as geometry or thickness of the central corneal. The Young's modulus corresponds to the elasticity of the corneal tissue. In addition, it is an important biomechanical feature that states different solutions for their extraction in vitro or in vivo through the use of numerical methods and/or modeling of complex systems of equations to simulate the real corneal tissue. There is also the obtaining of the elasticity of the cornea through processes of images of the human eye, which shows the action of the cornea at the time of using the tonometer. As a result, it demonstrate, in some cases, that the Young's modulus relates only to the intraocular pressure, but also in other cases, it changes due to the increase of other factors such as age of people, corneal resistance, the central cornea thickness, etc.

Keywords: Corneal biomechanics · Young's modulus · Intraocular pressure · Biomechanical parameters · Central corneal thickness

M. Botto-Tobar et al. (Eds.): ICAETT 2019, AISC 1066, pp. 219–228, 2020.
https://doi.org/10.1007/978-3-030-32022-5_21

1 Introduction

Eyes are the most valuable and sensitive organs in the human body, because through them people have the ability to see and interact with the environment surrounding us [6]. Part of this process is a delicate, transparent and external structure called the cornea, which is exposed to multiple conditions, caused by the deformation [2].

There are studies [1, 2, 3], which mentioned the different diseases directly related to the cornea, for example cases of Keratoconus, keratitis, hypermyopia, myopia and indirectly glaucoma. It is very common to find devices in the market that would diagnose these diseases such as the ocular response analyzer (ORA) and Oculus Corvis. These devices provide rates of bio-mechanical behavior of the eye globe [4, 5, 9], some of those behaviors are the corneal resistance factor (CRF), central thickness of the cornea (CCT), intraocular pressure (IOP), hysteresis (CH) among others [7]. However, the results obtained through these biomechanical indexes instruments are not independent parameters that also influences geometry corneal, IOP and the thickness of the cornea [10] hampering the diagnosis of mild conditions as a corneal ectasia subclinical [11].

Through the study of corneal biomechanics and behavior, obtaining biomechanical indices is possible to analyze the balance and the deformation of the cornea subjected to forces such as intraocular pressure (IOP) in the rear and external forces as the atmospheric pressure or the action of the tonometer or rubbing with your hands on the front [1, 2].

It is important to know the properties of intrinsic material corneal biomechanical since using these, you can know the state of the structure corneal and pathologies which may occur effectively [12], as well as allowing to calculate the elasticity, Young's modulus, and the viscosity of the cornea.

The most frequent application from the point of view of the eye is the corneal refractive surgery to correct small errors of graduation where the light rays are not focused directly on the retina, but by the cornea and lens, which results in blurred vision. The success of this surgery depends on the decision of the specialist and this is based on the correct analysis of the tests carried out clinical [1, 3].

2 Modeling of the Cornea and the Module of Elasticity

Cornea material shows different behavior by what the authors of the works mentioned present different mathematical models for the structure corneal.

In some cases, like [12, 13, 14] presents a model of all the eyeball, in which Cao et al. the takes 10 patients and measures IOP through a method not invasive as the Oculus Corvis, takes into consideration three forces that apply on it: the ambient air pressure, the pressure of the internal fluid which has the eye and the iris of the eye, Fig. 1.

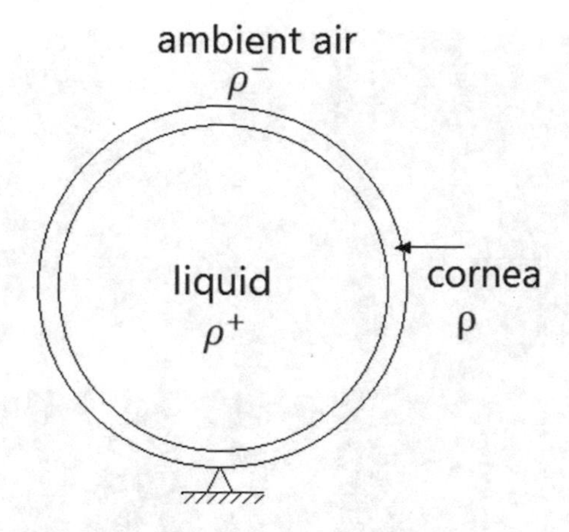

Fig. 1. The dynamic model of the eyeball [13].

Air environment and the liquid inside the eyeball is modeled using the Euler equations, applying these equations in the Helmholtz function; modified aid in the simplification of those derived from the Eq. (1) raised initially [13].

$$\frac{(\mathrm{E}+\mathrm{T}).\mathrm{t}^3}{12(1-V^2)}\nabla^4 w - T.t\nabla^2 w - [p] = \rho.t\frac{\partial^2 w}{\partial t^2} = 0 \quad (1)$$

Where:

E, is Young's module
T, is the tension in the diaphragm of the eyeball
T, the thickness of the diaphragm
V, is the Poisson's ratio
ρ, is the density of the diaphragm
[p], is the difference in pressure between the two sides of the diaphragm.

This model assumes that the distribution of the pressure produced by the big air of the tonometer has a Gaussian distribution three dimensional, which consider the Poisson ratio of 0.49 [8].

While Joda et al. [14], mentions in his article the modeling of the eyeball which used parameters such as corneal thickness, corneal topography, material properties, conditions of border at the edge of the cornea, and also it is considered the density of the mesh and the method of desertification.

This article is made use of finite elements (FE) to simulate the pressure of air in the eye by Corvis, the model includes the cornea, the sclera and the aqueous and vitreous humor, seeking to be as real as possible to the response of the cornea before this action.

On the other hand, other authors made the model via image processing to have the distance of displacement of the cornea at the time of applying jet of air, Fig. 2, as he presents in his work L Wang et al., this is to determine the elasticity and the viscosity of the cornea [9].

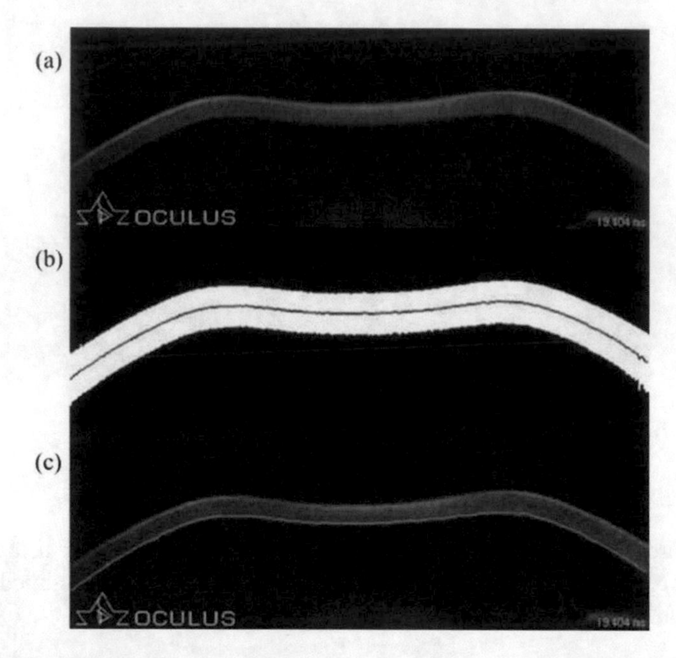

Fig. 2. Examples of image identification from Corvis ST, (a) represents the image of Corvis urinal (b) is the region of interest represented on the centerline, (c) limits superior and inferior corneal respectively [18]

The author performed the data collection of 59 patients, which of whom, 25 had Keratoconus and 34 had no pathologies for the force of the breath of air that shoots the Oculus Corvis, it adapted a precise load cell at a distance of 11 mm with respect to the nozzle of the Corvis. Furthermore, a pressure of Gaussian distribution had a peak of 36 mN constant stream value, Fig. 2 [18].

Using Hooke's law, takes into account that the cornea has as characteristic non-linearity on the elasticity coefficient. The data to be used for this modeling were CH, Corneal (FRC), and PIO resistance factor, coming to the viscosity equation and elasticity that are based on the potential energy (load, download and absorbed) for force-displacement relationship.

It is considered of great importance that the corneal rigidity given that it is which supports the intraocular pressure by keeping the shape and geometry of the cornea, as well as the vision in their normal ranges, some eye conditions and surgical interventions are closely linked with the biomechanical properties [9].

This study used 10 artificial corneas silicones each of which had different rigidity and 5 corneas pig, these were placed vertically fastened by a few clamps in order to take pictures of the cross section through tomography optical coherence, (OCT) (Fig. 3) which undergo treatment of images to improve their sharpness and brightness [9].

The greater compressions of corneal biomechanics method are through a test of a strip of thin cornea in vitro based on the stress-strain relationship in uniaxial, further more, by segmenting the cornea into thin strips links the structure of the corneal tissue.

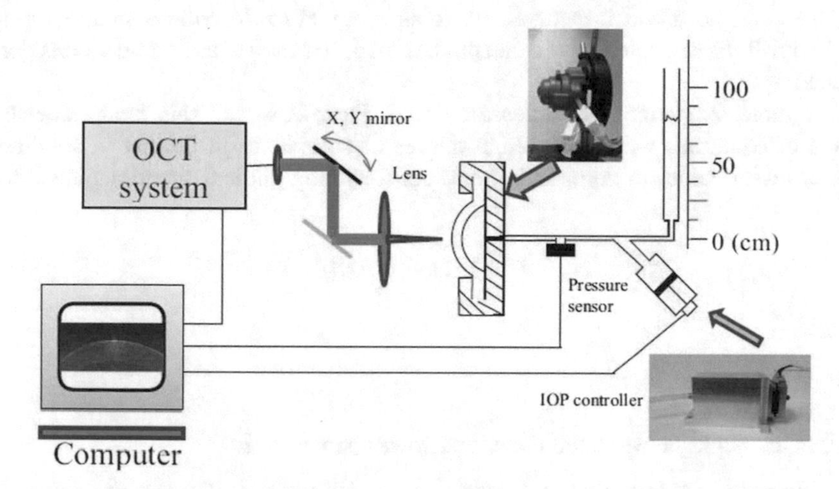

Fig. 3. Scheme of the treatment of the artificial and porcine corneas and the application of the OCT to obtain the required images [9].

Porcine corneas, as well as the artificial corneas, were subjected to inflation by Hooke's law and the theory of thin shell when T/R < 0.1, is modelled the relationship of tension- displacement of the PIO being as a result (2):

$$D = \frac{pR^2}{2ET}(1 - \nu) \tag{2}$$

Where:

D, elevation of the vertex,
p, intraocular pressure,
R, ratio of curvature,
E, Young's modulus,
T, central thickness of the cornea,
v, ratio of Poisson.

Similar to J. Cao et al. to's case, Wang et al. also considered a Poisson's Ratio of 0.49, assuming that the cornea is almost incomprehensible.

Artificial corneas showed them that they are almost completely elastic, while in porcine corneas, CH, it depends on the speed of load, and it is possible to distinguish values significantly higher than Young's modulus.

Shih et al., mentions the use of team clinical management to obtain the data in the state of the eyeball, from the data obtained properties biomechanical, which may vary by factors such as the corneal thickness or its geometry, which hinders the effective diagnostics in cases in which the cornea damage slight as subclinical ectasia. By which, he considers important to obtain the properties biomechanical of the cornea from Young's modulus, which can be used to facilitate the diagnostic physician in cases where the data delivered by the team which is not sufficient to diagnose pathology.

Therefore, he raised it using Shell Taber's model to describe a static warp shell filled with fluid subjected to a concentrated load. To model the cornea, consider the sequence:

1. dimple, 2. stretch under pressure and 3. Flex; however, this model develops a system of equations with a very high degree of complexity, resulting in solution set variables with values of zero, taking into consideration clinical situations, resulting in (3):

$$P - 2\pi(1 - \cos\alpha)p \tag{3}$$

Where:

P, is the applied force
p, is the IOP
α, is the angle between the upper and lower cover on 2.

Furthermore, shows that it is possible to deduce the Young's modulus, from geometric deformations and IOP, this value is implicit in the standard force (4).

$$P = \frac{pR^2}{Et^2} \tag{4}$$

Where:

R, is the radius of curvature
E, is Young's modulus
t, is the corneal thickness

According to Shih et al., the CCT minimally influence in the Young's module, and presumably does not depend on the geometry or external forces, therefore, Young's modulus slightly higher.

3 Discussion

At first, there are several devices that allows to obtain characteristics of corneal biomechanics, however, none of these give direct biomechanical parameters, so that each author presents a different method to model the cornea, coming to present similar results of the modulus of elasticity, that is to say, the Young's modulus, being necessary for the calculus to Poisson's ratio, to this, in most cases the authors take fixed and determined value of 0.49.

There are different ways to reach Young's modulus, through mathematical processes involving Eqs. (1) (2) and (3), image processing, all resulting in an average of 0.2 MPa, each method presents a degree of complexity fairly high. However, the

Table 1. Results presented in the calculation of cornea's Young's modulus.

Parameter	Range			Value	J. Cao et al.	J. Keraut ret et al.	Shih et al.
Age (years)	0–14	15–64	>64	–	x	–	–
Thickness (um)	–	550.3	–	–	x	–	–
PIO (mm Hg)	15.19	14.7	17.66	18.7	x	x	x
AL1 (mm)	1.78	1.77	1.81	–	–	–	x
AV1 (m/s)	0.14	0.15	0.12	–	–	–	x
AL2 (mm)	1.73	1.72	1.81	–	–	–	x
AV2 (m/s)	−0.35	−0.37	−0.32	–	–	–	x
PD (mm)	3.91	3.79	3.56	–	–	–	x
Radius (mm)	7.11	7.17	7.63	–	–	–	x
DA (mm)	1.02	1.07	1.03	–	–	–	x
CCT (um)	537.54	543.10	561.68	–	–	–	x
AT1 (mseg)	7.45	7.38	7.76	–	–	–	x
AT2 (mseg)	21.76	21.77	21.21	–	–	–	x
CRF (mm Hg)	–	–	–	6.8	–	–	x
CH (mm Hg)	–	–	–	6.9	–	x	–
Young's Modulus (MPa)	0.243	0.29852	0.187	0.209	x	x	x

PIO: intraocular pressure; AT1 and AT2: applanation times; AL1 and AL2: applanation diameters; AV1 and AV2: applanation velocities; PD: peak distance; DA: maximum deformation amplitude; CCT: central corneal thickness; *CH: corneal Hysteresis; CRF: corneal resistance factor*

method with higher process and difficulty presents Shih, he says that the biomechanical parameter should not be linked to values as the CCT or the geometry of the cornea.

In almost all cases, the data obtained porcine corneas in vivo, also, Shih proposes this study starting from synthetic corneas which give results that the cornea is almost purely elastic, it is considered that age is also a factor that influences in Young's modulus.

Wang et al. and J. Keraut ret et al., considered great importance of corneal rigidity and provides data of CH, Corneal (FRC), CCT, and PIO resistance factor to perform its calculations coming to the equation of viscosity and elasticity.

While A. Joda et al., data involving to the mathematical model considered is CCT, the properties of the material, corneal topography, boundary conditions at the edge of the cornea slightly contradicting Shih method et al., take as important data CCT and conditions border (corneal geometry).

Data from each of the authors are presented in Table 1. Where the visible features being used and the results obtained in Young's modulus.

Results indicate that the Young's modulus has an average value of 0.25 MPa, it can also be seen that this varies with age, having an average value of 0.187 MPa in people older than 64 years.

J. Kerautret et al., on the contrary, does not consider age as a relevant factor for the calculation of the modulus of elasticity. However, the corneal hysteresis and the corneal resistance factor have the same results in the calculation of modulus of elasticity which indicates that each of the proposed models is valid to represent the modulus of elasticity. As shown in Table 1.

4 Future Work

Considering the difficulties found in the reviewed bibliography, new alternatives that can collaborate in the ophthalmological diagnostic efficacy and, a three stage investigative work is proposed.

The first step, using the revised bibliography, to find the most reliable method for the determination of the Young's Modulus and the Poisson's Ratio, biomechanical properties that indicate the elasticity and viscosity of the cornea, respectively. For which, the use of equations proposed by Wang et al. and Shih et al. is considered viable, whose results agree in spite of being developed with different methodologies, the combination of these methods defines the unknowns not found with the other method, besides, the variables with which they start their models can be found in analyzes carried out on the Corvis and Pentacam equipment. The methods applied by Cao et al. and Joda et al. are not viable in this topic, because they are more complex and integrate a greater number of variables, including border conditions and image processing.

As a second stage, perform the modeling of the cornea, where it is sought to integrate the own geometry and characteristics that this one presents, in addition, the properties of a material of behavior similar to the corneal tissue to simulate its tension and deformation, as can be: Blatz-Ko, Mooney-Rivlin, Hart-Smith, Neo-Hookean, materials that have hyperelastic characteristics, resembling the flexible nature of the cornea, additionally, the incorporation of the Young's Module, the Poisson's Ratio and density, which are necessary biomechanical properties to define the model.

Finally, performs the simulation applying a pressure similar to that exerted by the air jet of the tonometer, to obtain maps of the pressure distribution and the tension exerted on the cornea surface, this way the zones of fragility can be analyzed in the corneal tissue.

5 Conclusions

According to the analysis of modeling methods for the determination of the modulus of the elasticity of the cornea, it is concluded that the elasticity of the cornea is an important biomechanical parameter that allows to analyze in more detail the structure of the corneal tissue. Since, through the cornea's elasticity and the study of the pressure exerted on this, facilitates the compression of the structural behavior of the collagen fibers in healthy tissue or pathological, which facilitates the diagnosis of corneal tissue resistance in difficult cases and avoids surgical intervention in cases with high risk of developing postoperative ecstasies.

The different methods seeked efficiently to determine Young's module, through mathematical models or treatment of images of the cornea, looking to include factors not contained in other models, among them, the geometric shape of the cornea and density, in addition, not to include factors which are variable and make precise determination difficult.

It has been considered essential for the calculation of each of the models, the value of the IOP, however, other parameters that are used are variants, in some cases, the authors use the CCT or thickness of the cornea, in other cases, it includes a greater number of variables, among them, applanation speed, the corneal hysteresis, the DA, among others, which leads to a model of greater complexity.

Results of the exposed investigations indicate that the Young's cornea module has an average value of 0.20–0.25 MPa, if further research were to be conducted where a different method is applied to look for this Young's module, and this one reaches an approximate value to it, can be said that the method has been successful in its objective. Although it is indicated that it may vary due to age the higher the age of the patient, the lower the elasticity module that presents itself, which has an average value of 0.187 MPa in patients older than 64 years.

References

1. Calabuig Goena, M.: Estudio de la variación del grosor de la cornea por capas en la facoemulsificación de la catarata mediante tomografía de coherencia óptica del dominio espectral, Universidad de Valladolid (2015)
2. Del Buey Sayas Maria de los Angeles y Peris Martines Cristina: Biomecánica y Arquitectura Corneal, Editorial S.I. Elsevier (2014)
3. Del Buey Sayas Maria de los Angeles: Estudio de la biomecánica corneal: relación entre las propiedades biomecánicas corneales, determinadas mediante el Analizador de Respuesta Ocular ORA y la patología ocular., Universidad de Zaragoza, Zaragoza (2013)
4. Gatzioufas, Z., Seitz, B.: Determination of corneal biomechanical properties in vivo: a review. Mater. Sci. Technol. **31**, 188–196 (2015)
5. Hon, Y., Lam, A.K.C.: Corneal deformation measurement using Scheimpflug noncontact tonometry. Optom. Vis. Sci. **90**, E1–E8 (2013)
6. Dorronsoro, C., Pascual, D., Perez-Merino, P., Kling, S., Marcos, S.: Biomed. Opt. Express **3**, 473–487 (2012)
7. Luce, D.A.: Determining in vivo biomechanical properties of the cornea with an ocular response analyzer. J. Cataract Refract. Surg. **31**, 156–162 (2005)
8. Tanter, M., Touboul, D., Gennisson, J.L., Bercoff, J., Fink, M.: High-resolution quantitative imaging of cornea elasticity using supersonic shear imaging. IEEE Trans. Med. Imaging **28**, 1881–1893 (2009)
9. Wang, L., Tian, L., Huang, Y., Huang, Y., Zheng, Y.: Assessment of corneal biomechanical properties with inflation test using optical coherence tomography. Ann. Biomed. Eng. **46**(2), 247–256 (2018)
10. Shih, P.-J., et al.: Estimation of the corneal Young's modulus in Vivo based on a fluid- filled spherical-shell model with scheimpflug imaging. J. Ophthalmol. **2017**, 1–11 (2017)
11. Vellara, H.R., Patel, D.V.: Biomechanical properties of the keratoconic cornea: a review. Clin. Exp. Optom. **98**(1), 31–38 (2015)

12. Matteoli, S., Virga, A., Paladini, I., Mencucci, R., Corvi, A.: Investigation into the elastic properties of ex vivo porcine corneas subjected to inflation test after cross linking treatment. J. Appl. Biomater. Funct. Mater. **14**(2), 163–170 (2016)
13. Cao, H.-J., Huang, C.-J., Shih, P.-J., Wang, I.-J., Yen, J.-Y.: A method of measuring corneal Young's modulus. In: Goh, J., Lim, C.T. (eds.) 7th WACBE World Congress on Bioengineering 2015, vol. 52, pp. 47–50. Springer, Cham (2015)
14. Joda, A.A., Shervin, M.M.S., Kook, D., Elsheikh, A.: Development and validation of a correction equation for Corvis tonometry. Comput. Methods Biomech. Biomed. Eng. **19**(9), 943–953 (2016)
15. Elsheikh, A., Wang, D.: Numerical modelling of corneal biomechanical behaviour. Comput. Methods Biomech. Biomed. Eng. **10**(2), 85–95 (2007)
16. Kohlhaas, M.: Effect of central corneal thickness, corneal curvature, and axial length on applanation tonometry. Arch. Ophthalmol. **124**(4), 471 (2006)
17. Han, Z., et al.: Air puff induced corneal vibrations: theoretical simulations and clinical observations. J. Refract. Surg. **30**(3), 208–213 (2014)
18. Wang, L.-K., Tian, L., Zheng, Y.-P.: Determining in vivo elasticity and viscosity with dynamic Scheimpflug imaging analysis in keratoconic and healthy eyes. J. Biophotonics **9**(5), 454–463 (2016)
19. Kerautret, J., Colin, J., Touboul, D., Roberts, C.: Biomechanical characteristics of the ectatic cornea. J. Cataract. Refract. Surg. **34**(3), 510–513 (2008)

Proposal for a Tool for the Calculation of Toric Intraocular Lens Using Multivariate Regression

Johanna Castillo-Cabrera[1], Nataly Pucha-Ortiz[1], Eduardo Pinos-Velez[2](✉), William Ipanque[3], and Carlos Luis Chacón[4]

[1] GIIATA, Research Group on Artificial Intelligence and Assistive Technologies, Universidad Politécnica Salesiana, Cuenca, Ecuador
{jcastilloca,npucha}@est.ups.edu.ec

[2] GIIATA, Research Group on Artificial Intelligence and Assistive Technologies and GIIB, Research Group on Biomedical Engineering at Universidad Politécnica Salesiana, Cuenca, Ecuador
epinos@ups.edu.ec

[3] PhD of Engineering Computer Science and Control, Universidad de Piura, Piura, Peru
william.ipanaque@udep.pe

[4] Board Member of the EcuadoRian Society of Glaucoma, Clinical Santa Lucía, Quito, Ecuador
clchacon@clinicasantalucia.com.ec

Abstract. A common cause of loss of vision for humans is lens opacity, or cat aracts, which prevent patients from having a full field of vision and lower their visual acuities. Cataracts can be corrected for by replacing the opaque lens with a toric intraocular lens. The calculation for the placement of the toric lens is exceedingly important for the patient's recovery of vision. Some houses, take date form medical equipment and perform the calculation, but more depends on the experience of the treating physician to ensure the desired effectiveness and efficiency, For this reason, in this investigation, the proposal is presented to perform the calculation, in addition, the approach to simulation of the placement of the lens inside the eye, especially in its axial length and depth in the anterior chamber, determinants for short, medium and long distance with an automatic focus and appropriate, in like manner, the results are exposed to the specialists of the clinic "Santa Lucia" in the city of Quito.

Keywords: Toric intraocular lens · Keratometry · Axial length · Anterior chamber · Biometric power

1 Introduction

The development of cataract surgery, especially the implantation of toric intra- ocular lenses, was a major breakthrough in the field of ophthalmology. Cataracts are the leading cause of blindness in the world, responsible for a 48% of cases. Most of these cases are related to age, and there is no way to prevent them from developing. The

M. Botto-Tobar et al. (Eds.): ICAETT 2019, AISC 1066, pp. 229–236, 2020.
https://doi.org/10.1007/978-3-030-32022-5_22

prevalence and importance of cataracts as a problem of public health are significant [1]. In the absence of a pharmacological treatment for cataracts, the standard treatment is the surgical removal of the opacified lens and the insertion of a toric intraocular lens, which presents a double challenge at the time of the surgery.

The first step, is to remove and remove the cloudy lens through techniques such as laser keratorefractive (LASIK) and keratorefractive photo refractive surgery (PRK). These dominate the ametropias rates and moderate. Ametropias and patients with corneas of thick insufficient for the treatment with laser, the technique of implantation of toric intraocular lenses [2, 3]. The second step, is to place a toric intraocular lenses that leave the patient as close to emmetropia as possible.

For the correction of astigmatism during cataract surgery, are used incisions relaxing limbic and toric intraocular lenses, the success of surgery with toric intraocular lens is measured by reference to the ability to reduce the patient´s astigmatism and maintain the lens´s stability in the capsular bag, taking as a priority align the cylinder shaft of the intraocular lens-ring with the shaft curve more and more of the cornea, in order to ensure the maximum correction [2] For this reason, it is important the measurements of the ocular structures such as: axial length, the depth of the anterior chamber, keratometry and biometric formulas are very important for the accuracy of the refractive results [4].

Now while you have the main keratometric and biometric parameters and for the calculation of toric intraocular lens and prediction of effective position of the toric intraocular lens, the most used formula is the 4th generation (Gerry Rafferty Suite, Holladay II, Olsen) which considers the axial length and the exact calculation of an anterior chamber depth in proportion to the same next to keratometry [5].

The calculation is possible made through statistical studies, such as multiple regression using the variables employed being the axial length, the anterior chamber depth and keratometry intended for the calculation of power by means of numerical methods [6].

The toric lens are intraocular lenses that allow you to correct small and large astigmatisms. Historically, intraocular lenses did not correct astigmatism and at times it was necessary to perform a second surgical procedure for astigmatism. This is a major surgical complexity, since the lens has to be aligned to a particular axis, specifying exactly its final position, which is achieved by marks on the lens that helps its orientation. The manufacture of these lenses is complex and often work with custom orders these may be multifocal and monofocal, or toric trifocal. The corneal surface should be symmetrical and regular in their curvatures; astigmatism occurs and prevents a clear focus of the objects [7].

2 Methods and Proposal

Currently, the surgical implantation of toric intraocular lenses is in evolution, to improve the quality of vision of people with corneal irregularity at the same time as astigmatism. Still, this antecedent a problem under study which originates in the different methods that the specialists applied at the time of cataract surgeries to determine the power of toric intraocular lenses was not always the same, often resulting in errors in calculation.

A study was conducted by comparing data obtained by the IOL Master device, using the formula of 4th generation of HAIGIS SUITE and the numerical method, multiple regression based on a sampling of data consisting of un-operated patients of cataract in the period in the year 2018 of the Clinic "Santa Lucia" located in the city of Quito. The information was obtained through the revision of history, specifically the results of the review through the IOL Master device of the patients to diagnostic astigmatism, all the information was entered into a database using the statistical program Microsoft Excel as variables for the calculation is used the axial length, depth of the anterior chamber, half keratometric.

Patients admitted to our data base will be randomly allocated. We have a group of 62 patients, with a sample of 119 eyes, 62 Rights and 57 lefts, whose keratometric and biometric calculations were carried out by the IOL Master device, To calculate the power of the toric intraocular lenses made the mathematical operation taking into account the values and parameters mentioned above in Fig. 1 shows the proposal for calculation of LIO-toric using multivariate regression.

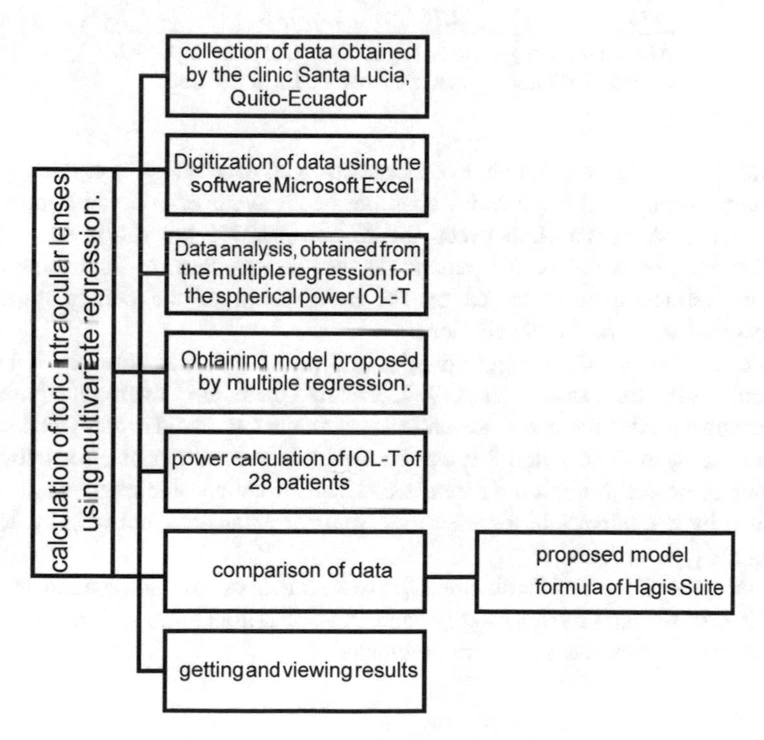

Fig. 1. Proposed graphical calculation of toric intraocular lenses using logistic regression multivariable.

3 Results

For the calculation of the power of toric intraocular lenses, the study sample was formed of 28 patients diagnosed with cataract and astigmatism. Of these 15 men (53.57%) and 13 women (46.43%) that have not been surgically, according to laterality included 27 eyes right and 26 left and power was calculated on the basis of the parameters set out. Comparing the values, obtained through the formula of Haigis Suite with the IOL-Master and the calculated using the mathematical model is given by the regression, the values calculated are approximate, which does not influence the selection of the type of lens.

Table 1. Population data

AL(mm)	RE	LE	Average Km(D)	Average ACD(mm)
<22	1	1	45,79	2,96
22–25	24	22	43,69	3,24
>25	2	3	42,76	3,48

AL: Axial Length. RE: Right Eye. LE: Left Eye. KM: Keratometric average. ACD: Anterior Chamber Depth.

In Table 1 are the axial length, eye diagnosed with cataract and astigmatism, mean keratometry average and the anterior chamber depth average, you can appreciate the 86.79% of normal eyes (AL between 22–25 mm), 3.77% are short (AL < 22) and 0944% are long eyes (>25). All patients should be operated on for cataract and to correct the refractive error caused by the deformation of the cornea through the introduction of the toric intraocular lens.

The calculation of the correct spherical power of the toric intraocular lens is a fundamental step for cataract surgery. Currently, there are several methods under certain parameters that require an accurate measurement, to avoid errors when selecting the type of lens, in Figs. 2 and 3 it can be seen that the margin of error between the calculation of power through the formula of Haigis Suite and the mathematical model determined by a multivariate regression is minimal what does not affect a lot when selecting the type of lens.

The analysis of the results obtained for two models of toric intraocular lenses, the error is 0.0086 for the Physiol Ankorys and Physiol Trifocal/fine vision is 0.0097, for the values with which the tests were conducted.

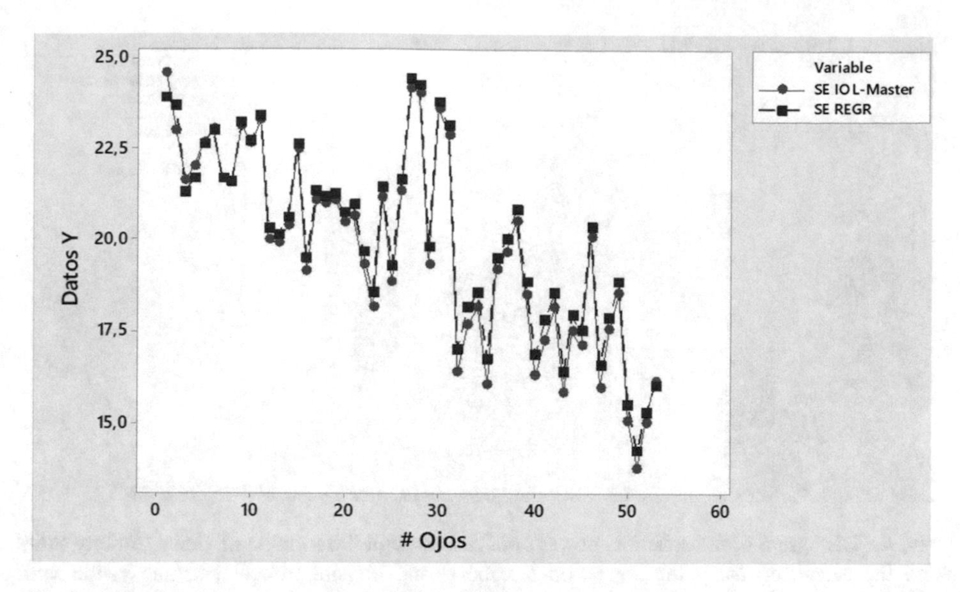

Fig. 2. Spherical power graph obtained through the formula of Haigis Suite and the calculated using the formulation for the toric lens Physiol Ankorys

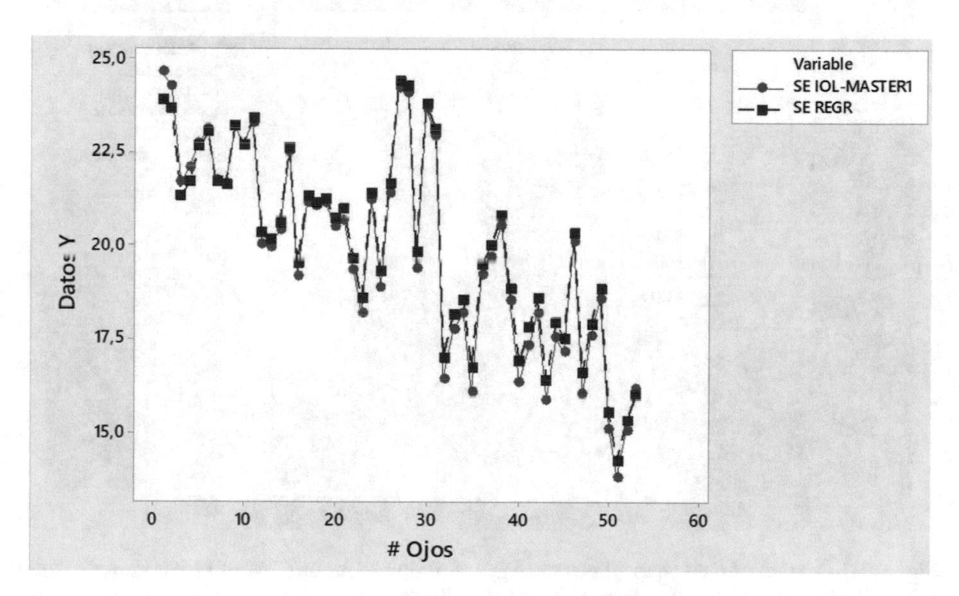

Fig. 3. Spherical power graph obtained through the formula Haigis Suite and the calculated using the formulation for the toric lens Physiol Trifocal/fine vision

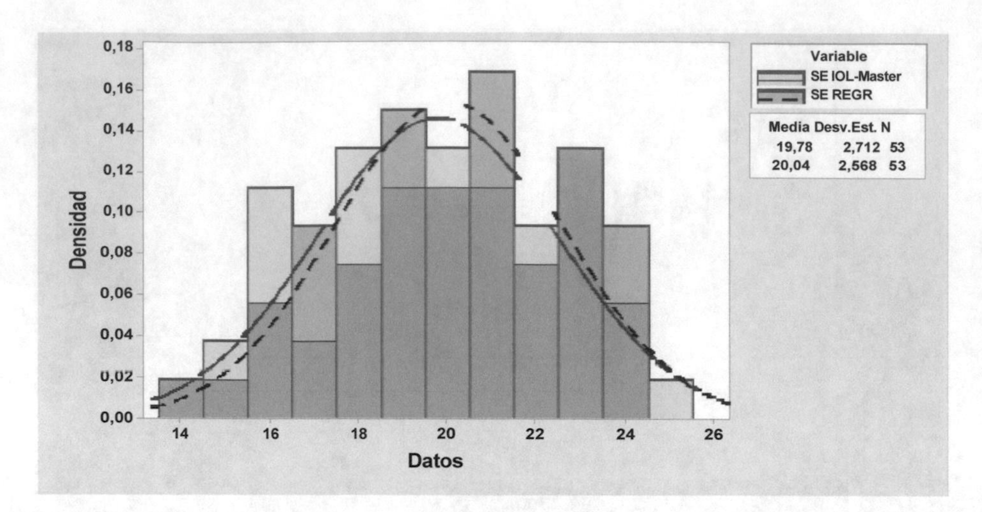

Fig. 4. Histogram of the spherical power obtained through the formula of Gerry Rafferty Suite and the calculated using the regression for the o-ring Physiol trifocal lens/fine vision with adjusted distribution.

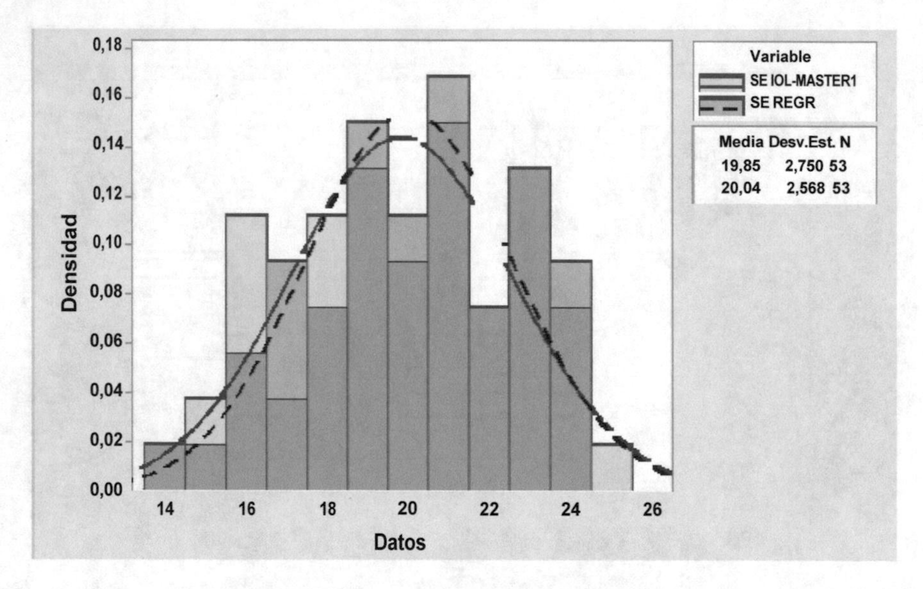

Fig. 5. Histogram of the spherical power obtained through the formula of Gerry Rafferty Suite and the calculated using the regression for the Toric Physiol trifocal lens/fine vision with distribution set

In Figs. 4 and 5 shows the distribution of the spherical power calculated by the formula of Haigis Suite and the obtained with the mathematical model analyzed on the basis of the parameters set out, in addition shows us the distribution set for each group (Table 2).

Table 2. Model summary

S	R-square.	R-Square (Tight)	R-Square (pred)
0,586271	96,59%	96,50%	96,28%

4 Discussion

The surgical intervention of implantation of intraocular lenses in cataract surgery is equally important the improvement and development of calculation techniques, so we have the power of the IOL-toric which will be implanted in the procedure of surgery, it is fundamental the correct calculation of power for which we have biometric data and keratometric obtained by the device IOL-master.

Cataract surgery with toric IOL implants is a common procedure in people with astigmatism. The objective of implanting toric IOLs is to reduce the degree of dependence on glasses, in addition to recovering visual acuity by accurately calculating the power of the toric IOL with low induced astigmatism, made it necessary to use efficient methods of calculating the spherical power for this type of lens, and thus obtain better refractive results post-surgery.

According to studies conducted by the Buenos Aires Vision Institute in 2014, confirmed that the implantation of toric IOLs in patients with cataracts is effective for various levels of astigmatism [7], it is important to know the different methods and calculation formulas to achieve the goal of implanting this type of lens in cataract surgery. One of the most used devices for toric IOLs is the IOL-Master 700 because it is one of the most precise devices that besides being a biometer contains a toric calculator that uses software with several formulas for the calculation, one of them is the Haigis Suite.

he evaluation of the mathematical model for the calculation of the spherical power, from the axial length, the keratometric mean and depth of the anterior chamber, for which it is determined that the so well adjusted model is in 96.59%.it is essential to know and have the knowledge to interpret the methods, formulas, biometric data and keratometrics to be able to understand the calculation carried out and to achieve the objective raised having this to provide a better visual quality of the patient.

One of the limitations of the study performed is the number of patients (28 patients), which would have 56 eyes however, there are patients who have one of the two eyes with problems of cataracts and astigmatism so you have 53 eyes however the results obtained coincide with published reports on this subject, which show that the correct calculation of power of intraocular toric lenses is a fundamental parameter in cataract surgery, so it is argued that the implantation of IOL-T are a good and safe option for patients.

Well, in addition to complying with the proposed objective target to design an application that enables to focus not only on the power of the lens, but also in the position and location of this, based on certain parameters obtained from the ocular biometrics, with the aim of reducing refractive errors to obtain precise surgical results when implementing the toric intraocular lens in patients with severe astigmatism.

Considering, The need on the part of the Doctors surgeons to use software to determine the type of lens and its location, it is estimated to provide an application such as calculation tool providing a soluble and didactic interface for specialist, in addition to incorporating the calculation of power spherical, taking into account that in the existing software power is a parameter that must be entered for the calculation of the lenses and many of these values are given by doctors based on their criteria.

References

1. Mura, C.J.I.: Cirugía actual de la catarata. Revista Médica Clínica Las Condes **21**(6), 912–919 (2010)
2. Tamez-Pena, A., Nava-García, J.A., Zaldívar-Orta, E.L., Lozano-Ramírez, J.F., Cadena-Garza, C.L., Hernández-Camarena, J.C., Valdez-García, J.E.: Efecto clínico de la rotación post- operatoria de los lentes intraoculares tóricos. Revista Mexicana de Oftalmología **89**(4), 219–224 (2015)
3. Moreno, N.R., Srur, A.M., Nieme, B.C.: Cirugía refractiva: indicaciones, técnicas y resultados. Revista Médica Clínica Las Condes **21**(6), 901–910 (2010)
4. Hernández, L.T.: Estudio de Técnicas Biométricas y cálculo de la lente intraocular, 6–7 (2010)
5. Gomez Lara, F.: Comparación de fórmulas biométricas en el cálculo de lentes intra- oculares mediante el uso biometría óptica (121), 20–21 (2013)
6. Prado-Serrano, A., Nava-Hernández, N.G.: Calculo del poder dióptrico de lentes intraoculares ¿Cómo evitar la sorpresa refractiva? Revista Mexicana de Oftalmologia **83**(5), 272–280 (2009)
7. Carracedo, M., Mengual, M., Córdoba Delgado, S.: Lentes multifocales: una buena opción en la cirugía de catarata. Revista Cubana de Oftalmología **30**(3), 1–14 (2007)
8. Haigis, W.: Corneal power after refractive surgery for myopia: contact lens method. J. Cataract Refract. Surg. **29**, 1397–1411 (2003)

An Algorithm Oriented to the Classification of Quinoa Grains by Color from Digital Images

Moisés Quispe[1], José Arroyo[1], Guillermo Kemper[1(✉)], and Jonell Soto[2]

[1] Universidad Peruana de Ciencias Aplicadas, Av. Prolongación Primavera 2390, Santiago de Surco, Lima, Peru
{u911675,u201020649,pcelgkem}@upc.edu.pe
[2] Instituto Nacional de Innovación Agraria, Av. La Molina 1981, La Molina, Peru
jonellsjeri@gmail.com

Abstract. The present work proposes an image processing algorithm oriented to identify the coloration of the quinoa grains that make up the different samples obtained from the production of a crop field. The objective is to perform quality control of production based on the statistics of grain coloration, which is currently done manually based on subjective visual perception. This generates results that totally depend on the abilities and the particular criteria of each observer, generating considerable errors in the identification of the colors and tonalities. The problem is further complicated by the nonexistence, at present, of a pattern or standard of coloration of quinoa grains that specifically defines a referential color map. In this sense, through this work, an algorithm is proposed oriented to classify the grains of the acquired samples by their color via digital images and provide corresponding statistics for the quality control of the production. The algorithm uses the color models RGB, HSV and YCbCr, thresholding, segmentation by binary masks, erosion, connectivity, labeling and sequential classification based on 8 colors established by agronomists. The obtained results showed a performance of the proposed algorithm of 91.25% in relation to the average success rate.

Keywords: Image processing · Quinoa · Color classification

1 Introduction

In recent years, the agricultural sector faces a new challenge to satisfy the constant national and foreign demand for quinoa (Chenopodium quinoa Willd), belonging to the subfamily Chenopodioides de amaranthaceae. This grain has a great nutritional value and an exceptional balance between proteins, fats and carbohydrates. Also, it is a rich source of amino acids, among which are lysine, arginine and histidine, essential for human development. However, one of the difficulties presented by the production of these grains is the little information that is available about their agro-botanical, phenological characterization and about the reaction to biotic and abiotic factors. Quinoa

M. Botto-Tobar et al. (Eds.): ICAETT 2019, AISC 1066, pp. 237–247, 2020.
https://doi.org/10.1007/978-3-030-32022-5_23

has become a flagship product of global consumption and export thanks to its nutritional value and agricultural versatility.

In cultivation, quinoa is normally exposed to genetic changes due to cross pollination. The resulting varieties are a mixture of different types of quinoa, so there is no genetic purity proper in its use within Peruvian trade. [1].

One of the main distinguishing characteristics of the types of quinoa seed is the color of these, many of them released by the INIA (Instituto Nacional de Innovación Agrícola).

The classification by color is given in a generic way, that is, it focuses only on the identification of colors by type of quinoa but not tones.

The evaluation consists of a quick review by visual inspection by producers or specialists, who analyze samples of approximately 200 g.

The process itself presents subjectivity, since it depends on the visual perception and the particular experience of each evaluator. Likewise, the influence of external factors alters the perceived tonalities and generates high variability in the results.

As a consequence of all this, erroneous classifications are generated and there is a high percentage of error in the evaluation. In addition, there is currently no standard that defines the tonalities and reference colorations for the classification of the different types of quinoa. In the literature, some proposals have been presented to solve this problem.

Montalvo proposes in [2] a study on color descriptors for the identification and classification of different types of quinoa seeds, evaluating the algorithms of image processing by computational vision. The technique for obtaining images without distortion and shadow was that of sampling under controlled image capturing conditions. The images were selected by seed types, which were validated as good or bad and were taken as positive or negative. Filters were applied to correct and quantify the color of the images in order to highlight the differences in the characteristics of each type of quinoa seed such as MedianBlur, GuisarBlur, Blur, and contrast, erosion and dilation techniques. The descriptors studied were Haar and SVM, the first one, allowed to classify by means of the histograms and, the second one, by a pixel to pixel analysis. However, this process validated the state of the seed, not its differentiated type by tonalities.

On the other hand, Montalvo et al. propose in [3], an algorithm of automatic classification of quinoa seed types through color descriptors. The study is based on being able to select the seeds properly for cultivation without mixing them with other varieties of quinoa. In this case three types of quinoa seed are considered: the Sacaca or yellow, the Pasankalla or red and the Salcedo through the use of color descriptors. The process involves the characterization, segmentation (using a Gaussian blur filter and Otsu method) and extraction of characteristics. The results show average accuracy of 66% in the classification of quinoa seed types.

In [3, 4] and [5] quality and color detection methods based on neural networks and SVM (Support Vector Machines) are proposed. The percentages of correct detection in these works are in the range of 68% to 94%.

The algorithm proposed in this work is focused mainly on the detection of quinoa color grains and the obtaining of a statistic that allows the determination of the quality of the product based on the color uniformity. This means that the quality of the production is greater when all the grains present the same color.

According to the "Catalog of Commercial Varieties of Quinoa in Peru" developed by INIA and FAO (The Food and Agriculture Organization of The United Nations) there are 13 commercial varieties of quinoa.

The last variety released in 2013 considers 4 colors: cream-white, orange-yellow, dark and black-wine. Subsequently a translucent grain color was added.

With these 5 colors several tests were carried out obtaining 26 colors which finally were grouped and classified in 8 final colors: "chullpi" (translucent), orange, black, red, dark orange, cream or white.

The proposed algorithm was therefore developed with the objective of detecting and classifying the quinoa grains considering these 8 colors.

Subsequently, the algorithm must be implemented on a small single-board computer to operate on a portable equipment. Therefore, to achieve an adequate estimate of the production quality, a percentage of correct color detection above 90% with low computational complexity is required.

The results finally showed a percentage of correct color detection of 91.25% using the proposed algorithm.

In the following sections the details of the proposed algorithm are described.

2 Proposed Algorithm

The proposed algorithm is made up of 2 processing stages: segmentation stage of grains and stage of classification by color. The first stage generates a binary mask that indicates the pixels that make up the quinoa grains (of any color). The second stage analyzes each segmented grain from the first stage and classifies it according to the color that prevails in the pixels that comprise it. Both stages are described in the following sections.

2.1 Grains Segmentation

For the segmentation of the grains in the image, different aspects were analyzed, so that the process can segment each object of interest without cutting it, excessively dilating it or introducing undesired artifacts. In this sense, given the different colors of quinoa grains that are presented, several colors of platforms were analyzed to position the grains and verify contrasts and the best segmentation.

After the experimental tests, it was the white platform that gave the best results.

Another important aspect that was analyzed was the type of lighting for segmentation effects. In this case the backlighting (lighting below the white platform that holds the grains) was what allowed obtaining the best results in the "chullpi" type grain and direct lighting in the other cases.

Finally, the most suitable spatial resolution for the acquisition of the images was analyzed. In that sense it was aimed at concentrating a minimum amount of pixels in each grain, in order to have reliable and sufficient information to classify it by color. In this case the camera was configured in a resolution of 12 megapixels and color model RGB 24 bits (true color - 8bits per primary component).

For the choice of the indicated spatial resolution, the focal length of the sensor of the camera and the corresponding distance to the objective (4 cm) were also taken into account.

With this configuration it was possible to concentrate on average 2000 pixels for each grain of quinoa.

It is important to indicate that the different types of illumination and configurations for images acquisition were deeply studied. This was done in order to obtain images that helped the low-complexity algorithm to reach a low error rate, and for it to be implemented without problems on a small single-board computer.

The Fig. 1 shows the block diagram of the segmentation algorithm. In this case we define the input image with backlight and RGB color format with primary components $I_{RR}(x, y), I_{RG}(x, y)$ e $I_{RB}(x, y)$ respectively (see Fig. 2).

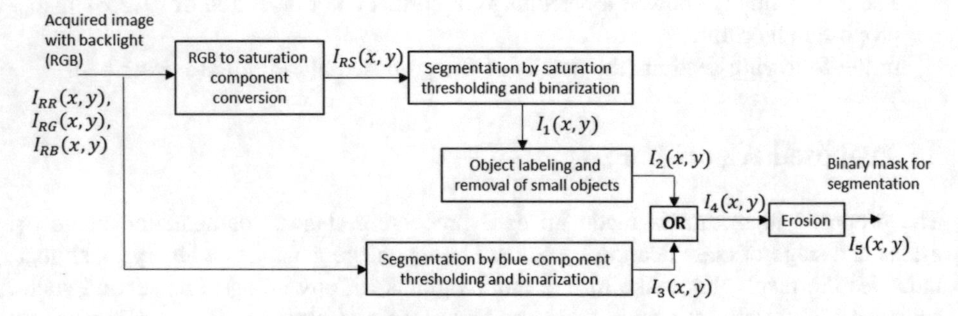

Fig. 1. Block diagram of the segmentation algorithm

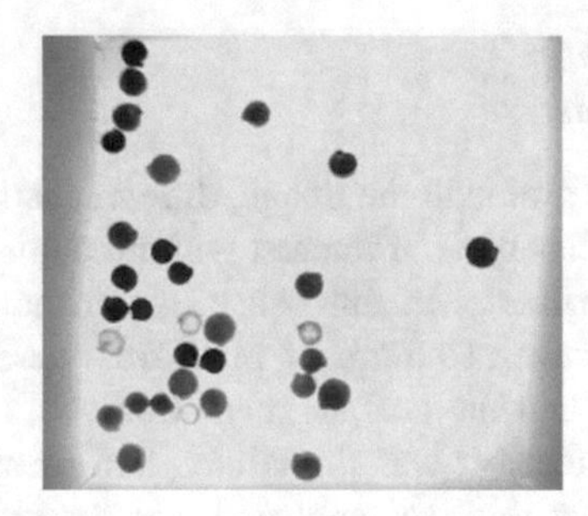

Fig. 2. Original acquired image with backlight

The algorithm can be described in steps:

Step 1: The saturation component $I_{RS}(x, y)$ is obtained from the primary input components RGB [4]:

$$I_{RS}(x,y) = 1 - \frac{3 \times min(I_{RR}(x,y), I_{RG}(x,y), I_{RB}(x,y))}{I_{RR}(x,y) + I_{RG}(x,y) + I_{RB}(x,y)} \tag{1}$$

Where $d = min(a, b, c)$ is a function that returns in d the minimum value between the values a, b and c.

Note that the saturation component is restricted to the range [0,1], where the value 1 indicates maximum saturation of the pixel and therefore the highest purity in color [4].

In this case, saturation was chosen because it is the only component able to detect without problems the pixels that make up the quinoa grains of the "Chullpi" type (semi-transparent tonality). However, for the other colors, the saturation did not yield the best results, so it was necessary to apply the procedure in step 4.

Step 2: The saturation values are subjected to thresholding in order to obtain a binary mask to indicate the presence of "Chullpi" type grains. This process can be expressed as:

$$I_1(x,y) = \begin{cases} 0 \ , & I_{RS}(x,y) < 0.12 \\ 1 \ , & other\ case \end{cases} \tag{2}$$

The threshold with value 0.12 yielded the best segmentation results based on the experimental tests.

Step 3: The binary image $I_1(x, y)$ it is then subjected to a process of 8-connectivity and tagging of objects [7] (pixels with value "1"), in order to segment each object and determine its area in number of pixels. This process allows the removal of objects (pixels taken to zero value) whose area is less than a threshold of 1000 pixels (based on the experimental tests carried out), since they are considered as noise and unwanted information. The binary image resulting from this process is $I_2(x, y)$.

Step 4: The values of the blue component $I_{RB}(x, y)$ of the acquired image are subjected to thresholding in order to obtain a complementary binary image that also constitutes a color presence mask. This is done in order to properly segment the other colors of grains that were not correctly segmented by saturation. This process can be expressed as:

$$I_3(x,y) = \begin{cases} 1 \ , & I_{RB}(x,y) < 100 \\ 0 \ , & other\ case \end{cases} \tag{3}$$

The threshold value of 100 provided the best segmentation results based on the experimental tests.

Step 5: The OR operation is performed between the two binary masks obtained to complete the final mask. However, an erosion procedure is applied to the resulting image in order to separate grains that were joined by the thresholding effect. As a structuring element for erosion, a disc with 2 radio pixels was used [8]. The process can be expressed with matrix operations in the following way:

$$\mathbf{I}_4 = \mathbf{I}_2 \ \mathbf{OR} \ \mathbf{I}_3 \tag{4}$$

$$\mathbf{I}_5 = \mathbf{I}_4 \ominus \mathbf{D}_1 \tag{5}$$

Where $\ominus$ expresses the erosion operation and $\mathbf{D}_1$ the disk type structuring element.

Finally the result of the segmentation algorithm is the binary mask $I_5(x, y)$ (represented matrix as $\mathbf{I}_5$).

2.2 Classification by Color

Figure 3 shows the block diagram of the classification algorithm of quinoa grains by color. Note that in this case the images acquired with direct lighting (see Fig. 4) and feedback are also used, in order to have more accurate information regarding the color of the grains.

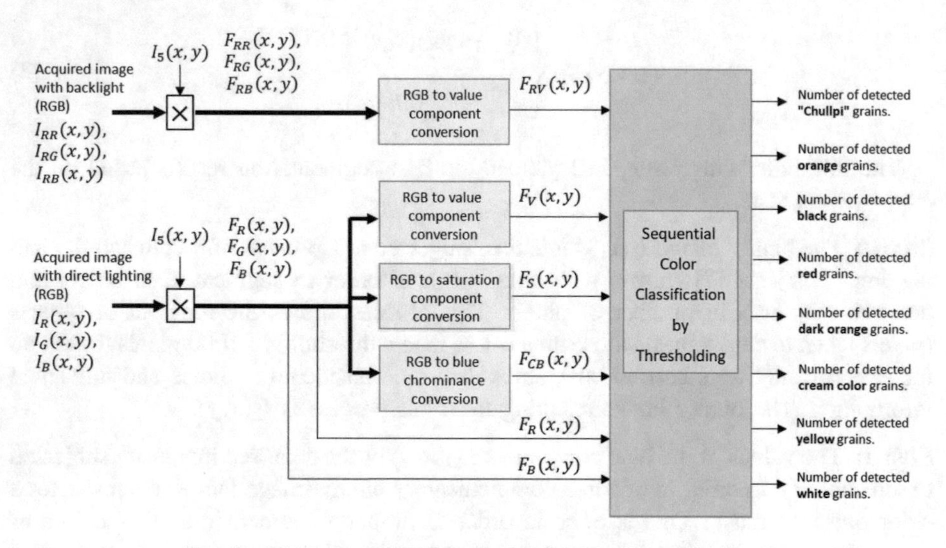

Fig. 3. Block diagram of the color classification algorithm

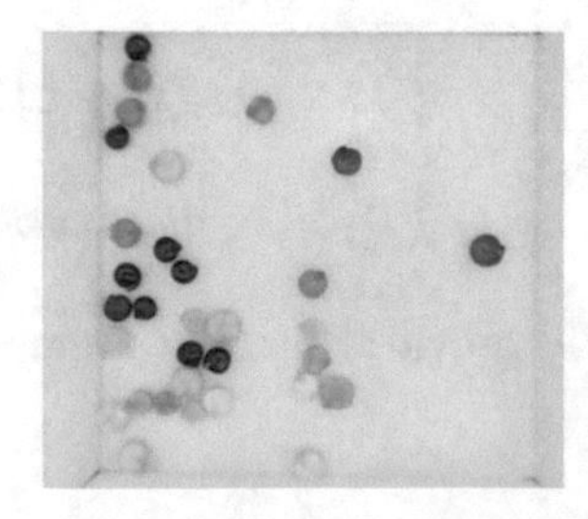

Fig. 4. Original acquired image with direct lighting

As in the case of the segmentation algorithm, we will describe the steps of the classification algorithm.

Step 1: Segment the pixels of the grains by multiplying the mask $I_5(x,y)$ with each of the RGB primary components of the original acquired images (with direct lighting and with feedback). The procedure can be expressed as:

$$F_R(x,y) = I_R(x,y).I_5(x,y) \tag{6}$$

$$F_G(x,y) = I_G(x,y).I_5(x,y) \tag{7}$$

$$F_B(x,y) = I_B(x,y).I_5(x,y) \tag{8}$$

$$F_{RR}(x,y) = I_{RR}(x,y).I_5(x,y) \tag{9}$$

$$F_{RG}(x,y) = I_{RG}(x,y).I_5(x,y) \tag{10}$$

$$F_{RB}(x,y) = I_{RB}(x,y).I_5(x,y) \tag{11}$$

Where $I_R(x,y), I_G(x,y)$ e $I_B(x,y)$ constitute the primary components of the original image acquired with direct illumination. Figure 5 shows the images obtained.

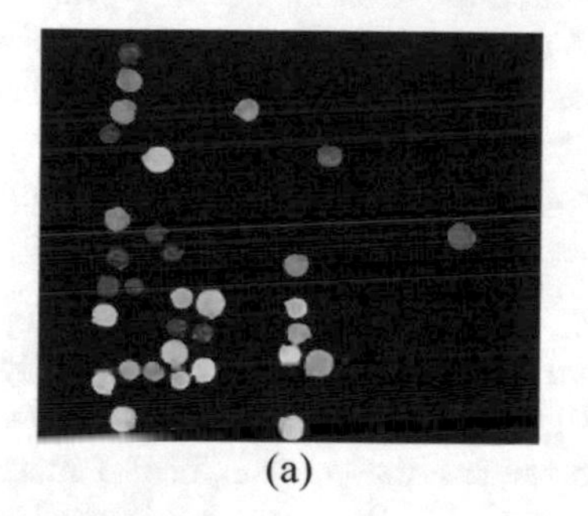
(a)

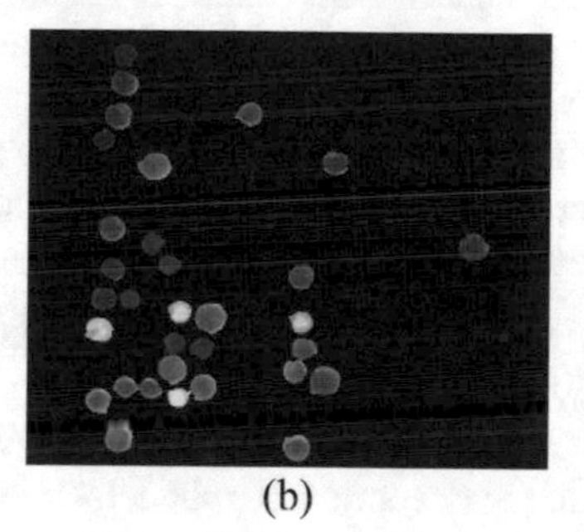
(b)

Fig. 5. Resulting images of segmentation: (a) Direct lighting, (b) Backlight

Step 2: From the segmented RGB images, the images corresponding to the value components are obtained $F_{RV}(x,y)$ y $F_V(x,y)$ and saturation $F_S(x,y)$ of the HSV color model (Hue, Saturation and Value) [7]:

$$F_{RV}(x,y) = \frac{max(F_{RR}(x,y), F_{RG}(x,y), F_{RB}(x,y))}{255} \tag{12}$$

$$F_V(x,y) = \frac{max(F_R(x,y), F_G(x,y), F_B(x,y))}{255} \tag{13}$$

$$F_S(x,y) = 1 - \frac{3 \times min(F_R(x,y), F_G(x,y), F_B(x,y))}{F_R(x,y) + F_G(x,y) + F_B(x,y)} \tag{14}$$

Where $d = max(a, b, c)$ is a function that returns in d the maximum value between the values a, b and c.

The values of the resulting color components $F_{RV}(x, y)$, $F_V(x, y)$ y $F_S(x, y)$ are normalized to the interval [0,1].

Step 3: From the segmented RGB image with direct illumination, the blue chromium is obtained $F_{CB}(x, y)$ of the color model YCbCr [9]:

$$F_{CB}(x, y) = 128 - \frac{37.945F_R(x, y)}{256} - \frac{74.494F_G(x, y)}{256} + \frac{112.439F_B(x, y)}{256} \quad (15)$$

In this cases $F_{CB}(x, y)$ it is restricted to the entire range [16, 240].

All the obtained color components will be used together with the primary components in the process of sequential classification by thresholding that is described in the next steps.

Step 4: The mask $I_6(x, y)$ is obtained by assigning the value of "1" to the pixels where $F_{RV}(x, y) \in [0.78, 1]$ and "0" the rest of pixels. The objects in this mask which are less than 1000 pixels of area are eliminated and brought to zero since they are not considered grains. This last procedure is applied to all the following steps, after the generation of the corresponding mask.

The objects that remain in the mask are counted as "Chullpi" quinoa grains (one grain per object). However, in certain cases if the area of an object (in number of pixels) exceeds the value of 1715, then a number of grains equal to the whole quotient resulting from the division of the object area between the value is counted by that object. 1715 (this last procedure is applied for all the following steps, with different values of thresholds).

Step 5: All the resulting objects in the mask $I_6(x, y)$ from the previous step are removed from the component $F_G(x, y)$. Then we get a binary mask $I_7(x, y)$ assigning the value of "1" to the pixels of $F_G(x, y) \in [113, 160]$ and "0" the rest of pixels. The filtering procedure is applied by area and the objects that remain in the mask are counted as orange quinoa grains (one grain per object). Then the procedure of accounting for grains in the same object is applied, using in this case a threshold of 2116.

Step 6: All the resulting objects in the masks $I_6(x, y)$ e $I_7(x, y)$ of the previous steps are removed from the component $F_S(x, y)$. Then you get a binary mask $I_8(x, y)$ by assigning the value of "1" to the pixels of $F_S(x, y) \in [0.01, 0.2]$ and "0" the rest of pixels. The filtering procedure is applied by area and the objects that remain in the mask are counted as black quinoa grains (one grain per object). Then the procedure for counting grains in the same object is applied, using in this case a threshold of 1677.

Step 7: All the resulting objects in the masks $I_6(x, y)$, $I_7(x, y)$ e $I_8(x, y)$ of the previous steps are removed from the component $F_R(x, y)$. Then you get a binary mask $I_9(x, y)$ by assigning the value of "1" to the pixels of $F_R(x, y) \in [180, 255]$ and "0" the rest of pixels. The filtering procedure is applied by area and the objects that remain in the mask are counted as red quinoa grains (one grain per object). Then the procedure of counting grains in the same object is applied, using in this case a threshold of 1460.

Step 8: All the resulting objects in the masks $I_6(x, y)$, $I_7(x, y)$, $I_8(x, y)$, e $I_9(x, y)$ of the previous steps are removed from the component $F_V(x, y)$. Then you get a binary mask $I_{10}(x, y)$ by assigning the value of "1" to the pixels of $F_V(x, y) \in [0.5, 0.8]$ and "0" the rest of pixels. The filtering procedure is applied by area and the objects that remain in the mask are counted as dark orange quinoa grains (one grain per object). Then the procedure of counting grain in the same object is applied, using in this case a threshold of 2286.

Step 9: All the resulting objects in the masks $I_6(x, y)$, $I_7(x, y)$, $I_8(x, y)$, $I_9(x, y)$ e $I_{10}(x, y)$ of the previous steps are removed from the component $F_{CB}(x, y)$. Then we get a binary mask $I_{11}(x, y)$ by assigning the value of "1" to the pixels of $F_{CB}(x, y) \in [70, 95]$ and "0" the rest of pixels. The filtering procedure is applied by area and the objects that remain in the mask are counted as cream color quinoa grains (one grain per object). Then the procedure for counting grains in the same object is applied, in this case using a threshold of 2700.

Step 10: All the resulting objects in the masks $I_6(x, y)$, $I_7(x, y)$, $I_8(x, y)$, $I_9(x, y)$, $I_{10}(x, y)$ and $I_{11}(x, y)$ of the previous steps are removed from the component $F_{CB}(x, y)$. Then we get a binary mask $I_{12}(x, y)$ by assigning the value of "1" to the pixels of $F_{CB}(x, y) \in [40, 56]$ and "0" the rest of pixels. The filtering procedure is applied by area and the objects that remain in the mask are counted as yellow quinoa grains (one grain per object). Then the procedure of accounting for grains in the same object is applied, using in this case a threshold of 2624.

Step 11: All the resulting objects in the masks $I_6(x, y)$, $I_7(x, y)$, $I_8(x, y)$, $I_9(x, y)$, $I_{10}(x, y)$, $I_{11}(x, y)$ e $I_{12}(x, y)$ of the previous steps are removed from the component $F_B(x, y)$. Then we get a binary mask $I_{13}(x, y)$ by assigning the value of "1" to the pixels of $F_B(x, y) \in [144, 255]$ and "0" the rest of pixels. The filtering procedure is applied by area and the objects that remain in the mask are counted as white quinoa grains (one grain per object). Then, the grain accounting procedure is applied to the same object, using in this case a threshold of 2166.

The sequence, intervals and thresholds defined in these steps were established based on several experimental tests analyzing each color with agronomists and always seeking to maximize the level of success of the algorithm.

The specialists in Andean grains also showed their conformity with the classification based on the 8 colors evaluated, thus establishing, for the first time, a reference of colors for the case of quinoa.

3 Results

For the validation of the proposed algorithm, 25 groups of 35 grains each were selected (Total: 875 grains). The colors of the grains that made up each group were varied and random. Each group was previously classified by visual inspection by 3 observers who used magnifying lenses. The observers had as reference the 8 possible colors that could present the quinoa grains: “chullpi”, orange, black, red, dark orange, cream, yellow and white.

The groups selected as reference for validation were those where the 3 observers showed a coincidence in the determination of the color of each grain.

The percentage of success rate per color was obtained from the following expression:

$$\%SR_{color} = \frac{NS_{color}}{NT_{color}} \times 100 \tag{16}$$

Where NS_{color} constitutes the number of hits in the detection of a certain color and NT_{color} the total number of grains evaluated with that color (this involves all the groups of grains evaluated).

Table 1 shows the obtained results, based on the success rate. In this case, an average success rate percentage of 91.25% was obtained.

Table 1. Percentage of success rate per grain color.

Grain color	$\%SR_{color}$
“Chullpi”	100%
Orange	80%
Black	100%
Red	90%
Dark orange	90%
Cream	95%
Yellow	80%
White	95%

4 Discussion

As can be seen, the results show variations in the percentage of success rate, according to the analyzed color. In that sense, confusions between the color orange and the color yellow, the color red and the dark orange and finally the cream with the color white were verified. However, the results were very satisfactory and sufficient for the agronomists, since they consider that the proposed algorithm will allow, with a high level of reliability, to determine the quality of the production from the uniformity of color of the grains produced in a harvest.

Percentages can be improved in future works by evaluating new illumination methods, an adequate choice and configuration of the sensor and improvement of processing algorithms adapted to the limitations of a small single-board computer.

5 Conclusions

In this research work it was possible to develop a simple algorithm of low complexity that could achieve error rate percentages accepted by agronomists.

It was decided to build an image acquisition scenario that could have different types of stabilized lighting and an adequate sensor configuration to guarantee fixed thresholds for the eight grain colors.

Although classifiers and artificial intelligence algorithms could be applied, we prefer to maintain a low computational load and achieve low error rates with a simple and iterative algorithm.

In fact future improvements can be implemented using artificial intelligence algorithms by evaluating the improved results and the computational load involved.

References

1. Marca, S., Chaucha, W., Quispe, J., Mamani, W.:Comportamiento actual de los agentes de la cadena productiva de quinua en la región puno. Gobierno Regional Puno, Dirección Regional Agraria Puno, Puno, Peru (2011)
2. Montalvo, J.: Procesamiento de imágenes digitales utilizando descriptores de color para la identificación y clasificación de diversos tipos de semilla de quinua. Tesis de Titulación en Ingeniería de Sistemas, Universidad Señor de Sipán, Facultad de Ingeniería, Arquitectura y Urbanismo, Pimentel, Peru (2017)
3. Montalvo, J.C., Tuesta, V.A., Arcila, J.C., Mejía, C.: Clasificación automática de tipos de semilla de quinua a través de descriptores de color. Revista Científica de Ingeniería. Ciencia Tecnología e Innovación **4**(2), 1–6 (2017)
4. Sharma, D., Sawant, S.: Grain quality detection by using image processing for public distribution. In: International Conference on Intelligent Computing and Control Systems (ICICCS), pp. 1118–1122. IEEE (2017)
5. Pabamalie, L.A.I., Premaratne, H.L.: A grain quality classification system. In: International Conference on Information Society, pp. 57–61. IEEE, London (2010)
6. Kiratiratanapruk, K., Sinthupinyo, W.: Color and texture for corn seed classification by machine vision. In: International Symposium on Intelligent Signal Processing and Communications Systems (ISPACS). IEEE, Chiang Mai (2011)
7. Gonzalez, R., Woods, R.: Digital Image Processing, 2nd edn. Prentice Hall, New Jersey (2002)
8. Solomon, C., Breckon, T.: Fundamentals of Digital Image Processing: A Practical Approach with Examples in MATLAB, 1st edn. Wiley-Blackwell, Chichester (2011)
9. International Telecommunication Union: ITU-R RecommendationBT.601–7: Studio encoding parameters of digital television for standard 4:3 and wide screen 16:9 aspect ratios (2011)

Computer Vision for Image Understanding: A Comprehensive Review

Luis-Roberto Jácome-Galarza[1], Miguel-Andrés Realpe-Robalino[1], Luis-Antonio Chamba-Eras[2], Marlon-Santiago Viñán-Ludeña[3](✉), and Javier-Francisco Sinche-Freire[3]

[1] Escuela Superior Politécnica del Litoral, Facultad de Ingeniería en Electricidad y Computación, CIDIS, 09-01-5863, Guayaquil, Ecuador
lrjacome@unl.edu.ec, mrealpe@fiec.espol.edu.ec

[2] Universidad Nacional de Loja, Grupo de Investigación en Tecnologías de la Información y Comunicación (GITIC), Carrera de Ingeniería en Sistemas, Loja, Ecuador
lachamba@unl.edu.ec

[3] Universidad Nacional de Loja, Carrera de Ingeniería en Sistemas, Loja, Ecuador
{marlon.vinan, javier.sinche}@unl.edu.ec

Abstract. Computer Vision has its own Turing test: Can a machine describe the contents of an image or a video in the way a human being would do? In this paper, the progress of Deep Learning for image recognition is analyzed in order to know the answer to this question. In recent years, Deep Learning has increased considerably the precision rate of many tasks related to computer vision. Many datasets of labeled images are now available online, which leads to pre-trained models for many computer vision applications. In this work, we gather information of the latest techniques to perform image understanding and description. As a conclusion we obtained that the combination of Natural Language Processing (using Recurrent Neural Networks and Long Short-Term Memory) plus Image Understanding (using Convolutional Neural Networks) could bring new types of powerful and useful applications in which the computer will be able to answer questions about the content of images and videos. In order to build datasets of labeled images, we need a lot of work and most of the datasets are built using crowd work. These new applications have the potential to increase the human machine interaction to new levels of usability and user's satisfaction.

Keywords: Computer vision · Deep Learning · Image understanding · CNN · Scene recognition · Object classification

1 Introduction

Computer vision is a field of computer science that works on enabling computers to see, identify and process images in the same way that human vision does [1]; that indeed is a very complex task. With the exponential growth of the computing power

M. Botto-Tobar et al. (Eds.): ICAETT 2019, AISC 1066, pp. 248–259, 2020.
https://doi.org/10.1007/978-3-030-32022-5_24

and the increased number of cameras that are installed all over the cities, now it is possible to build automated systems that accomplish with computer vision tasks [2]. However, complex tasks (HIT = Human Intelligence Task [3]) are still under investigation. Among these difficult tasks, we have image understanding in which the computer is able to describe the image in a similar way a human being would do it. Due to its complexity and its potential, it is a worth researching area for new applications. Table 1 presents a list of applications of Image Understanding [4], in which its contribution may change deeply those areas.

Table 1. Image Understanding Applications [4].

Area	Example
Inspection tasks	- Checking the results of casting processes for impurities - Screening of medical images, screening of plant samples
Remote sensing	- Cartography - Monitoring of traffic along roads, docks, and at airfields - Management of land resources such as water, forestry, soil
Making computer power more accessible	- Management information systems that have communication channels wider than systems that are work by typing or pointing - Document readers, design aids for architects, engineers
Military applications	- Tracking moving objects, automatic navigation based on passive sensing, target acquisition and range finding
Aids for the partially sighted	- Systems that read a document and say what was read - Automatic "guide dog" navigation systems

1.1 Theoretical Framework

Event Recognition. An event can be seen as a semantically meaningful human activity, taking place within a selected environment and containing a number of necessary objects. It can also be defined as a descriptive interpretation of the visual world for the blind. In the other hand, for best understanding of images we can use the 5Ws questions: who, where, what, when and how. With event recognition, 3 of the 5 questions can be answered [5]: what? - The event label, where? - The scene environment label, who? - A list of the object categories.

It can be said that event recognition is composed of scene recognition + classification. The SUN dataset [6] (Scene Understanding) is an example of event recognition effort in which all the pictures are organized in hierarchical categories.

Scene Recognition. In scene recognition, algorithms learn global statistics of the scene categories. In order to get better results and depending of the application, it is necessary to distinguish indoor scenes from outdoor scenes [7].

Fine-Grained Recognition. Is the task of distinguishing between visually very similar objects such as the species of a bird, the breed of a dog or the model of an aircraft [8].

Object Category Recognition. Classifying images can be defined as a collection of regions, describing only their appearance and ignoring their spatial structure. For object categorization there are generative and discriminative models. Image similarity metrics are also used for object recognition, for example, distance metrics: Dssd, Dwarp, Dshift [9]. Object classification can be binary or multiclass classification.

BOW (Bag of Words). Bag of visual words is a vector of occurrence counts of a vocabulary of local image features. A codebook represents an image as sequence of appearance words. BOW can be treated as a supervised or unsupervised task [10]; the scene classification is a supervised task and the object discovery is unsupervised.

1.2 Related Works

Many related works refer the use of Convolutional Networks for analyzing visual imagery, because CNN identify parts of objects in its convolution stages. [11]. For using CNN we need large amount of training data. Table 2 enumerates examples of popular specialized Image Datasets that are used for training a Deep Learning model; we have selected them in order to highlight the diverse fields of application.

Table 2. Specialized image databases for machine learning training.

Database	Description	Task	Content
VOC12	Pascal VOC 2012	Object image classification/segmentation (20 classes)	Person, animals, vehicles, indoor objects
MIT67	MIT 67 Indoor Scenes	Scene image classification (67 categories)	Indoor places
VOC11s	PASCAL VOC 2011	Object Category Segmentation/action classification (10 action classes + other)	Person, animals, vehicles, indoor objects
200Birds	UCSD-Caltech 2011-200 Birds dataset	Fine-grained recognition (200 categories)	Bird species
102Flowers	Oxford 102 Flowers	Fine-grained Recognition (102 categories)	Flower categories
H3Datt	H3D poselets human 9 attributes	Attribute Detection (150 categories)	Poselets for person
LFW	Labeled faces in the wild	Metric learning/face recognition	Famous people in different poses
Oxford5k	Oxford 5k buildings dataset	Instance retrieval (11 landmarks)	Oxford landmarks
Paris6k	Paris 6k Buildings [12]	Instance retrieval	Paris landmarks
Sculp6k	Oxford Sculptures Dataset	Instance retrieval (10 objects to detect)	Sculptures by Henry Moore and Auguste Rodin

(*continued*)

Table 2. (*continued*)

Database	Description	Task	Content
Holidays	INRIA holidays scenes dataset	Instance retrieval (500 image groups)	INRIA personal holiday photos
UKB	Uni. of Kentucky Retrieval Benchmark Dataset	Instance retrieval (2550 classes)	Animals, plants, household objects
IAPR TC-12	IAPR TC-12 Benchmark	Image captioning in English, German and Spanish	Sports and actions, people, animals, cities, landscapes
Flickr 30k	30k Flickr images with captions	Image captioning	Images of different types of objects
MSCOCO	Microsoft Common Objects in Context	Object detection, segmentation, image captioning	Person, animals, vehicles, furniture

Other interesting image datasets are [13] for video surveillance, human health monitoring, human pose, etc., in [14] they have 22.210 fully annotated images with objects and many with parts, in [15] there is a database of 360° panoramas, in [16] there is a database of 400,000 spoken captions for natural images (Places 205 dataset).

In the other hand, in [2] they highlight the use of deep and wide networks like VGGNet [17] and GooLeNet [18].

A very interesting work is [19]; they study human behavior-recognition algorithms to understand transit scenes. They present datasets and implementation details. They highlight that there is still a big gap in analytical skills between a typical security guard and state of the art in image processing algorithms.

There are also many image understanding reviews that are applied to medicine, like [20], in which they introduce the state of the art in image understanding for iris biometrics. They present datasets, applications and conditions that may affect the iris.

Moreover, in [21], they present a survey for video tracking. It consists from simple window tracking to complex models that learn shape and dynamics.

After doing this preliminary review we come up with a research question: The use of image understanding may assist workers to get a better performance compared with traditional methods? What are popular tools for working with image understanding? What limitations are there in the field of image understanding?

2 Materials and Methods

In order to make this paper, we worked in two parts: documental and practical. For the documental part, we included mostly the papers that work with Deep Learning because this technology has given the best results in many computer vision tasks [11]. We also gave importance to papers that use Natural Language Processing methods because it is necessary to generate the text that describe the images. Finally, our last criterion for selecting the relevant papers was its topic. We chose papers in the fields of Medicine, urban traffic, outdoor and indoor places, human actions and common object detection. In the other hand, we excluded papers that work in other fields like Agriculture, industry and other specialized areas, because we had to limit the depth of the study, considering that image understanding may be applied to any field of knowledge. In order to get the latest scientific information, we looked for academic databases like IEEE Xplore, Google Scholar, Research Gate, etc.; for all these academic databases we used the search string "image understanding". In the other hand, we found papers from 10 or 15 years ago that best describe the basic theory behind computer vision techniques and papers from recent years which describe the latest techniques.

For the practical part of this study, we used the github (public projects repository) to get the code that allows us to work with image understanding. We also followed the Kaggle's tutorials for deep learning in order to have a better knowledge of the key concepts of this technology. We use a machine with the following features: Alienware with Intel Core i7 processor, CPU 2.80 GHz, 64 bits architecture, 16 GB of RAM memory, NVIDIA GeForce GTX 1070 graphic card. Table 3 describes the experiments that we conducted.

Table 3. Image understanding in different IDE with datasets.

IDE	Library	Dataset	Task
Matlab	Deep Learning toolbox	AlexNet	Object recognition
Matlab	Deep Learning toolbox	VGG16, VGG19	Object recognition
RStudio	Keras	MNIST	Digit recognition
RStudio	Keras	Fashion MNIST	Clothes classification
RStudio	Keras	ImageNet [22]	Object recognition

We also tried platform solutions available online for image understanding. Table 4 describes them.

Table 4. Platform solutions for image understanding.

Solution	Vendor	Task
Watson [23]	IBM	Image captioning
Caption bot [24]	Microsoft	Image captioning
"Dog breed prediction with Keras" Tensorflow kernel [25]	Kaggle	Dog breed recognition

3 Results

The most relevant works for image understanding are presented in the following lines.

In [26], they present a Bayesian hierarchical model to learn and recognize natural scenes with the advantage that the learning model needs minimal human intervention.

In [27], a visual dictionary in which the nouns of the English language are arranged and related by their semantic meaning [28] is introduced.

In [29], they use Deep Learning to build a model for scene recognition. They also have an online demo where an image can be uploaded and the algorithm describes the content of that image [30].

In [31], they present a model which is able to answer questions about the content of an image. The model uses Long Short-Term Memory (LSTM) to extract the question representation, a Convolutional Neural Network (CNN) to extract the visual representation and LSTM to store the linguistic context in an answer; they use a fusing component to combine the information of the other 3 components. They also show the FM-IQA dataset with 150,000 images and 310,000 question-answer pairs. The model is able to answer questions like “Where is the cat?” giving the answer “On the table”

In [32], they propose neural networks and visual semantic embeddings to predict answers to simple questions about images. They also present a question generation algorithm which converts image descriptions into Question-Answer form.

In [33], they introduce a Region Proposal Network (RPN) that shares full-image convolutional features that enable low-cost region proposals. The RPN is a fully-convolutional network that simultaneously predicts object bounds and object scores at each position. They use RPN and the Fast R-CNN algorithm for training the convolutional features.

Scene labeling is a challenging computer vision task which requires the use of local discriminative features and global context information. In [34], they adopt a deep recurrent convolutional neural network (RCNN) for this task. They use the back-propagation through Time algorithm for an easy and simple training.

In [35], they present a multimodal Recurrent Neural Network (m-RNN) for generating sentence descriptions in order to explain the contents of images. The model consists of a deep RNN for sentences and a deep CNN for images. The model is validated on IAPR TC-12, Flickr 8K and Flickr 30K datasets.

In [36], they introduce an encoder-decoder pipeline that learns a multimodal joint embedding space with images and text. The encoder allows ranking images and sentences while the decoder generates descriptions of images. They also use LSTM to encode sentences.

In [37], they present a model that generates natural language descriptions of images and their regions. They use Convolutional Neural Networks for image regions, bidirectional Recurrent Neural Network architecture to generate descriptions of image regions. The experiments are done with Flickr 8k, Flickr 30k and MSCOCO datasets.

In [38], they trained a large, deep convolutional neural network to classify 1.3 million high-resolution images in the Large Scale Visual Recognition Challenge LSVRC-2010 ImageNet dataset into 1,000 different classes. The neural network has 60 million parameters and 500,000 neurons and 5 convolutional layers, some are connected to max-pooling layers, and 3 fully-connected layers with a final 1000-way softmax. They use the dropout method to reduce overfitting.

In [39], they address the task of learning new visual concepts, and their interactions with other concepts, from few images with sentence descriptions. Their method is able to conceive the semantic meaning of the new words and add them into its word dictionary, so the new concepts will be used to describe images.

In [40], they investigate the effect of the convolutional network depth on its accuracy in the large-scale image recognition setting. They evaluate very deep convolutional networks, with up to 19 weight layers. It is claimed that representation depth is beneficial for the classification accuracy.

In [41], they present a generative model based on a deep recurrent architecture that combines recent advances in computer vision and machine translation in order to generate natural sentences that describe an image. This is challenging because a descriptor must capture objects and it has to express how the objects interact with each other; moreover the semantic knowledge has to be put in a language like English.

In [42], they introduce an attention based model that automatically learns to describe the content of an image. They train the model using backpropagation techniques and maximize the variational lower bound (ELBO Evidence Lower Bound). They evaluate the model with Flickr 8k, Flickr 30k and MSCOCO.

In [43], they propose the use of visual denotations of linguistic expressions to define denotational similarity metrics. To compute denotational similarities they construct a denotation graph (a subsumption hierarchy over constituents and their denotations, based on a large corpus of 30K images and 150K descriptive captions).

In [44], they present a dataset to improve object recognition models by placing the question of object recognition in the context of scene understanding. They label objects with pre-instance segmentations to aid precise object localization. The dataset contains pictures of 91 object types which are easily recognizable by a 4 year old, having a total of 2.5 million labeled instances in 328k images. They also use a Deformable Parts Model for improving the results of bounding box and segmentation. Finally, they use MSCOCO dataset for training the model.

In [45], they present a simple model that is able to generate descriptive sentences when they give a sample image. The model has a strong focus on the syntax of the descriptions. They train a bilinear model that learns a metric between image representation and phrases used to describe them. The model is then able to infer phrases from a given image sample. They also propose a simple language model that is able to produce relevant descriptions for a given test image using the inferred phrases.

In [46], they propose a method for automatically answering questions about images by bringing together recent advances from NLP and computer vision. They combine discrete reasoning with uncertain predictions by a multiworld approach that represents uncertainty about the perceived world in a Bayesian framework. The system is trained from question-answer pairs. Table 5 shows the use of the proposed method.

In [47] and [48], they propose a method for NLP is which they group words of a human language into a corpus that is indexed in vectors for representing similarity and semantic meaning. In this approach (word embeddings) complex meaning and association can be represented.

Table 5. The proposed multiword approach (I = Individual, S = Set) [46].

Type	Description	Template	Example
I	Counting and colors	How many {color}{object} are in {image_id}?	How many gray cabinets are in image 1?
I	Room type	Which type of the room is depicted in {image_id}?	Which type of the room is depicted in image 1?
I	Superlatives	What is the largest {object} in {image_id}?	What is the largest object in image 1?
S	Negations type 1	Which images do not have {object}?	Which images do not have sofa?
S	Negations type 2	Which images are not {room_type}?	Which images are not bedroom?
S	Negations type 3	Which images have {object} but do not have a {object}?	Which images have desk but not have a lamp?

In [49], they present the Neural Image Caption (NIC) model that generates natural sentences describing a model. It uses Convolutional Neural Networks for computer vision and Recurrent Neural Networks for language generating.

In [50], they present the Novel Visual Concept (NVC) dataset, in this project, the model learns novel visual concepts and their interactions, from a few images with sentences descriptions.

In [51], they propose a multimedia analysis framework to process video and text jointly for understanding events and answering queries. The model produces a parse graph that represents the compositional structures of spatial information (objects and scenes), temporal information (actions and events) and casual information (causalities between events and fluent) in the video and text. The knowledge is represented in a S/T/C-AOG graph (spatial-temporal-causal And-Or Graph).

In [52], they compare the accuracy of models in many tasks like object classification, segmentation, etc., using different image datasets like Pascal VOC 2007, MIT 67 Indoor Scenes, etc.

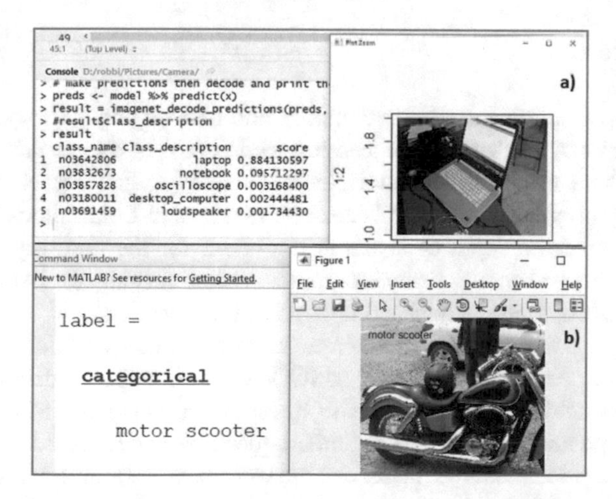

Fig. 1. Tests of image classification. (a) RStudio + Keras + imageNet (b) Matlab + Deep Learning toolbox + AlexNet

Finally, we tested some of the models using (a) RStudio with keras package, (b) Matlab with the Deep Learning Toolbox and AlexNet model, having good results as we can see in Fig. 1.

4 Discussion

Despite computer vision and image understanding don't get the same results as the human vision does, they have gotten an outstanding progress. Deep Learning models have achieved the best performance so far. Researchers can use pre-trained models like VGGNet or GooLeNet in order to skip the training process and start testing and using these models. For different problems, researchers also can benefit from transfer learning that is a technique that uses most of the layers of a pre-trained model but changes the last layers for a particular classification.

In order to answer our research question: "The use of image understanding may assist workers to get a better performance compared with traditional methods?", we can say that there are many fields like Medicine, security, traffic, industrial process that are experimenting with image understanding giving promising results that may benefit workers to perform much better and get brilliant results.

Popular tools for working with image understanding are the Python/R language in combination with Tensorfow library and Keras framework. Matlab with its deep learning toolbox is also an easy to start tool.

In the other hand, the limitations that we have with deep learning for image understanding are the need of a lot of data and computer power in order to train new models. To overcome these problems, we can build our own specialized datasets but this initiative takes a lot of work. To cope with the need of computer power, we can use cloud platforms like Microsoft cognitive services, IBM Watson, Google cloud services, Amazon cloud services, etc.

5 Conclusions

Describing content of an image is a complex and challenging task which joins computer vision and natural language processing techniques. In this context, for computer vision, researchers use Convolutional Neural Networks, and for Natural Language Processing they often use Recurrent Neural Networks and Long Short-Term Memory. We also found that the depth of layers is important for the accuracy of the classification. Researchers also use many computer vision techniques like segmentation, image classification, scene understanding, and color detection.

For training models we need a lot of human work, so many related works are based on crowd work. Among others the MSCOCO is a popular dataset. For image description, researchers train models with question-answer pairs. Models are able to infer phrases from images. So far, very limited questions can be asked to the model. For designing the questions that the model can answer, they use templates with the actions that the model can detect.

Newcomers and developers can use pre-trained deep learning models like VGGNet or GooLeNet in order to perform tasks like object identification, image segmentation, image captioning, etc. They can also use transfer learning to easily adapt those models to new datasets. In the other hand we can use languages like Python, Matlab or R language in conjunction with libraries like Tensorflow, OpenCV, Keras to work with image understanding tasks. Finally, web portals like Kaggle, allow us to program with Python or R with Tensorflow in the cloud, we do not need to install them.

For future work, we plan to test some of the models that we described in the present work in order to figure out the best techniques for image understanding. In the other hand, our future goal is to train and test models and datasets of urban traffic applications.

References

1. Techopedia: https://www.techopedia.com/definition/32309/computer-vision. 03 May 2019
2. Szegedy, C., Vanhoucke, V., Ioffe, S., Shlens, J., Wojna, Z.: Rethinking the inception architecture for computer vision. In: Proceedings of the IEEE Conference on Computer Vision and Pattern Recognition, pp. 2818–2826 (2016)
3. Eickhoff, C., de Vries, A.: How crowdsourcable is your task. In: Proceedings of the Workshop on Crowdsourcing for Search and Data Mining (CSDM) at the Fourth ACM International Conference on Web Search and Data Mining (WSDM), pp. 11–14 (2011)
4. Draper, R., Hunt, D.: Smart Robots, A Handbook of Intelligent Robotic System. Springer, Heidelberg (1985)
5. Li-Jia, L., Fei-Fei, L.: What, where and who? Classifying events by scene and object recognition. In: Proceedings/IEEE International Conference on Computer Vision (2007)
6. SUN dataset: https://groups.csail.mit.edu/vision/SUN/hierarchy.html. 26 Mar 2019
7. Fei-Fei, L., Iyer, A., Koch, C., Perona, P.: What do we see in a glance of a scene? J. Vis. 7(1), 10, 1–29 (2007). http://journalofvision.org/7/1/10/. https://doi.org/10.1167/7.1.10
8. Coursera, Université nationale de recherche, École des hautes études en sciences économiques. https://www.coursera.org/learn/deep-learning-in-computer-vision/home/welcome. 12 Mar 2019
9. Fei-Fei, L., Fergus, R., Torralba, A.: Recognizing and learning object categories. In: Short Course CVPR, International Conference on Computer Vision (2007)
10. Recognizing and Learning Object Categories course. http://people.csail.mit.edu/torralba/shortCourseRLOC/index.html. 25 Mar 2019
11. Sharif Razavian, A., Azizpour, H., Sullivan, J., Carlsson, S.: CNN features off-the-shelf: an astounding baseline for recognition. In: The IEEE Conference on Computer Vision and Pattern Recognition (CVPR) Workshops, pp. 806–813 (2014)
12. The Paris Dataset. http://www.robots.ox.ac.uk/~vgg/data/parisbuildings/. 24 Mar 2019
13. VisLab – Computer and Robot Vision Laboratory. http://vislab.isr.ist.utl.pt/datasets/#hda. 25 Mar 2019
14. ADE20 K dataset. http://groups.csail.mit.edu/vision/datasets/ADE20K/. 26 Mar 2019
15. SUN360 panorama database. http://people.csail.mit.edu/jxiao/SUN360/index_high.html. 24 Mar 2019
16. The Places Audio Caption Corpus. https://groups.csail.mit.edu/sls/downloads/placesaudio/index.cgi. 25 Mar 2019

17. Simonyan, K., Zisserman, A.: Very deep convolutional networks for large-scale image recognition. arXiv preprint arXiv:1409.1556 (2014)
18. Szegedy, C., Liu, W., Jia, Y., Sermanet, P., Reed, S., Anguelov, D., Erhan, D., Vanhoucke, V., Rabinovich, A.: Going deeper with convolutions. In: Proceedings of the IEEE Conference on Computer Vision and Pattern Recognition, pp. 1–9 (2015)
19. Candamo, J., Shreve, M., Goldgof, D.B., Sapper, D.B., Kasturi, R.: Understanding transit scenes: a survey on human behavior-recognition algorithms. IEEE Trans. Intell. Transp. Syst. **11**, 206–224 (2010)
20. Bowyer, K.W., Hollingsworth, K., Flynn, P.J.: Image understanding for iris biometrics: A survey. Comput. Vis. Image Underst. **11**, 281–307 (2008)
21. Trucco, E., Plakas, K.: Video tracking: a concise survey. IEEE J. Oceanic Eng. **31**, 520–529 (2006)
22. Imagenet large scale visual recognition challenge 2013 (ilsvrc2013): http://www.imagenet.org/challenges/LSVRC/2013/. 13 Mar 2019
23. IBM Watson demonstration website. https://www.ibm.com/watson/services/visual-recognition/demo/#demo. 10 May 2019
24. Microsoft Caption Bot. https://www.captionbot.ai/. 10 May 2019
25. Kaggle's "Dog breed identification" kernel. https://www.kaggle.com/kerneler/starter-dog-breed-identification-0c8eb184-8. 10 May 2019
26. Fei-Fei, L., Perona, P.: A Bayesian hierarchy model for learning natural scene categories. In: CVPR (2005)
27. Torralba, A., Fergus, R., Freeman, W.: 80 million tiny images: a large dataset for non-parametric object and scene recognition. IEEE Trans. Pattern Anal. Mach. Intell. **30**(11), 1958–1970 (2008)
28. Tiny Images dataset. http://groups.csail.mit.edu/vision/TinyImages/. 01 Mar 2019
29. Zhou, B., Lapedriza, A., Xiao, J., Torralba, A., Oliva, A.: Learning deep features for scene recognition using places database. Advances in Neural Information Processing Systems (NIPS) 27 (2014)
30. Cross-Modal Places database. http://projects.csail.mit.edu/cmplaces/. 23 Feb 2019
31. Are You Talking to a Machine? Dataset and Methods for Multilingual Image Question Answering (2015)
32. Ren, M., Kiros, R., Zemel, R.: Exploring models and data for image question answering. In: Conference Paper at NIPS (2015)
33. Ren, S., He, K., Girshick, R., Sun, J.: FasterR-CNN: towards real-time object detection with region proposal networks (2016)
34. Liang, M., Hu, X., Zhang, B.: Convolutional neural networks with intra-layer recurrent connections for scene labeling. In: Proceeding NIPS'15 Proceedings of the 28th International Conference on Neural Information Processing Systems, vol. 1, pp. 937–945 (2015)
35. Mao, J., Xu, W., Yang, Y., Wang, J., Yuille, A.L.: Explain images with multimodal recurrent neural networks. NIPS DeepLearning Workshop 201 (2014)
36. Kiros, R., Salakhutdinov, R., Zemel, R.S.: Unifying visual-semantic embeddings with multimodal neural language models. TACL (2015)
37. Karpathy, A., Fei-Fei, L.: Deep visual-semantic alignments for generating image descriptions. In: CVPR (2015)
38. Krizhevsky, A., Sutskever, I., Hinton, G.E.: Imagenet classification with deep convolutional neural networks. In: NIPS (2012)
39. Mao, J., Wei, X., Yang, Y., Wang, J., Huang, Z., Yuille, A.L.: Learning like a child: fast novel visual concept learning from sentence descriptions of images. arXiv preprint arXiv: 1504.06692 (2015)

40. Simonyan, K., Zisserman, A.: Very deep convolutional networks for large-scale image recognition. In: ICLR (2015)
41. Vinyals, O., Toshev, A., Bengio, S., Erhan, D.: Show and tell: a neural image caption generator. In: CVPR (2015)
42. Xu, K., Ba, J., Kiros, R., Cho, K., Courville, A., Salakhudinov, R., Zemel, R., Bengio, Y.: Show, attend and tell: neural image caption generation with visual attention. arXiv preprint arXiv:1502.03044 (2015)
43. Young, P., Lai, A., Hodosh, M., Hockenmaier, J.: From image descriptions to visual denotations: new similarity metrics for semantic inference over event descriptions. In: ACL pp. 479–488 (2014)
44. Lin, T.-Y., Maire, M., Belongie, S., Hays, J., Perona, P., Ramanan, D., Dollár, P., Zitnick, C. L.: Microsoft coco: common objects in context. arXiv preprint arXiv:1405.0312 (2014)
45. Lebret, R., Pinheiro, P.O., Collobert, R.: Simple image description generator via a linear phrase-based approach. arXiv preprint arXiv:1412.8419 (2014)
46. Malinowski, M., Fritz, M.: A multi-world approach to question answering about real-world scenes based on uncertain input. In: Advances in Neural Information Processing Systems, pp. 1682–1690 (2014)
47. Mikolov, T., Karafiát, M., Burget, L., Černocký, J., Khudanpur, S.: Recurrent neural network based language model. In: INTERSPEECH, pp. 1045–1048 (2010)
48. Mikolov, T., Sutskever, I., Chen, K., Corrado, G.S., Dean, J.: Distributed representations of words and phrases and their compositionality. In: NIPS, pp. 3111–3119 (2013)
49. Show and Tell: Lessons learned from the 2015 MSCOCO Image Captioning Challenge
50. Mao, J., Wei, X., Yang, Y., Wang, J., Huang, Z., Yuille, A.: Learning like a child: fast novel visual concept learning from sentence descriptions of images. arXiv preprint arXiv:1504. 06692 (2015)
51. Tu, K., Meng, M., Lee, M.W., Choe, T.E., Zhu, S.-C.: Joint video and text parsing for understanding events and answering queries. MultiMedia IEEE **21**(2), 42–70 (2014)
52. Benchmark of Deep Learning Representations for Visual Recognition. http://www.csc.kth. se/cvap/cvg/DL/ots/. 23 Feb 2019

Accessible eHealth System for Heart Rate Estimation

Víctor Santos[1,2](✉), María Trujillo[1], Karla Portilla[1], and Andrés Rosales[1]

[1] Departamento de Automatización y Control Industrial, Escuela Politécnica Nacional, Av. Ladrón de Guevara, E11-253, 170525 Quito, Ecuador
{victor.santos,maria.trujillo01,karla.portilla, andres.rosales}@epn.edu.ec
[2] Departamento de Física, Escuela Politécnica Nacional, Av. Ladrón de Guevara, E11-253, 170525 Quito, Ecuador

Abstract. Nowadays, increase of medical infrastructure in Ecuador implies high monetary cost to acquire new medical technologies. Population longevity and the substantial increase in patients with chronic diseases (mostly cardiac nature) force them to maintain frequent contact with the medical system to make a continuous monitoring of their pathologies. In order to reduce the concurrence of people towards medical centers for a primary diagnosis about the state of their cardiovascular system an eHealth system to monitoring heart rate is proposed. This system is focused on the estimation of this vital sign, incorporates diagnostic assistance service and support for continuous care, using computational algorithms that allow, through the application of mathematical techniques for the blind source separation and diagonalization of own matrices, all of the above is subsequently synthesized within a reduced plate computer using as peripheral devices a webcam and an smartphone; to finally be compared in a set of 100 people using two commercial medical devices (pulse oximeter and digital tensiometer), obtaining satisfactory results in around 70% of the samples. An early diagnosis and a more effective treatment are important tools for planning and optimization of resources. The proposed system will benefit most of the population in several aspects, including people with limited access into the health care system through accessible technology aiming to eliminate geographical dependence using wide area network links.

Keywords: eHealth · Heart rate · Hospital care · Signals decorrelation · WAN links

1 Introduction

Nowadays, information inclusion and communication technologies (ICT) within health field given a rise to new concepts among which is telemedicine. Although in 1924 Radio News magazine featured on its cover the radio doctor, who treated a child through a radio and various monitoring devices (see Fig. 1) [1], it was not until the 1960s that due to the constant and exponential evolution in ICT, the term telemedicine was finally

M. Botto-Tobar et al. (Eds.): ICAETT 2019, AISC 1066, pp. 260–269, 2020.
https://doi.org/10.1007/978-3-030-32022-5_25

defined as the use of systems that allow immediate access of medical experts to patient information regardless of the location of this or its relevant information [2].

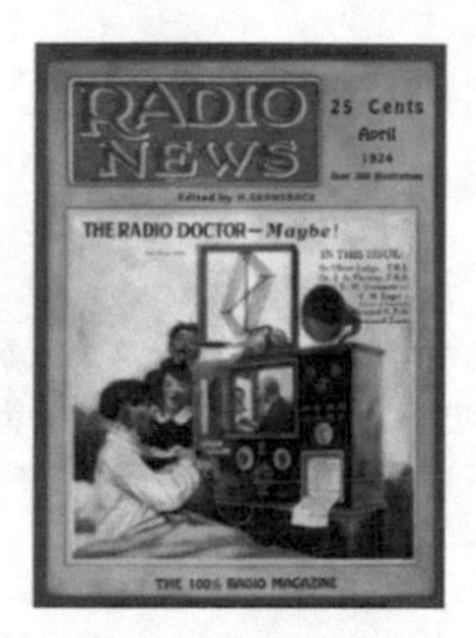

Fig. 1. Radio News magazine (April, 1924).

Although, there are many other definitions all agree in common three factors: medical services, distance between the parties and use of ICT; even in recent years, use of new technologies and electronic applications has been replaced by eHealth, which complements the concept of telemedicine by linking the market of clinical services with patients in a commercial way [3].

In Ecuador, the increase of health facilities has become more relevant in recent years. In one hand, the Public Health Ministry (MSP) manages 1.674 units of first level care, that represent 54% of public health institutions in the country. On the other hand, the number of facilities with hospital capacity is managed by the private sector with 74% [4]. As well in 2017, the number of hospital discharges in the public sector was 780.208, and in the private sector it was 363.557 with an average stay of 4, 3 days for the inpatient services. The statistics were performed without considering patients going to emergency services [5]. In the same way, it is important to note that diseases related to the cardiovascular system such as: hypertension, heart failure, cerebrovascular accidents or ischemic heart, are the main causes of mortality in Ecuadorians with approximately 24% of main global [6]; it is a main vital sign that can be diagnosed in a primary way by monitoring heart rate, and its alteration outside the normal resting range is an indicator of irregularities within the cardiovascular system [7].

Under these considerations, the implementation of an eHealth system that incorporates diagnostic assistance services and constant support to continuity of care focused on the monitoring of heart rate in patients was proposed. The goal is to maintain basic standards of telemedicine, such as reliability in performance and a possible market expansion, guarantee accessibility by optimizing resources and processes, reduce dependence on attending a health facility for monitoring heart rate.

The system is based on indicators and studies conducted in Italy and the United States [8] in which it was possible to reduce by up to 89% the emergency care and 72% the admission of patients. The system bases its operation on the digital processing of videos using recursive algorithms for blind separations of sources [9], based on the temporary change of coloration in the skin, as a result of variations in blood pressure;

these videos can be obtained using a local device, consisting of a reduced-board computer with a webcam or using a mobile device working as an IP video camera. These devices communicate with a central processing station using WAN links.

2 Methodology

2.1 Video Acquisition

One of the available methods for obtaining in patients is a local device that consist of a single board computer (Raspberry Pi 3 Model B) and a web video camera, whose only requirement is to have a minimum resolution of 360 × 480 pixels (480p) and a rate of 30 frames per second, (see Fig. 2).

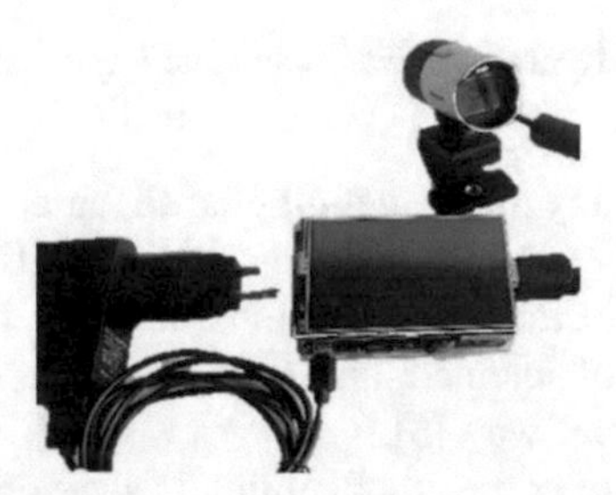

Fig. 2. Components for operation of local device.

As well, the possibility of obtaining video file using a mobile device was also enabled; currently a smartphone is often used by most people, using the microcomputer incorporated in the phone we will transform the regular camera into a camera that allows transmission of images using its own IP address, for which it is necessary to install an application in the mobile device that allows this configuration, maintaining the minimum resolution of 480p and also remembering that the video must be acquired with the phone in an upright position. Distance between the face of patient and any of these two devices must not exceed 70 cm in any case (see Fig. 3), within which the reliability of the measurements is guaranteed, patient must remain in this position during an interval of 10 s while devices acquire the images.

Fig. 3. Location of patients and acquisition devices.

Similarly, for both devices an identification code must be entered for the acquired video, which must be previously registered in the central processing station to have a historical record of the measurements made in each of the patients, information that will be available for its consultation when medical specialist or his team needs it.

2.2 Communication

It is also necessary to guarantee fast and reliable communication for data transmission between the acquisition devices and the central processing station, so a client-server architecture was chosen using a transmission control protocol (TCP/IP) through links of wide area network (WAN) under IEEE 802.11 standard (see Fig. 4).

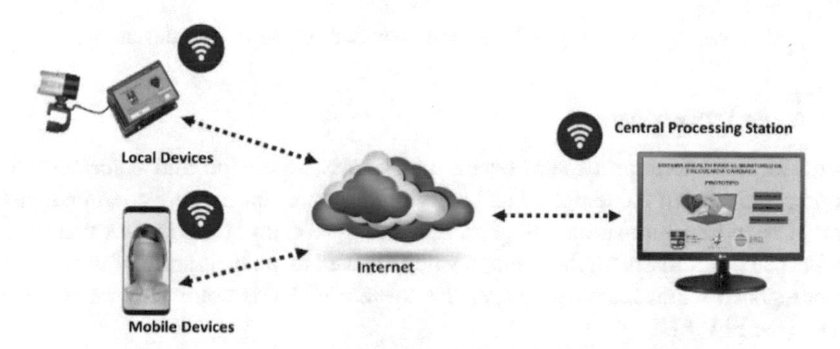

Fig. 4. Implemented network for data transmission.

In this way the scalability of the system is maintained, with the possibility of incorporating more acquisition devices to network and allowing its connection to the Internet through IP addresses that are visible from an external network to the network where the central processing station is located, eliminating in this way the limitation to certain geographical spaces.

The WLAN network is established by IP addresses, which must be previously configured within the local device (Raspberry Pi), which, via wire or wireless connection, must necessarily be visible in any Internet network through its static public IP address. In the same way, the remote device (smartphone) must necessarily access to the Internet using static public IP address provided by the telephony service provider and not access it using other data networks (Wi-Fi).

It also employs a network architecture of simple implementation and easy scalability, with cooperative processes (servers) that offer services to users (clients), avoiding excessive processing in protocols based on the model OSI using a simple communication format request-response type.

As mentioned above, public IP addresses will allow access to devices (local/remote) from any Internet network; these network addresses must be subsequently entered in the interface of the central processing station to start the process of acquiring the video file, for specific case of the remote device (see Fig. 5), it will be necessary to observe the IP address provided by the smartphone when starting the acquisition of video through the IP webcam application.

Fig. 5. Static public IP address provided by the remote device.

2.3 Image Processing

Change in the coloration of skin occurs due to the expansion and contraction of an artery, specifically of the temporal and carotid artery present in the face of patients. For a correct heart rate estimation will proceed to analyze each of the frames that make up the video, in order to obtain the average value, as well as its histogram of red, green and blue components that form the image, if a system of RGB color components is considered (see Fig. 6).

In an ideal scenario, three video cameras would be needed to monitor the change in each of color channels, transmit the image using a single monitor element, as is the case, the problem of quantifying these changes is transformed into finding a linear representation, most statistically independent of each of its red, green and blue components.

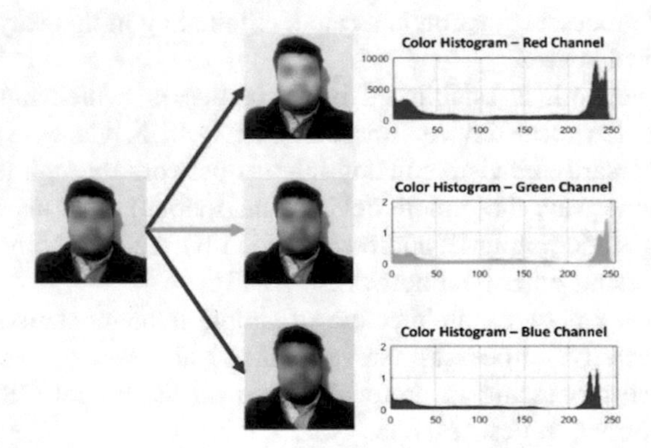

Fig. 6. Color histograms in red, green and blue channels of an RGB image.

Using a computational method that allows to find hidden factors in a set of variables, called approximate joint diagonalization of the own matrices, diagonalizes the matrix with the eigen values of combined signals, based on the statistical calculation of high order [9], finding a matrix of values so that, when the system reads the samples of obtained signals make it possible to extract output signals similar to the original sources.

In an analogous way, this system can be represented as indicated in (1). Where *S* represents samples vector of the obtained signals, *A* the matrix of combinations with dimensions *P x Q*, *F* the samples vector of the source signals and *R* the noise vector associated with components statistically independent of sources.

$$S = AF + R \tag{1}$$

System is then considered, as a problem of blind sources separation, a specific case of the independent components analysis that essentially consists of estimating a group of source signals, without knowing their nature, from a mixture of overlapping signals made by a set of sensors, separating their main adjacent components [9], this process can be summarized in two main stages (see Fig. 7).

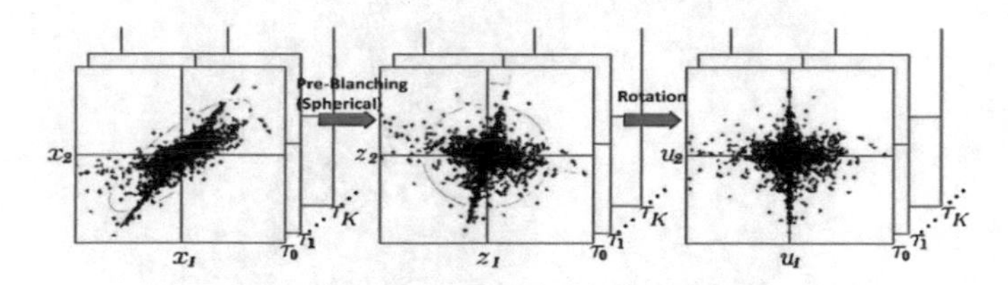

Fig. 7. Diagonalization steps within the decorrelation method for temporary signals.

Subsequently, a process known as whitening of the signal is performed, which consists of obtaining a base of eigenvalues for the covariance matrix, which indicates the degree of joint variation of two variables in relation to their average values, thus determining the dependence between both; as defined in (2), where *W* is the pre-bleaching or spherical matrix and *B* is a unitary matrix known as bleached blending matrix.

$$B = W \times A \tag{2}$$

In the final stage each of the rotations of the plane is a rotation applied to a pair of coordinates, that is, in a two-dimensional plane. If *G* is a vector of dimensions *n x 1*, the rotation of the plane *(i, j)* at an angle $_{ij}$ changes the coordinates *i* an *j* according to (3), while the other coordinates remain unchanged. Once the unitary matrix B is obtained, the sources Y can be estimated using (4).

$$\begin{bmatrix} G_i \\ G_j \end{bmatrix} \leftarrow \begin{bmatrix} cos(\theta_{ij}) & sin(\theta_{ij}) \\ -sin(\theta_{ij}) & cos(\theta_{ij}) \end{bmatrix} \begin{bmatrix} G_i \\ G_j \end{bmatrix} \tag{3}$$

$$Y = B^T \times W \times S \tag{4}$$

These stages are part of a recursive algorithm using Jacobi techniques, which is an iterative optimization technique on a set of orthonormal matrices, where the orthonormal transform is obtained as a sequence of flat rotations [10].

3 Results and Conclusions

The eHealth implemented system provides two connection methods for the data acquisition, either through the local device or the mobile device, medical staff will decide which of them is operative to initiate heart rate monitoring.

In the specific case of using mobile device as an image acquisition system it is necessary to specify the IP address that is generated in the smartphone (see Fig. 8), in this way we can visualize, the real-time image of the own camera of the mobile device, thus eliminating geographical limitation between patient and medical staff.

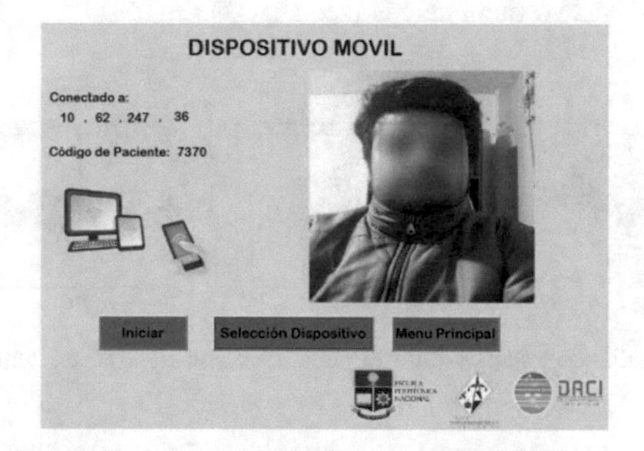

Fig. 8. Image acquired in real time using mobile device.

To verify the correct functioning of implemented system, it was decided to perform heart rate measurements in a group of 100 people, chosen completely at random within an age range of 18 to 29 years, the same as those residing in Quito-Ecuador and perform their activities as students and university professors.

The tests, in their entirety, were performed in an internal environment with artificial light, similarly to perform a quantitative analysis two commonly used commercial medical instruments were used to contrast the information obtained using the proposed eHealth system; during video acquisition the participants carried a digital blood pressure monitor (Panasonic EW-BU04) and a pulse oximeter (MD300C21C) with an accuracy of ± 5% and ± 2%, respectively.

In a first analysis, the implemented eHealth system requires a total of 22 s from the beginning in the images acquisition until the presentation of the obtained signals and the estimation of heart rate (see Fig. 9), thus reducing in 21.42% the measurement time compared to the digital tensiometer and 45% compared to manual measurement methods.

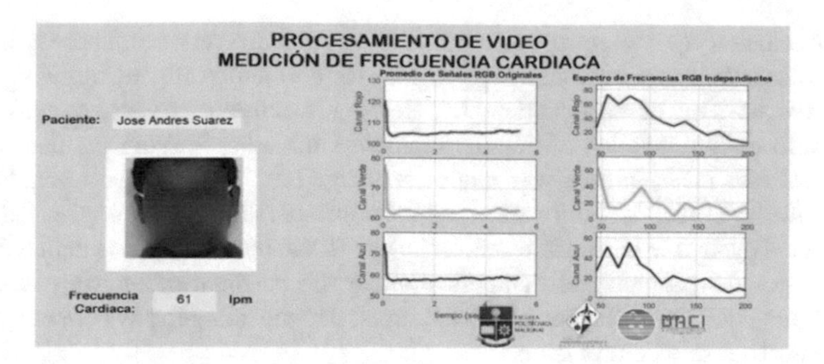

Fig. 9. Signal processing and hear rate estimation.

Contrarily, within test carried out in validation stage it could be noted that pulse oximeter, when using photoelectric principle to measure the saturation of the erythrocyte by the hemoglobin in the capillary beds, has errors or even does not perform measurement of heart rate if the fingernail has some type of varnish; in the same way any sudden movement made by participant will change the measurement made by the dig- ital tensiometer due to its oscillometric principle using the built-in electronic pressure transducer; these reasons explain thc variability between the measurements obtained using the two commercial instruments reached a maximum of 36% on several occasion. Considering the aforementioned variation and adding the fact that cardiovascular system, and consequently heart rate, can be altered by several reasons such as emotional state (mood), physical exercise, age, among other reasons, it was decided that the system will consider a correct measurement to be one that does not exceed 20% of that obtained by any of the two commercial instruments; thus within the set of 100 measurements made, 70% effectiveness was obtained as indicated in Table 1, with a sample of 10 participants, completely randomly selected from a total sample of about 100 people.

Table 1. Contrast of a sample in obtained measurements.

Participant	Heart rate using (Pulse Oximeter)	Heart rate using (Digital Tensiometer)	Heart rate using (Implemented eHealth System)
1	66	65	65
2	63	67	61
3	68	71	72
4	61	57	57
5	73	71	65
6	93	85	87
7	68	67	68
8	91	89	95
9	68	62	57
10	83	80	87

In conclusion the system incorporates diagnostic care services, such as heart rate monitoring, as well as providing a support tool for continuity of care, allowing medical staff to record data of each patient by creating electronic medical records, which contain information detailed of the measurements taken, specifying the date and time when heart rate measurement was made (see Fig. 10). The presented benefits, completed with the use of information and communication technologies, allow the implemented system to be called eHealth. Its use in the medical care system will reduce waiting times for patients and its application in the constant monitoring of this vital sign will prevent its mobilization to medial units, decongesting the system and allowing a more effective service provision.

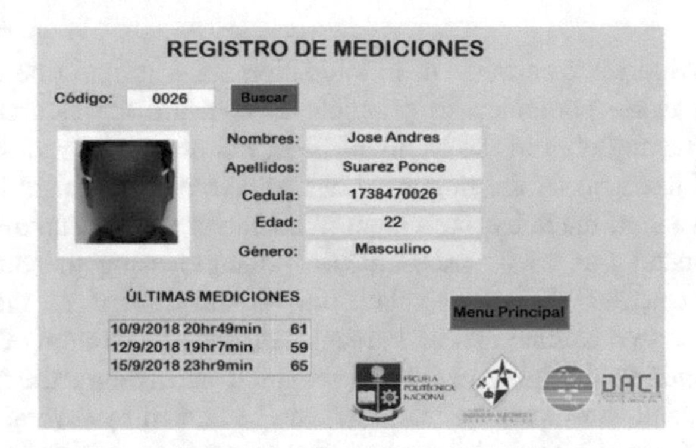

Fig. 10. Display of measurement history.

Finally, although the eHealth system implemented has advantages over methods frequently used for the same purpose, such as the easy-access technology used, the low cost of implementation and the use of non-invasive techniques that allow maintaining comfort and tranquility in the patient, it could include more benefits that make it even more competitive in the medical equipment market, such as the monitoring of other important vital signs, for example, the respiratory rate or modify the support to the continuity of care of medical personnel, with the inclusion of other relevant data in the electronic medical records.

References

1. Carrión, P., Ródenas, J., Rieta, J., Sánchez, C.: Telemedicina: Ingeniería Biomédica, 6th edn. Universidad de Castilla, La Mancha (2009)
2. Marimón, S.: La sanidad en la sociedad de la información: sistemas y tecnologías de la información para la gestión y la reforma de los servicios de salud, 1st edn. Díaz de Santos, Madrid (1999)
3. Della Mea, J.: What is e-health(2): the death of telemedicine. J. Med. Internet Res. **3**(2), e22 (2001)
4. ISAGS UNASUR. http://isags-unasur.org/es/sistema-de-salud-en-ecuador. Accessed 21 Mar 2019

5. INEC ECUADOR. http://www.ecuadorencifras.gob.ec/documentos/web-inec/Estadisti-cas_Sociales/Camas_Egresos_Hospitalarios/Cam_Egre_Hos_2017/Presentac-ion_CEH_2017.pdf. Accessed 21 Mar 2019
6. Lucio, R., Villacrés, N., Henríquez, R.: Sistema de salud de Ecuador. Journal **53**(2), 177–187 (2011)
7. Jinich, H., Lifshitz, A., García, J., Ramiro, M.: Síntomas y signos cardinales de las enfermedades, 6th edn. Manual Moderno, México (2013)
8. López, M., De la Torre, I., Herreros, J., Cabo, J.: Mejora de la calidad asistencial mediante la telemedicina y teleasistencia, 1st edn. Díaz de Santos, Madrid (2014)
9. Portilla, K., Santos, V., Trujillo, M., Rosales, A.: Non-invasive heart rate monitor applying independent component analysis in videos. In: Luján, S., Moscoso, O. (eds.) International Conference on Information Systems and Computer Science 2017, INCISCOS, pp. 121–127. IEEE, Quito (2017)
10. Ziehe, A.: Blind source separation based on joint diagonalization of matrices with applications in biomedical signal processing. Ph.D. dissertation. Postdam University, Berlin (2005)

The Use of New Technologies for Mindo Birdwatching

Genaro Sulca[1], Alexandrino Gonçalves[2], Anabela Marto[2,3], Nuno Rodrigues[2], and Rita Ascenso[2](✉)

[1] School of Technology and Management, Polytechnic Institute of Leiria, Leiria, Portugal

[2] School of Technology and Management, Computer Science and Communication Research Centre, Polytechnic Institute of Leiria, Leiria, Portugal
rita.ascenso@ipleiria.pt

[3] INESC – TEC, Porto, Portugal
https://www.ipleiria.pt/

Abstract. The growing acceptance of new technologies, along with trends focused on reducing the use of paper and facilitating access to information, make mobile applications increasingly important. In order to solve current demands, we want to create an attractive way to share information and offer a tool to educate people at the same time.

The main objective of this work is to provide tourists and users in general with a mobile application that provides information about the birds that inhabit the town of Mindo. This mobile application will provide multimedia content and with the use of Augmented Reality the user will be given the option to interact between the virtual objects and the environment, it also will provide the option to share this experiences with other people via social networks to enrich user experience when practicing birdwatching.

To develop the prototype application, we used native android, ARCore as the AR engine, geolocation services with Google's maps API, cloud services with Firebase to store the data and, in the mobile device, internal storage and SQLite to work offline.

Keywords: Birdwatching · Augmented reality · Mobile application · ARCore

1 Introduction

For Ecuador, tourism has become one of its priorities and a great and important source of income [1]. With programs promoted by the central government, it is intended to promote tourism in Ecuador worldwide [2]. Ecuador is the third country worldwide with more species of birds observed, with 1164 species of birds sighted [3], which makes the birdwatching activity ideal for tourists and mandatory for ornithologists.

Considering this and that the use of smartphones is very common nowadays, leads us to pursue new, innovative and attractive ways to instruct people about the different species of birds that inhabit the ecosystem (in Mindo) and raise awareness about the risks to which it is exposed.

M. Botto-Tobar et al. (Eds.): ICAETT 2019, AISC 1066, pp. 270–279, 2020.
https://doi.org/10.1007/978-3-030-32022-5_26

Related to tourism, several ways to utilize new technologies have been explored, such as Augmented Reality, to attract attention and offer new ways to present information to tourists, complementing the previous analogous forms, enhancing tourists' experiences. Currently there are mobile applications that offer information about birds, with images and sounds, but, in Ecuador, none with augmented reality is found.

This is one of the reasons why this work proposes the development of a mobile multilanguage application, for Android, that helps the user in the study of birds in the area of Mindo, Ecuador. This application will provide information about those birds, risks to which they are exposed, images, videos, sounds and, in specific cases, virtual models of them using Augmented Reality (AR).

The relevance of this work is that similar applications focused on birdwatching in Ecuador do not use augmented reality and also that the AR engine (ARCore) for Android used for this project is relatively recent (released on February 23, 2018) [4].

2 Related Work

Currently, there are mobile applications for birdwatching in Ecuador for other countries too, and also applications with augmented reality to learn about animals, but we haven't found any application having both features simultaneously, as proposed by this work.

The first example is an application on birds, with over 11400 species, divided into families. It provides information about birds and multimedia content (connection to the internet needed) with sounds, images and videos [5].

There is an application that offers information about birds in Ecuador, scientific information, range, subspecies and other information about more than 1600 species with the option to download photos, maps and sounds [6].

Considering applications that uses AR in this context, we can mention a project developed to support the teaching of zoology. This application uses the regular school cards as targets for augmented reality, whereas has other interesting features such as animal sounds and two modes of operation to make the learning activity more interactive [7].

Another example is the one for the University museum, in Göteborg. The application has 3 modes of use, the first one offers static information about animals, the second one uses AR with image targets to create virtual objects of the corresponding animal and the third, uses a book with illustrations that activate the AR and offers interactive playable activities aimed at learning [8].

3 Methodology

First, a digital format survey was conducted to both foreign and local Mindo tourists to evaluate the acceptance this work might have. We used some variables of the UTAUT model to get a survey that provides us with accurate information [9], and the number of respondents was defined considering the Central Limit Theorem, which indicates that if you have a large group of independent variables and all of them follow the same distribution model, its sum is distributed according to a normal distribution [10].

Then we define the structure's application, namely: information (text) in Firebase Cloud Firestore and multimedia files stored in Firebase Cloud Storage. The Android application will store all multimedia and textual information from the database to allow offline operation and will interact with the user as shown in Fig. 1.

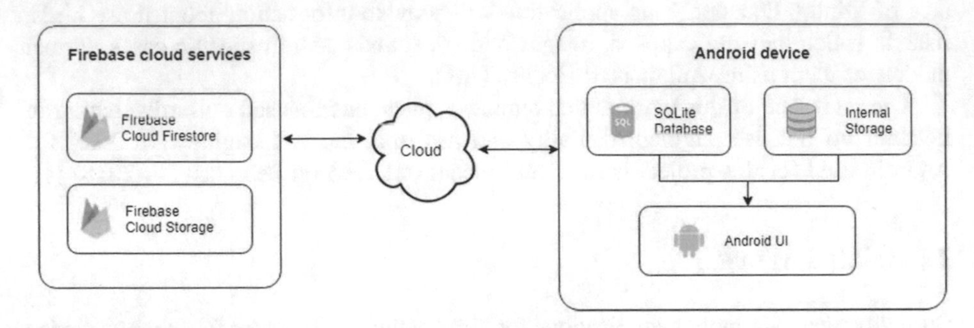

Fig. 1. App diagram

3.1 Android

The application is being developed for Android, version 7.0 and higher (API level 24), since it is from this version that OpenGL ES 3.2 is incorporated to render 2D and 3D graphics, as well as the SurfaceView object that allows content to be presented asynchronously with better performance and reduced battery consumption, features that allows to enable the use of the Google AR engine, ARCore [11].

SQLite. In order to allow the application to update most of the content on runtime and work offline, all this data need to be stored in an internal database. SQLite allows to save and manage information into the device, with a powerful engine that provides high-reliability and powerful data processing [12].

Google Material Design for Android. Google Material is the standard of Google to design interfaces for Android. Its API provides the facility to create interfaces according to this standard and with only a few lines of code, it is fully integrated with Java and can modify the UI in runtime. It also allows to thematize the application to change the colors used in the application [13].

3.2 Firebase

Firebase provides services for data storage with high availability and efficiency in data transmission. Its API is simple to integrate with native Android and, in addition to its free version, it offers plans with affordable costs. For this work we used the Blaze plan (pay per use), which allows us to integrate the project with Google Cloud Platform [14].

3.3 Augmented Reality

To develop the AR, feature we used Google AR engine, named ARCore. With its API, ARCore facilitates the integration with native Android and through the Sceneform plugin, allows to render and import virtual objects to a Visual Studio project. ARCore is available from Android version 7.0 and is limited to some devices. Notwithstanding, this list is being continuously updated, since number of devices that are compatible with this technology is increasing rapidly [15].

Google ARCore has 3 essential characteristics to integrate the virtual content with the real scenario, which are: motion tracking, environmental understanding and light estimation. With this, we can create an AR experience as similar as possible to reality [16].

- **Motion Tracking.** Allows us to render virtual content, birds in our case, with correct relation and perspective with the real world, placing it in the corresponding spot and maintaining its position even while the device is in movement.
- **Environmental understanding.** Allows us to detect flat surfaces in which to place the virtual object, these surfaces can be horizontal or vertical and can be used at convenience to present virtual objects.
- **Light estimation.** ARCore detects the ambient lighting and also the income light direction, in order to render the virtual object with the corresponding light and also, if it exists, generate a shadow in accordance with it, to increase the realism to the virtual object.

3.4 Geolocation

We use GPS to present the geolocation of the Hotspots by using google maps. This is also convenient to determine the user's location in the app integrated map, by using the Google Maps API.

For the functionality "Discover hotspot" that shows specific content for certain points of interest depending on the user's location, as well as to send notifications of entry to these points, we use the awareness Google API, with geofences.

3.5 Prototypes

Prior to the development of the application, prototypes were developed with Adobe XD to define an interface that is intuitive and, at the same time, pleasant for the user [17]. The main menu screens were inspired by the interface of Adobe Portfolio [18], for app layout. Side menu and bottom menu are using the standards of Google Material and color theming was selected from Google Material color tool (Fig. 2).

3.6 App Development

The development focused primarily on the main features: main menu, hotspots, routes and bird library with location and maps, reading/saving files from/to internal storage, multi-language, storing and reading information from SQLite.

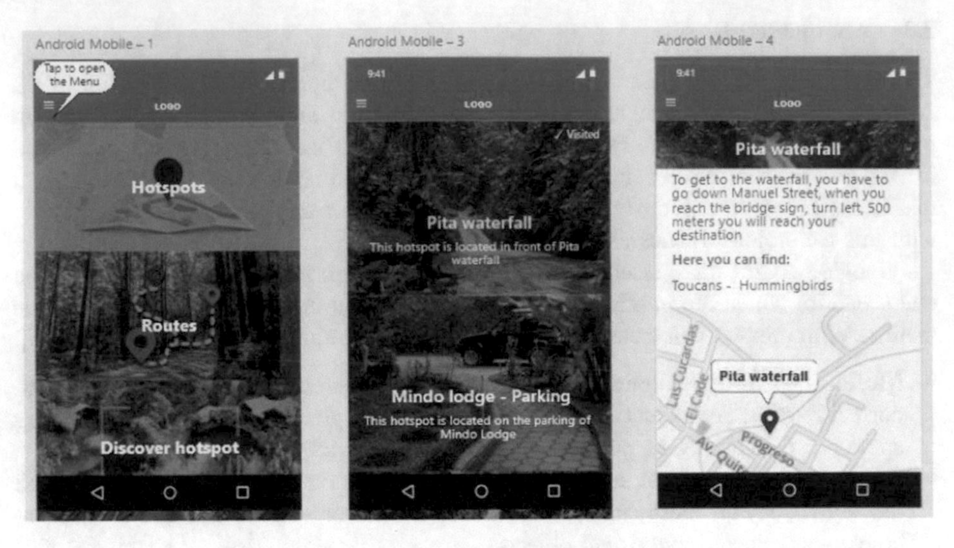

Fig. 2. Prototypes made in Adobe XD

After this process, we proceeded to implement the Geofences with notifications and the augmented reality option with plane detection and with virtual objects in. Finally, for the option to share or save a photo, we used PixelCopy [19] to take the screenshot (SurfaceView object) and transform it into bitmap, to allow us to treat it as desired (share or save into device).

4 Case Study

Since Mindo has a great diversity of bird species, although there are places where you can observe birds in captivity, as well as tours to observe birds in their natural state, it is very complicated to observe them all.

Even though there is information about these birds in books and the Internet, they do not offer an interactive experience for the user and, in addition, the places and tours for birdwatching are paid in their vast majority, which could be a reason for tourists to deprive themselves of living the experience of birdwatching.

The case study presented in this article is focused on the development of an application to provide information about birds to the user, with photos, videos and, for some bird species, digital 3D models through AR and audio. In addition, it offers information about places where these birds can be observed, as well as trails (or routes) in which several species of birds can be observed.

Considering that in Mindo there may be places where there is no GSM signal, one of the requisites of our app is that it should work on an offline mode, *i.e.* without an internet connection.

5 Results

Once the surveys where carried out, there were several answers that have influenced the realization of this work, however, the ones that motivated us to carry out this work were those presented in Figs. 3 and 4. The results indicate that 75% of the people surveyed don't have a good knowledge about birds (Fig. 3) and that 77.5% are interested in a smartphone or tablet application that allows them to learn about birds (Fig. 4).

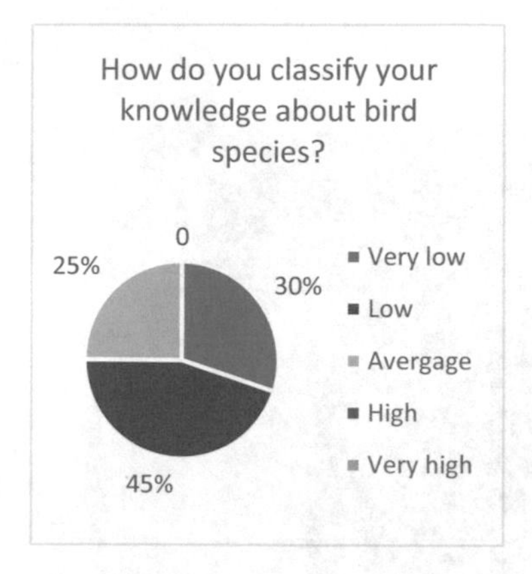

Fig. 3. Question: How do you classify your knowledge about bird species?

Fig. 4. Question: Would you like to use an application for cell phone/tablet to learn more about birds using AR?

In addition, another important fact, gathered from the surveys, is that 77% of users have or use a smartphone or tablet with Android as operative system, so it is adequate to develop the work for that platform (Fig. 5).

Since this is a work still in progress, the tests have been carried out only by the development team, however the prototype presented is a functional prototype, with all the features working already. Below is an analysis of these features (Fig. 5).

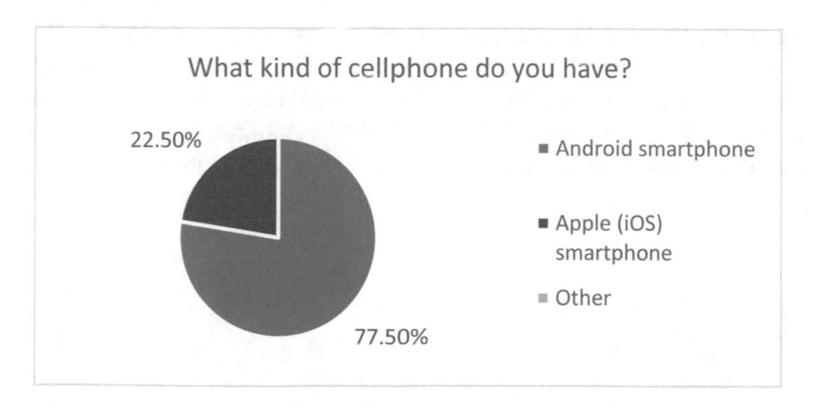

Fig. 5. Question: What kind of cellphone do you have?

Load Database. To save in local database, each time the app is opened it checks if there is a new version of the database, if it is the case, a modal is shown notifying this and asking if the user wants to update the local database. If accepted, it will ask for Wi-Fi connection, if is not available, to download the new database version. On the first run the Wi-Fi connection will be checked to create the database automatically.

Main Menu. The main menu always shows the three main options of the app, *Hotspots, Routes and Bird Library*, as shown in Fig. 6(a), when entering a geofence the option *Discover hotspot* is displayed, as shown in Fig. 6(b).

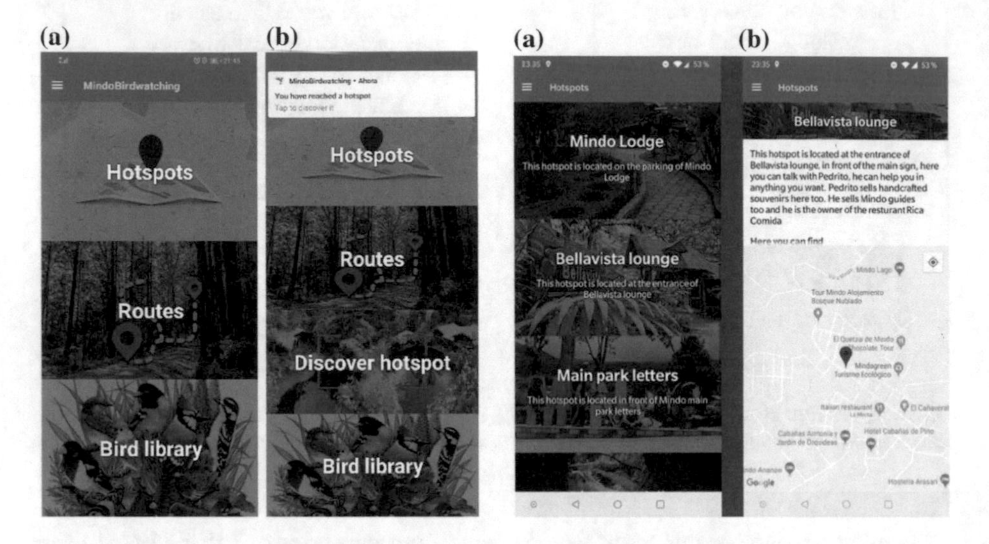

Fig. 6. (a) Main menu, (b) Geofence entrance **Fig. 7.** (a) Hotspot list, (b) Hotspot description

For geofences, a notification is displayed every time the user enters a geofence. When the user interacts with it, the application is opened in "Discover hotspot" feature and the notification is dismissed. If the notification is dismissed by the user, it is deleted when the user leaves the geofence.

Hotspot. In the Hotspot menu (as in the Routes menu), a list of available Hotspots is displayed, with an image and a brief description of it, as shown in Fig. 7(a). In the Hotspot description, the respective information and a map with the location of the Hotspot is displayed, as shown in Fig. 7(b). If the Hotspot marker is touched, the name is displayed and options to open in maps and to obtain directions.

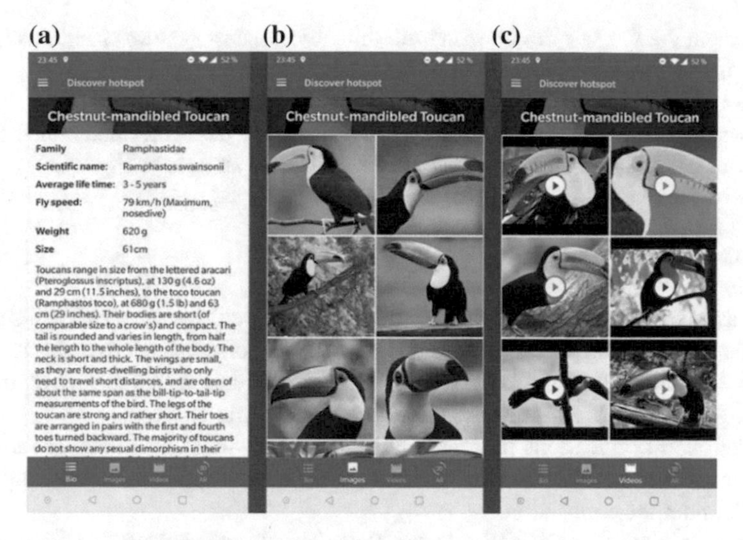

Fig. 8. (a) Bird bio, (b) Bird image gallery, (c) Bird video gallery

Bird Library/Discover Hotspot. When entering to Bird library or Discover hotspot, it is possible to navigate through the bottom bar between Bio Fig. 8(a), Images gallery Fig. 8(b), Video gallery Fig. 8(c), and AR option. This last option will be available only when entering from the "Discover hotspot" option.

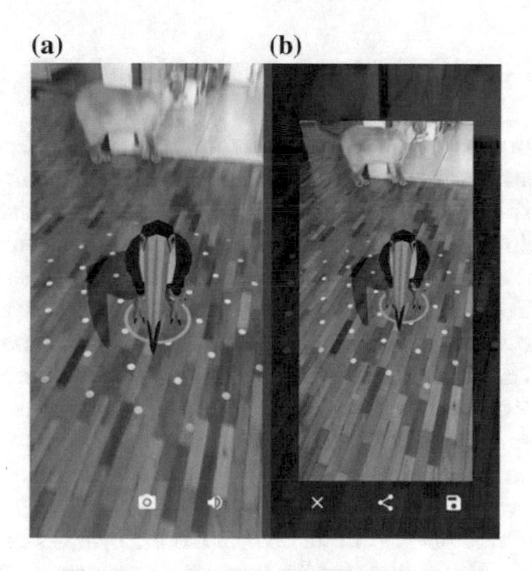

Fig. 9. (a) AR view, (b) AR screenshot

AR Feature. To activate the AR we use plane detection, for this, once a flat surface is detected, it shows several white points, which, when touched, overlap the virtual object on the camera view. User can interact with this object using gestures to increase its size, decrease it and rotate it. Other options in this section are: play a bird sound and take a screenshot, as shown on Fig. 9(a).

The screenshot is taken excluding the buttons of the user interface and then shows options to save in local storage, share or discard it, as shown in Fig. 9(b).

6 Discussion

As well as Marto et al. [20], we used the UTAUT model to carry out the surveys. It reveals that 75% of the people surveyed don't have a good knowledge about birds and that 77.5% are interested in a smartphone or tablet application that allows them to learn about birds, which have influenced the realization of this work.

Unlike other applications presented in Sect. 2, ours will use a database to store all the content in the mobile device, in order to make the application fully operational without an internet connection.

For the main menu, unlike Bird Data - Ecuador [6], we used eye-catching interfaces to capture the user's attention.

For library, differing for example, from the application South American Birds Sounds [5], we use a different approach, based on a bottom bar for the design of the bird library. We use this method-ology also in Discover hotspot feature

To use the AR feature, unlike to the work of Becerra [7] that uses image targets to activate the AR, in this application we use plane detection, whereby, once a flat surface is detected the user have some places to place the virtual object.

7 Conclusions and Future Work

The need for new tools to promote tourism and instruct users, offers the opportunity to create mobile applications that are striking and practical, with current technology and interactive elements. For this, following the standards and using the APIs of Google, it is possible to develop user friendly applications with good performance, besides the AR engine, ARCore together with Sceneform, it is easy to use and integrate in a common Activity. To develop the prototype mobile application, we used a phone with Android 9, but the application was developed for Android from version 7.0. The application worked correctly on the phone with android 9, but when testing on other phones with Android 7 and 8, there were problems, in the request for permissions, performance and geolocation. By this reason, to facilitate development, it should be tried on smartphones with the base operating system to which the application is being developed.

For future work, it is planned to implement more Hotspots, as well as 3D models for more bird species. Also, because at the moment changes on the information can just be made by a person with knowledge in firebase, it is planned to implement a web page with friendly interface, in which a common user is allowed to make these changes.

Acknowledgement. This work was supported by national funds through the Portuguese Foundation for Science and Technology (FCT) under the project UID/CEC/04524/2019.

References

1. Ingreso de divisas por turismo crece en 46% durante el primer trimestre de 2018 – Ministerio del turismo. https://www.turismo.gob.ec/ingreso-de-divisas-por-turismo-crece-en-46-durante-el-primer-trimestre-de-2018/. Accessed 26 Mar 2019
2. 'All you need is Ecuador' es la campaña que está en el mundo - Ministerio de Turismo. https://www.turismo.gob.ec/all-you-need-is-ecuador-es-la-campana-que-estara-en-el-mundo/. Accessed 26 Mar 2019
3. Global Big Day – eBird. https://ebird.org/globalbigday. Accessed 26 Mar 2019
4. GitHub ARCore repository. https://github.com/google-ar/arcore-android-sdk/releases. Accessed 27 Feb 2019
5. South American Birds Sounds - Birding by families and countries. Find and learn about over 11,800 bird species from around the world – luminousapps. https://www.luminousapps.net/. Accessed 27 Mar 2019
6. Bird Data – Ecuador. https://play.google.com/store/apps/details?id=com.birdphotos.user.gbd.ecuador. Accessed 27 Feb 2019
7. Becerra, D.A.I.: Augmented reality applied in the design of learning. Universidad Nacional de San Agustín, Arequipa
8. Berzén, E.: Use of augmented reality to increase learning in museums. Malmö University, Malmö (2018)
9. Venkatesh, V., Morris, M.G., Davis, G.B., Davis, F.D.: User acceptance of information technology: toward a unified view. MIS Q. **27**, 425–478 (2003)
10. Gutiérrez, H., De la Vara, R.: Análisis y Diseño de Experimentos, 2nd edn. McGraw-Hill Interamericana, México (2008)
11. Android 7.0 for Developers—Android Developers. https://developer.android.com/about/versions/nougat/android-7.0.html. Accessed 28 Feb 2019
12. SQLite. https://www.sqlite.org/index.html. Accessed 28 Feb 2019
13. Design – Material Design. https://material.io/design/. Accessed 28 Feb 2019
14. Firebase. https://firebase.google.com/. Accessed 26 Mar 2019
15. Supported Devices—ARCore—Google Developers. https://developers.google.com/ar/discover/supported-devices. Accessed 27 Mar 2019
16. Fundamental concepts—ARCore—Google developers. https://developers.google.com/ar/discover/concepts. Accessed 28 Feb 2019
17. Adobe XD. https://www.adobe.com/products/xd.html. Accessed 28 Feb 2019
18. Adobe Portfolio. https://portfolio.adobe.com/. Accessed 28 Feb 2019
19. PixelCopy—Android Developers. https://developer.android.com/reference/android/view/PixelCopy. Accessed 27 Mar 2019
20. Marto, A., Gonçalves, A., Martins, J., Bessa, M.: Applying UTAUT model for an acceptance study alluding the use of augmented reality in archaeological sites. In: 14 International Joint Conference on Computer Vision, Imaging and Computer Graphics Theory and Applications, VISIGRAPP 2019, vol. 2, pp. 111–120 (2019)

A MATLAB-Based Tool for 3D Reconstruction Technologies for Indoor and Outdoor Environments

Kevin Jaramillo[1], David Ponce[1], David Pozo[1](✉), and Luis Morales[2]

[1] Facultad de Ingeniería y Ciencias Aplicadas, Universidad de Las Américas, Quito, Ecuador
{kevin.jaramillo,david.ponce.herrera,david.pozo}@udla.edu.ec
[2] Departamento de Automatizacíon y Control Industrial, Escuela Politécnica Nacional, Quito, Ecuador
luis.moralesec@epn.edu.ec

Abstract. This paper presents a 3D-Reconstruction MATLAB-based tool for indoor and outdoor environments using the Kinect V2, Kinect V1, RPLIDAR A1 sensors and a ZED 2K stereo camera. The 3D reconstruction tool allows to obtain the data generated by any of the sensors and create a point cloud that allows to represent an environment in three dimensions. Likewise, through this tool it is possible to manipulate the point cloud by applying axis and distances filters, as well as spatial resolution modification of the acquired data. The aim of this work is to provide the user a tool capable to obtain 3D information of several devices to facilitate its use and avoiding the need of a different software for each device. Finally, several 3D images with each sensor are tested in order to manipulate them with the tool and to get a 3D point cloud with established bounds.

Keywords: 3D-Reconstruction · MATLAB · Kinect V2 · Kinect V1 · ZED 2K stereo camera

1 Introduction

A 3D reconstruction could be performed from an object, a scene, several images or from a video and is being applied in areas of study such as: infographics, medicine, architecture, cartography, archeology, among others. The need to abstract elements of the physical world to digitize them is a topic of relevance in the field of research; however, in some cases the development of three-dimensional models is a complex and expensive process [1].

Currently, new applications require real-time interaction, which leads to the development of models and tools that simplify the processing of large amounts of data, with a relatively low cost. This is done from digital information, obtained through cameras or range sensors such as: LIDAR, infrared, laser, which achieve an accurate representation of the 3D scene through a point cloud.

M. Botto-Tobar et al. (Eds.): ICAETT 2019, AISC 1066, pp. 280–290, 2020.
https://doi.org/10.1007/978-3-030-32022-5_27

Devices such as RPLIDAR A1, ZED 2K stereo camera, Kinect V1 and V2 are technological solutions that allow the development of low-cost applications that can also be used in research field. However, the management of different technologies bring inside a level of complexity in knowledge, due to the specialized software that each one requires for its operation. In addition, it must be considered that the type of technology used will directly influence the development of the reconstruction [2].

On the other hand, the need to reproduce the ability of human vision to interpret and recognize objects and environments in three dimensions has motivated research in this field to take great importance thanks to its infinity of applications oriented in different study areas. In medicine, 3D reconstruction systems focus mainly on the study of extremities, bones and internal organs [3,4]. In Criminalistics, the scanning of crime scenes is extremely important for documentation and case resolutions [5].

Likewise, in the field of robotics and industry applications such as the autonomous guidance of robots for rescue, quality control, autonomous vehicles and way assistance [6,7], the reconstruction of objects for its later modeling [8] and detection of nearby objects [9] are being developed.

In fields such as architecture, cartography and archeology the representation of structures, models [10,11], elaboration of three-dimensional maps [12] and reconstruction of archaeological structures [13] are having great acceptance.

All these applications and others in general [14], have been developed to facilitate delicate, repetitive or highly complex tasks for people [15]. In addition, analyzes between different technologies have been carried out, so in [16] and [17] comparisons of performance between sensors used for 3D reconstruction are presented.

In this context, a tool for working with several devices used for point cloud acquisition and treatment through a unique and user-friendly software is developed. The technologies that are used in this work are: RPLIDAR A1, stereo vision camera ZED 2K and Kinect sensors V1 and V2. The methodology and the mathematical foundations could be found in the previous work [16].

2 Background

Among the multiple technologies that can be used for the reconstruction of objects and environments we have techniques based on multiview images [18], RGB-D acquisition devices [19], ultrasound imaging technology [20], LIDAR sensors [21], among others.

LIDAR (Light Detection and Ranging), is a system that allows to obtain the distance to an object or surface by means of Time of Flight (TOF) of laser pulses [22]. This technology can be used in systems for reconstructing scenes or environments in 2D and 3D, by obtaining a high-resolution point cloud. Currently, LIDAR is widely applied to the elaboration of oriented mapping systems for autonomous vehicles, where the cooperative recording of point clouds is collected by the LIDAR sensors arranged in each automobile [23].

Table 1. Kinect V1 and V2, ZED 2K stereo camera and RPLIDAR A1 characteristics

	Kinect V1	Kinect V2	ZED 2K	RPLIDAR A1
Price	Very low	Low	Medium	Very low
Applications	Indoors	Indoors	Indoors Outdoors	Indoors Outdoors
Interface	USB 2.0	USB 3.0 PC Adapter	USB 2.0	USB 2.0 UART
Precision depth (cm)	1	1	1	≤0.5 or 1% of distance
Point cloud data type	RGB-D x, y, z points	RGB-D x, y, z points	RGB-D x, y, z points	x, y, z points
Field of view	57° × 43.5°	70° × 60°	max 110°	360° on 2D plane
Operation range (m)	0.8–3.5	0.5–4.5	0.5–20	0.2 a 6

Kinect V1 is a device developed by Microsoft, originally designed for the Xbox 360 platform. The objective of the peripheral is the interaction based on the movements of the user. It has four fundamental parts such as: 3D depth sensors, an RGB camera, microphone array and a tilt motor; while the design and technology were created by the Israeli company PrimeSense. The depth perception system consists of three components; the infrared laser projector, the CMOS projector and a microchip in charge of information processing.

Kinect v2 sensor technology developed by Microsoft has two cameras, a RGB resolution of 1920 × 1080 pixels and an infrared (IR) camera, also has three IR projectors which allow to triangulate the information of depth and distance of the scene [24,25]. This device is capable of generating depth maps in real time and even in color, called "RGB-D image" with more than 300 thousand points and with a refresh rate of 30 Hz [26,27].

The ZED 2K stereo camera, based on stereoscopic vision, it has a set of two cameras whose operation is based on human vision. The device allows the triangulation of interest points in three dimensions of the observed scene, thus allowing the space and depth of objects to be recognized [28,29]. In addition, it allows Motion Tracking and is widely used in 3D mapping applications. Moreover, is recommended for its use a computer with minimum 2Gb of ram memory and a graphic card [17].

In Table 1 is presented a features comparison among the devices that are going to be managed through the MATLAB-based tool for 3D reconstruction. Based on it, is appreciated that the simplicity of the hardware requirements of the RPLIDAR made it an option in embedded applications with low process capabilities. However, if a point cloud with RGB-D is desired, the Kinect V2 and ZED stereo camera would be the best options.

3 MATLAB Based Tool Functionalities

In order to obtain a three-dimensional reconstruction, a new tool was created capable of processing and manipulating the different captures made by the Kinect v1, Kinect v2, RPLIDAR A1 and ZED 2K stereo camera in order to deploy a point cloud of a 3D reconstruction environment. The GUI facilitates the use of the tool for users who want to work with 3D point clouds in research activities without the disadvantage of the use of different software for each device. In Fig. 1 is shown the main GUI and in the Fig. 2 is shown the GUI for data acquirement and treatment. In Table 2 are presented the functionalities of MATLAB-based tool.

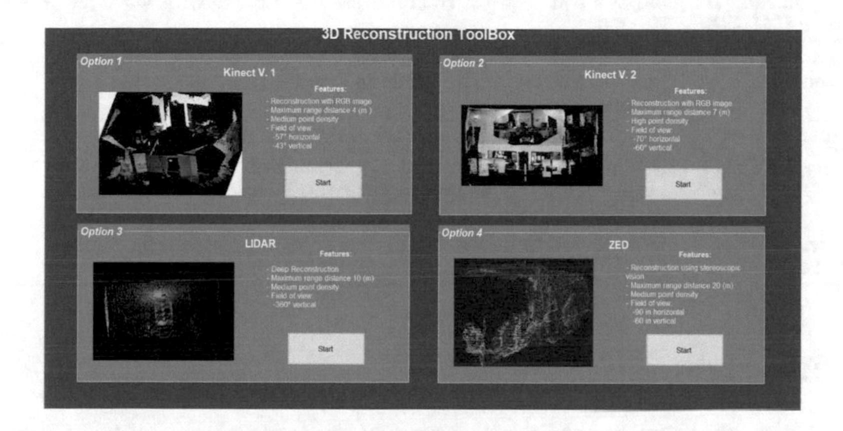

Fig. 1. Main GUI interface

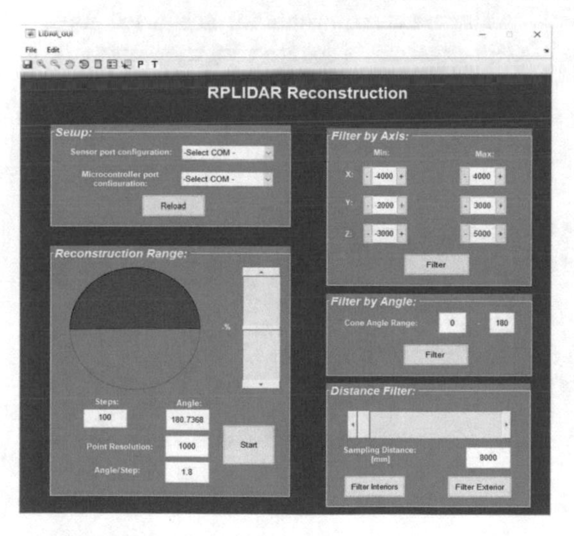

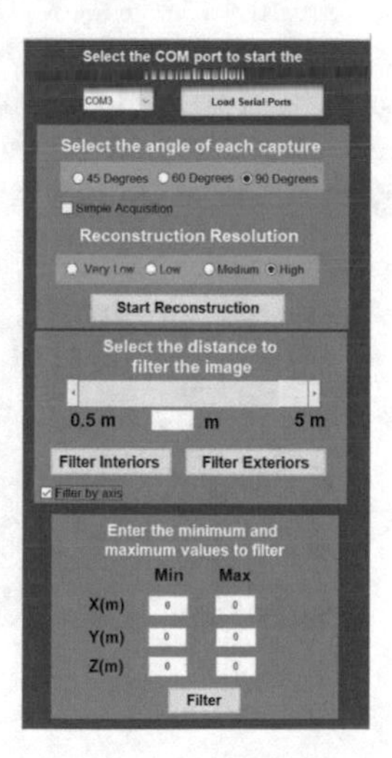

(a) (b)

Fig. 2. 3D reconstruction interfaces. (a) RPLIDAR A1 interface. (b) Kinect v1, Kinect V2 and ZED 2K stereo camera interface.

Table 2. Kinect V1 and V2, ZED stereo camera and RPLIDAR A1 characteristics

Functionality	Description
Save	Allows to save the original reconstruction previously obtained in a ".mat" file with the name provided by the user or to export the final result as a ".ply" file
Load file	Displays a menu, which allows you to select the resolution of the previously stored reconstruction. Later, it shows the point cloud loaded
Port management	Displays the available ports for communication between the devices and the interface
Platform rotation angle	Defines the angle at which the platform rotates at each sensor captures. These are 45°, 60°, 90° and 1.8° for the LIDAR sensor
Reconstruction range (LIDAR)	Allows you to select the reconstruction range with values from 1.8° to 360°
Reconstruction resolution	Defines the quality of the reconstruction obtained, allowing a greater fluidity of the result in devices with low processing resources
Filter by axis	Obtain the point cloud according to the limits in each axis entered
Distance filter	Eliminates the points that are lower than the minimum value and higher than the maximum value established, both inside and outside
Filter by angle (LIDAR)	Allows you to delete points depending on a set angular range

4 Test and Results

The developed tool allows the management of all the proposed devices in version MATLAB 2016. In order to test the tool, an experimental environment was used as is shown in Fig. 3. Moreover, for the next experimental scenarios must be considered that the 3D- reconstruction system is located in the center of the test room to perform the point cloud acquisition. In the next tests, all the 3D - reconstructions are shown with an upper view in order to appreciate the general geometrical form of the original test environment (see Fig. 4).

Fig. 3. Test environment

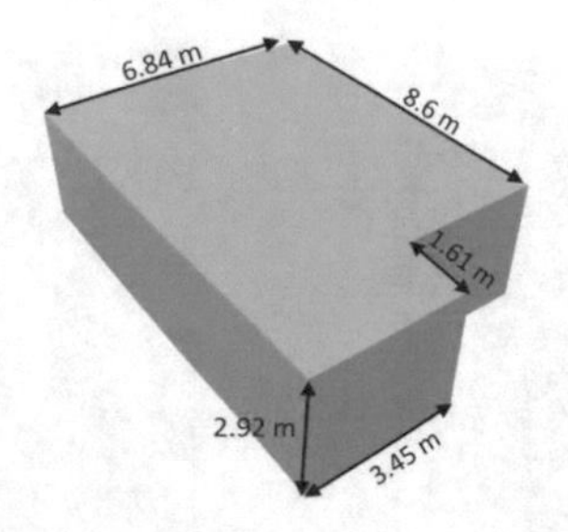

Fig. 4. Test environment

In Fig. 5 are shown the different point clouds obtained by the different devices. These point clouds are a three-dimensional representation of the scenario in 360°. In this reconstruction it can be clearly observed the existence of atypical points referring to noise, which reduce its quality. Red marks in Fig. 5 are shown in order to compare the filtered point cloud with the Fig. 6. Likewise, in Fig. 5a it can be seen that there are cutouts in the corners with respect to the real test environment, this is due to the fact that the environment exceeds the range of maximum acquisition range of the Kinect V1.

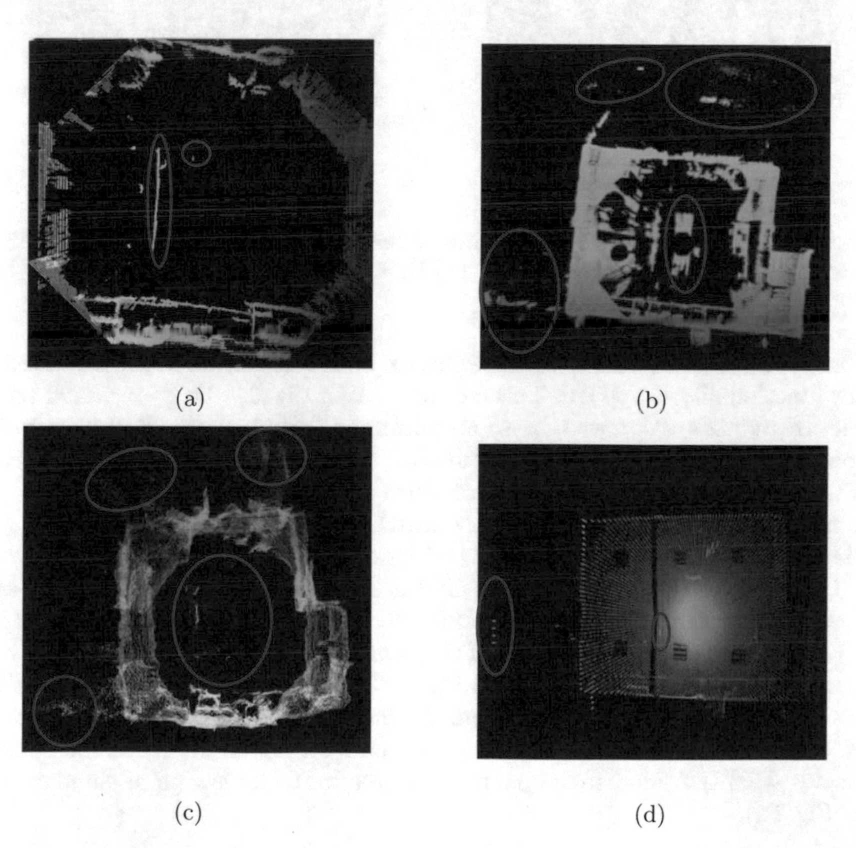

Fig. 5. Raw 3D reconstruction. (a) Kinect v1. (b) Kinect V2. (c) ZED 2K stereo camera. (d) RPLIDAR A1.

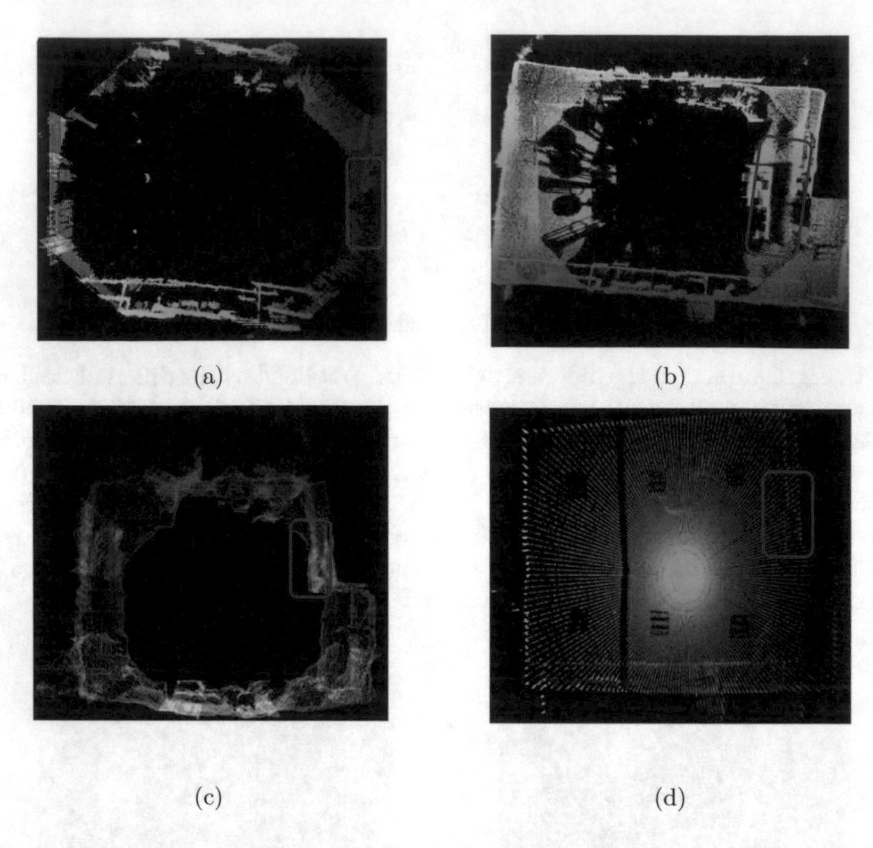

Fig. 6. Minimum and maximum distance filtered reconstruction. (a) Kinect v1. (b) Kinect V2. (c) ZED 2K stereo camera. (d) RPLIDAR A1.

In order to carry out a better analysis of the reconstruction, the distance filter functionality is applied. This can be seen in Fig. 6, where unwanted reconstruction points are removed, based on minimum and maximum limits applied to the axis (x, y, z). In this case any points less than 1.5 m were removed for Kinect v1, v2, ZED and RPLIDAR A1 device. Similarly, everything that is greater than the maximum value entered was eliminated, being 5 m used for Kinect v1, v2, ZED 2k stereo camera and RPLIDAR A1 device.

Thanks to the implementation of GUI it is possible to apply the filter by axis, which allows to segment the point cloud according to the ranges of coordinates (x, y, z) entered. In this way, portions of the point cloud are discarded focusing on a certain area of interest as is shown in Fig. 7. This interest area was obtained from the point cloud surrounded by red marks on Fig. 6. As can be seen, Kinect V1 and V2 shown a better performance in comparison with ZED 2K stereo camera. However, RPLIDAR A1 shows poor resolution due to its low point cloud density (see Fig. 7d).

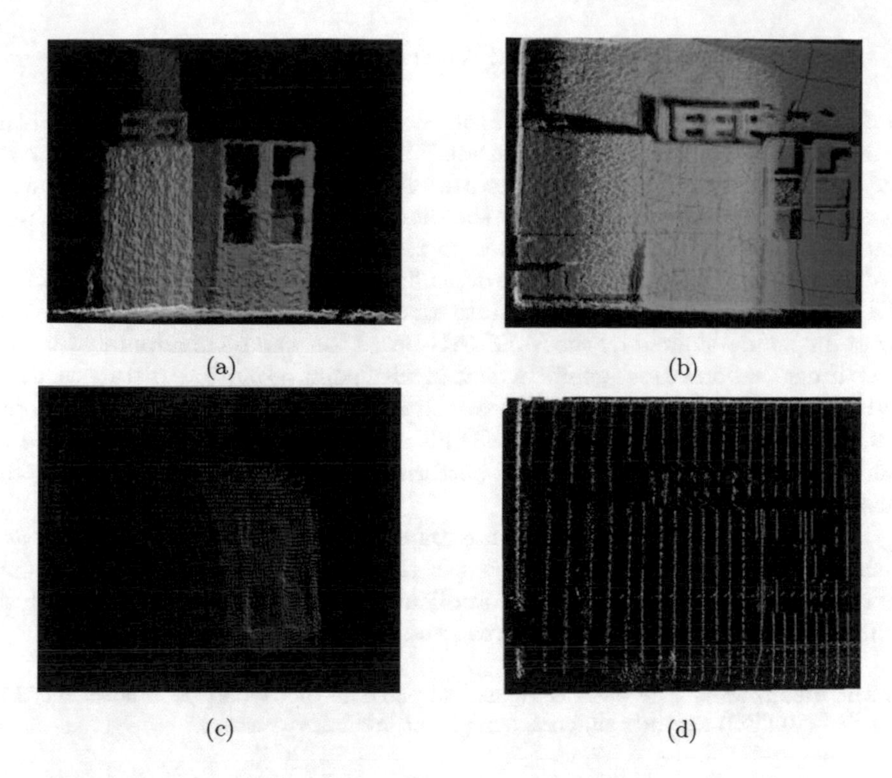

Fig. 7. Filter by axes. (a) Kinect v1. (b) Kinect V2. (c) ZED 2K stereo camera. (d) RPLIDAR A1.

Table 3. Data point cloud density

X, Y, Z points	Kinect V1	Kinect V2	ZED 2K	RPLIDAR A1
Raw 3D reconstruction	643230	8786241	73728	400000
Minimum distance filtered	596842	7629561	64850	187071
Maximum distance filtered	643230	8180334	71347	396047
Maximum and minimum distance filter	596842	7023654	62469	183118
Filter by axes	102124	884520	13370	13949

In Table 3 is shown the total points (x, y, z) plotted in Figs. 5, 6 and 7, where it could be seen that the point cloud density is reduced depending of each applied filter.

5 Conclusions and Future Works

In this document a MATLAB-based tool was developed in order to manage the data acquisition of four different sensors: Kinect V1, V2, ZED 2K stereo camera and RPLIDAR A1. The four devices used for the development of the tool were chosen due to its high availability in the market, its low costs and its wide range of applications both commercially and in research.

The tool is capable of acquire the depth information of each sensor in terms of global coordinates (x, y, z) and generate an environment 3D reconstruction. The 3D point cloud obtained by the MATLAB-based tool can be manipulated by the user through several functionalities related with point resolution, distances filters and write and read file options. However, due to the basic operating characteristics, especially the Kinect V2 and ZED 2K stereo camera, it is recommended to use this tool in computers with high performance and preferably with a graphic card.

For future work it is proposed to expand the devices that can be managed with the tool, in order to facilitate the user the acquisition of point clouds for further analysis, comparisons or future applications. Moreover, a usability analysis could be performed in order to improve the user interface.

Acknowledgment. The authors thanks to the staff of Unidad de Innovación Tecnológica (UITEC) for their support, equipment and infrastructure.

References

1. Cheng, Y., Wang, G.Y.: Mobile robot navigation based on lidar. In: 2018 Chinese Control and Decision Conference (CCDC), pp. 1243–1246 (2018)
2. MoBiVAP Group Research: Reconstrucción 3D, p. 15 (2013)
3. Vogiatzis, G., Hern, C.: Practical 3D reconstruction based on photometric stereo. In: Computer, pp. 313–345 (2012)
4. Pan, Z., Tian, S., Guo, M., Zhang, J., Yu, N., Xin, Y.: Comparison of medical image 3D reconstruction rendering methods for robot-assisted surgery. In: 2017 2nd International Conference on Advanced Robotics and Mechatronics, ICARM 2017, vol. 2018-January, pp. 94–99 (2018)
5. Bostanci, E.: 3D reconstruction of crime scenes and design considerations for an interactive investigation tool, pp. 1–9 (2015). http://arxiv.org/abs/1512.03156
6. Bobkov, V.A.: 3D reconstruction of underwater objects with autonomous underwater vehicle, pp. 3–7 (2017)
7. MIT: Mechanical engineers develop an intelligent co-pilot' for cars (2012). http://news.mit.edu/2012/mechanical-engineers-develop-intelligent-car-co-pilot-0713
8. Ilbay, L.: Reconstrucción activa de objetos 3D mediante escaneo láser y generación de la vista por medio del software matlab, pp. 1–126 (2014)
9. Navas, D., Vargas, J., Morales, L.: The nearest object localization through 3D lidar reconstruction using an embedded system. XXVII Jornadas de Ingeniería Eléctrica y Electrónica, Escuela Politécnica Nacional **27**, 32–38 (2017)

10. Cai, L., Wu, K., Fang, Q., Zheng, R.: Fast 3D modeling Chinese ancient architectures base on points cloud. In: 2010 International Conference on Computational Intelligence and Software Engineering, CiSE 2010, no. 200809128, pp. 4–6 (2010)
11. Yang, S.-C., Fan, Y.-C.: 3D building scene reconstruction based on 3D LiDAR point cloud. In: IEEE International Conference on Consumer Electronics, pp. 127–128 (2017)
12. Miao, R., Song, J., Zhu, Y.: 3D geographic scenes visualization based on WebGL. In: 2017 6th International Conference on Agro-Geoinformatics, Agro-Geoinformatics 2017 (2017)
13. Hernández, I.: métodos fotogramétricos (2015)
14. Ramírez-Pedraza, A., González-Barbosa, J.J., Ornelas-Rodríguez, F.J., García-Moreno, A.I., Salazar-Garibay, A., González-Barbosa, E.A.: Detección de Automóviles en Escenarios Urbanos Escaneados por un Lidar. RIAI - Revista Iberoamericana de Automatica e Informatica Industrial **12**(2), 189–198 (2015)
15. Navarro, F.: Aplicaciones de la Reconstrucción 3D: odometría visual e Integración con la realida virtual, pp. 1–64 (2017)
16. Pozo, D., Jaramillo, K., Ponce, D., Torres, A., Morales, L.: 3D reconstruction technologies for using in dangerous environments with lack of light: a comparative analysis. In: II Congreso Internacional de Sistemas Inteligentes y Nuevas Tecnologías: Tendencias Interdisciplinares en Comunicación (2019, Submitted)
17. Gupta, T., Li, H.: Indoor mapping for smart cities—an affordable approach: using kinect sensor and zed stereo camera. In: 2017 International Conference on Indoor Positioning and Indoor Navigation (IPIN), pp. 1–8, September 2017
18. Boora, S., Sahu, B.C., Patra, D.: 3D image reconstruction from multiview images. In: 2017 8th International Conference on Computing, Communication and Networking Technologies (ICCCNT), pp. 1–7, July 2017
19. Valgma, L., Daneshmand, M., Anbarjafari, G.: Iterative closest point based 3D object reconstruction using RGB-D acquisition devices. In: 2016 24th Signal Processing and Communication Application Conference (SIU), pp. 457–460, May 2016
20. Qinghua, H., Zhaozheng, Z.: A review on real-time 3D ultrasound imaging technology. BioMed. Res. Int. **2017**, 20 (2017)
21. Caminal, I., Casas, J., Royo, S.: SLAM-based 3D outdoor reconstructions from lidar data. In: 2018 International Conference on 3D Immersion (IC3D), pp. 1–8, December 2018
22. Schwarz, B.: LIDAR: mapping the world in 3D. Nat.Photon. **4**(7), 429–430 (2010). https://doi.org/10.1038/nphoton.2010.148
23. Li, B., Yang, L., Xiao, J., Valde, R., Wrenn, M., Leflar, J.: Collaborative mapping and autonomous parking for multi-story parking garage. IEEE Trans. Intell. Transp. Syst. **19**(5), 1629–1639 (2018)
24. Lachat, E., Macher, H., Landes, T., Grussenmeyer, P.: Assessment and calibration of a RGB-D camera (Kinect v2 Sensor) towards a potential use for close-range 3D modeling. Remote Sens. **7**(10), 13070–13097 (2015)
25. Lachat, E., Macher, H., Mittet, M.A., Landes, T., Grussenmeyer, P.: First experiences with kinect V2 sensor for close range 3D modelling. In: International Archives of the Photogrammetry, Remote Sensing and Spatial Information Sciences - ISPRS Archives, vol. 40, no. 5W4, pp. 93–100 (2015)
26. Popescu, C.R., Lungu, A.: Real-time 3D reconstruction using a kinect sensor. Comput. Sci. Inf. Technol. **2**(2), 95–99 (2014)
27. Wasenmuller O., Stricker D.: Comparison of kinect V1 and V2 depth images, vol. 10118, September 2017. https://doi.org/10.1007/978-3-319-54526-4

28. StereoLabs: ZED Stereo Vision (2018). https://docs.stereolabs.com/overview/getting-started/introduction/
29. Kirsten, E., Inocencio, L.C., Veronez, M.R., Da Silveira, L.G., Bordin, F., Marson, F.P.: 3D data acquisition using stereo camera. In: IGARSS 2018 - 2018 IEEE International Geoscience and Remote Sensing Symposium, pp. 9214–9217, July 2018

Measuring Surfaces in Orthophotos Based in Color Segmentation Using K-means

F. Enriquez, G. Delgado(✉), F. Arbito, A. Cabrera, and D. Iturralde

University of Azuay, Cuenca, Ecuador
efejaramillo@es.uazuay.edu.ec, {gabrieldelgado, apcabrera, diturralde}@uazuay.edu.ec, teslamillingec@outlook.com

Abstract. This research presented a method to perform the measurement of orthophotography areas. An unsupervised segmentation was developed using K-means for each pixel to reconstruct an image with fewer colors to perform an area measurement through each given color tone. The Calinski-Harabasz index (CHI) was also studied; which allows a better choice of the color space in which the algorithm will be executed. The results of the execution times for each color space are presented, obtaining an error of approximately 0.08% in the measurement of the area.

Keywords: K-means · Areas · Clustering · Orthophotography · Color spaces

1 Introduction

Measurements of particular urban or rural areas help in several research projects in civil engineering, mining engineering, high voltage cable routing, biology, natural reserves, agriculture, etc. These measurements are usually taken using satellites, which have limitations to determine patterns due to the low quality of satellite images and its big content shades of grey. However, due to new techniques that apply artificial intelligence algorithms it is possible to obtain better measurements [1].

Image segmentation is a fundamental process, and at the same time is a classic problem in the majority of computer vision applications. Usually, image segmentation is done using a greyscale image, which means that only information related with intensity is used. On the other hand, color image segmentation offers more levels of discrimination that reach millions [2].

The most used methods for image segmentation are based in histograms and association of colors. The method of color association includes two branches: supervised and unsupervised. Supervised algorithms are simple because they work with input data and the expected (output) data vector. However, they have a big disadvantage that consists in the loss of information related to color [2]. By contrast, unsupervised methods need training data, which consist in input vectors without a specific goal. Instead of searching a specific data, the unsupervised algorithm autonomously finds a pattern inside the data,

M. Botto-Tobar et al. (Eds.): ICAETT 2019, AISC 1066, pp. 291–300, 2020.
https://doi.org/10.1007/978-3-030-32022-5_28

resulting in a small error rate and an improvement in the rate of success in segmentation, though with a bigger computational cost [3].

This kind of classification systems help to develop algorithms capable of search and categorize patterns by colors in photographs, and, in this context, orthophotographs.

2 Methodology

This research is focused on obtaining a system able to measure areas using machine learning and unsupervised algorithms such as the one shown in (Fig. 1). The algorithm needs as input an orthophotographic reconstruction of the area to classify each pixel according to its color components to group them and create masks based on color.

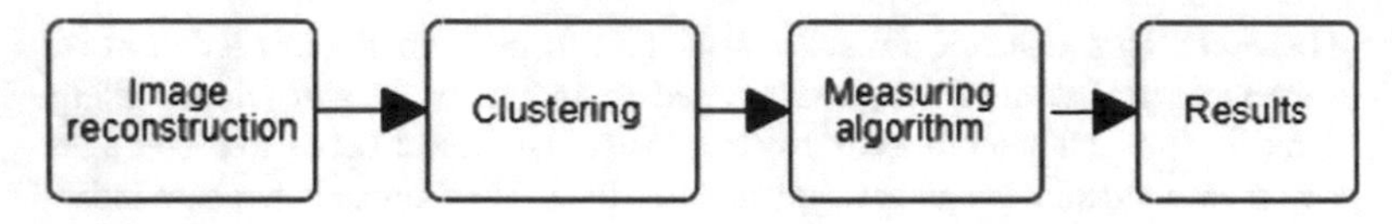

Fig. 1. Methodology

2.1 Orthophotography

Previously to the implementation of the algorithm, orthophotographs with different spatial resolutions are needed [5, 6]. A result of an orthophotograph using “Agisoft PhotoScan” [4], is shown in Fig. 2, as a result of the process of 250 aerial images covering an area of 2.1 ha with a spatial resolution of 1.78 cm/px.

Fig. 2. Orthophotography. 2.1 ha.

2.2 Clustering Using K-means

K-means is one of the most used unsupervised learning algorithms for cluster analysis. This denomination was used for the first time by MacQueen, [7]. The goal of this algorithm is the solid clustering of data in k groups using the nearest mean, giving as a result a partition of the data space into Voronoi cells [8, 9].

The result given by the k-means algorithm is such that it groups the data with high similarity and at the same time, each group is very different to the other clusters [12]. The similarity is decided by the median value of a group with respect to its centroid.

The algorithm uses Eq. (1) with the method of least squares.

$$J = \sum\nolimits_{j=1}^{k} \sum\nolimits_{i=1}^{n} \left\| X_I^{(J)} - C_J \right\|^2 \quad (1)$$

Where $\left\| X_I^{(J)} - C_J \right\|^2$ is the distance between a data point $X_I^{(J)}$ and its centroid C_J. K-means algorithm is shown in Fig. 3.

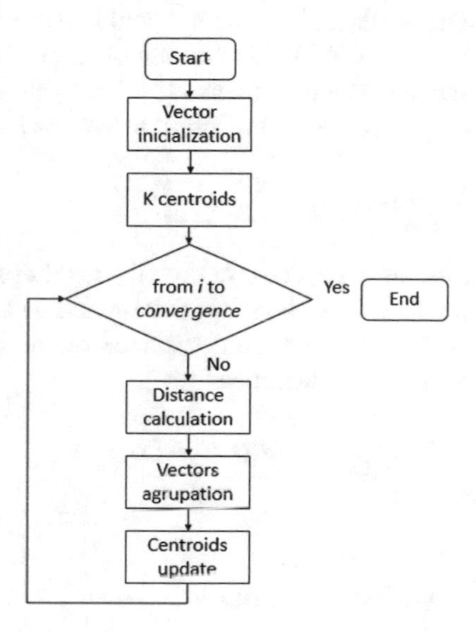

Fig. 3. Flow chart of the K-means method

2.3 Image Reconstruction

Image reconstruction consist of recreating the original image with a limited number of colors previously obtained by the k-means algorithm. To perform this procedure, a bidimensional zeros matrix is constructed, whose size is given by the width (w) and height (h) of the original photograph, these values are also used to create a nested loop that runs through each vector and at the same time it replaces it with the value of the corresponding centroid. Then the image is reconstructed by replacing each pixel with the centroid value of its group.

2.4 Measuring Algorithm with Binary Images

Area measurement consists of counting pixels of the same color that afterwards are converted in units of area using the scale given by the orthophotographic procedure (cm^2, m^2, km^2).

In order to obtain the number of pixels, the system uses a threshold mask to generate a binary image (two levels of intensity). Equation (2) determines the threshold used to classify the pixels into the two groups, where LS is the highest color value, LI is the lowest color value, P_i is the original pixel value, and P_0 is the final pixel value [12]

$$P_o = \begin{cases} 0\, si\, LI < P_i(i,j) < LS \\ 1\, si\, LI \leq P_i(i,j) \geq LS \end{cases} \tag{2}$$

To eliminate the noise in the binary image, basic morphologic transformations are applied, such as closing (dilation and erosion) that helps to fill in the interstices of an image [12].

The particular shape of an object in an image could be obtained using thc measure of its moments. A zeroth moment M (0,0) is the sum of all pixels of a certain value, and Eq. (3) shows the total number of white pixels [13]. Thus, the area in square meters is given by Eq. (4), where D_p is the spatial resolution (cm^2/px)

$$M(0,0) = \sum x \sum y f(xy) \tag{3}$$

The measurement of the areas depends on both the height of the programmed flight and the quality of generation of the orthophoto determined by the software, in addition to the minimum flight speed, the minimum distance of the overlap of the images, resolution of the camera and its shutter speed [14].

$$Area = \frac{M(0,0) * (D_p)^2}{10000} \tag{4}$$

2.5 Quality Measure and Validity Index

In order to verify which color space gives the best result, the Calinksi-Harabasz Index (CHI) is used. This index is a variance relationship criterion that gives an idea for the data structure. This method is described in Eq. 5 [11].

$$CHI = \frac{SS_B}{SS_W} x \frac{N - k}{K - 1} \tag{5}$$

Where N represents the total number of data, k the total number of clusters, SS_b is the variance between clusters and SS_w is the variance of the cluster [11]. The CHI values show that, if the criterion is larger, the clustering is better.

3 Results

Analyzed data were obtained from an orthophotograph with spatial resolutions of 1.78 cm/px and 10 cm/px, from a space of 2 ha approximately, using a segmentation with k = 12.

Figure 4 shows the results obtained after the image reconstruction with the values given by the algorithm after the clustering applied to the orthophotographs in HSV color space. It is important to mention that among the 12 final clusters white is included, which is discarded in the masks because it belongs to the background of the image [10].

a) Spatial resolution of 1.78 cm/px.

b) Spatial resolution of 10 cm/px

Fig. 4. Orthophotograph reconstruction results (HSV) with different spatial resolutions and k = 12

As expected, the reconstruction values (centroids given by the k-means algorithm) from Fig. 4a and b are similar, although its spatial resolution is different.

With each centroid (color) obtained by the k-means algorithm, the area corresponding to each color is calculated. Next, the results are shown to the user, who decides which ones are of its interest.

The measurements of each segmentation with its HSV value are shown in Table 1.

Table 1. Area measurements of masks in HSV (1.78 cm/px)

HSV color	Measurement (m^2)
[130.24, 158.39, 157.33]	354.514
[120.93, 49.87, 57.16]	708.173
[104.14, 19.77, 214.25]	1531.60
[112.31, 23.22, 112.71]	3975.951
[113.07, 14.5, 153.39]	2258.753
[18.01, 24.95, 145.85]	2497.51
[173.89, 74.43, 152.11]	1584.269
[6.725, 116.47, 167.55]	725.649
[38.36, 63.62, 54.26]	3036.562
[41.74, 114.40, 66.29]	3626.169
[33.83, 67.94, 103.65]	3887.056

The same procedure is applied to segment the color space (in HSV) of the image with spatial resolution of 10 cm/px.

Table 2 shows the dimensions for the masks with its respective segmentation value.

Table 2. Area measurements of masks in HSV (10 cm/px)

HSV color	Measurement (m^2)
[127.80, 150.98, 156.83]	335.21
[103.787, 17.03, 214.59]	1616.41
[19, 19.93, 146.04]	2374.74
[106.38, 14.86, 143.77]	2797.71
[111.86, 26.30, 103.79]	3720.04
[6.26, 109.07, 168.84]	693.85
[173.17, 58.38, 146.36]	1988.89
[40.75, 128.71, 57.96]	1717.4
[39.39, 41.41, 68.23]	2368.67
[32, 63.17, 107.38]	3430.5
[41.28, 84.55, 65.03]	4162.23

The algorithm previously described is applied in the RGB color space, for comparison. The results for each RGB masks measurements applied to 1.78 cm/px image are shown in Table 3.

Table 3. Area measurements of masks in RGB

RGB color	Measurement (m^2)
[61.45, 90.45, 159.35]	233.372
[212.31, 220.94, 225.7]	597.777
[180.99, 185.86, 189.43]	1391.565
[155.44, 153.93, 153.43]	1873.262
[32.30, 36.92, 25.84]	2166.686
[156.68, 103.14, 104.85]	2207.307
[132.45, 130.98, 128.16]	2794.846
[94.92, 97.20, 82.26]	2834.28
[74.13, 82.26, 57.87]	3105.772
[54.05, 61.27, 41.45]	3265.765
[108.11, 110.35, 109.56]	3607.206

The procedure is repeated applying the algorithm to the 10 cm/px image with RGB color space. The result is shown in Table 4.

Table 4. Area measurements of masks in RGB.

RGB color	Measurement (m^2)
[66.25, 95.86, 159.32]	221.94
[220.86, 227.14, 231.21]	516.35
[184.93, 191.12, 194.56]	1164.46
[160.49, 160.38, 161.03]	1514.98
[157.45, 107.42, 107.73]	2028.77
[137.44, 136.05, 135.21]	2662.22
[37.78, 43.10, 28.94]	2669.1
[118.55, 117.16, 114.22]	3190.78
[81.63, 87.32, 66.40]	3503.69
[99.84, 102.06, 97.71]	3670.13
[59.76, 66.24, 46.56]	3973.83

A very important aspect to assess the proposed algorithm is the execution time, from the adjustment of the data samples, until the area measurement. Table 5 shows the execution time for the algorithm using masks, which is given by the addition of the execution time of the different sub processes to complete the algorithm.

Table 5. Execution time using masks in HSV color space

Cm/píxel	Execution time (s)				
	t Adjust.	*t Labeled*	*t Recons.*	*t Measur.*	Total
1,78	82,713	22,542	116,397	103,809	327,241
10	3,313	0,729	2,827	2,717	19,586

Similarly, Table 6 shows the algorithm execution times for the masks in RGB color space

Table 6. Execution time using masks in RGB color space

Cm/píxel	Execution time (s)				
	t Adjust.	*t Labeled.*	*t Recons.*	*t Measur.*	Total
1,78	85,385	18,965	265,753	62,127	434,01
10	5,546	0,729	2,91	2,816	22,001

Table 7 shows the index values obtained using the CHI criterion, being the best value the larger one.

Table 7. Index values for clustering using k-means for the orthophotographs

Cm/píxel	Calinski-Harabasz index	
	HSV	RGB
1.78	17847,223	10090,756
10	52981,274	29955,421

Table 8 shows the results of clustering selected by user (using green tones) to obtain the total area measurement.

Table 8. Comparisons between measurements and execution time.

Tone color	Cm/píxel	Measurements			
		HSV		RGB	
		Measurement	*Time*	*Measurement*	*Time*
Green tones	1.78	10549.79	327,241	8538.22	434,01
	10	11679.07	19,586	10146.62	22,001

In order to calculate the errors, measurement using polygons approximation is taken as the norm (Fig. 5). The resulting area of green space is 10 558.4 m^2.

Fig. 5. Area approximation using polygons.

4 Discussion

Analysis of Tables 1, 2, 3 and 4 is performed, observing that the centroid values change as the spatial resolution of the image changes because the amount of information contained in each pixel (color) is modified. The value of the centroids will also change due to the change in the color space, because HSV color space is a nonlinear transformation of the RGB color space.

The time difference to obtain the clustering result between RGB and HSV is very similar. However, the same cannot be stated about the measurements obtained by these two. The values in Table 7 show that the HSV color space has better clustering, expressed in a larger CHI value obtained compared to the RGB one.

Comparing the measurement values shown in Table 8 with the data obtained by "Agisoft PhotoScan" software, the error is 0.08% for the image with a spatial resolution of 1.78 cm/px, and 10.61% for the image with a 10 cm/px resolution. Thus, we can conclude that the error is proportional to the orthophotographs spatial resolution.

Acknowledgement. This research was supported by Vicerrectorado de Investigaciones of Universidad del Azuay. We thank our colleagues from Department of Electronics Engineer who provided insight and experience that greatly assisted the research.

References

1. Lear, C.: Digital orthophotography: mapping with pictures. IEEE Comput. Graph. Appl. **17**(5), 12–14 (1997)
2. Li, J., Wang, J., Mao, J.: Color moving object detection method based on automatic color clustering. In: Proceedings of the 33rd Chinese Control Conference, pp. 7232–7235 (2014)
3. Dehariya, V., Shrivastava, S., Jain, R.: Clustering of image data set using k-means and fuzzy k-means algorithms. In: 2010 International Conference on Computational Intelligence and Communication Networks, pp. 386–391 (2010)
4. Agisoft LLC: Agisoft (2018). http://www.agisoft.com/
5. Szeliski, R.: Image alignment and stitching. Found. Trends Comput. Graph. Vis. 43–61 (2006)
6. Brown, M., Szeliski, R., Winder, S.: Multi-image matching using multi-scale oriented patches. In: 2005 IEEE Computer Society Conference on Computer Vision and Pattern Recognition, CVPR 2005, pp. 510–517 (2005)
7. Arabie, P., De Soete, G.: Clustering and Classification. Word Scientific, Singapore (1998)
8. Agarwal, S., Yadav, S., Singh, K.: K-means versus k-means++ clustering technique. In: 2012 Students Conference on Engineering and Systems, pp. 1–6 (2012)
9. Day, W., Edelsbrunner, H.: Efficient algorithms for agglomerative clustering methods. J. Classif. **1**, 7–24 (1984)
10. Ganesan, P., Rajini, V.: Assessment of satellite image segmentation in RGB and HSV color space using image quality measures. In: 2014 International Conference on Advances in Electrical Engineering (ICAEE), pp. 1–5 (2014)
11. Łukasik, S., Kowalski, P. Charytanowicz, M., Kulczycki, P.: Clustering using flower pollination algorithm and Calinski-Harabasz index. In: 2016 IEEE Congress on Evolutionary Computation (CEC), pp. 2724–2728 (2016)

12. Alegre, E., Pajares G., de la Escalera, A.: Conceptos y Métodos en Visión por Computador, España: Comité Español de Automática (CEA), p. 169 (2016)
13. Domínguez, S.: Analisis de imágenes mediante el metodo de los momentos usando funciones de base continuas a intervalos (PCBF). Revista Iberoamericana de Automática e Informática industrial **12**, 69–78 (2015)
14. Hernández, L., Rodríguez, E., Martínez, A., Álvarez, H., Kharuf, S.: Levantamiento Fotogrametrico De La Ubpc "Desembarco Del Granma" Utilizando Aviones No Tripulados, Solución De Bajo Costo Para La Agricultura Nacional. In: VII Edición de la Conferencia Científica Internacional sobre Desarrollo Agropecuario y Sostenibilidad, pp. 1–14 (2016)

Biomechanics of Soft Tissues: The Role of the Mathematical Model on Material Behavior

Carlos Bustamante-Orellana[1], Robinson Guachi[3,5], Lorena Guachi-Guachi[1,2](✉), Simone Novelli[3,4], Francesca Campana[3], Fabiano Bini[3], and Franco Marinozzi[3]

[1] Yachay Tech University, Hacienda San José, Urcuquí 100119, Ecuador
lguachi@yachaytech.edu.ec
[2] SDAS Research Group, Yachay Tech, Urcuquí, Ecuador
[3] Department of Mechanical and Aerospace Engineering, Sapienza University of Rome, Via Eudossiana 18, 00184 Rome, RM, Italy
[4] Institute for Liver and Digestive Health, University College London (UCL), Gower Street, London WC1E 6BT, UK
[5] Universidad Internacional del Ecuador, Av. Simon Bolivar, Quito 170411, Ecuador

Abstract. Mechanical properties of the soft tissues and an accurate mathematical model are important to reproduce the soft tissue's material behavior (mechanical behavior) in a virtual simulation. This type of simulations by Finite Element Analysis (FEA) is required to analyze injury mechanisms, vehicle accidents, airplane ejections, blast-related events, surgical procedures simulation and to develop and test surgical implants where is mandatory take into account the high strain-rate. This work aims to highlight the role of the hyperelastic models, which can be used to simulate the highly nonlinear mechanical behavior of soft tissues.

After a description of a set of formulations that can be defined as phenomenological models, a comparison between two models is discussed according to case study that represents a process of tissues clamping.

Keywords: Hyperelastic mathematical models · Soft tissues behavior · FEA

1 Introduction

A mathematical model may reproduce the mechanical behavior of soft tissues (tissues and organs) of animals and humans [1]. Nevertheless, the inherent complexity in the biological tissue's mechanical behavior and difficulties to acquire the parameters to describe the relationship stress-strain, have given place to an active area of research based on realistic modelling and simulation of tissue deformation.

Although, most of the works in the biomechanics domain are focused on understanding the fundamental properties of various tissues [2–4]. There is still a lack of a general mathematical model able to reproduce the mechanical behavior of soft tissues

M. Botto-Tobar et al. (Eds.): ICAETT 2019, AISC 1066, pp. 301–311, 2020.
https://doi.org/10.1007/978-3-030-32022-5_29

considering different load conditions as compression, axial tension, biaxial tension, shear, etc.

In literature, the effects of different testing conditions (tissues clamping, Influence of geometry, simple applied load and small deformation) [5–7] have been studied, which have demonstrated that the mission of choosing an appropriate mathematical model to describe the stress-strain response, when the tissue is subjected to different load conditions, is not straightforward.

The human organs and the majority of the soft tissues are inhomogeneous, anisotropic and often have inherently nonlinear viscoelastic mechanical behavior [7–9] because they are comprised of substantial amounts of interstitial fluid. Therefore, this mechanical behavior involves a response that changes with the time as a product of tissue relaxation, which implies different properties from traditional solid materials that are typically described with a linear elastic (LE) or hyperelastic behavior. All the above-mentioned characteristics together with a nonlinear stress-strain relationship make the characterization of biological tissues complicated. Due to these difficulties, some researches [8, 10–12] have introduced mechanical behavior approximations as linear elastic and hyperelastic material models, the applicability limit of these approximations is related to the strain rate that may affect the stress response [13, 14]. Therefore, it is also necessary to consider that hyperelastic models may replace a nonlinear visco-elastic model only if the simulation does not require load cycling as it happens in relaxation phenomenon [15]. This has recently been confirmed in bioengineering simulations [16], where visco-elastic properties are invoked in the relaxation step in analysis of breast tissues by considering time that tends towards infinity. It is important to mention that, in simulations by FEA is necessary to consider the sources of nonlinearities, such as the mechanical behavior and geometrical characteristics [17], with the aim to simulate the phenomenon in a similar way as it happens in the real world.

In this work we provide the description of a set of mathematical models also known as phenomenological hyperelastic models [18, 19] used to reproduce the mechanical behavior of soft tissues. In order to provide a point-of-view to choose an optimal mathematical model, their mathematical principle is evaluated, and the information of such models is summarized in a manner that it could be useful to experimentalists, doctors, and computational modelers.

The paper is organized as follows: Sect. 2 describes the most relevant behavior of soft tissues, particularly the phenomenological models, together with a comparison and discussion about the explored mathematical models. While, Sect. 3 presents a specific case study to give evidence of the difference between linear models and hyperelastic formulation during the Finite Element Analysis of an abdominal surgical procedure. Finally, some concluding remarks are presented in Sect. 4.

2 Phenomenological Models

Hyperelastic models were firstly addressed to describe stress-strain behavior of elastomers which are highly non-linear. From the solid mechanics point of view, the stress-strain curve of many hyperelastic models may be derived from the Strain Energy

Function (SEF), denoted as W, in function of three strain invariants I_1, I_2 and I_3, as shows Eq. (1):

$$W = f(I_1; I_2; I_3) \tag{1}$$

These three parameters are invariants of Cauchy-Green deformation tensor, which are expressed in terms of principal stretch ratios λ_1, λ_2 and λ_3. Other mathematical models use these stretch ratios to build their SEF as well. As elastomers are considered incompressible, models that use the three strain invariants can be simplified by using the assumption $I_3 = 1$, and therefore, W can be expressed as a function of two invariants only as is shown in Eq. (2):

$$W = f(I_1 - 3; I_2 - 3) \tag{2}$$

In addition, hyperelastic models use different material constants, which vary depending on the material that is being modeled and are usually denoted as C_{ij}.

In the case of phenomenological models W, is a model that express mathematically the results of observed phenomena through constant material parameters [20]. These material parameters are usually difficult to determine for such models, since they ask for ad-hoc experiments to be used in model calibration loops [21]. According to this, they can conduct to errors when used out of the deformation range in which their parameters were identified [18]. A brief description of the most discussed and applied phenomenological hyperelastic models is presented:

Money-Rivlin Model. [18] Initially known as Money model, it was firstly proposed by Money [22] in 1940 and then expressed in terms of invariants by Rivlin [23] in 1948. It is known by being one of the first hyperelastic models, which can handle very well strains lower than 100% [24]. However, it cannot capture the upturn (S-curvature) of the force-extension relation in uniaxial test and the force-shear displacement relation in shear test [12]. The form of the model for a compressible material is presented in Eq. (3):

$$W = C_{10}(I_1 - 3) + C_{01}(I_2 - 3) + \frac{1}{D_1}(J_{el} - 1)^2 \tag{3}$$

where:

C_{10} and C_{01} are material parameters. J_{el} is the elastic volume ratio. D_1 is a constant that defines the compressibility of the material, which can be estimated from volumetric test. I_1 and I_2 are the first and second invariants.

Considering an incompressible material, such as the rubber and soft tissues, the elastic volume ratio J_{el} becomes 1; and therefore, the Money-Rivlin model for an incompressible material is the one presented in Table 1. The Money-Rivlin model has some variants, which extend the basic model using different approaches. Some of these variants are (Table 1):

Full Polynomial Model. [13] This variant can be found in the literature with the same name "Money-Rivlin" [19] or as Full Polynomial model [13]. The model expresses a SEF as a series, which is often truncated to terms of second and third order [18], such truncations can constitute another model.

Biderman Model. [18] This model is a truncation of the series of Full Polynomial model but retaining only the terms for which i = 0 or j = 0. Only the first three terms for I_1 and one term for I_2 were considered.
Haines-Wilson Model. [18] This model is another truncation of the Full Polynomial model, choosing to retain only six terms of the series based in the comparison of invariants and principal stretches development of the SEF of Full Polynomial model.
Yeoh Model. [25] This model, also known as reduced polynomial, is based in the first invariant I_1 only. As previous mentioned models, it is also the result of selecting specific terms of the series of Full Polynomial model. It can capture the upturn of stress-strain curve of rubber, has good fit over a large strain range and can simulate various modes of deformation with limited data [13].
Neo-Hookean Model. [13] It is the simplest hyperelastic model, being first proposed in 1943. It is very similar to Hooke's law although it is a hyperelastic model, in fact, it is a particular case of Money-Rivlin (cannot capture the upturn of stress-strain curve) when C_{01} = 0. This model can be used when material data is limited, being able to make good approximations at small strains, up to 20% [26].

Ogden Model. [13] It is based on principal stretch ratios λ_1, λ_2, and λ_3 instead of invariants as the previous models. This model can capture upturn (stiffening) of stress-strain curve and model rubber accurately for large ranges of deformation, which can reach up to 700%. It is not advisable to use this model with limited laboratory test; the general form of the model can be seen in Table 1, where μ_n and α_n are material parameters. An excellent convergence between theoretical and experimental results is achieved when N = 3 [18]. This model is widely used for large strain problems, even if it is hard to determine the material parameters.

Gent-Thomas Model. [19] It is a two-parameter model similar to Money-Rivlin but using a scaled logarithm on I_1. However, this model has not proved to be more efficient than the basic Money-Rivlin model (Table 1).

Valanis-Landel Assumption. [18] This model defines its strain energy function in terms of principal stretch ratios instead of invariants in order for this function to be more efficient. The model proposes a form of W which depends on a function ω (Table 1) which in turn depends on principal stretch ratios, as shows Eq. (4):

$$\frac{\partial_\omega}{\partial \lambda} = 2\mu \ln \lambda \tag{4}$$

Yeoh-Fleming Model. [19] This model is based on Gent and Yeoh models, it proposes a new form for W which involves three new material parameters A, B and I_m. It is a consequence of the observation that the reduced Mooney stress tends to a constant value that does not depend on I_1 for large strains [18]. So, the form of W that Yeoh-Fleming proposed is shown in Table 1, where:

$$R = \frac{I - 3}{I_m - 3} \tag{5}$$

The Table 1 provides a generalized classification of the most used forms of phenomenological models, in terms of polynomial forms of the strain energy function and exponential and logarithmic forms. The first column shows the name of the model, and the second column presents the form of the strain energy function for each model.

Most of the models presented in Table 1 are based on the principal strain invariants of Cauchy-Green deformation tensor, with exception of the Ogden, Valanis-Landel and Martins models which are based on principal stretches. Besides, the models Full Polynomial, Biderman, Hynes-Wilson, Ogden, Ishihara, Swanson, Caroll, Hart-Smith, Yeoh-Fleming and Gent can predict deformations in uniaxial, biaxial, compression and pure shear loads; however, only Yeoh, Ogden, Isihara, Yeoh-Fleming and Gent can reproduce the S-curvature that hyperelastic materials present in such tests.

Table 1. Phenomenological models.

Polynomial forms of the SEF	Formula
Mooney-Rivlin model [13]	$W = C_{10}(I_1 - 3) + C_{01}(I_2 - 3)$
Full Polynomial model [19]	$W = \sum_{i,j=0}^{\infty} C_{ij}(I_1 - 3)^i (I_2 - 3)^j$
Biderman model [13]	$W = C_{10}(I_1 - 3) + C_{01}(I_2 - 3) + C_{20}(I_1 - 3)^2$ $+ C_{30}(I_1 - 3)^3$
Hynes-Wilson model [13]	$W = C_{10}(I_1 - 3) + C_{01}(I_2 - 3) + C_{11}(I_1 - 3)(I_2 - 3)$ $+ C_{02}(I_2 - 3)^2 + C_{20}(I_1 - 3)^2 + C_{30}(I_1 - 3)^3$
Yeoh model [20]	$W = \sum_{i=1}^{3} C_{i0}(I_1 - 3)^i$
Neo-Hookean model [19]	$W = C_{10}(I_1 - 3)$
Ogden model [19]	$W = \sum_{n=1}^{N} \frac{\mu_n}{\alpha_n}\left(\lambda_1^{\alpha_n} + \lambda_2^{\alpha_n} + \lambda_3^{\alpha_n} - 3\right)$
Swanson model [14]	$W = \frac{3}{2}\sum_{i=1}^{n} \frac{A_i}{1+\alpha_i}\left[\frac{I_1}{3}\right]^{1+\alpha_i} + \frac{3}{2}\sum_{j=1}^{n} \frac{B_j}{1+\beta_j}\left[\frac{I_2}{3}\right]^{1+\beta_j}$
Caroll model [14]	$W = aI_1 + bI_1^4 + c\sqrt{I_2}$
Exponential and logarithmic forms of SEF	Formula
Fung-Demiray model [23]	$W = \frac{\mu}{2b}\left[e^{b(I_1-3)} - 1\right], b > 0$
Veronda-Westmann model [20]	$W = C_1\left[e^{C_2(I_1-3)} - 1\right] - \frac{C_1C_2}{2}(I_2 - 3)$
Gent-Thomas model [14]	$W = C_1(I_1 - 3) + C_2 ln(I_2/3)$
Valanis-Landel model [13]	$W = \omega(\lambda_1) + \omega(\lambda_2) + \omega(\lambda_3)$
Yeoh-Fleming model [14]	$W = \frac{A}{B}(I_m - 3)(1 - e^{-BR}) - C_{10}(I_m - 3) * \ln(1 - R)$
Humphrey model [20]	$W = C_1(e^{C_2(I_1-3)} - 1) + C_3(e^{C_4(\lambda-1)^2} - 1)$
Martins model [20]	$W = C_1(e^{C_2(I_1-3)} - 1)$
Gent model [14]	$W = -\frac{E}{6}(I_m - 3)\ln\left[1 - \frac{I_1-3}{I_m-3}\right]$

3 Numerical Comparison Through FEA - Case Study

The surgical simulators and the preoperative planning based on FEA are tools that help surgeons in different medical applications such as improving dexterity or trying to understand the possible issues present in a complicate surgery. These simulators also help to develop surgical implants with innovative piezoelectric Nano-sensors [27, 28].

In order to determine how the stress changes depending on the choice of the mathematical model, this case study analyzes the interaction between the surgical clamp and the soft tissue during the soft tissues clamping. Particularly, this interaction is found in different surgical processes as: colorectal, bariatric, thoracic surgeries, etc. In this sense, we used de geometrical model of the colon tissues obtained by a segmentation process [16] (Fig. 1).

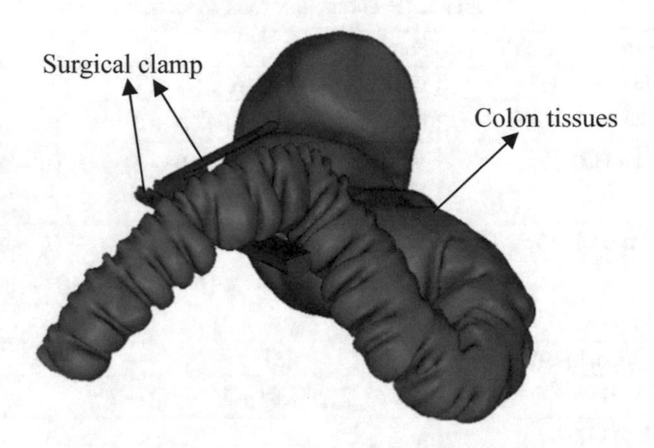

Fig. 1. Geometrical model of colon tissues.

Numerous studies performed with various tissue types (gastric [29], colorectal [30] and pancreatic [31]) provide data that demonstrates the importance of a surgeon's familiarity with tissue thickness and compressibility in order to optimize stapler (surgical clamp)-tissue interaction.

In fact, in the literature it is possible to find different values of colon thickness from 0.8 up to 2 mm [12, 32, 33], these values change not only by the different laboratory procedures in order to obtain the thickness value, but also by a wall thickening that can occur in pathological situations. Due to all these considerations a FEA based in the variation of the thickness has been done, considering both the linear elastic (LE) and Mooney Rivlin mathematical models (HE-MR).

Since there are some particular applications where an approximation of the mechanical behavior of the soft tissues (linear elastic) are required, some laboratory tests [10] have found an experimental approximation to the linear mechanical behavior corresponding to $E = 5.18$ MPa. Concerning the HE-MR model, the constant values are set to $C_{10} = 0.085$ MPa and $C_{01} = 0.0565$ MPa, according to [7]. These values are sufficient to define the hyperelastic Mooney-Rivlin model when the used materials are incompressible.

The load on the soft tissues have been applied by a rotation of the upper element of the surgical clamp (the blue part in Fig. 1). For colorectal surgery, staple cartridges are often used. When they are in the closing configuration, the staple height is 1.5 mm, so this value has been assumed as the final gap reached between the clamps in the final position of the FEA simulation.

FEA has been carried out in the HyperWorks environment through RADIOSS solver. This choice has been justified by the necessity of evaluating the adoption of material mathematical models within a commercial code that does not ask for specific implementations.

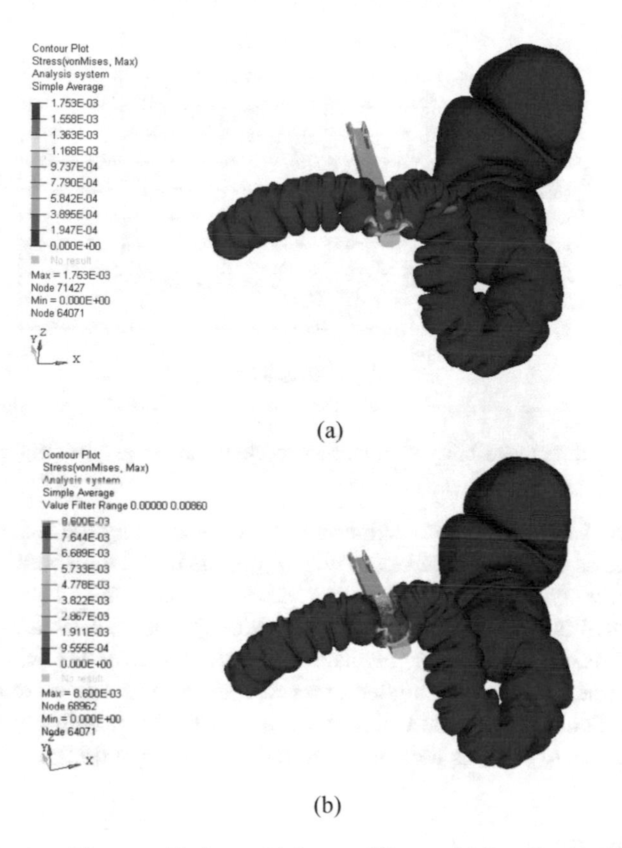

Fig. 2. Stress Contour-Map considering a thickness of 2 mm; (a) Results using Mooney-Rivlin model (GPa) (b) Results using Linear elastic model (GPa).

The Fig. 2 shows contour-plots to describe the stress distribution at the closure of the colorectal geometrical model. Particularly, Fig. 2(a) shows element's average length of 1 mm, thickness equal to 2 mm and Mooney Rivlin mechanical behavior. Figure 2(b) shows the same kind of stress distribution for a Linear Elastic behavior.

Although they have similar gradients, the final stress is different due to the mathematical model. In particular, the linear elastic formulation overestimates the stress 500% in condition of imposed strain.

Figure 3 highlights the comparison between the adopted mathematical models showing the maximum stress computed at the final closure of the clamp by changing initial thickness of the colon. The values are referred as the maximum values of the equivalent von Mises' stress that certainly occurs in the areas of stress concentration as shown in Fig. 2.

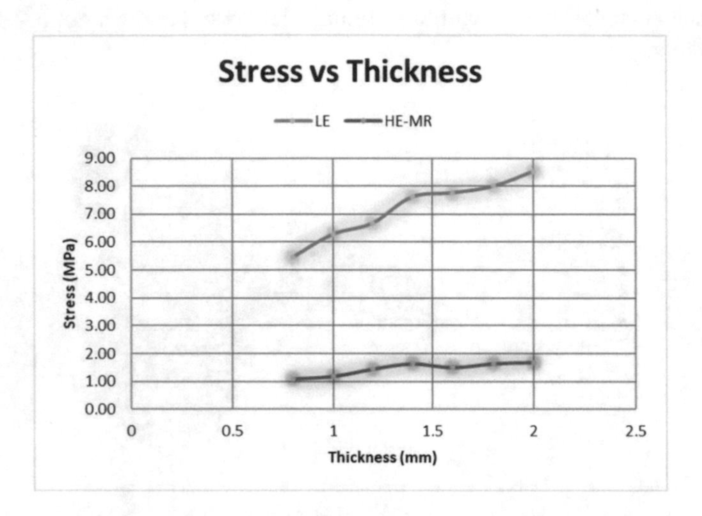

Fig. 3. Stress sensibility with respect to the thickness and material model.

In the case of Fig. 3, the overestimation of the stress using the LE is clearly shown and it increases as a consequence of the increase of the initial thickness. This increasing trend is partially connected to the fact that the FEA condition has an imposed motion on the clamp that is stopped at a specific gap-value (1.5 mm) related to clamp tighten. This means a maximum limit for the distance between clamps but with an increase of the initial thickness, it leads to higher stress in the tissue that is more compressed. In fact, a higher thickness means a higher volume with the possibility of absorbing a higher level of energy that is locally converted into stress and strain.

4 Conclusions

In different fields of medicine, which use the help of engineering to simulate surgical processes, the mechanical behavior of soft tissues could be well described using different mathematical models depending the type of simulated phenomenon. Therefore, type of the applied load imposes the selection of ad hoc mathematical model. They have to be calibrated with experimental test suitable to carry out the material constants. Since the complexity of the mathematical formulations may ask for many constants and thus complex experiments to find out them, in this paper a systematic discussion for looking up the most suitable mathematical model has been introduced.

Moreover, to give evidence of the peculiar hyperelastic behavior of the soft tissues a surgical example is provided comparing a linear elastic model with the phenomenological mathematical model derived from the Mooney Rivlin formulation. The LE model present higher values in terms of stress with respect to the HE-MR. This can lead to an overestimation of stress and produces a wrong prediction of the surgical tools path, or an erroneous value in terms of force-feedback when is necessary. This effect increases with the tissue initial thickness, producing inaccurate results when force reaction has to be computed according to the surgical problem to solve (e.g. thickening of the colon walls).

These preliminary results will help to researchers to understand the influence/importance of mechanical behavior giving the possibility to understand inaccuracy generated by model approximations.

The future works will be to identify the strains range, where each model fully fit in a good way to the stress-strain curve, and to identify the different mathematical models considering the tissues as homogeneous and anisotropic.

References

1. Wren, T., Carter, D.: A microstructural model for the tensile constitutive and failure behavior of soft skeletal connective tissues. J. Biomech. Eng. **120**(1), 55–61 (1996)
2. Fung, Y.C.: Biomechanics: Mechanical Properties of Living Tissues, 2nd edn. Springer Science, New York (1993)
3. Yamada, H.: Strength of Biological Materials, 1st edn. Williams & Wilkins, Baltimore (1970)
4. Marinozzi, F., Bini, F., Marinozzi, A., Zuppante, F., De Paolis, A., Pecci, R., Bedini, R.: Technique for bone volume measurement from human femur head samples by classification of micro-CT image histograms. Ann. I. Super. Sanità **49**(3), 300–305 (2013)
5. Ottensmeyer, M., Kerdok, A., Howe, R., Dawson, L.: The effects of testing environment on the viscoelastic properties of soft tissues. In: Cotin, S., Metaxas, D. (eds.) Medical Simulation, ISMS 2004. Lecture Notes in Computer Science, vol. 3078, pp. 9–18. Springer, Heidelberg (2004)
6. Guachi, R., Bini, F., Bici, M., Campana, F., Marinozzi, F.: Finite element model set-up of colorectal tissue for analyzing surgical scenarios. In: Tavares, J., Natal Jorge, R. (eds.) VipIMAGE 2017. Lecture Notes in Computational Vision and Biomechanics, vol. 27, pp. 599–609. Springer, Cham (2017)
7. Mayeur, O., Witz, J., Lecomte, P., Brieu, M., Cosson, M., Miller, K.: Influence of geometry and mechanical properties on the accuracy of patient-specific simulation of women pelvic floor. Ann. Biomed. Eng. **4**(1), 202–212 (2016)
8. Rubod, C., Brieu, M., Cosson, M., Rivaux, G., Clay, J.C., De Landsheere, L., Gabriel, B.: Biomechanical properties of human pelvic organs. Urology **79**(4), 968.e17–968.e22 (2012)
9. Chantereau, P., Brieu, M., Kammal, M., Farthmann, J., Gabriel, B., Cosson, M.: Mechanical properties of pelvic soft tissue of young women and impact of aging. Int. Urogynecol. J. **25**(11), 1547–1553 (2014)
10. Christensen, M., Oberg, K., Wolchok, J.: Tensile properties of the rectal and sigmoid colon: a comparative analysis of human and porcine tissue. Springerplus **4**(1), 142 (2015)

11. Gao, C., Gregersen, H.: Biomechanical and morphological properties in rat large intestine. J. Biomech. **33**(9), 1089–1097 (2000)
12. Liao, D., Zhao, J., Gregersen, H.: 3D Mechanical properties of the partially obstructed guinea pig small intestine. J. Biomech. **43**(11), 2079–2086 (2010)
13. Shahzad, M., Kamran, A., Siddiqui, M.Z., Farhan, M.: Mechanical characterization and FE modelling of a hyperelastic material. Mater. Res. **18**(5), 918–924 (2015)
14. Sasso, M., Palmieri, G., Chiappini, G., Amodio, D.: Characterization of hyperelastic rubber-like materials by biaxial and uniaxial stretching tests based on optical methods. Polym. Test. **27**(8), 995–1004 (2008)
15. Chung, T.: General Continuum Mechanics, 2nd edn. Cambridge University Press, Huntsville (2007)
16. Calvo-Gallego, J., Domínguez, J., Gómez Cía, T., Gómez Ciriza, G., Martínez-Reina, J.: Comparison of different constitutive models to characterize the viscoelastic properties of human abdominal adipose tissue. A pilot study. J. Mech. Behav. Biomed. Mater. **80**, 293–302 (2018)
17. Guachi, R., Bici, M., Guachi, L., Campana, F., Bini, F., Marinozzi, F.: Geometrical modelling effects on FEA of colorectal surgery. Comput.-Aided Des. Appli. **16**(4), 778–788 (2019)
18. Marckmann, G., Verron, E.: Comparison of hyperelastic models for rubber-like materials. Rubber Chem. Technol. **79**(5), 835–858 (2006)
19. Steinmann, P., Hossain, M., Possart, G.: Hyperelastic models for rubber-like materials: consistent tangent operators and suitability for Treloar's data. Arch. Appl. Mech. **82**(9), 1183–1217 (2012)
20. Thewlis, J.: Concise Dictionary of Physics, p. 248. Pergamon Press, Oxford (1973)
21. Broggiato, G., Campana, F., Cortese, L.: Parameter identification of a material damage model: inverse approach by the use of digital image processing. In: University of Parma/Ingegneria Industriale (eds.) 22nd Danubia-Adria Symposium on Experimental Methods in Solid Mechanics, DAS 2005, Parma, pp. 19–20 (2005)
22. Mooney, M.: A theory of large elastic deformation. J. Appl. Phys. **11**(9), 582–592 (1940)
23. Rivlin, R.: Large elastic deformations of isotropic materials. IV. Further developments of the general theory. Philos. Trans. Roy. Soc. Lond. Ser. **241**(835), 379–397 (1948)
24. Hamza, M.N., Alwan, H.M.: Hyperelastic constitutive modeling of rubber and rubber- like materials under finite strain. Eng. Technol. J. **28**(13), 2560–2575 (2010)
25. Martins, P., Jorge, R., Ferreira, A.: A comparative study of several material models for prediction of hyperelastic properties: Application to silicone-rubber and soft tissues. Strain **42**(3), 135–147 (2006)
26. Gent, A.: Engineering with Rubber, 2nd edn. Carl Hanser Verlag, Munich (2001)
27. Araneo, R., Bini, F., Rinaldi, A., Notargiacomo, A., Pea, M., Celozzi, S.: Thermal-electric model for piezoelectric ZnO nanowires. Nanotechnology **26**(26), 265402 (2015)
28. Araneo, R., Rinaldi, A., Notargiacomo, A., Bini, F., Marinozzi, F., Pea, M., Lovat, G., Celozzi, S.: Effect of the scaling of the mechanical properties on the performances of ZnO Piezo-semiconductive nanowires. In: Nanoforum 2013 on AIP Conference Proceedings, vol. 1603, pp. 14–22. Rome (2014)
29. Elariny, H., González, H., Wang, B.: Tissue thickness of human stomach measured on excised gastric specimens from obese patients. Surg. Technol. Int. **14**, 119–124 (2005)
30. Offodile, A., Feingold, D., Nasar, A., Whelan, R., Arnell, T.: High incidence of technical errors Involving the EEA circular stapler: a single institution experience. J. Am. Coll. Surg. **210**(3), 331–335 (2010)
31. Okano, K., Oshima, M., Kakinoki, K., Yamamoto, N., Akamoto, S., Yachida, S., Hagiike, M., Kamada, H., Masaki, T., Susuki, Y.: Pancreatic thickness as a predictive factor for

postoperative pancreatic fistula after distal pancreatectomy using an endopath stapler. Surg. Today **43**(2), 141–147 (2013)
32. Kester, E., Rabe, U., Presmanes, L., Tailhades, T., Arnold, W.: Measurement of mechanical properties of nanoscaled ferrites using atomic force microscopy at ultrasonic frequencies. Nanostruct. Mater. **12**(5–8), 779–782 (1999)
33. Egorov, V., Schastlivtsev, I., Prut, E., Baranov, A., Turusov, R.: Mechanical properties of the human gastrointestinal tract. J. Biomech. **35**(10), 1417–1425 (2002)

Image Segmentation Techniques Application for the Diagnosis of Dental Caries

Alfonso A. Guijarro-Rodríguez[1(✉)], Patricia M. Witt-Rodríguez[2], Lorenzo J. Cevallos-Torres[1], Segundo F. Contreras-Puco[1], Mirella C. Ortiz-Zambrano[1], and Dennisse E. Torres-Martínez[3]

[1] Faculty of Mathematical and Physical Sciences, University of Guayaquil, Guayaquil, Ecuador
{alfonso.guijarror,lorenzo.cevallot, francisco.contrerasp,mirella.ortizz}@ug.edu.ec

[2] Germany Faculty of Dentistry, University of Guayaquil, Guayaquil, Ecuador
patricia.wittr@ug.edu.ec

[3] Superior Technological Institute Vicente Rocafuerte, Guayaquil, Ecuador
detorres@itsvr.edu.ec

Abstract. The diagnosis of a patient with caries is a process performed by oral health professionals, who after auscultation on the dental surfaces, determine the degree of affectation, following the visual inspection protocols, which present international standards. In recent years, there has been a growing interest in developing new techniques that allow the establishment of medical diagnoses, supported by information technologies, specifically in the early detection of diseases to apply the respective treatments. This work presents a method to determine the level of affectation of caries in the oral cavity, which applies a classifier that consists of 5 phases: capture, preprocessing, segmentation, extraction of characteristics and classification of objects. The proposed methodology considers a bank of images of dental pieces, all extracted from private dental clinics, as well as from the Integral Clinic CIAM II, which pertains to the Odontology Pilot School of the University of Guayaquil. For the classification, a multilayer perceptron artificial neural network was used, while for the validation of the work, 2030 images were analyzed, finding 80% success in the results, which were corroborated following the norm of caries classification and the criteria exposed by experts.

Keywords: Computer vision · Classifier · ICDAS · Matlab · Image processing · Segmentation techniques

1 Introduction

Dental caries is an oral disease, which affects the majority of the population in the world [1, 10, 11, 14], considered a disease that occurs on the tooth structure [11, 14], is manifested by the presence of cavities in a tooth [1, 2, 6, 10, 11, 14, 15]. This disease, considered multifactorial, develops due to certain types of bacteria that produce an acid that destroys the enamel and dentin layer of the tooth [1, 14].

M. Botto-Tobar et al. (Eds.): ICAETT 2019, AISC 1066, pp. 312–322, 2020.
https://doi.org/10.1007/978-3-030-32022-5_30

Modern research refers to technological advances based on systems platforms for teledentistry [10], a term associated with the fusion of Information and Communication Technologies (ICTs) with oral health care. In [10], a system based on telemedicine is shown, which allows the diagnosis and monitoring of patients with dental caries, to avoid loss of their teeth, although the work is very interesting, has few details in its structure and does not specify how it solves the dental caries module, rather, suggests storing photos of the tooth, as evidence and diagnoses associated with a visual inspection. There are other works related to health issues, such as the one presented by [15], who proposes a model for automatic detection of edges in dermatoscopic images, which establishes a medical diagnosis, based on the execution of the segmentation techniques applied by the doctor. Mean Shift algorithm, which executed with a multilayer perceptron neural network, presents a great advance to the results obtained from this work, although the author maintains that there are several problems to be solved, it is still a great contribution for science, besides The segmentation technique used in this work could be considered to segment dental caries as future work.

In [2], the volume of affectation of caries by computerized tomography is determined, it uses segmentation techniques, with the k-mean grouping method and the threshold method, the latter being the most recommended. Among the techniques mentioned, models for the detection of borders appear, which have allowed to focus on diseases and establish more precise diagnoses avoiding subjectivities that are based on the experience of physicians, generating results with high percentages of acceptance in health areas [2, 13, 15]. The segmentation techniques have allowed establishing a preliminary diagnosis on the areas of affectation, based on the International Caries Detection and Assessment System (ICDAS), which contrasted with the visual inspection protocols performed by dentists, provide reliability to diagnosis and validity to the proposal.

There are multiple risk factors that are associated with this disease, ranging from poor oral hygiene, reduced salivary flow, periodontal disease, among others [6]. In addition, there are those who attribute the lifestyle that people lead, when they consume excessive amounts of alcoholic beverages, tobacco, and diets rich in sugars, as the main causes that cause conditions in the oral cavity and are the most important factor in the production of caries. Thus affecting the integrity and aesthetics of the tooth, presenting an important problem in oral health [6, 11, 14].

The oral cavity is a fundamental part of the body, having it healthy is important to be able to chew, speak and look good, but carelessness can affect the entire organism; For these reasons, the mouth is a window that allows a skilled dentist to make a general health assessment. Oral diseases, such as dental caries and periodontitis, are preventable [6, 11, 14]. Therefore, it will be applied the use of segmentation techniques [15] that is the process which divides an image into parts constituting one of the key elements in the automated image analysis [15], so that this is the stage where objects of interest are extracted for further processing, such as description and recognition [2, 13, 15, 17].

Segmented images are now routinely used in a multitude of different applications, such as diagnosis, treatment planning, pathology localization, anatomical structure study, integrated computer surgery, among others [2, 10, 13, 15, 17]. The use of a classifier system that discriminates the level of affectation of a dental piece will help us to know, what is the state of the oral cavity of the patient and thus be able to prevent chronic diseases, also, it allows an increasing evidence of the untreated oral diseases due to have systemic consequences that aggravate the course of no communicable diseases [6, 11].

This work presents a model that determines the level of involvement of a dental piece in the oral cavity, from the application of the model consisting of 5 phases [16]; capture, preprocessing, segmentation, feature extraction, and classification.

For this the structure of the document exposes the Sect. 2, where it gives to know the materials and methods, site where the model of the proposal is developed, then in Sect. 3 it gives to know the experimental results, Sect. 4 presents the conclusions.

2 Materials and Methods

The proposed methodology considers a bank of images of 2030 dental pieces, all of them obtained in private dental clinics, as well as in CIAM II of the pilot school of Odontology of the University of Guayaquil. The proposed model follows the phases presented in [16], however, for a better understanding, the phases of the model are illustrated in Fig. 1.

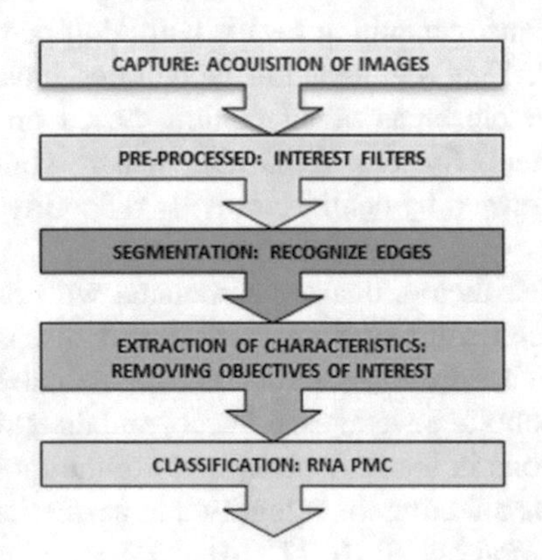

Fig. 1. Model of the proposal developed

2.1 Image Capture

The first phase of the methodology is aimed at capturing the images of different teeth. In the particular dental clinics, the images were captured by means of a Nikon Cool-pix L820 camera, this camera has a retro-illuminated CMOS image sensor that has 16 megapixels, which allows capturing exceptionally detailed images, and even when there is an environment with a shortage of light. On the other hand, the pilot faculty of dentistry of the University of Guayaquil, made the captures with the Nikon D200 camera of 10 Megapixels circular flash and a focal length of 30 cm, as shown in Fig. 2.

Fig. 2. Capture of images – University of Guayaquil

It is important to mention that the records of clinical cases are stored in magnetic media (CD-R), with their respective medical evaluations, which have a reserved use of the Odontology Pilot Faculty.

In total 2030 images were considered for this work, of which one part was used for training and another part for validation. In order to provide continuity to the work, the images were stored online, under a Dropbox repository, to perform tests from anywhere with the help of an internet connection, Fig. 3 shows some of the images captured belonging to the image bank.

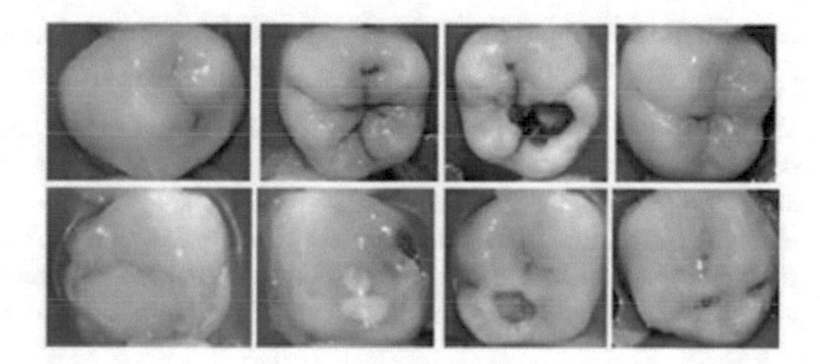

Fig. 3. Sample of images from the database

2.2 Pre-processed

In this phase, there were used a set of techniques with the aim of improving the quality of the image, from RGB plane, eliminating noise, smoothing the image and improving the quality of the edges [15], which improved on the analysis in the next segmentation phase, as corresponds in Fig. 1.

Among the techniques applied, there is the media filter that reduces the variations of intensity between neighboring pixels, which is discarded by the loss of sharpness affecting the segmentation, as well as the Gaussian filter. Another technique applied was the median filter, which returns a grayscale image, without noise and with greater clarity, totally applicable to segmentation, as shown in Fig. 4.

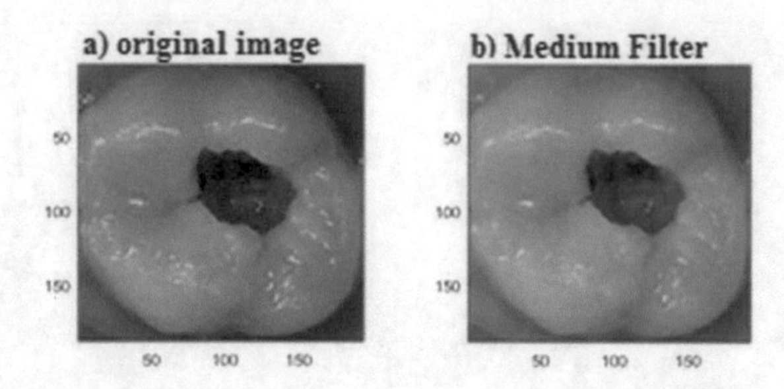

Fig. 4. Median filter.

2.3 Segmentation of Images

This phase divides the image into homogeneous blocks, to separate the objects of interest and perform the analysis of their characteristics between pixels, starting from their contour and connectivity, in order to discriminate one region from another, resulting in a segmented image. For the segmentation of the images, several segmentation techniques were used, the most significant being those shown in this work as:

Watershed Transform: Classify the pixels according to their spatial proximity, gray level gradient and the homogeneity of the textures.

Canny Transform: It is an algorithm used in edge detection of images or objects. It is characterized by avoiding the breaking of the edges of objects.

Sobel Transform: Shows how an image changes abruptly or softly in each point or pixel analyzed. Hence, it tells us how likely it is to represent an edge in the image, and also the orientation to which this edge will tend.

When applying the techniques of segmentation to the dental pieces, it was found that they obtain a similar visually result; however, Sobel presents better behavior for the segmented area affected by caries, and the total area of the tooth, as shown in Fig. 5.

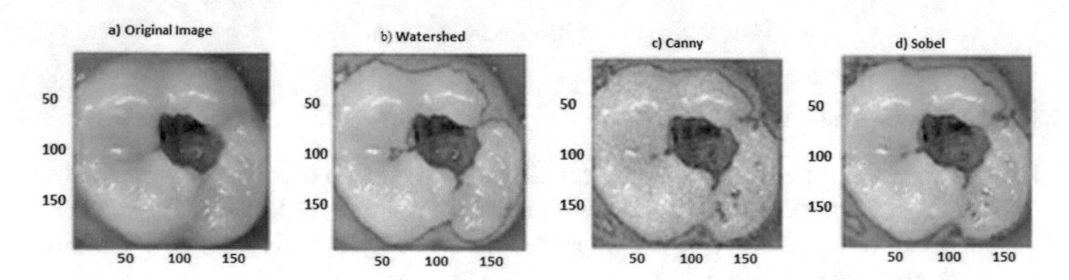

Fig. 5. Segmentation techniques on a dental piece

2.4 Feature Extraction

For this phase, it is considered about 8 types of features that have been extracted from the segmentation phase of each of the images analyzed of which are: healthy area, affected area, total area, entropy, contrast, correlation, energy and homogeneity. In addition, histograms were calculated, which presented an appearance of increasing order, finding two higher peaks [16], we proceeded to calculate the mean, variance [15]. At the morphological level, some characteristics that allow a better analysis is applied, such as those cited below:

Entropy: it allows to measure the uncertainty that exists in relation to the distribution of the pixels that the image in study has.

$$-\sum_{\mathrm{i}}\sum_{\mathrm{j}} p(i,j)\log(p(i,j)) \tag{1}$$

Contrast: it facilitates observing the illumination and being able to differentiate the intensity that the image in the analysis has.

$$\sum_{\mathrm{n}=0}^{\mathrm{N_{g-1}}} \mathrm{n}^2\left\{\sum_{i=1}^{N_g}\sum_{j=1}^{N_g} p(i,j)\right\}, |\mathrm{i},\mathrm{j}| = n \tag{2}$$

Correlation: Statistical method which allows knowing if there is any relationship between the variables; in other words, it allows to perform the displacement measurement or the pixels' deformations that make up the image under study.

$$\frac{\sum_{\mathrm{i}}\sum_{\mathrm{j}} (i,j)p(i,j) - \mu_{\mathrm{x}}\mu_{\mathrm{y}}}{\sigma_{\mathrm{x}}\sigma_{\mathrm{y}}} \tag{3}$$

Energy: it facilitates the sum of the square elements through the Gray Level Cooccurrence Matrix (GLCM), which is in charge of maximizing the large values and also decreasing the smaller values.

$$\sum_{\mathrm{i}}\sum_{\mathrm{j}} (\mathrm{i},\mathrm{j})^2 \tag{4}$$

Homogeneity: Formed by elements in which the pixels with equal value measure the existing proximity of the elements distribution of the GLCM, which means that it measures the equality or similarity, obtaining, as a result, a single value equal to 1, otherwise will be taken null, whose value will be 0. The formula of the equation is:

$$\sum_{\mathrm{i}}\sum_{\mathrm{j}} \frac{1}{1 = (\mathrm{i},\mathrm{j})^2} p(i,j) \tag{5}$$

2.5 Classification of Objects

Based on the variables obtained, there was established a multivariate analysis in the SPSS 23 application, using the Principal Components Analysis (PCA) method, in order to reduce the extracted characteristics and work with the most significant ones. Due to the presence of a matrix of 2030x8, and the excessiveness of the data, PCA was applied to the variables: healthy area, affected area, total area, energy, entropy, correlation, contrast, and homogeneity, as shown in Table 1, resulting the most relevant those that correspond to the healthy area and affected area, as shown in Table 2. After the analysis, it gives way to the training of the Artificial Neural Network.

Table 1. PCA communities

	Initial	Extraction
Healthy area	1,000	,961
Affected area	1,000	,961
Entropy	1,000	,782
Contrast	1,000	,761
Correlation	1,000	,011
Energy	1,000	,813
Homogeneity	1,000	,813
Exit	1,000	,721

Extraction method: analysis of the main Components.

Table 2. Component transformation matrix

Component	1	2
1	.855	,519
2	-,519	,855

Extraction method: analysis of the main components.
Rotation method: Varimax with Kaiser standardization

Artificial Neural Network (RNA) Multilayer Perceptron
For the creation of RNA, (see Fig. 6), it has been used Matlab platform as a tool, in which the training algorithm pertains to the retro-propagation conjugate gradient, with 7 neurons in the hidden layer and a sigmoidal hyperbolic tangent training function [15]. Thus, as an input of the network, there are variables such as Healthy Area, Affected area and bias, 8 hidden layer neurons and for the output the values that the ICDAS standard is set from 0 to 6, that is 7 values as shown by Table 4.

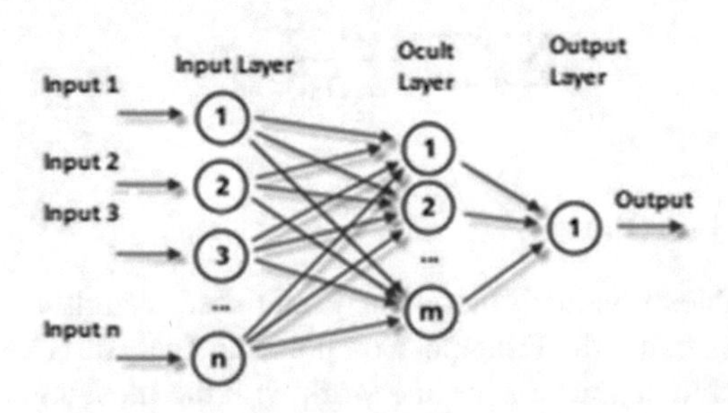

Fig. 6. Artificial Neural Network

As a first approximation of the work, there were considered 100 images from the total. It was observed that the retro-propagation algorithm is at a minimum value, which leads to safe learning, the mean square error graph is shown in Fig. 7, and it can be observed that in 3 iterations appears an almost immediate convergence. How ever, when the number of images increases, computational processing increases proportionally, and learning follows the same trend as in Fig. 7, another graphic that supports the exposed is found in Fig. 8 and produces 80% success in the results.

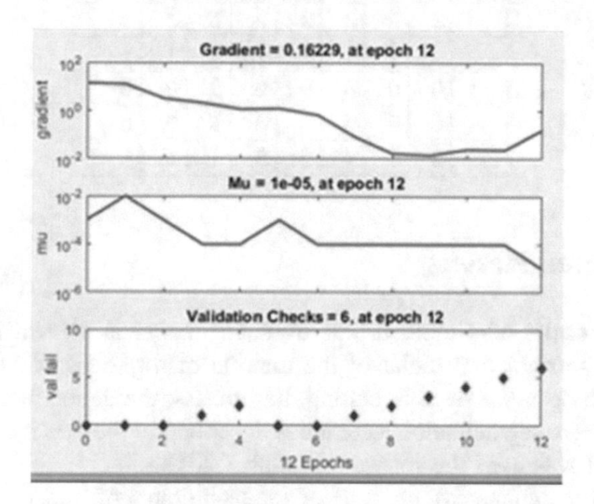

Fig. 7. Graph of mean square error vs. iterations.

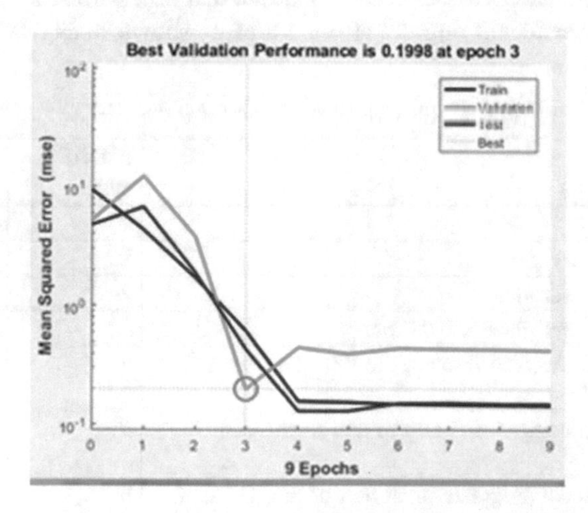

Fig. 8. Gradient graphs and validations.

To validate the work, 200 test images were chosen and classified according to Table 4, which supports the ICDAS standard. From these images, 7 were randomly selected so that doctors from the Odontology Pilot School, through a survey, comment on their diagnosis by the method of direct observation, finding very satisfactory the work done, as illustrated in Table 3.

Table 3. Consultation with experts

Photo	E1	E2	E3	E4	E5	E6	E7	RNA
1	3	3	4	3	4	3	3	3
2	3	4	4	3	4	3	4	4
3	2	2	2	2	2	2	2	2
4	1	1	1	1	1	1	1	1
5	0	0	0	0	0	0	0	0
6	6	6	6	5	6	5	5	6
7	5	5	4	5	5	4	5	5

3 Experimental Results

To display the results of a tooth it was used an image, as shown in Fig. 5, which corresponds to the right first molar of the maxilla of a patient of 13 years old. This molar has a tooth decay of stage 6. During the process, the dental piece was submitted by each of the image segmentation techniques described in the previous section. For the image process, it was used the software Matlab R2018a.

Once done the comparative analysis of the techniques of image segmentation and verifying the result that they showed when subjected dental imaging to tests, it was determined that Transform Sobel is the technique that best helps for the creation of the code responsible for identifying the affected area as shown in Fig. 5.

Table 4. Classification of caries by the ICDAS

Visual threshold	ICDAS code	Degree of affectation (%)
Healthy	0	0
White/brown spot dry glaze	1	0,5–4,3
White/brown stain wet glaze	2	4,31–9,69
White/brown stain wet glaze < 0.05 mm	3	9,7–11,35
Dark shadow dentin viewed through the wet glaze with or without micro cavity	4	11,36–13,9
Dentin exposure in cavity > 0.5 mm to half of the dental surface in dry	5	13,91–35,1
Dentin exposure in cavity greater than half of the dental surface	6	35,11 o mayor

Once the affected part is segmented, the area is calculated in relation to the total of steel and the affected part, resulting in the number of pixels analyzed, taking these results as a percentage, it is feasible to place them in the classification table. Caries given by the ICDAS.

As a result, reference is made to Fig. 5, the calculation of the area of the dental unit is made and contrasting the healthy and affected part thereof; results in the area of the molar 171686 px, healthy area 146041 px and affected area 25645.6 px. Therefore, the percentage of healthy area is 85.06% and the affected area corresponds to 14.94%. Finally, the degree of caries in this practice is shown in Grade 5, which corresponds to a dentine exposure in the cavity according to Table 4.

4 Conclusions

In the present work, a comparative analysis of image segmentation techniques has been carried out: Watershed, Sobel, and Canny. After carrying out the tests for each of the techniques and discarding those that presented false positives that could hinder the final result, it was determined that Sobel's technique was the one that has most approached the reality of what is presented in the analyzed dental piece highlighting elements such as: the area, contour, size among others.

The capture phase has used the oral cavity of several patients; however, the dental pieces have been separated detailing which dental piece is and the affectation that each of them presents.

In the segmentation phase, it was applied the Sobel technique in order to determine the limits that appear between the healthy part and the affected part. This technique gives more flexibility to the moment of image processing and accompanied by the preprocessing of images to perform the analysis.

In the phase of extraction of characteristics, the calculation of the healthy and affected area in percentage has been detailed, as well as the segmentation of the affected area.

From the calculation made in the test piece, the affected area is detailed, which is the most relevant when making the comparison with the table given by the ICDAS in order to determine the degree to which the caries is located, giving the dentist an indicator that possible dental treatment.

To apply new segmentation techniques such as the Mean Shift algorithm, to find the area of interest and help doctors establish their diagnosis of dental pieces.

Integrate this solution with teledentistry and benefit rural areas in Ecuador and other populations in the world.

References

1. Agarwal, R., et al.: A review paper on diagnosis of approximal and occlusal dental caries using digital processing of medical images. In: International Conference on Emerging Trends in Electrical Electronics & Sustainable Energy Systems (ICETEESES), pp. 383–385. IEEE (2016)
2. Ahmed, S., et al.: Identification and volume estimation of dental caries using CT image. In: 2017 IEEE International Conference on Telecommunications and Photonics (ICTP), pp. 48–51. IEEE (2017)

3. Gálvez, S.: Operadores de detección de bordes. Universidad de Chile Facultad de Ciencias Físicas Y Matemáticas **31**(144), 378–381 (2014)
4. González Sanz, Á.M., Aurora, B., Nieto, G., González Nieto, E.: Salud dental: relación entre la caries dental y el consumo de alimentos. Nutr. Hosp. **28**(4), 64–71 (2013)
5. Gonzalez, M., Ballarin, V.: Segmentación de imágenes utilizando la transformada Watershed: obtención de marcadores mediante lógica difusa. IEEE Lat. Am. Trans. **6**(2), 6 (2008)
6. Hidalgo Gato-Fuentes, I., Duque de Estrada Riverón, J., Pérez Quiñones, J.A.: La caries dental: Algunos de los factores relacionados con su formación en niños. Revista Cubana de Estomatología **45**(1) (2008)
7. Hu, Z., et al.: Teeth segmentation using dental CT data. In: 2014 7th International Conference on Biomedical Engineering and Informatics (BMEI), pp. 81–84. IEEE (2014)
8. Karlsson, L.: Caries detection methods based on changes in optical properties between healthy and carious tissue. Int. J. Dent. **2010**, 9 (2010)
9. Kim, H.-E., Kim, B.-I.: Early caries detection methods according to the depth of the lesion: an in vitro comparison. Photodiagn. Photodyn. Ther. **23**, 176–180 (2018)
10. Lancheros-Cuesta, D.J., Suarez, D.R., Arias, J.L.R.: Tele-dentistry information system for promotion, prevention, diagnosis and treatment of dental caries. In: 2016 11th Iberian Conference on Information Systems and Technologies (CISTI), pp. 1–7. IEEE (2016)
11. Luyo, A.G.P.: ¿ Es la caries dental una enfermedad infecciosa y transmisible? Re-vista Estomatológica Herediana **19**(2), 118–124 (2009)
12. Marín: Desarrollo de un sistema de ayuda a la decisión para tratamientos odonto-lógicos con imágenes digitales, p. 33. Universidad de Málaga – España (2015)
13. Molina, Á.V.: Segmentación de los pulmones usando maquinas de soporte vectorial en imágenes de tomografía computarizada
14. Núñez, D.P., García Bacallao, L.: Bioquímica de la caries dental. Revista Habanera de Ciencias Médicas **9**(2), 156–166 (2010)
15. Peláez, J.I., et al.: Un modelo de detección automática de bordes en imágenes dermatoscópicas. In: CISCI 2016-Decima Quinta Conferencia Iberoamericana en Sistemas, Cibernética e Informática, Décimo Tercer Simposium Iberoamericano en Educación, Cibernética e Informática, SIECI 2016-Memorias (2016)
16. Peláez, J.I., Vaccaro, G., Guijarro, A.: Un Modelo para la Categoriza-ción de Hormigones Mediante Procesamiento Digital de Imágenes. In: 19th World Multi-Conference on Systemics, Cybernetics-CISCI 2015 (2015)
17. Pinto-González, S.F., et al.: Validación mediante el Método PPI de un Algoritmo Computacional para la medición automática del área de afectación por Sigatoka ne-gra en imágenes de hojas de plátano del Departamento del Meta, Colombia, vol. 14, no. 1, pp. 24–28 (2017)
18. Pretty, I.A.: Caries detection and diagnosis: novel technologies. J. Dent. **34**(10), 727–739 (2006)
19. Riveros Guevara, A., Salas López, C.N., Solaque Guzmán, L.: Proximación a la navegación autónoma de una plataforma móvil, mediante visión este-reoscópica artificial. Ciencia E Ingeniería Neogranadina **22**(2), 111–129 (2013)
20. Shokouhi, E.B., et al.: Comparative study on the detection of early dental caries using thermo-photonic lock-in imaging and optical coherence tomography. Biomed. Opt. Express **9**(9), 3983–3997 (2018)

Assistive Technology in Ecuador: Current Status of Myoelectric Prostheses of Upper Limbs

Washington Caraguay[1(✉)], Marco Sotomayor[1], Christoph Schlüter[2], and Doris Caliz[3]

[1] Faculty of Electronic, Computer and Telecommunication Engineering, Universidad Espíritu Santò, Guayaquil, Ecuador
{wcaraguay,mvinicio}@uees.edu.ec

[2] Faculty of Mechanical Engineering, Karlsruher Institut für Technologie, Karlsruhe, Germany
updvv@student.kit.edu

[3] Institute of Telematics, Karlsruher Institut für Technologie, Karlsruhe, Germany
caliz@teco.edu

Abstract. Despite the academic research on myoelectric upper limb protheses conducted in Ecuador, there are no local companies manufacturing electromyography-controlled protheses. In addition, some local universities have developed their own prototypes. However, many of the components are expensive, need to be imported, and are difficult to acquire, such as the electromyography sensors. This literature survey covers relevant studies developed by the industry and the academia on the design of upper limb protheses, both locally and internationally, with the main objective of collecting information as a starting point for the manufacture of prostheses using low-cost materials in Ecuador. Document analysis techniques were applied, to the search of scientific databases. The results are justified and finally, the acquisition and advances of upper limb prostheses with electromyography sensors related to the local context is discussed.

Keywords: Electromyography · Upper limb · Sensor · Low-cost · Prostheses

1 Introduction

The Ecuadorian "Consejo Nacional para la Igualdad de Discapacidades" (National Council for Equality of Disabled People, CONADIS) presents statistics on disabled people in Ecuador. It states that 204,607 out of 14.48 million population are registered as physically disabled, without digging into what type of disability [12]. According to the last census in 2010, this makes up about 1.41% of the population. CONADIS also publishes the number of hospital discharges in

M. Botto-Tobar et al. (Eds.): ICAETT 2019, AISC 1066, pp. 323–334, 2020.
https://doi.org/10.1007/978-3-030-32022-5_31

Ecuador and the reasons for hospitalization. Statistics on upper-limb amputees from 2015, 2016, and 2017 are summarized in Fig. 1 [20–22]. Wrist and hand amputations are not mentioned in the graphic, but they are a very common, with over 600 amputations in Ecuador in each of the last three years.

In order to reduce upper limb disabilities, the current market offers several products, both prosthetic arms and devices for amplification and interpretation of EMG-signals. However, in addition to being expensive and including non-reusable elements, none of these solutions meet the expectations [10,11,23,46]. Developments of upper limb prostheses technology include aspects such as, intelligent control, brain-to-computer interface, and control sources of prosthetic hands [16,23,32].

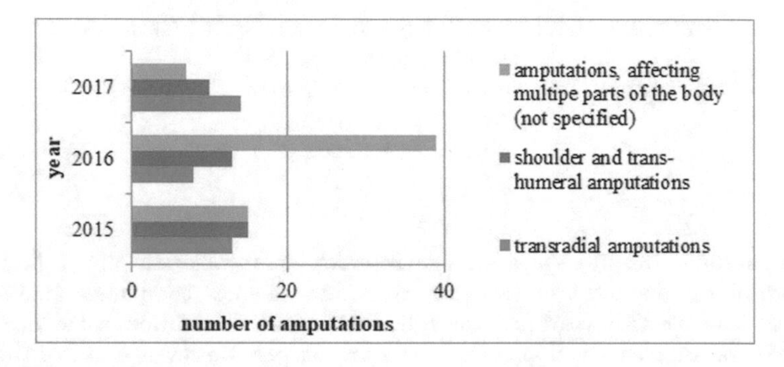

Fig. 1. Number of hospital discharges in Ecuador related to the reason of hospitalization. Taken and adapted from [20–22]

Although there are companies selling prostheses from recognized brands in Ecuador [37], the asked prices are too expensive when compared to the average salary of the working class [50], making them inaccessible to low-income people. With the purpose of identifying the appropriate technologies for the production of myoelectric upper limbs protheses in Ecuador, the following question arose: what are the barriers in Ecuador for people willing and needing to access upper limb protheses? This work starts with the biological structure of human muscles, focusing on the generation of EMG-signals. The control techniques are reviewed with a focus on those that require low-cost devices for their design and that are supplied by local companies, with the objective of designing a product using components available in Ecuador, implying the justification and need for this work.

2 Method

The method applied in this paper is structured as follows, and was taken and adapted from [2,28,35]. It then describes the literature review process.

2.1 Literature Review Process

A literature search was performed via SCOPUS, ACM and ISI Web of Science databases, using the following topics: Biomechanical model, EMG Control, Low-costs and Upper limb prostheses market. The search was carried out in June 2018, and limited to articles published from 2000 onwards. The search returned 420 references, classified in a number of topics as shown in Fig. 2. Also, 10 web pages of prostheses manufacturers and dealers were used. The selection criteria were composed of four searches considering the four topics. The short-listing strategy flow diagram is shown in Fig. 3.

For all the databases were used a set of simple search string and adding the outcome from each of the searches for each source:

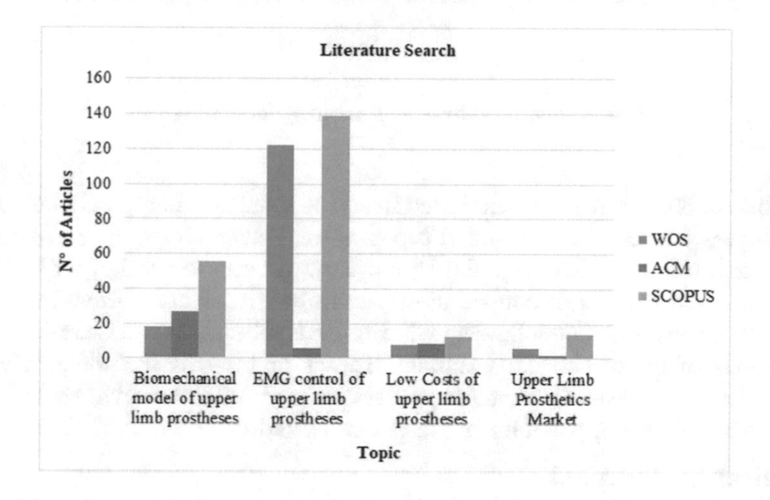

Fig. 2. Results of literature search by topic

(i) Biomechanical AND model AND upper AND limb AND prostheses.
(ii) EMG AND control AND upper AND limb AND prostheses.
(iii) Low AND costs AND upper AND limb.
(iv) Upper AND limb AND prosthetics AND market

The primary list of 420 articles using the four topics was filtered based on the relevance of the paper titles and date of publication. This returned 177 papers $(50+100+15+12)$. Then, the list was further screened based on the relevance of the content of the abstracts. This resulted in a list of 74 articles $(20+40+8+6)$. These papers were read and analysed, after what 25 articles were found to be relevant to the four topics that would respond to the research question: what are the barriers in Ecuador for people willing and needing to access upper limb protheses?

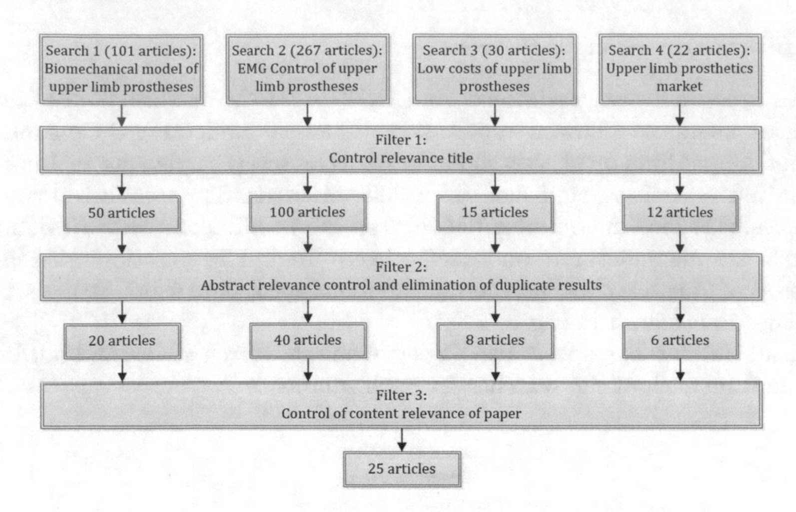

Fig. 3. Search refinement strategy flow diagram

A total of 395 studies were excluded from the total number of articles included in the survey process. The excluded topics were: Sensors using invasive methods for the extraction of EMG-signals, Upper limb prostheses design without costs estimation, and Prosthesis control of upper limbs with other sensors (other than EMG-signal sensors). The following will review the literature associated with the foundations of biological EMG-signals. Topics of interest include EMG-signal sensors, low-cost ideas, market for prostheses and an exhaustive review of the advances in robotic upper limb protheses in Ecuador.

Biological Background

Human muscles can be divided into three different types: skeletal, cardiac, and smooth muscle tissue. EMG measures skeletal muscle activity. Skeletal muscles consist of parallel-arranged muscle fiber bundles that can be divided into single muscle fibers to which motor endplates are attached [10]. Motor endplates are synapses that receive electrical signals from the nervous system. If a certain threshold is overcome, a potential action is triggered in the muscle fiber, and is passed along it in both directions. In resting muscle cells, there is a potential difference of about -80 mV between the inside and the outside; this is caused by an ion imbalance, called resting potential. The depolarization of the cell membrane leads to a potential action of about 30 mV, thus immediately starting a repolarization, bringing the potential back to the resting potential with a small phase of hyperpolarization. This whole process takes around 5 ms, after which the membrane can be activated again [29]. Observations show that the force of muscle contraction correlates approximately linearly with the EMG-signal's amplitude.

In the upper arm, there are two individually contractible muscle groups, the biceps and the triceps. For people with trans-humeral amputation or elbow exarticulation, these muscles often remain partially. As they cannot serve the function of exing the forearm anymore, they can be used to control a prosthetic arm [29,52]. A prosthetic arm needs an elbow articulation, a robotized hand, including a wrist, and links between those elements and to the shaft connected to the arm [36].

To control a prosthesis by myoelectric signals, the EMG-signal must be measured using an appropriate sensor. There is a wide variety of existing EMG sensors. They can be classified by the way they capture the EMG-signal or by their amplification strategy, distinguishing between active and passive electrodes. There are two general ways to measure EMG-signals. One is intrusive EMG (IEMG). Needle or fine wire electrodes are inserted into the skin and muscle. The advantage is that one can measure potential actions in specific muscle fibers. This is the only way to measure underlying muscles activity [38,45]. A second way is surface EMG, for which two surface electrodes are usually placed onto the skin over the muscle to be examined. The surface EMG sensor measures the potential difference between those two electrodes, which is the result of a superposition of all the underlying potential actions. This technique though shows accurate results only for superficially-aligned muscles [29].

EMG-Signals Sensors

Several universities, research centers, and companies are doing research on the improvement of EMG-signal acquisition for different purposes. Some researchers describe in their papers how they have developed low-budget EMG sensors. In 2013, for example, a comparably small sensor with dimensions of $44 \times 20.5 \times 11$ mm was presented and documented [19]. The use of implantable devices to get a better and cleaner EMG seems to be the answer to achieve a wide range of movement on a prosthetic arm or hand; that is why the use of implantable devices that record intramuscular EMG-signals has been proposed to overcome the constraint of disallowing simultaneous movement of multiple degrees of freedom (DOFs) [47]. Also, a wireless mobile platform for reception and communication of surface EMG-signals embedded in a complete mobile control system structure has been presented [6]. In this way, in the last decade there has been a significant development of EMG technology. Prosthetic hand control based on the acquisition and processing of EMG-signals is a well-established method that makes use of the electric potentials evoked by the physiological contraction processes of one or more muscles. [2,24].

Low-Cost Ideas

There has been a lot of research and brain-storming when it comes to achieving a low-cost prosthesis for disabled people [25,27]; thankfully, there are a lot of different ideas to achieve this goal. In one proposed design, a single DC motor actuates all five digits of the hand [15]. The hand is voice-controlled via a smartphone application and Arduino.

Another interesting idea arose from an investigation in which researchers designed a prosthesis for patients with transradial limb amputations. What is compelling is that it was shoulder-controlled and externally powered with an anthropomorphic terminal device. The user could open and close all five fingers and move the thumb independently. This allows making a prosthesis with an estimated cost of $300. The prototype was tested in a 13-year-old female patient with a transradial traumatic amputation [18].

A different approach was made by another group of researchers [26], who proposed that sensorless techniques can be used to reduce design complexities and costs, and to provide easier access to the electronics. To achieve this, a closing and opening finite state machine (COFSM) was developed to handle the actuated digit joint control state along with a supervisory switching control scheme, used for speed and grip strength control. Three torque and speed settings were created to be preset for specific grasps. The hand was able to replicate ten frequently used grasps and grip some common objects. The main cost would be that of a suitable DC motor, an Arduino (or equivalent printed circuit board, PCB), and guitar strings; this brought the product to under €100, excluding labor cost and manufacturing.

Market Conditions for Prostheses

Thanks to the fast-developing technical standards for the functionality and accuracy of electrical and mechanical components, several highly-developed prostheses may be bought nowadays. Prensilia S.r.l [41], is the manufacturer of IH2 Azzurra, a prosthesis with a price of €23,000; however, including warranties, assistance and support services for 36 months, it adds up to €33,000 [13]. IH2 Azzurra is a self-contained robotic hand, equipped with 24 sensors for exact determination of the finger position and the force applied to them [13,49]. This hand adapts to the object that is to be gripped and can be, but not necessarily, controlled by EMG-signals. The most developed market-available solutions are the I-Limb ultra [5], Bebionic hand [36], and Taska hand [43], these can be controlled by myoelectric sensors attached to the forearm and offer a wide variety of different grips and hand movements. I-Limb products consist only of hands, but are compatible with many wrist and arm shafts from other manufacturers. A unique feature of the newest product generation is the gesture control to change between the most important grips; however, they can also be changed via a smart-phone application with access to 36 different grips. The Bebionic hand lets users choose between 14 grips by moving the thumb to specific locations and pushing integrated buttons. It is able to hold objects with a weight of up to 45 kg. Both of these robotic hands are able to sense the slipping of gripped objects, and to adjust the applied force. Taska offers 23 grips to perform a broad variety of different tasks, is fully waterproof, and lifts a peak weight of 20 kg. The Defense Advanced Research Projects Agency (DARPA), requested the development of two prostheses. One was designed by The Research and Development Corporation (DEKA), and offers a robotized arm that can substitute the whole arm, including the shoulder joint. The product is now distributed by Mobius Bionics. DARPA's second project was designed by Johns Hopkins University's Applied Physics Laboratory. The result is one of the most advanced above-elbow prostheses, which so far serves for research purposes only. It focuses on neural control, the measurement of nerve and brain activity, as well as haptic feedback from the prosthesis back to the nervous system [14]. Ottobock and several different companies also offer simpler and therefore more inexpensive solutions.

The most basic ones are simple grippers consisting of two or three fingers or hooks. The London Prosthetic Center [42], reports that prosthetic care can be expensive and that all amputees or their providers are concerned about costs. It is difficult to tell someone how much their prosthesis is going to cost without an in-person evaluation. Therefore, the cost is approximate and varies greatly depending on the amputation [44], Table 1. However, low-budget solutions are offered by several projects that have emerged in recent years. Nearly all of them are based on 3D-printed designs, as this has turned out to be a good way to easily adapt models to clients. To mention just two of those projects, Bionicohand and exiii HACKberry offer open-source solutions with manuals on how to make your own prosthesis [4,30]. These all projects focus on lower arm devices, and there are rarely low-cost products for trans-humeral amputees. Most of these devices integrate EMG sensors from other companies.

Table 1. Upper limb high-end prosthesis prices. Taken and adapted from [13,44,48]

Company	Prosthesis	Price
Prensilia S.r.l	IH2 Azzurra	€33,000
Touch Bionics	I-Limb ultra	US$40,000
Ottobock	Bebionic hand	US$35,000
Taska Prosthetics	Taska hand	US$35,000

EMG sensors are the interface for human-machine communication in myoelectric prostheses. There are many different types available. Most consist of adhesive, non-reusable pre-gelled electrodes. Some of the cheapest solutions are the MyoWare Muscle Sensor [40], SX230 sensor [3], among others. Currently, there is only one product available in the market, the Myo Gesture Control Armband [31], which integrates EMG sensors into an armband. It also integrates an IMU and is developed to be sported at the forearm.

Advances in Ecuador

In 2012, a prototype of a robotic hand controlled by EMG-signals was successfully developed at Politécnica Salesiana University. The prototype had four DOFs, including the movement of three fingers and rotation of the wrist articulation. Each finger was equipped with a force sensor to control the pressure applied by the hand onto the objects [11]. In 2014, researchers from Técnica Particular de Loja University developed a prototype called "Hand of Hope", aiming to produce low-priced upper limb prostheses. The developed prosthesis made use of biosignals for its control and was made of plastic biopolymers [8,9]. Following the next year, a bionic prosthesis of seven DOFs was designed at the Fuerzas Armadas University using intelligent materials and myoelectric control, and it was adapted for various grip patterns. The main objective of this project was to enable transradial amputees to perform basic grips to perform daily tasks [33]. Later, control algorithms were applied and tests of movement, load, grip, and pressure were performed, validating the model by the means of statistical tests of Chi square independence [34]. In the same year, students of Católica de Cuenca University together with the Secretaría de Educación Superior, Ciencia, Tecnología e Innovación, within the framework of the Prometeo Project, developed biomechanical hand prostheses for children older than eight years and teenagers. Those weighed up to 200 g and worked with a speech recognition system that allowed to perform three important functions: cylindrical grip, pinch grip, and pointing with the index finger to enable the use of touch screens for instance. These hands do not use springs or tension wires, thus significantly reducing their weight [17]. In 2016 at Técnica Del Norte University, an academic work dealing with the construction of a robotic hand focusing on the control of finger movement was published. The movement of the fingers was done through wire ropes that acted as tendons. The full control of the structure was carried out through a glove. It also contains an accelerometer, which simulates the movement of the wrist to make the device perform the same movement as the glove [51]. In 2017, in order to demonstrate the feasibility of using 3D printing technology on the development of low-cost prototypes, the San Francisco de Quito University developed a hand-arm that could be controlled by exing any muscle in the user's arm through a sensor that collects EMG-signals from the muscle. The movement of the finger joints was done by one servomotor per finger, coiling a nylon thread [7].

3 Results

Statistics show that there are several trans-humeral amputations performed in Ecuador every year [20–22], which is a strong indicator for an existing demand for upper arm prostheses. As also shown, a great majority of disabled people are adults and are therefore full-grown, and only small adaptations need to be done once fitting a data-acquisition device. It has also been analysed that the number of wrist amputations is high. All in all, the market offers a variety of highly-developed prostheses for different purposes and EMG sensors. Especially in recent years, great advances have been achieved and several versatile, multifunctional prostheses have emerged [4,5,14,30,36,43]. Unfortunately, high manufacturing costs is a barrier to prosthetics access. This is confirmed by other researchers [1,24,39]. There are also different low-cost prostheses [4,30], but not many low-cost devices for signal acquisition [3,31,40]. The control of a prosthesis with EMG-signals is a promising approach, because of the easy attachment method and non-intrusive measurement of muscles that cannot serve another function anymore. Several universities are researching robotized hands, but there are no commercial EMG sensors manufactured or designed in Ecuador [7–9,11,33]. One reason is the lack of specialized electronic components in the local market. Another reason is that the local demand of EMG sensors is low and the infrastructure costs for manufacture are high.

4 Discussion and Future Directions

Many prostheses that have been developed in Ecuadorian universities share the use of the flexo-tensor biomechanical model using EMG devices for the acquisition of signals, which are amplified and then sent to servomotors for the main action of tension and exion of the phalanges. Although the use of 3D printing technologies has reduced the weight of prostheses, increasing the DOFs proportionally increases their weight and reduces their energy efficiency. Another research contributes with the use of artificial intelligence and the integration of other sensors to increase the number of possible combinations and achieve a greater functionality of the prostheses. However, these efforts have not been successful in real life situations due to the lack of robustness. These limitations include muscle crosstalk and the location of the electrode in relation to muscle activity. It is known that when the upper arm is available there are two individually contractile muscle groups, the biceps and the triceps, limiting to two states: open and close. The technology industry trends have made significant progress in the development of robotic prostheses. In addition to the use of EMG sensors, the integration of external sensors and wireless commands has contributed to improved strength and precision controls. One of the disadvantages when trying to access these prostheses is their high costs. A possible solution for this issue is the reduction of the gap between industry and university, with the purpose of allowing the fabrication of low cost robotics hands in Ecuador. Finally, although there are advances in the design and manufacture of robotic prostheses in Ecuadorian universities, many of the parts that are used are purchased and not locally manufactured, such as EMG sensors.

5 Conclusions and Future Works

A review of significant advances in the development of robotic prostheses of upper limbs in Ecuador has been carried out. Many prostheses have been designed using imported electronic devices at high costs. The parameters to consider are size, shape, weight, and functionality. With the use of 3D printing technologies, significant advances in size and shape have been achieved, but studies should be directed to improve the autonomous functionality of prostheses with affordable costs for users. Future research, have to include the reviewing of the best materials for its manufacture, as well as the optimization of energy for its operation.

Acknowledgments. This work was supported by the Research Center of the Espíritu Santo University in Ecuador.

References

1. Andrade, A.O., Pereira, A.A., Walter, S., Almeida, R., Loureiro, R., Compagna, D., Kyberd, P.J.: Bridging the gap between robotic technology and health care. Biomed. Sig. Process. Control **10**, 65–78 (2014)

2. Biddiss, E.A., Chau, T.T.: Upper limb prosthesis use and abandonment: a survey of the last 25 years. Prosthet. Orthot. Int. **31**(3), 236–257 (2007)
3. Biometrics: Surface EMG sensor, November 2018. http://www.biometricsltd.com/surface-emg-sensor.htm
4. Bionicohand: Opensource myohand, November 2018. https://bionico.org/
5. Bionics, T.: I-LIMB ultra, November 2018. http://touchbionics.com/products/active-prostheses/i-limb-ultra
6. Brunelli, D., Tadesse, A., Vodermayer, B., Novak, M., Castellini, C.: Low-cost wearable multichannel surface EMG acquisition for prosthetic hand control. In: 2015 6th International Workshop on Advances in Sensors and Interfaces, pp. 1–6. IEEE (2015). https://doi.org/10.1109/IWASI.2015.7184964
7. Cañizares, A., Pazos, J., Benítez, D.: On the use of 3D printing technology towards the development of a low-cost robotic prosthetic arm. In: 2017 IEEE International Autumn Meeting on Power, Electronics and Computing, pp. 1–6. IEEE (2017). https://doi.org/10.1109/ROPEC.2017.8261579
8. Calderon-Cordova, C., Ramírez, C., Barros, V.: Una mano de esperanza para todos. Perspectivas de Investigación **10**(1), 3 (2014)
9. Calderon-Cordova, C., Ramírez, C., Barros, V., Quezada-Sarmiento, P.A., Barba-Guamán, L.: EMG signal patterns recognition based on feedforward artificial neural network applied to robotic prosthesis myoelectric control. In: 2016 Future Technologies Conference, pp. 868–875. IEEE (2016). https://doi.org/10.1109/FTC.2016.7821705
10. Clauss, C., Clauss, W.: Humanbiologie kompakt, 2nd edn. Springer, Heidelberg (2017)
11. Collahuazo, J., Ordoñez, E.: Design and construction of a robot hand activated by electromyographic signals. In: International Symposium on Robotic and Sensors Environments, pp. 25–30. IEEE (2012). https://doi.org/10.1109/ROSE.2012.6402629
12. CONADIS: Estadísticas de discapacidad, October 2018. https://www.consejodiscapacidades.gob.ec/estadisticas-de-discapacidad/
13. Controzzi, M.: D5.20: first robot hand development. Report, CogLaboration (2012). http://www.coglaboration.eu/node/33
14. DARPA: Revolutionizing Prosthetics, November 2018. https://www.darpa.mil/program/revolutionizing-prosthetics
15. Foody, J., Maxwell, K., Hao, G., Kong, X.: Development of a low-cost underactuated and self-adaptive robotic hand. In: 38th Mechanisms and Robotics Conference, p. 9. ASME (2014). https://doi.org/10.1115/DETC2014-35075
16. Fougner, A., Stavdahl, O., Kyberd, P., Losier, Y., Parker, P.: Control of upper limb prostheses: terminology and proportional myoelectric control–a review. IEEE Trans. Neural Syst. Rehabil. Eng. **20**(5), 663–677 (2012). https://doi.org/10.1109/TNSRE.2012.2196711
17. Gámez, B., Flores, C., Cabrera, F., Cabrera, J.: Design of a biomechanics prosthesis for child. Ingeniería UC **23**(1), 58–66 (2016)
18. Gretsch, K., Lather, H., Peddada, K., Deeken, C., Wall, L., Goldfarb, C.: Development of novel 3D-printed robotic prosthetic for transradial amputees. Prosthet. Orthot. Int. **40**(3), 400–403 (2016). https://doi.org/10.1177/0309364615579317
19. Imtiaz, U., Bartolomeo, L., Lin, Z., Sessa, S., Ishii, H., Saito, K., Zecca, M., Takanishi, A.: Design of a wireless miniature low cost EMG sensor using gold plated dry electrodes for biomechanics research. In: 2013 IEEE International Conference on Mechatronics and Automation, pp. 957–962. IEEE (2013). https://doi.org/10.1109/ICMA.2013.6618044

20. INEC: Camas y egresos hospitalarios 2015, October 2015. http://www.ecuadorencifras.gob.ec/camas-y-egresos-hospitalarios-2015/
21. INEC: Camas y egresos hospitalarios 2016, October 2016. http://www.ecuadorencifras.gob.ec/camas-y-egresos-hospitalarios-2016/
22. INEC: Camas y egresos hospitalarios 2017. October 2018. http://www.ecuadorencifras.gob.ec/camas-y-egresos-hospitalarios/
23. Inglis, T., MacEachern, L.: 3D printed prosthetic hand with intelligent EMG control. Report, Carleton University (2013). http://www.doe.carleton.ca/Course/4th_year_projects/Am4_Inglis_Timothy_2013.pdf-.pdf
24. Iqbal, N.V., Subramaniam, K.: A review on upper-limb myoelectric prosthetic control. IETE J. Res. **64**(6), 740–752 (2018)
25. Jiang, Y., Sakoda, S., Hoshigawa, S., Ye, H., Yabuki, Y., Nakamura, T., Ishihara, M., Takagi, T., Takayama, S., Yokoi, H.: Development and evaluation of simplified EMG prosthetic hands. In: 2014 IEEE International Conference on Robotics and Biomimetics, ROBIO 2014, pp. 1368–1373. IEEE (2014)
26. Jones, G., Rosendo, A., Stopforth, R.: Prosthetic design directives: low-cost hands within reach. In: 2017 International Conference on Rehabilitation Robotics, pp. 1524–1530. IEEE (2017). https://doi.org/10.1109/ICORR.2017.8009464
27. Khanna, P., Singh, K., Bhurchandi, K., Chiddarwar, S.: Design analysis and development of low cost underactuated robotic hand. In: 2016 IEEE International Conference on Robotics and Biomimetics (ROBIO), pp. 2002–2007. IEEE (2016)
28. Kitchenham, B., Pretorius, R., Budgen, D., Brereton, O.P., Turner, M., Niazi, M., Linkman, S.: Systematic literature reviews in software engineering-a tertiary study. Inf. Softw. Technol. **52**(8), 792–805 (2010)
29. Konrad, P.: The ABC of EMG A Practical Introduction to Kinesiological Electromyography, 1st edn. Noraxon, Scottsdale (2006)
30. Koprnicky, J., Najman, P., Safka, J.: 3D printed bionic prosthetic hands. In: 2017 IEEE International Workshop of Electronics, Control, Measurement, Signals and their Application to Mechatronics, pp. 1–6 (2017). https://doi.org/10.1109/ECMSM.2017.7945898
31. Geryes, M., Charara, J., Skaiky, A., Mcheick, A., Girault, J.: A novel biomedical application for the Myo gesture control armband. In: 2017 29th International Conference on Microelectronics, pp. 1–4. IEEE (2013). https://doi.org/10.1109/ICM.2017.8268823
32. Mallik, S., Dutta, M.: A study on control of myoelectric prosthetic hand based on surface EMG pattern recognition. Int. J. Adv. Res. Sci. Eng. **6**(07), 635–646 (2017)
33. Monar, M., Murillo, L.: Diseño y construcción de una prótesis biónica de mano de 7 grados de libertad utilizando materiales inteligentes y control mioeléctrico adaptada para varios patrones de sujeción. Report, Universidad de las Fuerzas Armadas (2015). http://repositorio.espe.edu.ec/xmlui/handle/21000/10187
34. Morales, D.: Diseño e implementación del sistema de control de una prótesis biónica de 7 grados de libertad utilizando materiales inteligentes y control mioeléctrico adaptada para varios patrones de sujeción. Report, Universidad de las Fuerzas Armadas (2016). http://repositorio.espe.edu.ec/handle/21000/12449
35. Nacher, V., Cáliz, D., Jaen, J., Martínez, L.: Examining the usability of touch screen gestures for children with down syndrome. Interact. Comput. **30**(3), 258–272 (2018)
36. Ottobock: Armamputation und rehabilitation, October 2018. https://www.ottobock.de/prothetik/informationen-fuer-amputierte/von-amputation-bis-rehabilitation/leben-mit-armamputation/

37. Ottobock: Ottobock region andina - localidades, October 2018. https://www.ottobock.com.co/information-pages/locations.html
38. Pasquina, P., Evangelista, M., Carvalho, A., Lockhart, J., Griffin, S., Nanos, G., McKay, P., Hansen, M., Ipsen, D., Vandersea, J.: First-in-man demonstration of fully implanted myoelectric sensors for control of an advanced electromechanical arm by transradial amputees. J. Neurosci. Methods **244**, 85–93 (2015). https://doi.org/10.1016/j.jneumeth.2014.07.016
39. Pasquina, P.F., Perry, B.N., Miller, M.E., Ling, G.S., Tsao, J.W.: Recent advances in bioelectric prostheses. Neurol: Clin. Pract. **5**(2), 164–170 (2015)
40. Poveda, G., Trujillo Guerrero, M.F., Rosales, A.: Muscular biofeedback system for the rehabilitation of the upper extremity. In: 2018 International Conference on Information Systems and Computer Science, pp. 1–8 (2018). https://doi.org/10.1109/INCISCOS.2018.00008
41. Prensilia, S.: Self-contained robotic hand, November 2018. https://www.prensilia.com/portfolio/ih2-azzurra/
42. Prosthetic, L.: Cost, November 2018. http://thelondonprosthetics.com/consultation/cost/
43. Prosthetics, T.: Doing more build confidence, November 2018. http://www.taskaprosthetics.com/
44. Van der Riet, D., Stopforth, R., Bright, G., Diegel, O.: An overview and comparison of upper limb prosthetics. In: 2013 AFRICON, pp. 1–8. IEEE (2013). https://doi.org/10.1109/AFRCON.2013.6757590
45. Sella, G.: Clinical utilization of surface electromyography and needle electromyography: a comparison of the two methodologies. Biofeedback **35**(1), 38–42 (2007)
46. Sharmila, K., Sarath, T., Ramachandran, K.: EMG controlled low cost prosthetic arm. In: Distributed Computing, VLSI, Electrical Circuits and Robotics, pp. 169–172. IEEE (2016). https://doi.org/10.1109/DISCOVER.2016.7806239
47. Smith, L.: Real-time simultaneous and proportional myoelectric control using intramuscular EMG. J. Neural Eng. **11**, 2–14 (2014). https://doi.org/10.1088/1741-2560/11/6/066013
48. Stuff: Kiwis engineer the world's first waterproof prosthetic hand, January 2019. https://www.stuff.co.nz/business/innovation/94370766/kiwis-engineer-the-worlds-first-waterproof-prosthetic-hand
49. Townsend, W.T., Hauptman, T., Crowell, A., Zenowich, B., Lawson, J., Krutik, V., Doo, B.: Intelligent, self-contained robotic hand, January 2007. https://patents.google.com/patent/US7168748B2/en. US Patent 7,168,748
50. Trabajo: Ministerio del trabajo establece salario básico unificado 2019, December 2018. http://www.trabajo.gob.ec/incremento-del-salario-basico-unificado-2019/
51. Yacelga, P., Paul, H.: Construcción de una mano robótica enfocado al control del movimiento de los dedos. Report, Universidad Técnica del Norte (2016). http://repositorio.utn.edu.ec/handle/123456789/5697
52. Zuo, K., Olson, J.: The evolution of functional hand replacement: from iron prostheses to hand transplantation. Can. J. Plast. Surg. **22**(1), 44–51 (2014). https://doi.org/10.1177/229255031402200111

Design, Simulation and Comparison of Controllers that Estimate an Hydric Balance in Strawberry Plantations in San Pedro

Raúl Carrasco[1(✉)], Carolina Lagos[2], Eduardo Viera[3], Leonardo Banguera[4], Ginno Millán[3], Manuel Vargas[5], and Álvaro González[6]

[1] Facultad de Ingeniería, Ciencia y Tecnología, Universidad Bernardo O'Higgins, Santiago, Chile
raul.carrasco.a@usach.cl
[2] Pontificia Universidad Católica de Valparaíso, Valparaíso, Chile
[3] Departamento de Ingeniería Eléctrica, Universidad de Santiago de Chile, Santiago, Chile
[4] Facultad de Ingeniería Industrial, Universidad de Guayaquil, Guayaquil, Ecuador
[5] Facultad de Ingeniería y Tecnología, Universidad San Sebastián, Santiago, Chile
[6] Departamento de Ingeniería Industrial, Universidad de Santiago de Chile, Santiago, Chile

Abstract. This work has a great relevance in modern agriculture, because of the nowadays problematic of the hydric resources at national and world level. Which evaluates different control technics able to estimate an hydric balance in strawberry plantations. Through the PID controllers with neural networks and diffuse logic. Getting better results with neural networks in Adequation Index, Settling Time, Overshoot and Stability, with which the obtained results were validated.

Keywords: Controllers · Hydric balance · PID · Neural networks · Fuzzy

1 Introduction

The increasing interest for the sweet water consumption in the world and the increase of desertic zones due to climate change, has caused a legitimate concern in the scientific world for the efficient use of the water resource in modern agriculture [1,2].

It has been researched in the efficient use of the sweet water resource in the agriculture using predictive control methods, achieving greater efficiency and reducing significantly the water consumption [2,3], with a limitated supply of

M. Botto-Tobar et al. (Eds.): ICAETT 2019, AISC 1066, pp. 335–347, 2020.
https://doi.org/10.1007/978-3-030-32022-5_32

water and uncertain meteorologic conditions achieve in maintaining soil humidity above the wilting of crops [4], with good estimations of the demand of water in crops and the local time it is achieved to optimize supply and demand with the objective of optimizing the performance of the crops [5].

It has been used a multiplatform application of a PC and a mobile module for the watering of strawberries, taking into account the agroclimatic conditions, soil information and information of the hydric system obtaining information of the watering time required and achieving a save of water between an 11% and 33% [6].

A study was done about the hydric balance of several crops in dry area in Spain, finding that the (ETa) represented the biggest loss and the improve in the watering efficiency [7].

ETa and the unique coefficients of crops (Kc) has been investigated for the development phases of the coffee (Coffee arabica), through a superficial watering system, in the region of San Andrés, Cuba. It was obtained that the most demand of water is produced in the phase flowering-fructification, the ETa annual daily average was of 3.24 mm/d and the unique global coefficient of crop was of 0.86 [8].

It has been implemented an Artificial Neural Networks (ANN) to program the watering of strawberries using the humidity of the soil and its physical properties as supplies of the model. The system got a save of water of 20.5% and an energy saving of 23.9%. However, ANN requires great set of data for the training and are incapable of describing the physical dynamic of a system. This makes that its use is limited in the supporting tools of decisions in real time [9].

It was applied a diffuse logic model to program drip irrigation using ETa, humidity of the soil and the growth stage of the crops as entries to the mode [10].

It was applied a diffuse logic model using data of soil humidity, foliar moisture, and climatologic data as entries for the model and implement the decisions of watering programming. The system was able to maintain the soil moisture thresholds in the specified range [11].

It has been propose an efficient watering system based o ETa, using the diffuse logic with the purpose of avoiding the watering excess that affects the harvest, their results show that the diffuse model is a fast and accurate tool to calculate the ETa, as well as the net irrigation required [10].

It has been investigated about the water applied in the rice field, where it is recommended to install a drainage in the central line in the rice fields, a drainage installation eliminating excess water in the root zone in time [12].

It were determined the periods of need of water to maintain the hydric balance in the strawberry plantations in the commune of San Pedro, through the agroclimatic variables and the needs of water by evapotranspiration for strawberries plantations in the commune use the correlation analysis, principal component analysis and k-means [13].

2 Characteristics of the Zone of San Pedro

2.1 Irrigation

The commune of San Pedro, located in the Metropolitan region of Chile, is the most important area of the country in the strawberry cultivation [14]. Affected by the water shortage of the last decades product of the aquifers over exploitation.

The basin of the Yali estuarian is mainly located in the comunes of Santo Domingo and San Pedro. Being the last one, the commune with the greatest irrigated surface of e 4,800 hectare (ha) [15] from a total of 61,298 ha and of these 3,495 ha with system of micro watering, mainly thanks to the drip irrigation, according to the last Agricultural and Forestry Census 2007 [16].

The main source of water is of waterwheel and in second place of deep wells [17], since the commune lacks of water from the Andes Mountains that could provide water for irrigation through the rivers and water ways.

2.2 Evapotranspiration

The evapotranspiration s the total transfer of water from the vegetation cover to the atmosphere, by the process of evaporation of the water in its different stages from the earth surface to the atmosphere and the transpiration of the vegetation cover by the water absorbed by its roots that transfer the water mainly through the stomata located in its leaves [18].

Below is shown Eq. 1 relating to Evapotranspiration [19].

$$et_0 = \frac{0,408\Delta(R_n - G) + y\frac{900}{T+273}u_2(e_s - e_a)}{\Delta + y(1 + 0,34u_2)} \tag{1}$$

where :
et_0: reference Evapotranspiration [mm/d],
R_n: net radiation on the surface of the crop [MJ/(m^{2}d)],
G: flux density of soil heat [MJ/(m^{2}d)],
T: air average daily temperature at 2 mhigh [°C],
u_2: wind speed at 2 m high [m/s],
e_s: saturation vapor pressure [kPa],
e_a: real steam pressure [kPa],
$e_s - e_a$: saturation vapor pressure deficit [kPa],
Δ: vapor pressure curve of the slope [kPa/°C],
y: psychrometric constant [kPa/°C].

Evapotranspiration f the crop is obtained from:

$$et_c = K_c\ et_0 \tag{2}$$

where,
et_c: Evapotranspiration of the cultivation (mm/d),
K_c: Cultivation Coefficient,
et_0: Potential Evapotranspiration or of reference (mm/d).

3 Transfer Function

For measuring the hydric balance represented in Fig. 1, like [2,3]:

$$\dot{\theta}(t) = ir(t) + rf(t) + cr(t) - et_c(t) - dp(t) - ro(t) \tag{3}$$

where,
$\dot{\theta}$: Soil moisture,
$ir(t)$: Irrigation,
$rf(t)$: Rains,
$cr(t)$: Capillary rise,
$et_c(t)$: Evapotranspiration of plants and soil,
$dp(t)$: Deep drainage,
$ro(t)$: Water loss in irrigation.

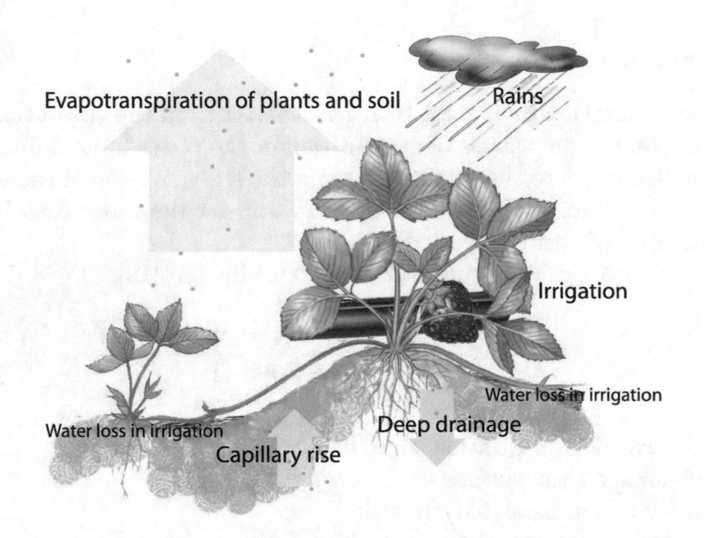

Fig. 1. Components of hydric balance in irrigation systems.

Considering, the Winter rains of Mediterranean climate, capillar rise by study zone lacking of wetlands or superficial groundwater; superficial spill that is minimized thanks to drip irrigation technology or tape that are despised in the model and the deep percolation in a watering system is clearly proportional to the moisture of the soil [20].

$$\dot{\theta}(t) = ir(t - \tau) - et_c - c_1\theta \tag{4}$$

et_c, is considered disturbance.

Finally the hydric balance model is represented in Eq. 5.

$$\dot{\theta}(t) = ir(t - \tau) - c_1\theta \tag{5}$$

3.1 The Transform of Laplace

Applying the transform of Laplace results:

$$\frac{\theta(s)}{ir(s)} = \frac{e^{-\tau s}}{(s + c_1)} \tag{6}$$

3.2 Characterization of the Model

To characterize the response of a dynamic system, this must be submitted to its levels of design, which in this case correspond to the estimative study of the hydric balance defined in Eq. 5. For which the response to the step, is a good estimator of the behavior of the system in this stage.

An entry signal of the step type allows to know the response of the system in the face of abrupt changes in its entry. So, it gives an idea of the time of the establishment of the signal, that is to say, how long does it take to the system in achieving its steady state.

$$\text{Response to the step} \rightarrow ir(s) = \frac{1}{s}e^{Ts} \tag{7}$$

Is obtained,

$$\theta(s) = \frac{1}{s}e^{Ts}\frac{e^{-\tau s}}{(s + c_1)} \tag{8}$$

To proceed to a first approach in the application of control strategies, the PID controller is the base alternative, which is possible to tune in through the Ziegler-Nichols method. In Table 1.

Table 1. Calculation of tuning of Ziegler-Nichols.

Controller	K_p	T_i	T_d
P	48.0	∞	0
PI	43.2	33.3333	0
PID	57.6	0.0417	0.0104

4 Feedback Control Systems

4.1 PID Controllers

According to Fig. 2 and performance data in Table 2. The PID controller gets a shorter Rise Time of 0:20:01, the P 0:23:54 and the PI is obtained at 0:26:38, the stabilization time or Settling Time the PI controller reaches with the shorter time of 4:39:13, then the P controller at 6:27:56 and the controller achieve stabilization at 10:17:28. Where it can be appreciated an output overrun related to the slogan or Overshoot, where PI controller with a 41.28%, the P controller obtains a 49.89% and the PID controller gets the biggest excess of 68.89%, analogously with a Peak PI = 28.4648, P = 29.9165 and PID = 33.7183 which occur Peak Time PI = 1:29:17, P = 1:33:36 and PID = 1:24:58.

So, PI controller gets the best results with the lower stabilization time 4:39:13 and lower Overshoot of a 41.28%.

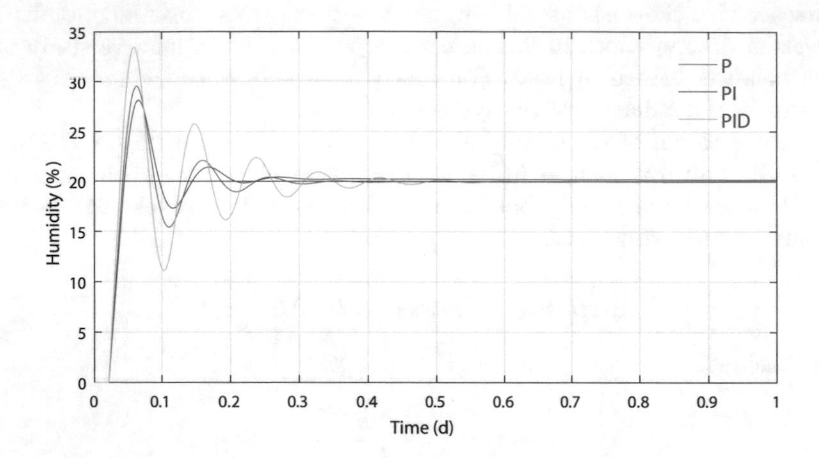

Fig. 2. Response PID controller.

Table 2. Summary of performance indicators.

Variable/controller	P	PI	PID
Rise Time	0:23:54	0:26:38	0:20:01
Settling Time	6:27:56	4:39:13	10:17:28
Settling Min	15.2289	17.1645	11.0122
Settling Max	29.9165	28.4648	33.7183
Overshoot	49.89%	41.28%	68.89%
Peak	29.9165	28.4648	33.7183
Peak Time	1:29:17	1:33:36	1:24:58

4.2 PID Controller with Disturbances

According to Fig. 3 and performance data in Table 3 in disturbances. The PID controller gets a lower Rise Time of 0:19:26, the P 0:23:28 and gets the PI at

0:26:04, the stabilization time or Settling Time the controller P, PI and PID do not achieve stabilization in the evaluation period. The lower excess or Overshoot is obtained by the PI controller with a 42.27%, the P controller gets a 50.64% and the PID controller gets the greatest excess of 69.73%, analogously with a Peak PI = 27.9688, P = 29.3660 and PID = 33.1689 with Peak Time PI = 1:34:11, P = 1:30:00 and PID = 1:24:06.

Similar results are obtained between the simulations with and without disturbances, except that the simulations with disturbances do not achieve a stabilization time of the system.

The PI controller gets better results with the lower Overshoot of a 42.27%.

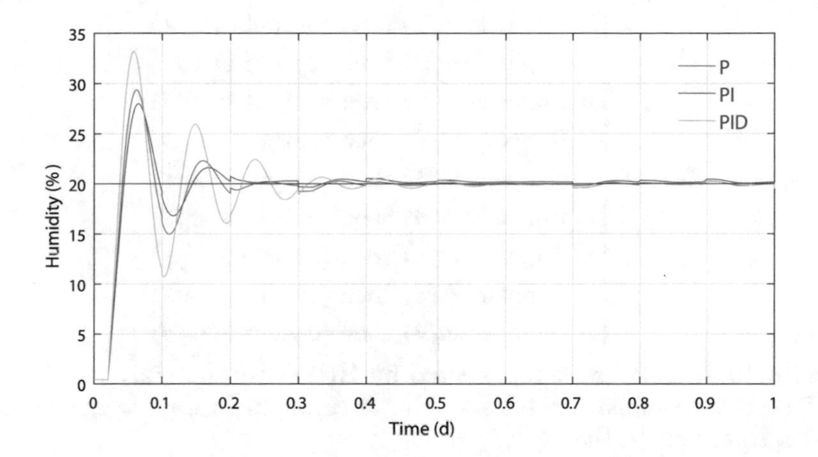

Fig. 3. Response system of control PID with disturbances.

Table 3. Summary of performance indicators under disturbances.

Variable/controller	P	PI	PID
Rise Time	0:23:28	0:26:04	0:19:26
Settling Time	1d	1d	1d
Settling Min	14.9818	16.8244	10.7448
Settling Max	29.3660	27.9688	33.1689
Overshoot	50.64%	42.27%	69.73%
Peak	29.3660	27.9688	33.1689
Peak Time	1:30:00	1:34:11	1:24:06

4.3 Fuzzy Controller

Basically, fuzzy controller was configured with nine specific rules, based on the following variable (data set):

$$\mathbf{A}(i) = \begin{cases} Irrigation \\ Evapotranspiration \\ Weather\ statics \\ Anual\ rain\ statics \end{cases} \tag{9}$$

These rules are summarized in nine rules:

$$\mathbf{X}(i) = \begin{cases} \text{if (input is } \mathit{mf1}\text{) then (output is } \mathit{mf1}\text{)} \\ \text{if (input is } \mathit{mf2}\text{) then (output is } \mathit{mf1}\text{)} \\ \text{if (input is } \mathit{mf3}\text{) then (output is } \mathit{mf5}\text{)} \\ \text{if (input is } \mathit{mf4}\text{) then (output is } \mathit{mf6}\text{)} \\ \text{if (input is } \mathit{mf5}\text{) then (output is } \mathit{mf6}\text{)} \\ \text{if (input is } \mathit{mf6}\text{) then (output is } \mathit{mf7}\text{)} \\ \text{if (input is } \mathit{mf7}\text{) then (output is } \mathit{mf7}\text{)} \\ \text{if (input is } \mathit{mf8}\text{) then (output is } \mathit{mf9}\text{)} \\ \text{if (input is } \mathit{mf9}\text{) then (output is } \mathit{mf9}\text{)} \end{cases} \tag{10}$$

In Eq. 10 the rules set is represented by $\mathbf{H}(i) = \{mf1, mf2, ..., mf9\}$ where the $\mathbf{H}(i)$ are the summarized set of the input variables for each "required" point of these, shown in the the Eq. 9.

The Fuzzy controller can not reach the set point but achieves the stability of the system, according to rule set applied in this case.

Regarding Fig. 4 and performance data in Table 4 the fuzzy controller reaches the Rise Time at 16:07:58 and at 17:09:36 with disturbances, the stabilization time or Settling Time is reached at 21:54:09 and in a day at 21:11:14 for the fuzzy controller with disturbances.

Table 4. Performance indicators in fuzzy controller.

Var/controller	Without disturbances	With disturbances
Rise Time	16:07:58	17:09:36
Settling Time	21:54:09	1d 21:11:14
Settling Min	17.4858	18.0490
Settling Max	19.4255	20.0362
Overshoot	0%	0.62%
Peak	19.4255	20.0362
Peak Time	1d 3:41:28	1d 12:00:00

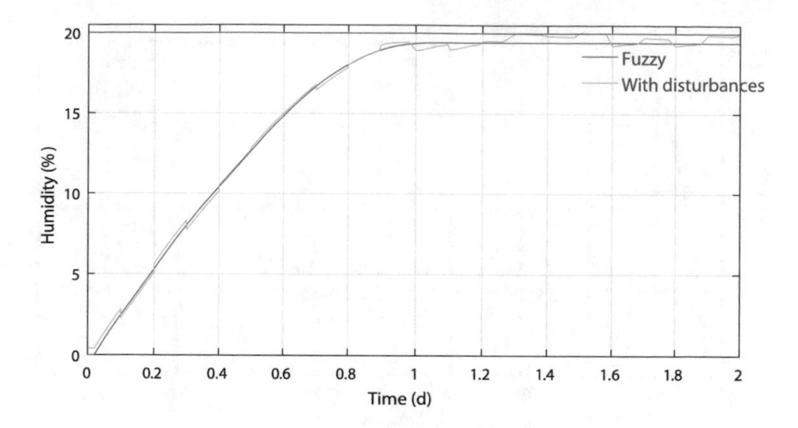

Fig. 4. Response fuzzy control with and without disturbances.

4.4 Neural Networks

The Artificial Neural Networks was configured in a "feedfordward" architecture [21,22], and trained with "back-propagation" algorithm where the input layer ($f(x)$) has three neurons to process the input variables these are:

$$f(x) = \begin{cases} Irrigation \\ Anual\ rain\ statics \\ Weather\ statics \end{cases} \tag{11}$$

On other hand hidden layer has ten neurons and the output layer has just one output. In the hidden layer ($y(x)$) ten neurons has the function to process the input variables. In the output layer just one layer represent the humidity % in the soil.

The response of the controller by neural network shows an overshoot to achieve the stability of the system, corresponding to a conventional FeedForward neural network with a layer of ten neurons.

According to Fig. 5 and performance data in Table 5 the controller of neural network achieves the Rise Time at 1:56:21 and at 1:59:14 with disturbances, the stabilization time or Settling Time is reached at 3:45:22 and in a day at 23:38:41 for the fuzzy controller with disturbances and a peak of 20.1548 without disturbances.

5 Performance Indices

According to Table 6 the RMS, RSD, AI, ISE and IAE performance measures respectively. The controller of neural network achieves the best adequation index or AI, instead the controller PI type achieves better results in all the indicators of error evaluation.

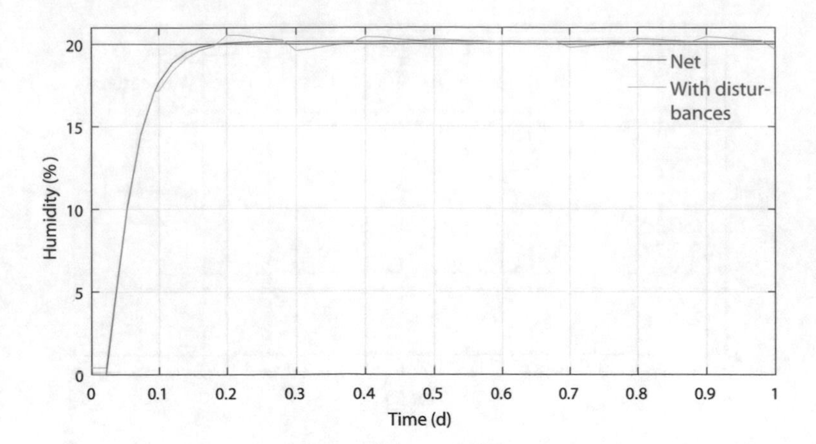

Fig. 5. Response neural network with and without disturbances.

Table 5. Indicators of neural network performance.

Var/controller	Without disturbances	With disturbances
Rise Time	1:56:21	1:59:14
Settling Time	3:45:22	23:38:41
Settling Min	18.8170	18.3368
Settling Max	20.1548	20.6515
Overshoot	0%	4.91%
Peak	20.1548	20.6515
Peak Time	1d	5:20:50

Table 6. Performance measures.

Controllers	RMS	RSD	AI	ISE	IAE
P	0.1988	3.9246	0.6090	13.9212	1.3657
PI	0.1924	3.8397	0.6017	13.3630	1.3859
PID	0.2241	4.4547	0.5629	17.7203	1.8343
Neural network	0.2184	4.2200	0.6793	17.4461	1.9531
Fuzzy	1.0287	11.5548	0.0916	106.7185	10.5885

5.1 Performance Evaluation

In Table 7, is summarized a general performance evaluation for each controller according to information developed in this work, getting better results the neural network in Adequation Index, Settling Time, Overshoot and Stability and secondly the controller of the PI type for its low performance index at error proof and its low implementation cost [23].

Table 7. Performance evaluation.

Criteria/controllers	P	PI	PID	Neural network	Fuzzy
Adequation Index				✓	
Error measures		✓			
Settling Time				✓	
Overshoot				✓	
Stability				✓	
Implantation costs	✓	✓	✓		

6 Conclusion

In this work it was established that the use of different control tools, allows an improvement of the hydric balance in strawberry plantations. Where the results obtained, could be validated through the performance indices shown in Table 6.

Highlighting the neural networks as the technique of best response when implemented in a simulation environment, getting a negligible error and with less response time than the rest of techniques.

Given the characteristics of hydric balance in the cultivation of strawberries, stands for being a slow system, to which humidity of soil can be controlled fast and efficiently from the dynamic of the system, allowing an optimal control and improving productivity and quality of the fruit with the use of neural network controllers.

This work can be implemented in different types of cultivations, where exist shortage of water, loss or potential decrease of it, at a national or international level.

Acknowledgement. The authors acknowledge the funding for the investigation to FVF Ingeniería y Consultoría Ltda.

References

1. Adeyemi, O., Grove, I., Peets, S., Norton, T.: Advanced monitoring and management systems for improving sustainability in precision irrigation. Sustainability **9**(353), 1–29 (2017). https://doi.org/10.3390/su9030353
2. Lozoya, C., Mendoza, C., Mejía, L., Quintana, J., Mendoza, G., Bustillos, M., Arras, O., Solís, L.: Model predictive control for closed-loop irrigation. IFAC Proc. Vol. **47**(3), 4429–4434 (2014). https://doi.org/10.3182/20140824-6-ZA-1003.02067

3. Lozoya, C., Mendoza, C., Aguilar, A., Román, A., Castelló, R.: Sensor based model driven control strategy for precision irrigation. J. Sens. **2016**(9784071), 12 (2016). https://doi.org/10.1155/2016/9784071
4. Delgoda, D., Malano, H., Saleem, S.K., Halgamuge, M.N.: Irrigation control based on model predictive control (MPC): formulation of theory and validation using weather forecast data and AQUACROP model. Environ. Model. Softw. **78**, 40–53 (2016). https://doi.org/10.1016/j.envsoft.2015.12.012
5. Saleem, S.K., Delgoda, D., Ooi, S.K., Dassanayake, K.B., Liu, L., Halgamuge, M., Malano, H.: Model predictive control for real-time irrigation scheduling. IFAC Proc. Vol. **46**(18), 299–304 (2013). https://doi.org/10.3182/20130828-2-SF-3019.00062
6. González, R., Fernández, I., Martin, M., Rodríguez, J., Camacho, E., Montesinos, P.: Multiplatform application for precision irrigation scheduling in strawberries. Agric. Water Manag. **183**, 194–201 (2017). https://doi.org/10.1016/j.agwat.2016.07.017
7. Merchán, D., Causapé, J., Abrahão, R., García-Garizábal, I.: Assessment of a newly implemented irrigated area (Lerma Basin, Spain) over a 10-year period. I: water balances and irrigation performance. Agric. Water Manag. **158**, 277–287 (2015). https://doi.org/10.1016/j.agwat.2015.04.016
8. Cisneros, E., Rey, R., Martínez, R., López, T., González, F.: Evapotranspiración y coeficientes de cultivo para el cafeto en la provincia de Pinar del Río. Revista Ciencias Técnicas Agropecuarias **24**(2), 23–30 (2015)
9. Karasekreter, N., Başçiftçi, F., Fidan, U.: A new suggestion for an irrigation schedule with an artificial neural network. J. Exp. Theor. Artif. Intell. **25**(1), 93–104 (2013). https://doi.org/10.1080/0952813X.2012.680071
10. Mousa, A.K., Croock, M.S., Abdullah, M.N.: Fuzzy based decision support model for irrigation system management. Int. J. Comput. Appl. **104**(9), 14–20 (2014)
11. Prakashgoud, P., Desai, B.: Intelligent irrigation control system by employing wireless sensor NetworksPatil. Int. J. Comput. Appl. **79**(11), 33–40 (2013)
12. Darzi-Naftchali, A., Mirlatifi, S., Shahnazari, A., Ejlali, F., Mahdian, M.: Effect of subsurface drainage on water balance and water table in poorly drained paddy fields. Agric. Water Manag. **130**, 61–68 (2013). https://doi.org/10.1016/j.agwat.2013.08.017
13. Carrasco, R., Soto, I., Seguel, F., Osorio-valenzuela, L., Flores, C.: Water balance in plantations of strawberries, in the commune of San Pedro. In: Proceedings of 2017 CHILEAN Conference on Electrical, Electronics Engineering, Information and Communication Technologies, CHILECON 2017, pp. 1–5. IEEE, Pucon (2017). https://doi.org/10.1109/CHILECON.2017.8229715
14. Sánchez, S., Gambardella, M., Henriquez, J., Diaz, I.: First report of crown rot of strawberry caused by Macrophomina phaseolina in Chile. Plant Dis. **97**(7), 996–996 (2013). https://doi.org/10.1094/PDIS-12-12-1121-PDN
15. Departamento de Recursos Hidricos, Facultad de Ingeniería Agrícola, U.d.C.: Diagnóstico de fuentes de agua no convencionales en el regadío inter-regional. Comisión Nacional de Riego, Ministerio de Agricultura, Gobierno de Chile **4**, 154–246 (2010)
16. Instituto Nacional de Estadísticas - Chile: Censo Agropecuario y Forestal (2007)
17. Flores, J., Rodríguez, R.: Resultados y experiencias del proyecto: "Incorporación de tecnología WMS (Web Map Service) en sistemas de acumulación de agua para la producción de frutillas en las comunas de Alhué, María Pinto, Melipilla y San Pedro". Technical report, Fundación para la Innovación agraria - (FIA), Ministerio de Agricultura, Santiago (2015)

18. Reckmann, O.: Demanda de agua por parte de los cultivos (chap. 2). In: Antúnez, A., Felmer, S. (eds.) Nodo tecnológico de riego en el secano región de O'Higgins. Fase II, pp. 51–72. Boletín INIA, num 190, Instituto de Investigaciones Agropecuarias, Santiago (2009)
19. Allen, R., Pereira, L., Raes, D., Smith, M.: Crop evapotranspiration-guidelines for computing crop water requirements-FAO Irrigation and drainage paper 56. Technical Report 9, FAO - Food and Agriculture Organization of the United Nations, Rome (1998)
20. Ooi, S.K., Mareels, I., Cooley, N., Dunn, G., Thoms, G.: A systems engineering approach to viticulture on-farm irrigation. IFAC Proc. Vol. **41**(2), 9569–9574 (2008)
21. Carrasco, R., Vargas, M., Alfaro, M., Soto, I., Fuertes, G.: Copper metal price using chaotic time series forecating. IEEE Lat. Am. Trans. **13**(6), 1961–1965 (2015)
22. Carrasco, R., Vargas, M., Soto, I., Fuentealba, D., Banguera, L., Fuertes, G.: Chaotic time series for copper's price forecast: neural networks and the discovery of knowledge for big data. In: Liu, K., Nakata, K., Li, W., Baranauskas, C. (eds.) Digitalisation, Innovation, and Transformation, vol. 527, pp. 278–288. Springer, Cham (2018). https://doi.org/10.1007/978-3-319-94541-5_28
23. Ogata, K.: Ingeniería de Control Moderna, vol. 53, 5th edn. Pearson Educación S.A., Madrid (2013)

Design and Proposal of a Database for Firearms Detection

David Romero[1(✉)] and Christian Salamea[1,2]

[1] Grupo de Investigación en Interacción Robótica y Automática (GIIRA), Universidad Politécnica Salesiana, Cuenca, Ecuador
{dromerom, csalamea}@ups.edu.ec

[2] Speech Technology Group, Universidad Politécnica de Madrid, Ciudad Universitaria, Av. Complutense 44, 40, 28040 Madrid, Spain

Abstract. Closed circuit television (CCTV) surveillance systems that implement monitoring operators have multiple human limitations, these systems usually don't provide an immediate response in different situations of danger like an armed robbery. To address this security gap, a firearms detection system has been developed through convolutional neural networks (CNNs). For its development a large database of images is necessary. This article presents the creation and characteristics of this database, which is made up of 247,576 images obtained from the web. This article addresses the application of different techniques for the creation of new images from the initial ones to increase the database, obtaining up to 22.7% relative improvement in the accuracy of the network after increasing the database. The database is structured into two classes. The first class is made up of people that have a gun and the second class of people not carrying a gun. The use of this database in the development of the detection system obtained up to 90% in "Precision" and "Recall" metrics in a convolutional neural network configuration based on "VGG net", through the use of grayscale images.

Keywords: Convolutional neural network · Database · Detection · Firearm

1 Introduction

Surveillance systems and remote monitoring devices are an important part of the technological infrastructure of transportation systems, education, government and commerce, making closed television systems (CCTV) an essential part of these infrastructures [1]. The operation of these surveillance systems involves monitoring operators, who through remote observation activate action protocols when a dangerous situation happens, such as an armed robbery. However, these monitoring systems have many human limitations, such as the number of computer screens that an operator can see at the same time or the limitations of the operator in diligently monitoring a computer screen. The development of detection systems through artificial intelligence, specifically deep learning, allow cameras to perform object detection in real time, which can provide a solution to human limitations in monitoring tasks. To address this security problem, we have proposed the development of a short firearms detection

M. Botto-Tobar et al. (Eds.): ICAETT 2019, AISC 1066, pp. 348–360, 2020.
https://doi.org/10.1007/978-3-030-32022-5_33

system through convolutional neural networks, in order to perform gun detections in dangerous situations such as an armed robbery. Being able to make gun detections in an automatic way and without the intervention of an operator provides a totally autonomous monitoring system, which would eliminate human limitations in the current monitoring systems. For most systems that use machine learning as an analysis strategy, it is necessary to have a large database to develop this type of systems. Due to the non-existence of an image database of people who have a firearm in real robberies or different situations that simulate a real robbery, we proceed to the creation of it.

This article presents the creation of an image database, which is composed of two classes, the first class is composed of images of people who have a gun and the second class of images of people without a gun. This database is composed of images where different types of factors were considered, including different types of luminosity, focus, image quality, firearm position and camera position. In this paper we also present the implementation results of the detection system using this database. The main contributions of this work are:

- Designing an image database of people who have a gun in real robberies and in situations that simulate (for the most part) a real robbery, considering factors like luminosity, image quality, firearm position and camera position.
- Finding the CNN architecture that achieves the best detection results with the new database.

2 Related Works

The problem of firearm detection in CCTV videos has been addressed in many different ways, firstly through the use of classic machine learning algorithms like "K-means clustering", to make color based segmentation and then combining it with algorithms like "SURF" (Speeded Up Robust Features), "Harris Interest Point Detector" and "FREAK" (Fast Retina Keypoint) to make the detection and localization of the gun [2, 3]. Also, algorithms like SIFT (Scale-Invariant Feature Transform) are used to extract different features of the image, combining it with "K-means clustering" and "Support Vector Machines" to decide if an image contains a gun or not [4]. The authors in [5] use algorithms like "Background" and "Canny edge detection" in combination with sliding window approach and neural networks to make the detection and localization of the gun. The disadvantage of these systems is that they use a database where the gun occupies most of the image, which does not represent areal robbery scenario. Therefore, these systems are not optimal for continuous monitoring, where the images extracted from CCTV videos have a high complexity due to the multiple factors involved, and where there are open areas with multiple objects around.

This problem has also been faced with more complex algorithms like deep CNN's. In [6] the authors use transfer learning with CNN's, using "Faster R-CNN" trained in a new database made by the authors. However, this database only has high quality and low complexity images, with good luminosity and focus. The authors show that the best system that was evaluated in well-known films had a low "Recall" produced by the frames with very low contrast and luminosity, which are the kind of scenarios that appear in real robberies. Also, several false positives occurred in the detection, produced by the objects in the background of the image, which occurs because all the areas of the image were analyzed with the sliding window and regions proposal approach to make the detection and localization of the gun. The authors in [7] face this problem using a symmetric dual camera system and CNN's, using also a database made by the authors. However, the most common cameras in CCTV systems are not dual cameras, therefore the use of this system would not apply to most commercial places [8]. The most common problem that we found is that the developed systems use small databases which don't represent the different scenarios that appear in real robberies, where the gun can appear in different positions and in different luminosities in the image, depending on the camera location.

This work addresses the problem of gun detection, using a deep CNN and a new database that has images with different types of luminosity, focus, firearm position, camera position and image quality.

3 Database

The database is composed of two classes, the first class consists of images of people who have a gun and the second class consists of images of people without a gun. The images of each class were obtained from the web, from "Google", "Instagram" and "YouTube". The database consists of images of people who have guns instead of images where only the gun is in the image for to two reasons. Firstly, this is done to provide the neural network with images that are similar to those that the network will face in its operation, where the firearm appears in complex environments with multiple objects around it. Secondly, in the security industry there are visibly armed personnel who are in charge of security in different types of places, such as security guards or the police, and in these cases it would not be appropriate for the system to make a detection. It is necessary that the system make a detection when a person has the gun in his hands, because these situations are considered dangerous. In this type of images, the gun is found in a small part of the image in a very complex environment. The gun will only be found next to the person, so only the image segments where people are located are the segments that are important to the system to search a gun. For this reason, the object detection and localization system "Yolo" [9] was applied to the images, to detect and locate people in the image. When these are located we segmented them in each of the images, and in this way we obtain only these segments of the image which will be included in the database. With these modifications it was possible to reduce images of large dimensions to smaller images, in which the most important information is included. The localization and segmentation are shown in Fig. 1.

3.1 Class A – Images with Firearms

The first class of the database is composed of images of people who have guns. The first type of images of this class correspond to images captured by CCTV cameras in real robberies. In this type of situation, the robber can show the gun in different types of positions. In [10] it is concluded that in the majority of cases the robbers had their weapons when entering the scene and tended to keep their weapons low at waist height or at shoulder height when posing the initial threat, being the most common case when a robber has a gun. The position in which the firearm is presented also depends on the camera position, in which the camera could be in a lateral position to the robber or above. This class covers all these types of situations, firstly, with images where the robber is aiming the gun at shoulder height and where the camera is in a lateral position to the robber. In this case the gun is shown in a lateral position. Also, in the case of images where the robber is aiming the gun at waist or shoulder height, but the camera is above the robber, these images show the top of the gun. And finally, with images where the robber is not aiming the gun, and the gun is at waist height, these images show the top or lateral side of the gun. These types of images are shown in Fig. 2. The second type of images of this class correspond to images of people that practice shooting with firearms. The third type of images of this class are images of people that have guns in other types of situations, different from a robbery or shooting practices. These types of images were chosen because these images are very similar to those that happen in real robberies scenarios and they present very similar positions to the previously described positions, which are common in robberies. An example of these types of images are shown in Fig. 3. This class is composed of 8,843 images, these images have just one type of firearm, which are handguns, the structure of this class is shown in Fig. 4.

Fig. 1. Localization and segmentation.

3.2 Class B – Images Without Firearms

The second class of the database is composed of images of people who don't have a gun. The first type of images of this class correspond to images of people who are in different positions and places, such as supermarkets and stores. The second type of images of this class are images that were also obtained from the images of people that are practicing shooting with firearms, but when the person does not have the gun. This class is conformed of 8,841 images, some of these images are shown in Fig. 5, and its structure is presented in Fig. 6.

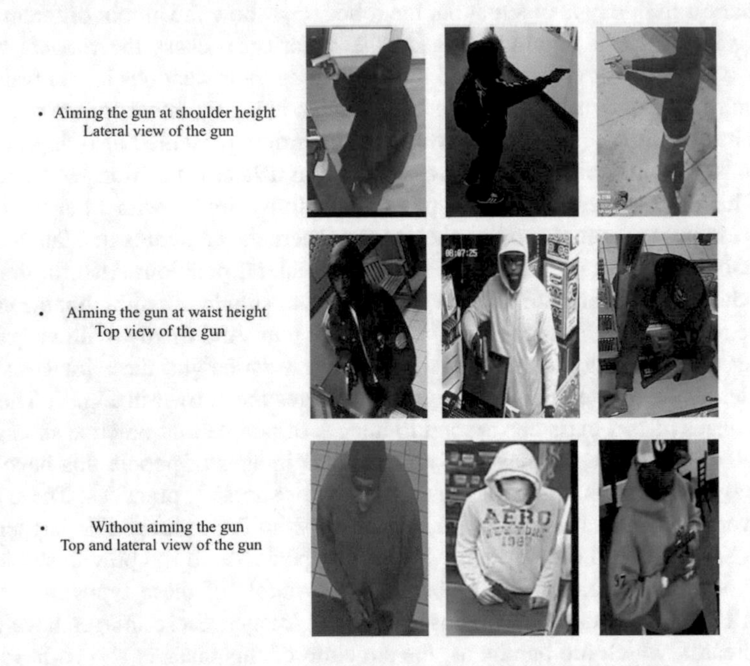

Fig. 2. Positions in which the firearm is shown – real robberies.

Fig. 3. Other type of scenarios in which the firearm is shown.

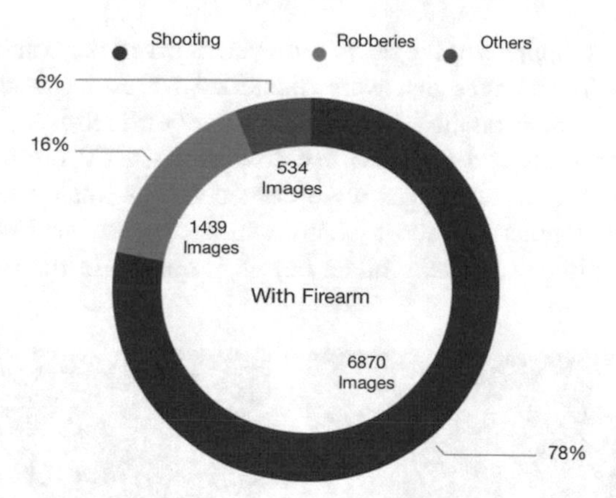

Fig. 4. Structure – Class A

Fig. 5. Images without firearms.

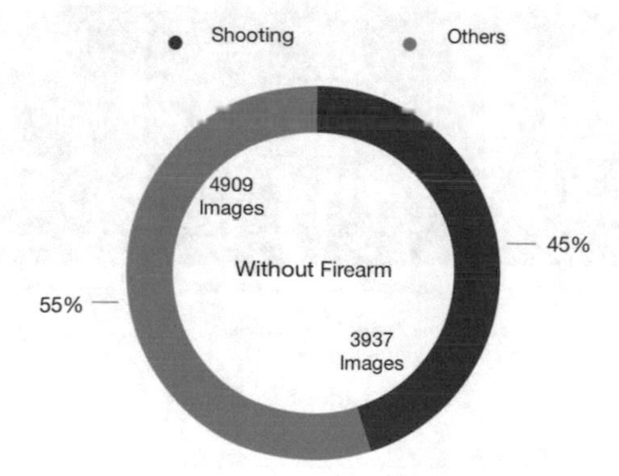

Fig. 6. Structure Class B

On the web there are a lot of images and videos of these types of situations, however, most of these images do not have an adequate quality. For this reason, we proceeded to choose the images that present an optimal quality to compose the database. The criterion followed to choose the appropriate quality of the images can be seen in Fig. 7, since image number 3 it is considered that the quality of the image is adequate for the development of the detection system given the conditions of focus and graphic resolution. The images that were discarded were images of very old CCTV videos that have very poor quality, where it was very difficult to differentiate between objects. Medium markets and business use low cost CCTV cameras that have low quality, however this type of images were considered as images that have sufficient quality for the development of the system, where despite the low quality of these images, it was not difficult to differentiate between objects in the image.

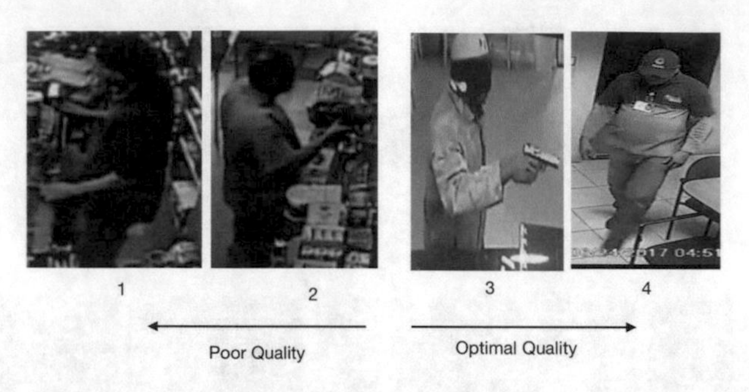

Fig. 7. Range of quality.

Luminosity in the image is another important factor in these types of situations because a robbery could happen at any time of the day. The database has images with different luminosity scales, this is shown in Fig. 8.

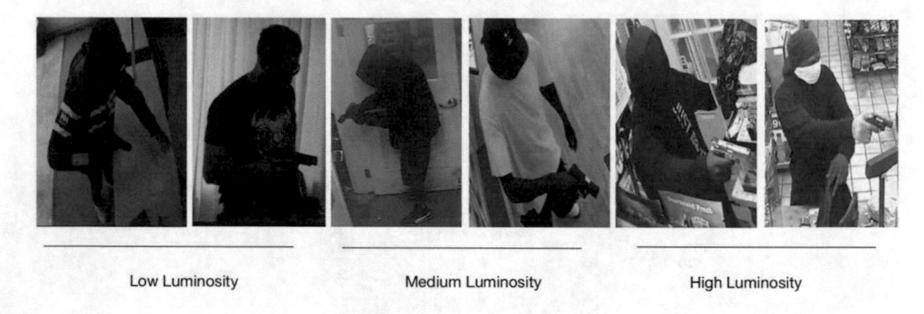

Fig. 8. Luminosity scale

3.3 Preprocessing and Database Augmentation

Before using the database, it is necessary to resize all the images to a fixed size because the input of the CNN needs to have images of the same size. Also, it is necessary to increase the number of images in the database. This is done by applying multiple techniques to modify the original images and create new ones to increase the database.

Image Size Normalization. The database is made up of images of multiple sizes and since the input of the CNN needs images of the same size, we proceeded to establish the image size to 224 × 224 pixels. This value was established after having done multiple tests, in which we were able to determine that the gun does not have a great distortion when we change the image size to 224 × 224 pixels. The objective was also to set the smallest possible image size to reduce the computational cost in the training phase of the CNN without causing a loss of information and a great distortion in the images.

Database Augmentation. The database originally consists of 17,684 images, due to the difficulty of detecting firearms in complex environments and that the structure of the CNN could have a large number of parameters due to the complexity of the problem, it was necessary to increase the size of the database to avoid overfitting the model. The increase has been carried out through the application of different techniques for the modification of the images initially obtained to create new images to increase the number of images in the database, these techniques are:

Flip. The first technique that was applied to increase the database was a "Flip" in the horizontal axis of the original images, with which other images are created, achieving in a certain way a change in the position of the firearm in the image.

Rotation. The second technique that was implemented to increase the database was the rotation of the images. Security cameras originally have an inclination in the places in which they are placed, the images that the cameras show have a certain degree of inclination. For this reason, it was concluded that a non-excessive rotation of the images would create new images that will be similar to those that the neural network will face in the real world, a rotation was made to the images at the angles of 10, 20, 30, −10, −20, −30.

With these techniques we increased the original database from 17,684 images to 247,576 images. An example of the application of these techniques to the images is shown in Fig. 9. This database will be available in: christiansalamea.info

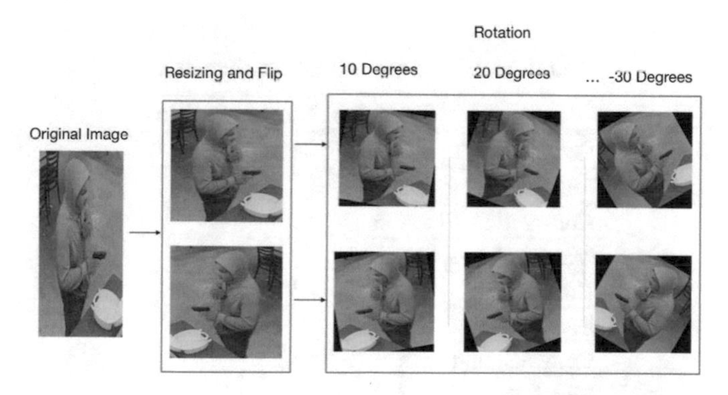

Fig. 9. Techniquesto process and increase the database.

4 Implementation and Analysis of Results

The development of the convolutional neural network was carried out after performing different tests using different configurations in the structure of the network. These configurations started from the idea of two types of network structures: VGG Net [11] and ZF Net [12].

The system was evaluated with different metrics in order to quantify the quality of the predictions of the detection system. The evaluation was made through the following metrics:

- Accuracy: In terms of a classifier is defined as the probability of predicting a class of an instance in a right way [13].
- Equal Error Rate: Equal Error Rate (EER) is the value where the proportion of false positives (FAR) is equal to the proportion of false negatives (FRR). This sets the detection sensitivity where the number of errors produced is minimized [14, 15].
- Precision: Precision measures that fraction of examples classified as positive that are truly positive [16].
- Recall: Recall measures that fraction of positive examples that are correctly labeled [16].
- Relative Improvement: Relative Improvement (RI) measures the existing improvement between two quantities, and is defined by the following equation:

$$\frac{Test\,B - Test\,A}{Test\,B} \tag{1}$$

Where Test B is the new measurement and Test A is the relative measurement.

To develop the detection system, the database was divided into 70% for training, 15% for evaluation and 15% for tests. The network that obtained the best results was the network structure based on “VGG net” using grayscale images, this configuration is shown in Table 1.

Table 1. CNN architecture - configuration

	Filter size	Number of filters
Convolution 1	3 × 3	64
Max-Pooling 2 × 2		
Convolution 2	3 × 3	128
Max-Pooling 2 × 2		
Convolution 3	3 × 3	256
Max-Pooling 2 × 2		
Convolution 4	3 × 3	512
Max-Pooling 2 × 2		
Convolution 5	3 × 3	512
Max-Pooling 2 × 2		
Neural network: 4096, 2 Neurons		

This architecture was tested in the different phases of the creation of the database, firstly, with the images that were originally obtained before the increase of the database (Test A), secondly, with the database that was obtained after the application of the described techniques to increase the database (Test B). In this test only the half of the database was taken with the objective of analyzing whether the techniques that were applied provide new information or just cause overfitting. Finally, the complete database was taken (Test C). The results obtained in these tests can be seen in Table 2.

Table 2. Tests with different phases of the database

	Training		Evaluation	
	Loss	Accuracy	Loss	Accuracy
Test A	0.30	0.781	0.42	0.701
Test B	0.20	0.850	0.26	0.890
Test C	0.17	0.855	0.21	0.908

It can be seen that in the first test "Test A" with the original database we obtained low accuracy and high loss values in the training and evaluation phases of the network. Also, this shows that there is an overtraining in the model. In the second test "Test B", the results show a higher accuracy with lower loss values, which indicates an improvement in the results after the increase of the database, obtaining a better accuracy in the evaluation of the network and reducing overfitting. Finally, in Test C the lowest loss and the highest accuracy values were obtained in the evaluation of the network. The relative improvements among the three tests are measured by the RI metric, taking the accuracy of the tests as the values to be compared, being the accuracy of the test A the relative measurement, the results of the relative improvement are shown in Table 3.

Table 3. Relative improvement

Tests	RI%
Test A – Test B	21.2%
Test A – Test C	22.7%

Afterwards, different evaluation metrics were obtained. Firstly, the EER metric and the proportions of false positives (FAR) and false negatives (FRR) were obtained using the complete database. The crossing between the proportions of FAR and FRR provide the EER value and corresponds to a value of 0.09 at a sensitivity of 0.52, this can be seen in Fig. 10.

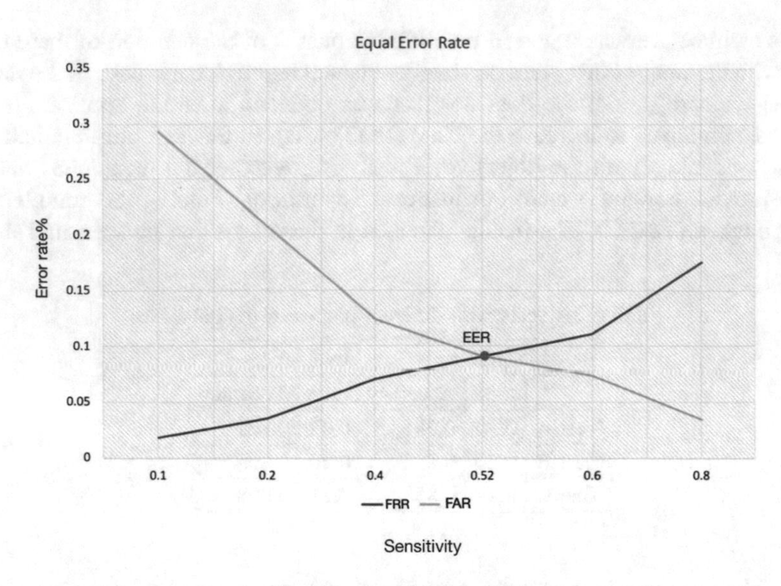

Fig. 10. Equal Error Rate

The confusion matrix with the most optimal sensitivity for the lowest EER value can be seen in Table 4.

Table 4. Confusion matrix

		Predictions	
		With firearm	Without firearm
Classes	With firearm	15.657	1.553
	Without firearm	1.561	15.649

Finally, the values of "Recall" and "Precision" obtained from the test set of the complete database are shown in Table 5.

Table 5. Metrics

Metrics	Results
Recall	0.9097
Precision	0.9093

5 Conclusions

In the development of the detection system, the database is the most important aspect, because the network will learn from these images. For its creation, different aspects were evaluated like the type of images to be included, the correct size of the images, types of luminosity, image quality, firearms position and camera position. According to the results obtained in the evaluation metrics it can be concluded that:

- The conformation of the database with images of people practicing shooting with firearms, as well as images in other types of situations in which people have firearms, allowed us to provide the network with images of situations that are very similar to real robberies and in this way provide the model with new information to have a better "generalization", which avoids overfitting.
- The segmentation of people in the images that were obtained from the web allowed us to obtain new images which will conform the database. In these images we have the most important information that is the person carrying the gun. In this way the complex environment was reduced, conforming the database only with the sections of the images that are important.
- The increase of the database through the implementation of different techniques provided complementary information to thc initial one, due to the fact that in the tests that were performed there wasn't a strong overfitting of the network. In addition, with the relative improvement metric, it was possible to analyze the improvement of the system in each stage of the creation of the database, obtaining an improvement of 22.7% in the accuracy of the network after increasing the database.
- The development of the detection systcm using grayscale images allowed us to obtain better results than with RGB images in "accuracy" and "loss" in the training and cvaluation phases.

References

1. Deisman, W.: CCTV: Literature review and bibliography. Royal Canadian Mounted Police, Ottawa (2003)
2. Tiwari, R., Verma, G.: A computer vision based framework for visual gun detection using SURF. In: 2015 International Conference on Electrical, Electronics, Signals, Communications and Optimization (EESCO), pp. 1–5, Visakhapatnam, India. IEEE (2015)
3. Tiwari, R., Verma, G.: A computer vision based framework for visual gun detection using Harris interest point detector. Procedia Comput. Sci. **54**, 703–712 (2015)
4. Halima, N., Hosam, O.: Bag of words based surveillance system using support vector machines. Int. J. Secur. Appl. **10**(4), 331–346 (2016)
5. Gaga, M., Lach, S., Sieradzki, R.: Automated recognition of firearms in surveillance video. In: 2013 IEEE International Multi-Disciplinary Conference on Cognitive Methods in Situation Awareness and Decision Support (CogSIMA), San Diego, CA, pp. 45–50. IEEE (2013)
6. Olmols, R., Tabik, S., Herrera, F.: Automatic handgun detection alarm in videos using deep learning. Neurocomputing **275**, 66–72 (2018)

7. Olmos, R., Tabik, S., Lamas, A., Pérez-Hernández, F., Herrera, F.: A binocular image fusion approach for minimizing false positives in handgun detection with deep learning. Inf. Fusion **49**, 271–280 (2019)
8. Reolink, http://reolink.com/cctv-camera-types/. Accessed 26 Apr 2019
9. Redmon, J., Farhadi, A.: Yolov3: an incremental improvement. arXiv (2018)
10. Mosselman, F., Weenink, D., Lindegaard, M.: Weapons, body postures, and the quest for dominance in robberies: a qualitative analysis of video footage. J. Res. Crime Delinq. **55**(1), 3–26 (2018)
11. Simonyan, K., Zisserman, A.: Very deep convolutional neural networks for large-scale image recognition. arXiv (2014)
12. Zeiler, M.D., Fergus, R.: Visualizing and understanding convolutional neural networks. In: Lecture Notes in Computer Science, vol. 8689, pp. 818–833 (2014)
13. Menditto, A., Patriarca, M., Magnusson, B.: Understanding the meaning of accuracy, trueness and precision. Accredit. Qual. Assur. **12**, 45–47 (2006)
14. Galdi, P., Tagliaferri, R.: Data mining: accuracy and error measures for classification and prediction. Encycl. Bioinform. Comput. Biol. **1**, 431–436 (2018)
15. Hammad, A.M., Elhadary, R.S., Elkhateed, A.O.: Multimodal biometric personal identification system based on Iris & Fingerprint. Int. J. Comput. Sci. Commun. Netw. **3**, 226–230 (2013)
16. David, J., Goadrich, M.: The relationship between precision-recall and ROC curves. In: Proceedings of the 23rd International Conference on Machine Learning, vol. 06, pp. 233–240 (2006)

A Proposed Architecture for IoT Big Data Analysis in Smart Supply Chain Fields

Fabián-Vinicio Constante-Nicolalde[1,2](✉), Jorge-Luis Pérez-Medina[1](✉), and Paulo Guerra-Terán[1]

[1] Intelligent and Interactive Systems Lab (SI2-Lab), Universidad de Las Américas, Quito, Ecuador
{fabian.constante, jorge.perez.medina, paulo.guerra.teran}@udla.edu.ec

[2] School of Technology and Management, Polytechnic Institute of Leiria, Leiria, Portugal
2162316@my.ipleiria.pt

Abstract. The growth of large amounts of data in the last decade from Cloud Computing, Information Systems, and Digital Technologies with an increase in the production and miniaturization of Internet of Things (IoT) devices. However, these data without analytical power are not useful in any field. Concentration efforts at multiple levels are required for the extraction of knowledge and decision-making being the "Big Data Analysis" an area increasingly challenging. Numerous analysis solutions combining Big Data and IoT have allowed people to obtain valuable information. Big Data requires a certain complexity. Small Data is emerging as a more efficient alternative, since it combines structured and unstructured data that can be measured in Gigabytes, Peta bytes or Terabytes, forming part of small sets of specific IoT attributes. This article presents an architecture for the analysis of data generated by IoT. The proposed solution allows the extraction of knowledge, focusing on the case of specific use of the "Smart Supply Chain fields".

Keywords: Big Data Analysis · Internet of Things · Data mining · Hadoop · Radio frequency identification

1 Introduction

The growth of data generated through interconnected IoT (Internet of Things) devices has played an important role, affecting all technology areas and businesses by increasing benefits for organizations and individuals in the Big Data landscape [1]. The collection of these data is generally available in a semi-structured and unstructured form, which makes it difficult to process them with database administration tools or data processing applications. Therefore, the adoption of Big Data in IoT applications is overwhelming [2]. In the IT (Information Technology) and business fields these two technologies have already been recognized. The increase in the amount of data and category due to the deployment of IoT; offers the opportunity for the application and development of Big Data Analysis (BDA) [3]. In addition, its applications accelerate the advances in research and IoT business models [2].

M. Botto-Tobar et al. (Eds.): ICAETT 2019, AISC 1066, pp. 361–374, 2020.
https://doi.org/10.1007/978-3-030-32022-5_34

IoT will continue to make considerable improvements in several fields. One of these domains are the "Smart Supply Chains" [2]. In Smart Supply Chains, IoT sensors continuously monitor the flow and their data is collected and stored on a server from which they can extract, transform and analyze their dispatch in real time [2]. Advanced analytical techniques, such as Machine Learning algorithms, to analyze IoT data are used to ensure efficient and efficient operation [2]. For example, Real-time analysis of data extracted and generated from IoT sensors not only to improve the efficiency in the logistics of the Supply Chains, but also for analysis of the demand and prediction of late shipments.

The control of the external environment and the execution of decisions in a Supply Chain are allowed through IoT and BDA. The IoT enabled factory equipment will be able to communicate within the data parameters and optimize performance by changing the equipment configuration or process workflow [4]. Another case of use presented in future Supply Chains within the IoT infrastructure is visibility in transit. The key technologies used for visibility within transit are the Global Positioning System (GPS) and RFID providing information about their location, identity and tracking [2]. The data collected through these technologies will allow Supply Chain managers to improve automatic delivery and accurate delivery information by predicting arrival time [2].

Software architecture manifest the design decision about a system, constitutes a graspable model for how a system is structured and how its elements work together [5]. In a computing system is the structure of the components, their interrelationships, the principles and guidelines governing the design and their evolution over time. The interrelationships involve a runtime mechanism to transfer control and data around a system. The main function is acts as a communication vehicle between stakeholders, being a basis for understanding, negotiation and consensus [5].

Big Data Management technologies and analysis techniques are integrated to effectively support analysis and autonomous decision-making. Therefore, this research focuses on the challenge of the real-time analysis of the large amount of high-speed data generated from the IoT devices and Information Systems that intervene in the Supply Chain. It provides a new integrated architecture of IoT and BDA [3]. The proposed architecture has three main components of integrated technology: IoT Sensors, Big Data Management and Data Analysis. The remain of the article is organized as follows: Sect. 2 presents some related works related to Big Data in IoT. Section 3, some technologies for Big Data Processing are introduced. In the Sect. 4, the proposed architecture is described. Section 5, explains a preliminary experiment. Finally, Sect. 6 presents the research conclusions and future works.

2 Related Works

Different architectures for BDA are considered due to the wide range of sources, varieties and data structures [2]. The BDA is classified as real-time and offline analysis. The real-time analysis is used mainly in electronic commerce and finance. Its main existing architectures include (a) parallel processing clusters that use traditional relational databases and (b) memory-based computing platforms [6]. Off-line analysis is used for applications without high response time requirements, for example, machine learning, statistical analysis and recommendation algorithms [6]. Performs the analysis when importing large data records in a special platform through data acquisition tools [2]. In this section, we present some works related to Big Data in IoT.

2.1 IoT+Small Data: Transforming In-store Shopping Analytics and Services

A Small Data approach based on analytical data from a retail store, the combination of sensor data, personal portable devices and store-deployed sensors and IoT devices, is used to create individualized services its named ShopMiner [7]. Key challenges include: Data Mining Sensor Mining and Judicious Unchaining.

The experiments were carried out with 5 smartwatch giving initial results 94% accuracy in identifying a gesture of item selection, 85% accuracy in identifying the location of the shelf from which the item was collected and 61% accuracy in the identification of the exact element collected [7]. Figure 1 illustrates the architecture.

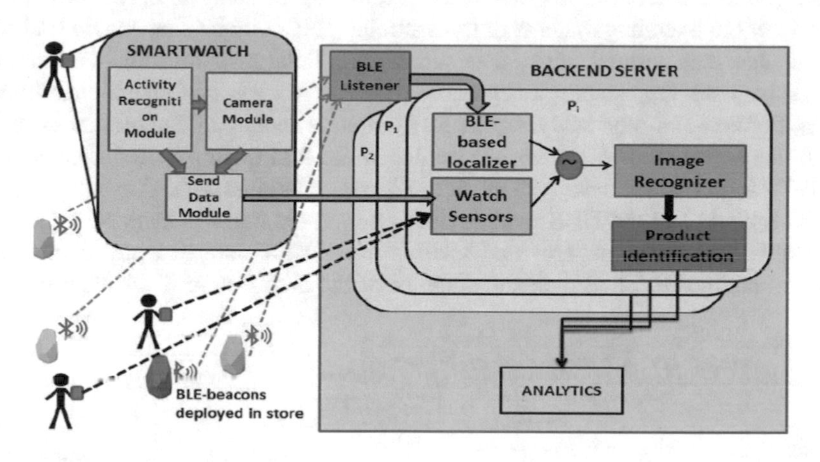

Fig. 1. System architecture of ShopMiner [7].

2.2 Development of a Smart City Using IoT and Big Data

The rapid growth of population density in urban areas requires a medium, smart city is a difficult task. Therefore, a system for the smart development of the city based on IoT using BDA has been proposed. The deployment of sensors was used, including smart sensors for the home, vehicle networks, meteorological and water sensors, smart parking sensors and surveillance objects [8]. Initially, a four-level architecture is proposed, which includes:

1. Lower level: Responsible for IOT resources, data generations and collections.
2. Intermediate level 1: Management of all types of communication between sensors, relays, base stations, Internet.
3. Intermediate level 2: Administration, data processing using the Hadoop

4. Higher level: Responsible for the application, use of data analysis and generated results. Real-time processing of data collected from the intelligent system allows smart cities to be obtained using Hadoop with Spark, VoltDB or Storm [8].

2.3 An Architecture for Big Data Driven Supply Chain Analytics

In [9] the authors have proposed a Big Data architecture and an analysis framework for Supply Chain Management (SCM) applications. Their data sources are represented in the lowest layer by entities such as: suppliers, manufacturers, warehouses, distributors/retailers and customers; which provide the input data in structured and unstructured formats generating the Big Data in the system.

Structured data are extracted by Extract, Transform and Load (ETL) mechanisms and filled in a Data Warehouse (DW), the HDFS and MapReduce systems of the Hadoop cluster take care of the unstructured part by storing it in the NoSQL database. An Operational Data Store (ODS) is implemented after the ETL operation for structured data entries, integrating data from multiple sources. A Real-Time Intelligence System (RTI) that accesses the data in DW, allowing its analysis in real time through direct access to operating systems or transactions. Technologies that enable RTI in real time include data virtualization, data federation, Enterprise Information Integration (EII), Enterprise Application Integration (EAI), and Service-Oriented Architecture (SOA). For non-real-time analyzes, the RTI output is sent to a Dimensional Data Store (DDS) that stores the DW data or the output of the RTI module in a different form to the OLTP format in the base systems of traditional relational data. The outputs of both the RTI module and the DDS module are incorporated into the Data Mining module responsible for finding patterns and relationships in the data that are of interest for decision-making [9]. The schematic diagram of the architecture is depicted in Fig. 2.

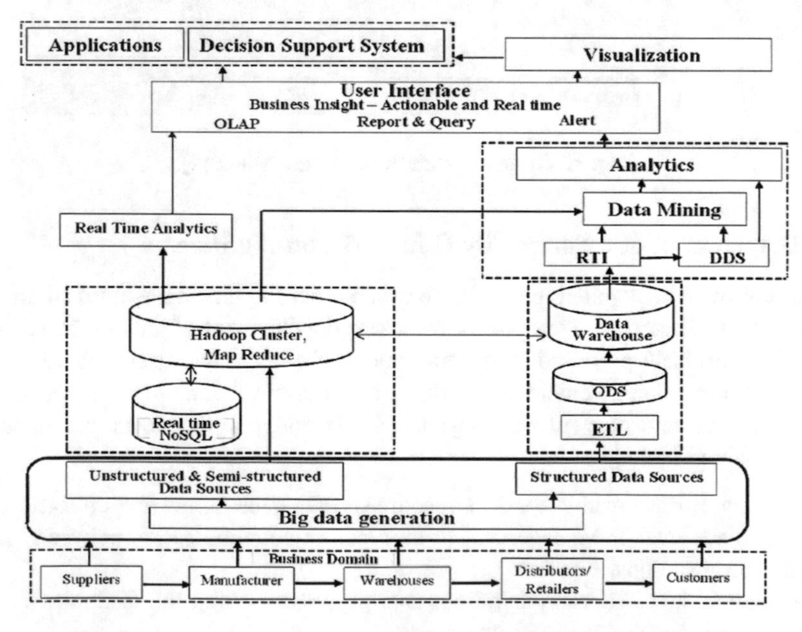

Fig. 2. A proposed architecture of Big Data Analytics for SCM [9].

2.4 Towards an IoT Big Data Analytics Framework: Smart Buildings Systems

This document presents an Integrated Big Data Analysis Framework (IBDA) for the extraction, storage and analysis of data in real time, corresponding to IoT sensors inside a smart building [10].

Python and Big Data Cloudera platform have been used for the development of the initial version of the IBDA framework which demonstrates with the help of a scenario that involves the analysis of smart building data in real time to automatically manage the level of oxygen, light and dangerous gases in different parts of the building [10]. The key to this work is the integration of BDA and IoT, as shown in Fig. 3.

The design of the IBDA framework for real-time data analysis in smart buildings uses IoT and Apache Hadoop sensors. Its scope is limited to data generation, data extraction, ingesting data in HDFS (Hadoop Distributed File System), data visualization, data analysis and real-time control [10].

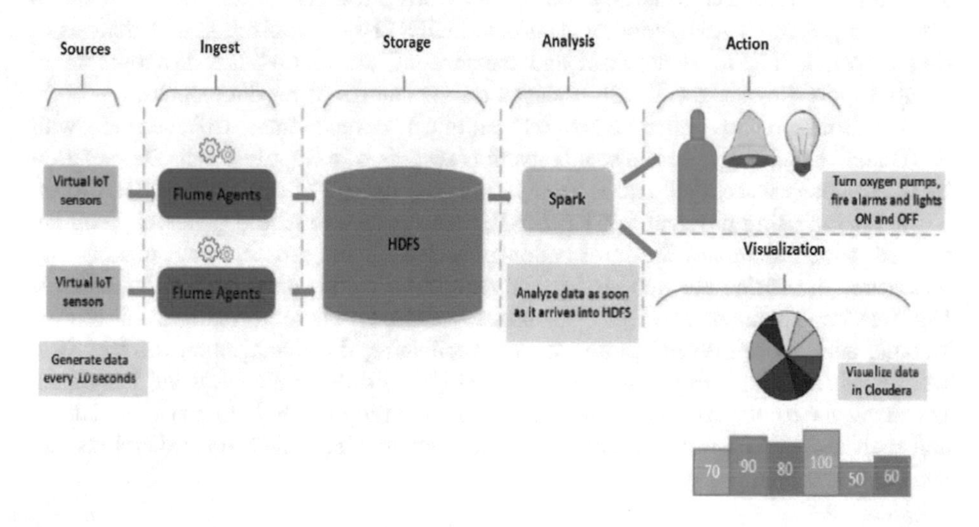

Fig. 3. IBDA architecture – ingest, store, analyze and actuate [10].

3 Technologies for Big Data Processing

The data massive volumes generated in the different contexts where the organization performs lead to Big Data consuming enormous computational resources, capable of handling over-demands of storage, processing, analysis and visualization of data Leading organizations to improve predictions, identify recent trends, find hidden information and make decisions [3].

Currently there are several tools for the development of Big Data projects. Among the most widespread are: MapReduce [11], Apache Spark [12] and Storm[1]. The R Language is an open source and statistical programming language. The syntax of R is little bit complex for learning but it consists of large number of machine learning packages and visualization tools [13].

Batch processing and flow processing are characteristics that most of the tools available are focused on. The batch processing tools are based on the Apache Hadoop infrastructure [14], such as Mahout [15] and Dryad [16]. Flow data applications are mainly used for real-time analysis. An example of a large-scale transmission platform is Splunk [17]. Dremel and Apache Drill [18] are Big Data platforms that support interactive analysis.

4 Proposed Architecture

The proposed architecture is inspired by the work presented in 2.3 and its scope is limited to generation, extraction, ingestion in HDFS [14], visualization and analysis of data corresponding to clickstream and transactions, facilitating the development of Smart Supply Chains, specifically focused on the control of product stocks.

The solution has three main technological components: IoT sensors with RFID/NFC technology, a Database that stores records of a Supply Chain Dataset used by the company DataCo Global for analysis, finally the use of open source BDA tools.

Instead of using physical IoT RFID/NFC sensors, a web record generator program is used, which simulates the information provided by the sensor, when making any transaction or visiting the web platform in real time. For this purpose, the Python code has been implemented. The CDH (Cloudera Distributed Hadoop) platform allows the storage, analysis and visualization of data facilitating the management of Big Data workflows from one end to the other. Tableau [19] and the language R are integrated, which allows to execute Algorithms of automatic learning that help to make predictions and soon to obtain knowledge for the good decision making. The proposed architecture for BDA is shown in Fig. 4.

4.1 Data Sources

The unstructured data used come from Clickstream, this data contains the routes that visitors choose when they browse a specific website, products seen, items searched [20]. Simulating its generation from Information Systems focused on Supply, Python code was used, a record is generated every five seconds and sent by TCP (Transmission Control Protocol) in the port 2181. On the other hand, the structured data come from IoT sensors and are stored in an RDBMS (Relational Data Base Management System) for the management of transactions involved in the management of Supply Chain.

[1] http://storm.apache.org/index.html.

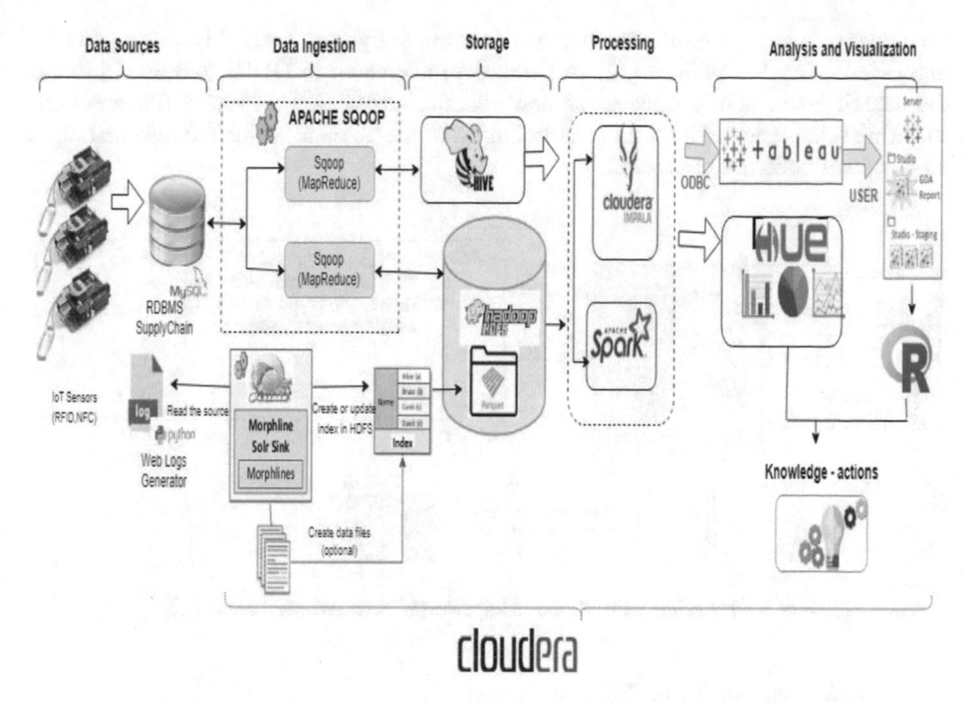

Fig. 4. BDA architecture proposed for Smart Supply Chains.

4.2 Data Extraction

The extraction of unstructured data generated from Clickstream, is executed through an Apache Flume Agent [14] is con figured listening to TCP port 2181, to which the Python program is sending the data. The Apache Flume agent is configured with HDFS as the receiver and stores the source data in the HDFS as soon as the data arrives. The RDBMS (Relational Database Management System) data are processed within the CDH environment used Apache Sqoop [14], which allows to automatically load relational data from MySQL to HDFS, conserving its structure. Then the relational data is loaded directly into a form ready to be consulted by Apache Impala using the Apache Avro file format [21], for the workload in the cluster.

The executed command launches MapReduce [11] jobs to extract the data and writes them to HDFS; they are distributed through the cluster in the Parquet file format. Create tables to represent the HDFS files in Impala/Apache Hive with the corresponding schema.

4.3 Data Ingestion into HDFS

The structured data is ingested in HDFS using Apache Sqoop, then this tool imported the tables in Apache Hive and Impala was used to consult this data. Hive works by compiling SQL queries in MapReduce jobs, which makes it very adaptable and easy to use, while Impala executes queries by itself. Clickstream records are ingested in real time using Apache Flume which reads incoming Pyhton data, this information is processed using Morphlines [22]. A Collection is created in HDFS and within this is the search index that is updated as new records arrive [22]. Figure 5 illustrates the Indexing scheme with Flume and the definition of the fields in which the information of Clickstream becomes structured.

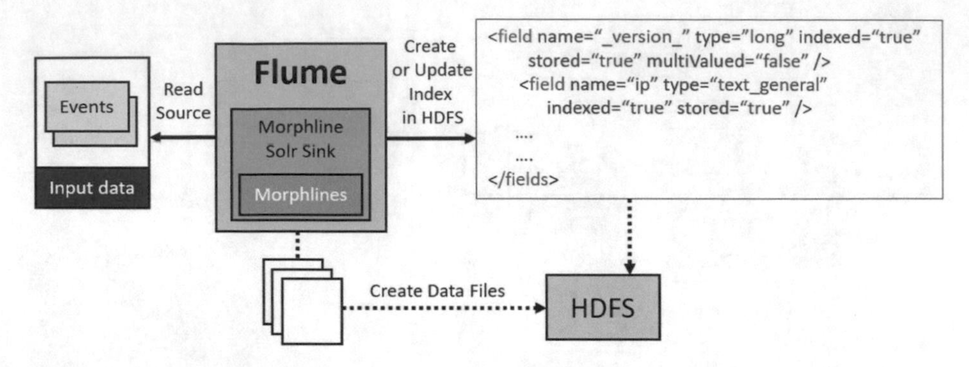

Fig. 5. Indexing with Flume using Morphlines. Adapted from [22].

4.4 Processing and Data Storage

Data processing requires the creation of tables in Apache Hive and Impala defined by means of a schema over the existing files with the use of 'CREATE EXTERNAL TABLE' statements, similar to the Relational Databases, which allowed obtaining reports [21]. "Spark-on-YARN" was used for more advanced data processing, in order to perform a Relationship Strength Analysis. MapReduce and Spark [12] share the same resource manager, facilitating the management of resource sharing among many users. As a result, the components were positioned together to generate a strong client base with Apache Spark and Language Scala producing a list of items purchased more frequently in a very short time.

4.5 Data Visualization

Once the data is stored in HDFS, the visualization is done with the HUE tool that is available in the CDH environment, as shown in Fig. 6. The connection between CDH and Tableau was made for the visualization of data. Cloudera's multi-faceted and performance DWs make it easy to store and query Big Data. Tableau allowed to quickly and easily find valuable information in vast data sets from Hadoop [19]. Figure 5 shows four Clickstream Data Visualization Dashboards, corresponding to a period of three months.

The "action" panel presents in a pie chart, the number of actions performed by user clicks. The actions considered are: checkout, add product to cart, view product, view department, support and income without any action. The "request" panel contains in a bar graph, the number of visits made to each page of the website in a general way. In the "department" panel, the number of clicks corresponding to visits made to a specific department is displayed in a pie chart. The departments observed are: apparel, footwear, bookshop, fan shop, discs shop, pet shop, technology, health and beauty, fitness and outdoors. The "request_date" panel is reflected in a time series, the sequence of observations or clicks made in each hour, day and ordered chronologically.

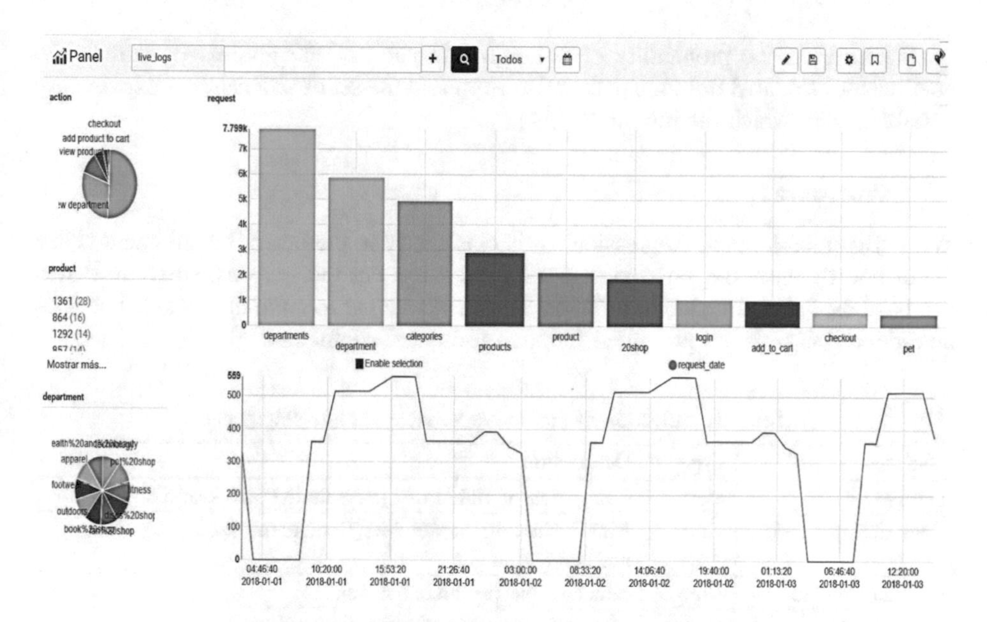

Fig. 6. Real-time Clickstream Analysis Dashboard.

5 Preliminary Experiment

In this section, we present the results obtained in the implementation of architecture with an experiment performed.

5.1 Materials

The experiment was configured a virtual machine with the following technical characteristics of Hardware: 4 processors, 64 GB of Hard Disk (SCSI) and 9 GB of RAM. The CDH environment and the R Programming Language were used for storage and processing. The Dataset used: DataCoSupplyChainDataset.csv, the detailed description of their fields are described in the DescriptionDataCoSupplyChain.csv file which are found in [23], with its attached description.

5.2 Method

Logistic regression was used to adjust a regression curve, $y = f(x)$ when "y", is a categorical variable, and can be predicted by a set of predictor variables x_i, which can be categorical, continuous, or a combination of both. The Logistic Distribution function is given by:

$$P_i = \frac{1}{1 + e^{-(\beta_0 + \beta_1 X_1 + \ldots + \beta_i x_i)}} \quad (1)$$

Where P_i is the probability of belonging to each of the groups, β_0 and β_i are coefficients estimated from the data and x_i represent the set of independent explanatory variables, with which fits the model [24].

5.3 Procedure

With "Binomial Logistic Regression", it is classified and predicted if a shipment is late (1) or not (0), since the variable to predict is binary. For the analysis, software R was used and the detailed variables in Table 1 were selected as input to the model. The 95% confidence intervals are presented for each estimated coefficient.

Table 1. Selection of predictive variables of late shipments.

Variable	Type	Description
Delivery Status	<fct>	Categorical variable that presents the status of a shipment
Late_delivery_risk	<int>	Binary categorical variable (Dependent) indicating states 0: not late and 1: late
Product Card Id	<int>	Indicates the product code sent
Order Item Quantity	<int>	Shows the number of units per product
Order Item Product Price	<dbl>	Presents the price per product
Customer Segment	<fct>	Categorical variable that shows the customer segment
Order Status	<fct>	Categorical variable indicating the status of the order
Shipping Mode	<fct>	Categorical variable indicating how the order was sent

Figure 7, shows the assembly and adjustment of the Logistic Regression model adjusted for prediction. Variables with triple asterisks are statistically significant. No regularization techniques were used since only the initial model was readjusted considering the existing correlation between the variables. For more information about the description of each of the variables that make up the used dataset, check the DescriptionDataCoSupplyChain.csv file located in [23].

To analyze the deviation of the model, the ANOVA() function was executed, in which it was observed that by adding the variable "Shipping Mode", the residual deviation was significantly reduced as indicated in the Fig. 8.

```
> summary(model)
Call:
glm(formula = Late_delivery_risk ~ ., family = binomial(link = "logit"), data = train)
Deviance Residuals:
    Min       1Q   Median       3Q      Max
-2.4768  -0.9789   0.3105   1.2434   1.3963
Coefficients:
                              Estimate Std. Error  z value Pr(>|z|)
(Intercept)                  2.989e+00  3.392e-02   88.114   <2e-16 ***
Product.Card.Id              3.007e-05  2.093e-05    1.437    0.151
Shipping.ModeSame Day       -3.173e+00  3.903e-02  -81.302   <2e-16 ***
Shipping.ModeSecond Class   -1.817e+00  3.465e-02  -52.428   <2e-16 ***
Shipping.ModeStandard Class -3.491e+00  3.242e-02 -107.665   <2e-16 ***
---
Signif. codes:  0 '***' 0.001 '**' 0.01 '*' 0.05 '.' 0.1 ' ' 1
(Dispersion parameter for binomial family taken to be 1)
    Null deviance: 198848  on 144419 degrees of freedom
Residual deviance: 164437  on 144415 degrees of freedom
AIC: 164447
Number of Fisher Scoring iterations: 5

> confint(model , level=0.95)
2.5 %          97.5 %
(Intercept)                  2.923169e+00  3.056170e+00
Product.Card.Id             -1.095307e-05  7.109202e-05
Shipping.ModeSame Day       -3.250153e+00 -3.097136e+00
Shipping.ModeSecond Class   -1.885168e+00 -1.749310e+00
Shipping.ModeStandard Class -3.555018e+00 -3.427896e+00
```

Fig. 7. Assembly and adjustment of the Logistic Regression model.

```
> anova(model, test="Chisq")
Analysis of Deviance Table
Model: binomial, link: logit
Response: Late_delivery_risk
Terms added sequentially (first to last)

                Df Deviance Resid. Df Resid. Dev Pr(>Chi)
NULL                           144419     198848
Product.Card.Id  1        1    144418     198847    0.396
Shipping.Mode    3    34411    144415     164437   <2e-16 ***
---
Signif. codes:  0 '***' 0.001 '**' 0.01 '*' 0.05 '.' 0.1 ' ' 1
```

Fig. 8. Analysis of the Variance of the Logistic Regression model.

5.4 Evaluation of the Predictive Ability of the Model

The Dataset was splitted into 2 parts: for training, 70% of random data was chosen and test set 30%. Functionality of the model is evaluated by predicting "y" that has the values of the variable "*Late_delivery_risk*" in a new data set. The ROC curve is plotted as shown in Fig. 9 and the AUC (area under the curve) is calculated to measure the performance of the binary classifier. The decision limit is 0.5, if $P(y = 1 \mid X) > 0.5$ then $y = 1$ otherwise $y = 0$. The model showed an Accuracy = 0.700000000000001. A model with good predictive capacity must have an AUC closer to 1 [24]. In the model evaluated, it results in AUC = 0.7307086. The model is in an acceptable range.

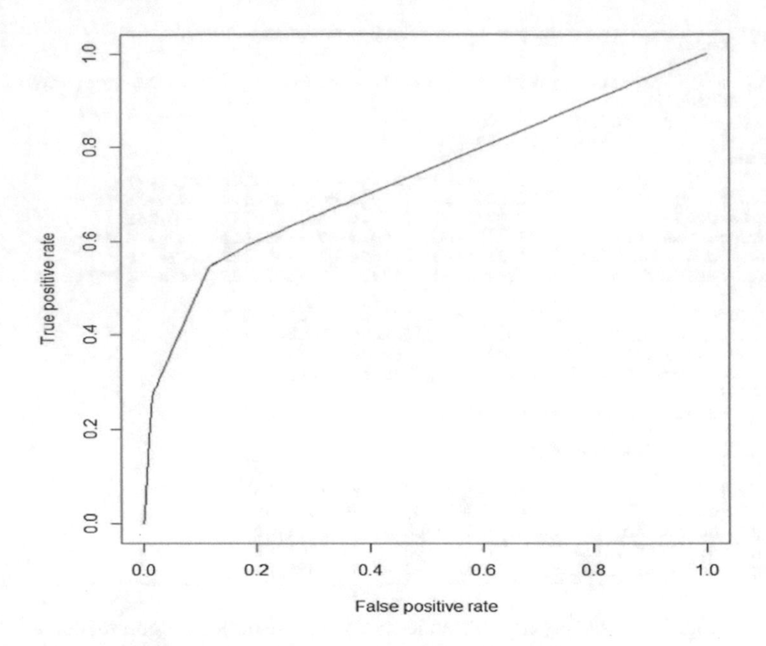

Fig. 9. ROC Curve of binary classifier performance

6 Conclusions and Future Work

In this work, we have implemented and evaluated an architecture for the Smart Supply Chains stage. The architecture allowed the generation, extraction, and ingestion of data from IoT and the RDBMS in HDFS, its visualization and real-time analysis with Machine Learning models, generating knowledge and good decision-making. The CDH platform allowed the storage, analysis, visualization and exploration of data in real time, using Flume, Solr and Morphlines, facilitating the management of Big Data workflows from end to end. Tableau and R were integrated allowing executing machine learning algorithms that ultimately helped to make predictions.

The main contribution of the design was to make an innovative architecture proposal that presents characteristics such as: Variety, Flexibility, Low Costs and Faster decision making. Variety since it allowed processing unstructured data as structured data. Flexibility of integration with several technologies that helped to obtain data and manage their information. Low in costs since using the CDH platform generates an advantage in storage costs. Faster decision making referring to Hadoop, its speed and information analytics, combined with the ability to analyze new data sources, helps companies to have immediate information (either by way of summary or as data specific ones that are required) and in this way make decisions based on what has been learned.

As a future work, we will present some experiments in architecture that will be carried out through the application of automatic learning algorithms in order to determine: detection of fraud, market basket analysis, demand analysis and groupings in sales and profits for client.

Acknowledgments. This work was possible thanks "Universidad de Las Américas" from Ecuador for the financing of research and Leiria Polytechnic Institute from Portugal.

References

1. Puranam, K., Tavana, M.: Handbook of Research on Organizational Transformations through Big Data Analytics, p. 109 (2012)
2. Marjani, M., et al.: Big IoT data analytics: architecture, opportunities, and open research challenges. IEEE Access **5**, 5247–5261 (2017)
3. Constante Nicolalde, F., Silva, F., Herrera, B., Pereira, A.: Big data analytics in IOT: challenges, open research issues and tools. Adv. Intell. Syst. Comput. **746**, 775–788 (2018)
4. Rowe, S., Pournader, M.: How big data is shaping the supply chains of tomorrow. KPMG, Supply Chain Big Data Series, no. March, pp. 1–16 (2017)
5. Kazman, R., Bass, L., Clements, P.: Software Architecture in Practice. Addison-Wesley Professional, Westford (2012)
6. Lele, A.: Big data: related technologies, challenges and future prospects. Smart Innov. Syst. Technol. **132**, 155–165 (2019)
7. Radhakrishnan, M., Sen, S., Vigneshwaran, S., Misra, A., Balan, R.: IoT+Small Data: transforming in-store shopping analytics & services. In: 2016 8th International Conference on Communication Systems and Networks, COMSNETS 2016, no. January (2016)
8. Ganesh, E.N.: Development of SMART CITY Using IOT and BIG Data. Int. J. Comput. Tech. **4**(1), 36–37 (2017)
9. Biswas, S., Sen, J.: A proposed architecture for big data driven supply chain analytics. SSRN Electron. J., 1–24 (2016)
10. Bashir, M.R., Gill, A.Q.: Towards an IoT big data analytics framework: smart buildings systems. In: Proceedings of 2016 IEEE 18th International Conference on High Performance Computing and Communications. IEEE 14th International Conference on Smart City; IEEE 2nd International Conference on Data Science and System, pp. 1325–1332 (2016)
11. Dean, J., Ghemawat, S.: MapReduce: simplified data processing on large clusters. In: Proceedings of 6th Symposium on Operating Systems Design and Implementation, pp. 137–149 (2004)
12. Deploying Spark, HPE Elastic Platform, Big Data Analytics. Spark – a modern data processing framework for cross platform analytics Deploying Spark on HPE Elastic Platform for Big Data (2014)
13. Prakash, M., Padmapriy, G., Kumar, M.V.: A review on machine learning big data using R. In: Proceedings of International Conference on Inventive Communication and Computational Technologies. ICICCT 2018, no. Icicct, pp. 1873–1877 (2018)
14. Nagdive, A.S., Tugnayat, R.M.: A review of Hadoop ecosystem for bigdata. Int. J. Comput. Appl. **180**(14), 35–40 (2018)
15. Ingersoll, G.: Introducing apache mahout: scalable, commercial friendly machine learning for building intelligent applications. White Paper, IBM Developer Works, pp. 1–18 (2009)
16. Isard, M., Budiu, M., Yu, Y., Birrell, A., Fetterly, D.: Dryad: distributed data-parallel programs from sequential building blocks. In: ACM SIGOPS Operating Systems Review, pp. 59–72 (2007)
17. Chen, C.L.P., Zhang, C.-Y.: Data-intensive applications, challenges, techniques and technologies: a survey on big data. Inf. Sci. **275**, 314–347 (2014)
18. Apache Software Foundation: Apache Drill Brings SQL-Like, Ad Hoc Query Capabilities to Big Data (2014). https://drill.apache.org/faq/. Accessed 03 Mar 2018

19. Tableau Software: Build your big data platform with Tableau and Cloudera (2018). https://www.tableau.com/tableau-and-cloudera. Accessed 10 Mar 2019
20. Rajagopalan, N.: Big Data Analytics with Clickstream. https://www.ness.com/big-data-analytics-with-clickstream/
21. Apache Software Foundation: Apache Impala Overview. https://www.cloudera.com/documentation/enterprise/5-9-x/topics/impala_intro.html#impala_cdh. Accessed 10 Mar 2019
22. Cloudera. Introducing Morphlines: The Easy Way to Build and Integrate ETL Apps for Hadoop (2019). https://www.cloudera.com/documentation/enterprise/5-9-x/topics/search_morphline_example.html. Accessed 31 Jan 2019
23. Constante, F., Silva, F., Pereira, A.: DataCo SMART SUPPLY CHAIN FOR BIG DATA ANALYSIS. Mendeley (2019). http://dx.doi.org/10.17632/8gx2fvg2k6.5. Accessed 12 Mar 2019
24. DataSciencePlus: How to Perform a Logistic Regression in R. https://datascienceplus.com/perform-logistic-regression-in-r/

A New Approach to Interoperability in Disaster Management

Marcelo Zambrano Vizuete[1,4](✉), Francisco Pérez[2], Ana Zambrano[3], Edgar Maya[1], and Mauricio Dominguez[1]

[1] Universidad Técnica del Norte, 100110 San Miguel de Ibarra, Ecuador
omzambrano@utn.edu.ec
[2] Universitat Politècnica de València, 46022 Valencia, Spain
[3] Escuela Politécnica Nacional, 170525 Quito, Ecuador
[4] Instituto Tecnológico Superior Rumiñahui, 171103 Sangolquí, Ecuador

Abstract. This paper proposes an architectural scheme for developing an interoperability platform between diverse information systems involved in emergency management. The proposal consists of a distributed communications infrastructure, based on an EDXL-DE model for the exchange of information in emergency settings as a single data model within it. A middleware layer works to adapt services and information offered by each system integrated into the platform according to the EDXL-DE standard, in order to access and/or publish information from any node or system within the infrastructure. To validate the proposal, the interoperability platform was implemented with three communications nodes located in three different locations under simulated emergency conditions. Information was exchanged between the three proprietary systems, each linked to one of the communications nodes of the platform. The staff of two response and humanitarian aid agencies participated in interoperability testing, and they confirmed the usability and functionality of the platform. The results verified the advantages and benefits of the platform, and uncovered some issues that need to be dealt with in the short and medium terms. The main contribution of this work is to provide an interoperability solution for independent information systems involved in emergency management, and to lay the groundwork for a coordinated and collaborative response between the agencies involved.

Keywords: Disaster management · Information systems · Interoperability

1 Introduction

Throughout history, humanity has had to deal with different types of disasters that commonly result in emergency situations in which the response efforts of local governments were often overwhelmed, leading to social instability and dangerous conditions for survivors, tactical responders, and humanitarian aid workers [1–3]. To improve the effectiveness of response efforts, multiple protection forces and public security organizations (firefighters police, medical assistants, Red Cross workers, etc.) must work together in a coordinated and orderly manner, responding together to control the situation and fulfill the critical needs of all agents involved: the government, humanitarian aid groups, first responders, victims, etc. But these widespread and

M. Botto-Tobar et al. (Eds.): ICAETT 2019, AISC 1066, pp. 375–388, 2020.
https://doi.org/10.1007/978-3-030-32022-5_35

diverse resources and skills, indispensable for a response that is comprehensive, are also the main obstacle to effective interoperability, since each group or agency has its proprietary technology, policies, objectives, and procedures.

Interoperability can be defined as the ability of two or more systems to exchange information and use it to meet their objectives [4]. Within the disaster management field, interoperability is indispensable to achieving accurate situational awareness in the operational environment that is shared by all agents involved, one that facilitates timely and realistic decision-making and actions [5]. Agencies such as Federal Emergency Management Agency (FEMA) [6], the Technical Committee for Societal Security of the International Organization for Standardization (ISO/TC 223) [7], the United Nations Office for Disaster Risk Reduction (UNISDR) [8], NATO's Multilateral Interoperability Programme (MIP) [9], etc., have all published standards and recommendations, acknowledging the importance of interoperability in complex networks that address disaster response—environments in which multiple subsystems must interact with themselves and with outside agents effectively.

This paper proposes an alternative approach to disaster response. It includes the architecture for an interoperability platform that facilitates real-time information exchange between all organizations responding to a particular disaster. It is based on a shared communications infrastructure and a common data model that supports interaction and data sharing between the pertinent agencies' information systems (IS). The architecture's functionality was tested in an emergency simulation in which information was exchanged between three proprietary systems (WebGIS, Ne.on, and Relief-Web) deployed in three different locations in Spain (Valencia, Cuenca, and Madrid).

This article is divided into five sections. First is an introduction to our proposal that describes the methodology we used. Second, we review prior solutions and their contribution to this project. Third, we lay out the platform architecture and each of its components. Fourth, we describe our test scenario and the results we obtained. Finally, we present out conclusions.

2 Methodology

The following data models were referenced to design and develop our interoperability platform architecture; all are focused on distributing and exchanging information between heterogeneous systems:

- The Joint Consultation, Command and Control Information Exchange Data Model (JC3IEDM), created by MIP to support multinational operations and facilitate command and control information exchange for tactical operations. (See Fig. 1) The IEDM is based on a common database model that allows transparent information exchange between different information systems (IS's) involved in disaster response operations during a crisis [10, 11].

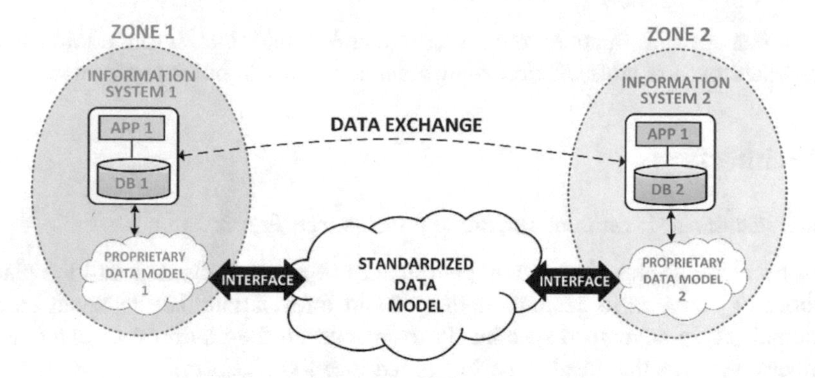

Fig. 1. JC3IEDM schema.

- The Emergency Data Exchange Language–Distribution Element (EDXL-DE) is a set of XML-based technical messaging rules designed to facilitate information exchange between emergency management IS's. Data is transferred in packets using a common XML format that is independent of the platform, programming language, and/or data model used by the IS's (Fig. 2). It was developed by OASIS (International Open Standards Consortium) in collaboration with the U. S. Department of Homeland Security (DHS) [12, 13].

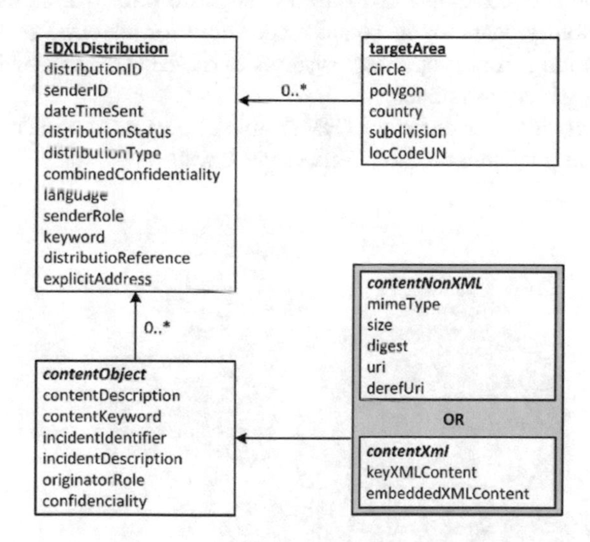

Fig. 2. Generic EDXL-DE schema.

The architecture combines the two data models for developing the platform described above with an innovative PUBLICATION/SUBSCRIPTION function that supports the distribution of information, and its corresponding metadata, among all the IS's integrated into the system. Information is distributed in a form independent of the data models used by the IS's, and without interfering with the privacy protocols used by agencies involved. The main contribution of our proposal lies in the methodology used to manage information; our system enables real-time information exchange implemented on a reliable, flexible, and scalable interoperability platform.

3 Architecture

The architecture rests on three main components; see Fig. 3:

- Information systems, that is, any equipment or system integrated into the platform whose objective is to generate and/or obtain information. The IS's can be either internal (those developed specifically to operate on the platform) or external (proprietary systems that need to be integrated into the platform).
- The communications subnet that provides connectivity between the nodes that enable access to the platform. The subnet topology, as well as the technology deployed in communications switching, have a transparent middleware layer. Thus new nodes can be added more easily, while subnet scalability and reliability increase.
- A middleware layer that serves as a transceiver between the communications subnet and the IS's. This layer has the following components:
  - Communications nodes: these are portals into the platform that, among their main functions, control platform security and privacy, publish information, display the services available to users, and store information, as necessary.
  - Interoperability adaptors: these are responsible for adapting services and information coming from proprietary systems to the EDXL-DE standard defined for the communications subnet.
  - The Human-Machine Interface (HMI): this allows users to access services and information available on the platform via a web navigator.

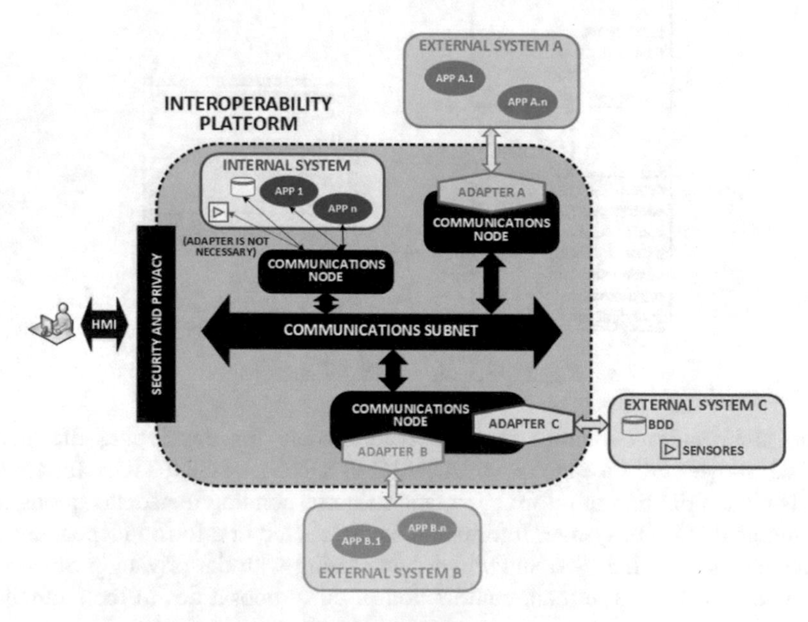

Fig. 3. Architecture.

3.1 Communications Nodes

These are the core components of the architecture; they are subdivided into two layers:

- The user layer, which is responsible for managing the information and services available to users.
- The operations layer that handles management of platform functions, such as the user and systems register, communications with other nodes, data management functions, etc.

Each node has its own independent database, thereby different data models can be used. Data formats are validated and transformed at the node level. Besides storing the data that is published on each of the corresponding nodes, the databases store the metadata to trace the access routes of all available data and services. They also store a copy of the register of platform users and systems. Developers can implement as many nodes are required, and each node can serve more than one IS, depending on the platform's range, the node processing capacity, and the particular IS's requirements, in terms of resources, security, availability, and/or reliability.

There is also a suite of transverse services responsible for monitoring and securing all processes executed on each node.

Figure 4 (below) depicts the high-level internal architecture of the communications nodes.

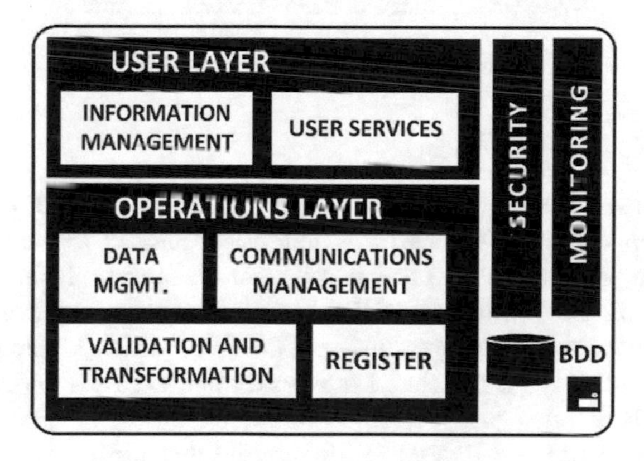

Fig. 4. Communications nodes.

3.2 Interoperability Adaptors

The adaptors combine information and services streaming from each of the external IS's according to the EDXL-DE standard. Each adaptor has two interfaces: one for the external IS and one to handle data from the communications node. Between these two interfaces, there are two sublayers, transformation and communications that convert data formats in both directions.

Requirements for proprietary IS's can be gathered by using any type of web service that has been defined and implemented on the IS's in question: SOAP, REST, HTTP, etc. However, requirements for the communications nodes, in accord with the platform design and the proposed standardization guidelines, only use SOAP [14] for their requirements.

Every external IS must have its own adaptor, in align with the data format that it manages, one for each proprietary IS that is integrated into the platform. Figure 5 represents a general, high-level view of the interoperability adaptor.

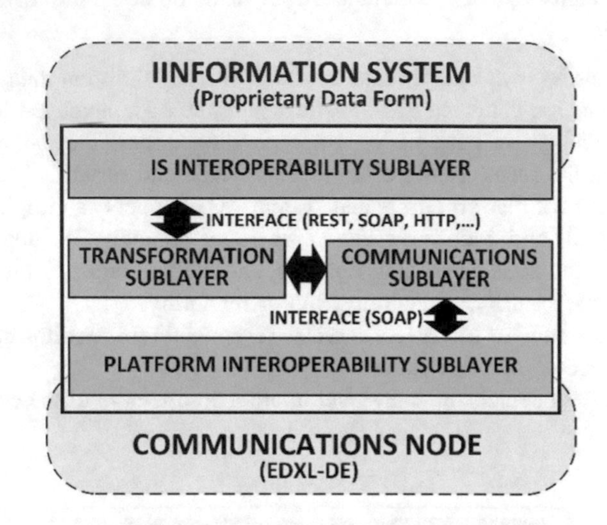

Fig. 5. Interoperability adaptors.

3.3 Human-Machine Interface

The HMI allows any user registered on the platform to connect to a communications node using a web navigator. The user can then easily interact with every IS integrated into the platform, and both obtain and publish information on it. This interactivity is made possible with the CLIENT/SERVER architecture described in Fig. 6.

The HMI server was developed using NODE.JS, which efficiently manages all recurring requests through a JavaScript interpreter that executes all tasks involved.

To address how often web services offered by the communications nodes are engaged, the server uses an EXPRESS.JS module that facilitates robust and secure communication within the operations layer of each node. The HMI client side uses Angular as its JavaScript framework, which in turn uses an MVC (Model View Controller) to offer a flexible application with a light footprint—developed with HTML5 and CSS3—to the client web navigator. Using our application's GUI (Graphic User Interface), the user can then access the services offered.

Connections to the nodes can be either local or remote, depending on the access configured for each node. We assume that the user will access the HMI from their web navigator and, through it, the services available on the platform.

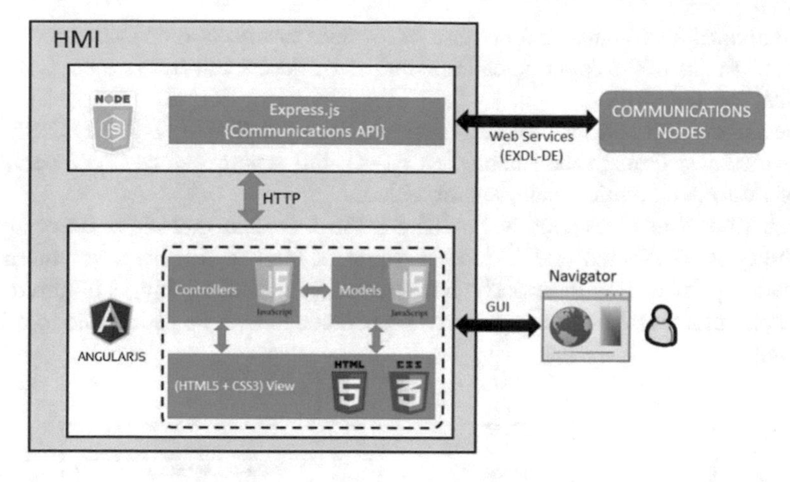

Fig. 6. Human-Machine Interface.

Figure 7 represents the navigation scheme that the GUI offers the user via a web navigator.

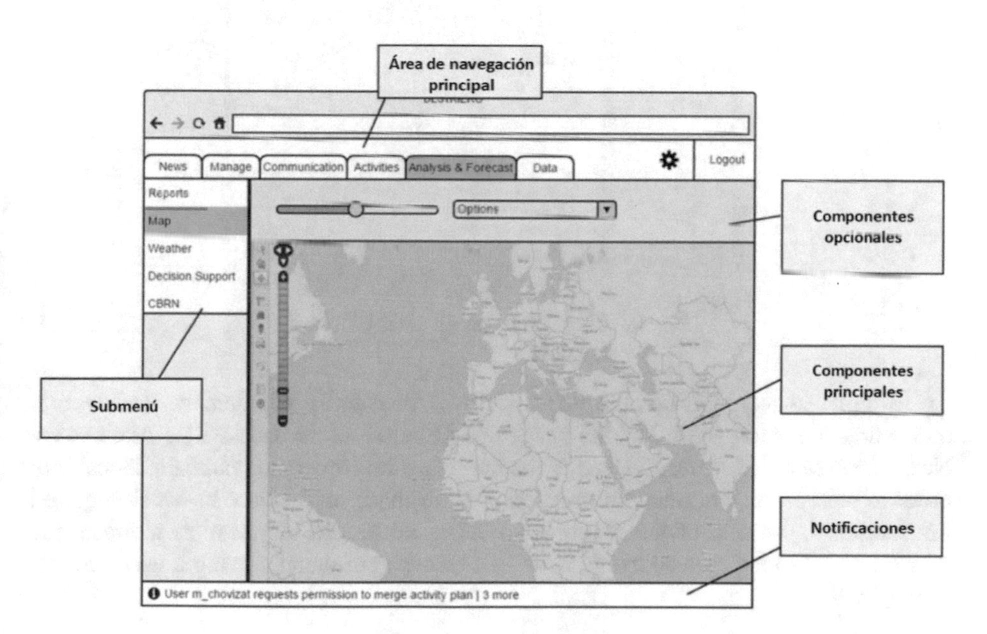

Fig. 7. HMI navigation schema.

3.4 Information Management

The PUBLICATION-SUBSCRIPTION function handles information exchange and facilitates real-time data distribution, including the corresponding metadata, throughout the communications subnet. Information or services that need to be shared must first be published on one of the communications nodes, then users can access them from any of the nodes via the HMI.

The publication and its data requirements form a REQUEST-RESPONSE that is always mediated through the established EDXL-DE scheme, since it can contain any type of information: voice, data, and/or video.

Each time a new resource is published, the metadata that alerts users about its availability is distributed and its access route is tracked. Figure 8 summarizes the information publication process. The information published by any IS is stored locally (on the parent node), so that the resource metadata can later be distributed to all nodes in the subnet.

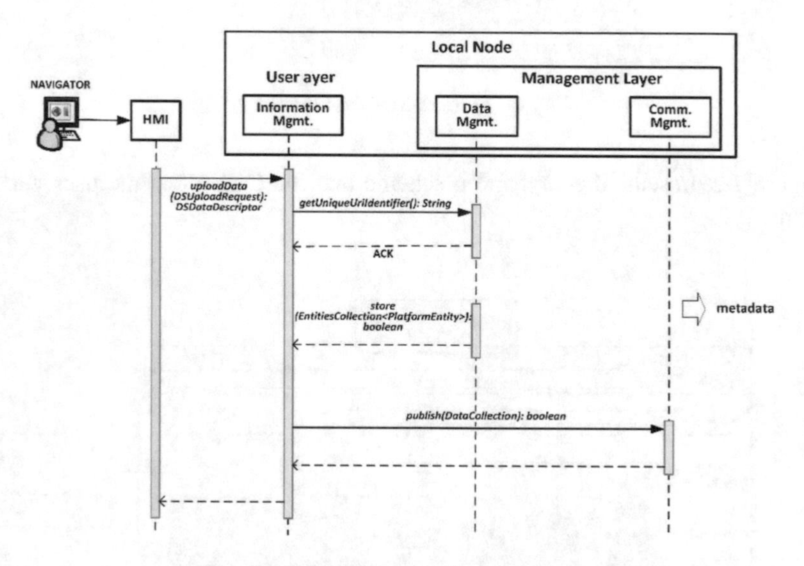

Fig. 8. Information publication process.

In terms of requirements, the first priority is to identify whether the user who is requesting the information has the necessary permissions to access files or services. Next, the system has to verify whether the requested information is published locally in order to receive the requested information as a response. In the case in which required information is on one of the platform's remote nodes, the application redirects the user's request to the corresponding node in a transparent manner using a *wire* service. (See Fig. 9).

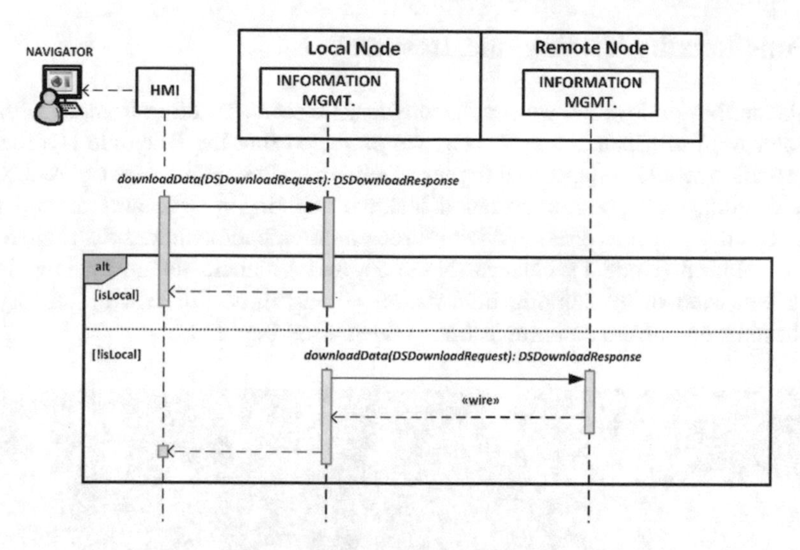

Fig. 9. Information request process.

3.5 Security

Platform accessibility and privacy is managed by assigning profiles and permissions to users. Communications security is managed through security protocols such as HTTP or TLS. Since the architecture is distributed, each node operates independently; therefore, the register of platform users and systems must be replicated.

Figure 10 presents an example of one of the platform's possible topologies with its integrated IS's.

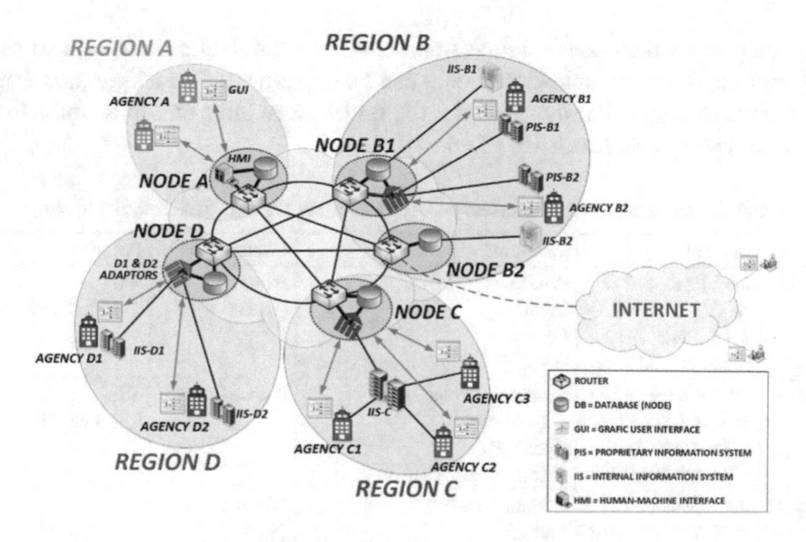

Fig. 10. Example of one possible topology.

4 Functionality Testing and Results

To validate the architecture, we simulated a disaster scene resulting from an earthquake in greater Madrid (Spain). Specifically, we projected that the Bolarque Hydroelectric Power Station would collapse and trigger a flood that affected the José Cabrera Nuclear Power Station, with possible radiation leakage affecting a wide area around it. We implemented a platform prototype with three communications nodes distributed in the cities of Madrid (Node 1), Cuenca (Node 2), and Valencia (Node 3). Functionality testing consisted of exchanging information among three different proprietary IS's, each hosting one of the communications nodes. (See Fig. 11).

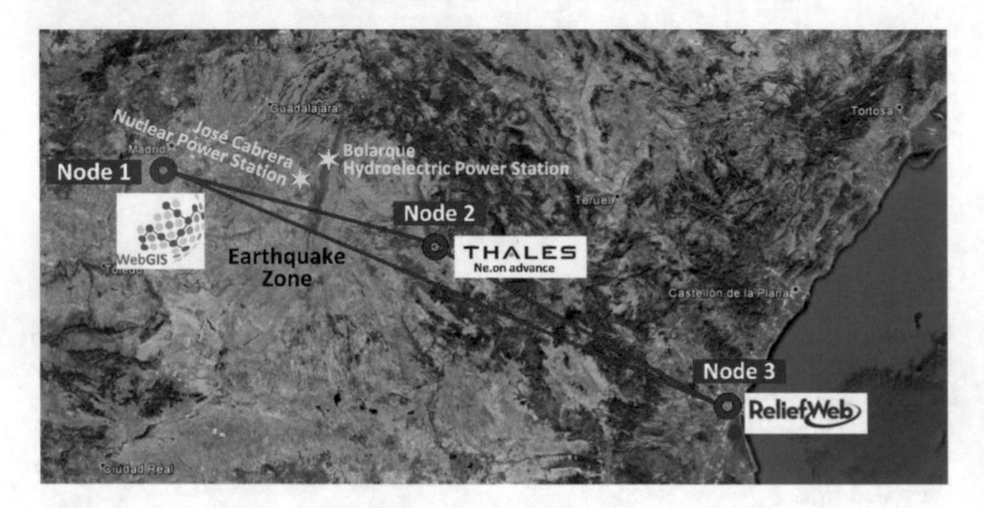

Fig. 11. Test scenario.

The platform's distributed architecture distributes the data processing and storage load; therefore, the communications nodes can be implemented on a light hardware and software infrastructure. For the test case, the nodes were built on top an infrastructure with the following characteristics (Table 1):

Table 1. Infrastructure implemented on each node for the functionality testing.

Node	Hardware	Software	Connectivity	Security
1 Madrid	Intel Processor i7 16 GB DDR4 RAM 2 TB Disk Network 100/1000	Windows 10 PRO MySQL OpenVPN WebGIS (e-GEOS, map server)	Internet (VPN)	SSL Certified VPN 1
2 Cuenca	Intel Processor i7 16 GB DDR3 RAM 1 TB Disk Network 100/1000	Windows Server 2018 SQL Server OpenVPN Ne.on (Thales-CBRN)	Internet (VPN)	SSL Certified VPN 2
3 Valencia	Intel Processor i7 16 GB DDR3 RAM 1 TB Disk Network 100/1000	Ubuntu 14.04 MySQL OpenVPN ReliefWeb App	Internet (VPN)	SSL Certified VPN 3

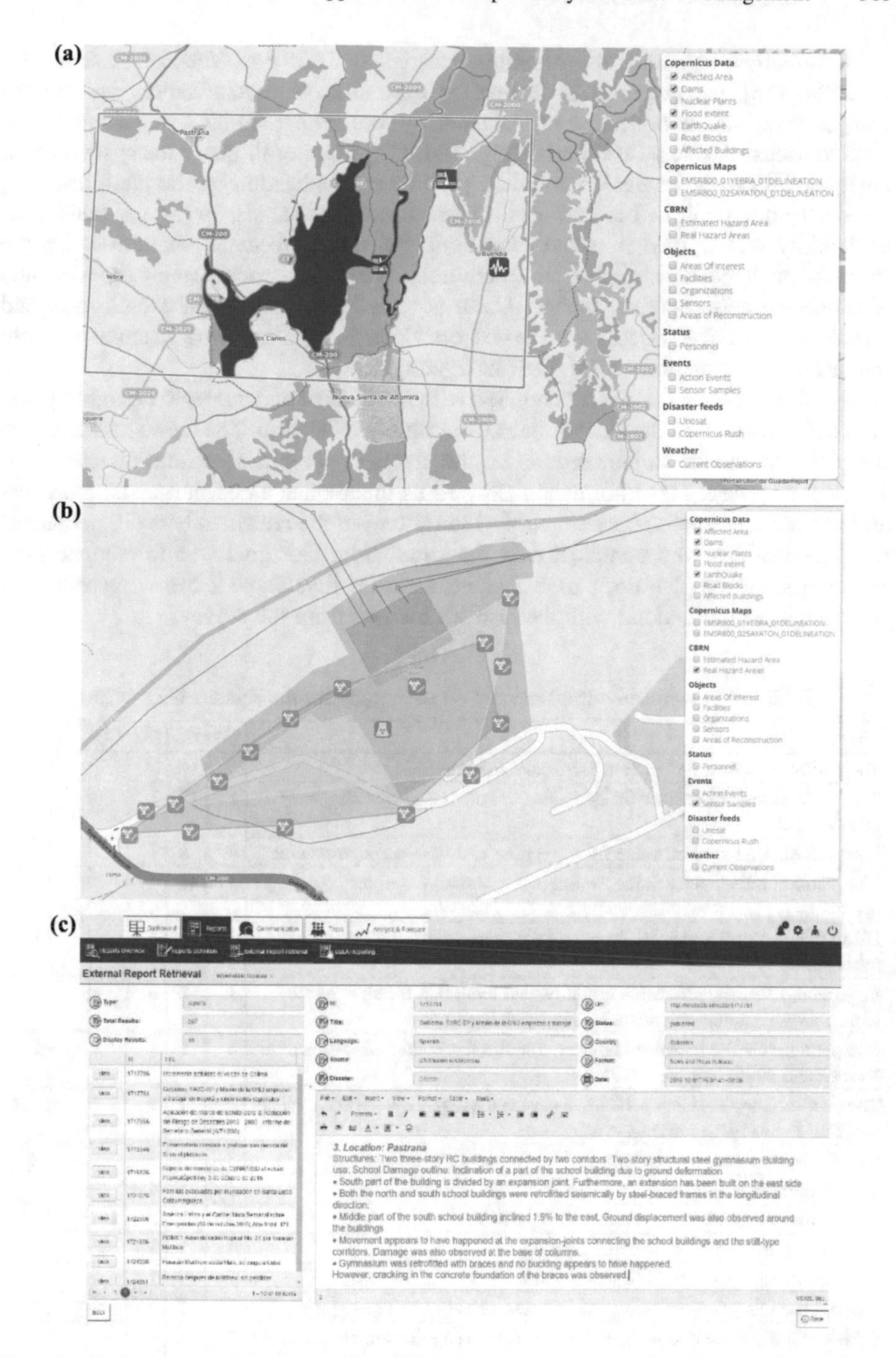

Fig. 12. (a) Perimeter of flood damage - WebGIS. (b) Perimeter of radiation leakage - Ne. on Advance. (c) ReliefWeb reports available via the platform.

Information regarding the flooding was taken from the WebGIS system (See Fig. 12a) [15], the projected perimeter for the radiation contamination was created using tools available on the Ne.on Advance system (Fig. 12b) [16], and Node 3 was used to access the Reliefweb portal (Fig. 12c) [17] in order to query the system about emergencies and humanitarian aid via a proprietary application on the platform.

Each node published reports in different formats (doc, xls, jpg, mpeg) and their availability was verified from the other nodes on the platform. In addition, using the information services available on each node, the user made remote queries and obtained the corresponding reports. Users accessed the HMI to make both local and remote queries. Web access was hosted on Node 3, thus allowing external users to interact with the platform using their navigators.

For the test scenario, we collaborated with the tactical and strategic firefighting and civil defense crews of the Autonomous Community of Madrid. The crews had access to the platform and its various reports on the status of the crisis and the disaster management response. This information allowed us to simulate a decision-making process and coordinate response operations. We then surveyed the crewmembers (15 people) of the organizations who participated in the simulation. Our goal was to tabulate their perceptions of the platform's usability and functionality. Table 2 below recreates the statements we used, along with the results obtained from the survey.

Table 2. Infrastructure implemented on each node for the functionality testing.

Issue	(1)	(2)	(3)	(4)	(5)	NA
The platform's visual design is user-friendly and intuitive	4	10				1
The platform has the characteristics needed to collaborate on emergency management	1	12	1			1
The platform offers added value for supporting real-time decision-making	4	8	3			
The platform offers added value for improving response and recovery operations	1	10	4			
Using this platform allows me to improve response times for emergency management	3	6	5			1
In general, I like the platform's usability, and think that it helps reduce unnecessary communications traffic and operating costs	1	8	6			
The platform can be integrated flexibly and effectively into all information sources and third-party applications	4	9	1			1
The platform supports interoperability between the organizations and systems involved in emergency management	3	7	5			
I would recommend this platform	3	5	5			2
The platform is easy to use	3	8	4			
The platform's functionality meets my expectations	2	3	8	1		1
I would use this platform in a real situation	2	7	5			1
I would use this platform for training and field training exercises	2	11	1			1
I would use this platform during the response phase of an emergency	2	7	5			1
I think that my organization would benefit from using the functions offered by this platform	2	8	4			1

Scale: (1) Completely agree - (5) Completely disagree

In general, 100% of the people surveyed were in favor, more or less, of the way we implemented platform usability features. 93% said they would use the platform for either training or in real-time situations. In addition, 87% of the people would recommend our platform, and 93% believed that their agencies would benefit by using our platform's functionality.

5 Final Notes

The interoperability platform described in this article enables responders to exchange information effectively in a crisis situation, and lays the foundations for obtaining and sharing an accurate and timely picture of the operational environment of any organizations involved in disaster response.

During the test phase, we were able to verify the capacity of the platform to meet the tactical and strategic goals of the personnel involved, including their ability to obtain information from any information system (IS) integrated into the platform. Shared information was utilized for timely decision-making in accord with the nature of the disaster, and to plan response operations.

One of the key characteristics of our architecture is its distributed nature, which facilitates the addition of new nodes and the segmentation of data processing and storage functions, thus improving the scalability and reliability of the platform.

For follow up research, we plan to add real-time communications tools to the system, such as instant messaging and Voice Over IP (VoIP) communications.

The architecture presented in this document serves as a basis for the development of the interoperability platform proposed for Project DESTRIERO in Europe, in which UPV plays an active role. The DESTRIERO project ended this past year (2018), and was part of the 7th European Union Framework Programme for Research and Technological Development (FP7). The goal of this program was to improve factors of interconnectivity, coordination, and collaboration in the management of crises [18].

References

1. Wayne, B.: Guide to emergency management and related terms, definitions, concepts, acronyms, organizations, programs, guidance, executive orders & legislation. FEMA (2008)
2. Alto Comisionado de las Naciones Unidas para los Refugiados, «Manual para situaciones de emergencia», ACNUR (2012)
3. Federal Emergency Management Agency: Course IS-0230.d: fundamentals of emergency management (2015)
4. IEEE: IEEE Standard Computer Dictionary: a compilation of IEEE standard computer glossaries. Institute of Electrical and Electronics Engineers, New York (1990)
5. Williams, A.P.: Agility and Interoperability for 21st Century Command and Control, vol. 4. CCRP (2010)
6. Department of Homeland Security US: «Federal Emergency Management Agency». http://www.fema.gov/. Accessed 25 July 2016
7. International Organization for Standardization: «ISO/TC 223 Societal security». http://www.isotc223.org/. Accessed 20 July 2016

8. United Nations: «The United Nations Office for Disaster Risk Reduction». http://www.unisdr.org/. Accessed 1 June 2016
9. NATO: «Multilateral Interoperability Programm». https://mipsite.lsec.dnd.ca/pages/whatismip_3.aspx. Accessed 15 May 2016
10. de Souza, L.C.C., Pinheiro, W.A.: An approach to data correlation using JC3IEDM model, pp. 1099–1102. IEEE, Tampa (2015)
11. NATO: Overview of the joint C3 information exchange data model. North Atlantic Treaty Organization – MIP (2012)
12. Pack, D., Coleman, C.: Assessing interoperability in emergency management standards, pp. 334–339. IEEE, Huntsville (2008)
13. DHS: Emergency data exchange language. Suit of Standards, Department of Homeland Security (2014)
14. Bloebaum, T.H., Johnsen, F.T., Brannsten, M.R., Alcaraz-Calero, J., Wang, Q., Nightingale, J.: Recommendations for realizing SOAP publish/subscribe in tactical networks. IEEE, Brussels (2016)
15. e-geos: «e-geos Satellite Data», 6 June 2016. http://www.e-geos.it/satellite-data.html
16. Thales Group: «Thales Command and Control Systems», 6 June 2016. https://www.thalesgroup.com/en/spain/what-we-do-defence-solutions/command-control-systems
17. United Nations Office for the Coordination of Humanitarian Affairs: «Reliefweb», 6 June 2016. http://reliefweb.int
18. FP7 European Union: «DESTRIERO», 29 July 2106. http://www.destriero-fp7.eu

Vitiligo Detection Using Cepstral Coefficients

Christian Salamea[1,2] and Juan Fernando Chica[1,2(✉)]

[1] Universidad Politécnica Salesiana, Cuenca, Ecuador
jchicao@ups.edu.ec
[2] Grupo de Investigación en Interacción, Robótica y Automática,
Cuenca, Ecuador

Abstract. Vitiligo is a pathology that causes the appearance of macules achromic (white spots) in the skin. Besides, generates a negative emotional burden in the people that have it, what make necessary to develop suitable methods to identify and treat it properly. In this paper we propose a novel system formed by two stages: The Front End where the principal characteristics of the image are extracted using the Mel Frequency Cepstral Coefficients (MFCC) and i-Vectors (techniques widely used in speech processing) and the Back End, where these characteristics are received and through a classifier is define whether and image contains or not vitiligo. Artificial Neural Networks and Support Vector Machines were selected as classifiers. Results shows that both MFCC and i-Vectors could be used in the field of image processing. Although, the i-Vectors allows us to decrease more the dimensionality of a feature vector and without losing the characteristics of the high dimensionality, this was reflected in their performance with an accuracy of 95.28% to recognize correctly images.

Keywords: Dimensionality reduction · Feature extraction · i-Vectors · MFCC · Medical image processing

1 Introduction

In humans, skin is by far the largest organ and its main function is to protect internal organs from external agents. When a beam of light hits the skin, only a small fraction is directly reflected (4–7%) [1], the rest enters into the skin and follows a paths through the first layers of the skin, until it is dimmed by chromophores found in the skin, like melanin (absorbents of electromagnetic energy to protects skin); or comes back out. Thus, the information of the skin is found in the total reflection (resulting light), information that can be recorded in a digital image.

Occasionally, there may be problems with the functionality of the skin. Among the skin pathologies, vitiligo is one of the most common pigmentary pathology in the world, affecting up to 1% (approximately) of the world population without distinction of sex, race or age [2]. Vitiligo is characterized by provoking the absence of melanin in the skin, which results in the appearance of achromic macules that can change in shape or spread to other areas of the body. Therefore, when a beam of light interacts with the skin, due to the absence of melanin, the light will not be dimmed, causing the light in this area to have a different characteristic than healthy skin.

M. Botto-Tobar et al. (Eds.): ICAETT 2019, AISC 1066, pp. 389–398, 2020.
https://doi.org/10.1007/978-3-030-32022-5_36

On the other hand, this pathology generates a negative emotional burden on the person that suffers from it; which in turn may even make their situation worse [3]. Therefore, it is necessary to develop suitable methods to identify and treat it properly. However, this task is complex because the skin color varies (both for healthy skin, and skin with vitiligo) depending on the type of skin of an individual person [4].

Das [5] proposes the use of different techniques of feature extraction in combination with Support Vector Machines to identify between different skin pigmentary pathologies: vitiligo, viz, leprosy and tineaversicolor. Above all the techniques, the Local Binary Pattern (LBP) technique was the one that got a better performance in the recognition process with a precision score of 89.65%.

Moreover, just as in [5], there are additional works where other vectors with helpful characteristics were individualized in the identification of vitiligo [1], [6–8]. Nevertheless, none explore the information that could be present in the frequency domain and that may be useful to the identification. Also, due to the variability in the skin types of the people, information more condensed and robust is necessary which becomes complex due to the variability in the time domain. Thus, in this paper we explore to what extent the process used in the speech processing could be used in the detection of this pathology.

Although there are several techniques used in the speech processing, among them, the Mel Frequency Cepstral Coefficients (MFCC) is the most widely used since it allows the extraction of the most relevant characteristics of a signal through Cepstrals Coefficients [9]. In addition, in similar works this technique has been used to extract characteristics from images, obtaining quite satisfactory results as shown in [10] and [11]. On the other hand, there are the i-Vectors [12], that allow to represent signals of high dimensionality in a format of low dimensionality, while retaining features of the high.

In this context, we propose the use of a novel method to identify the vitiligo pathology in digital images through the use of the techniques of MFCC and i-Vectors, which allow the extraction and the use of information contained in the frequency domain, that may be used later as input in a classifier that determines whether or not vitiligo is present in an image. Additionally, the results of the techniques are compared to a baseline technique known as Gaussian Mixtures Model (GMM) that had been used in other works and has proven to be able to model the skin [1].

The paper is organized as follows: Sect. 2 which shows the structure of the proposed method. Section 3 which shows the methodology used in experimentation. Section 4 which shows the results of the proposed system; and finally, in the Sect. 5 the conclusions are shown.

2 Proposed System

The proposed system is formed by two stages: The Front-End and the Back-End (As seen in Fig. 1). We used two techniques of feature extraction and two techniques to classification task. The main idea is to identify the best combination of these techniques.

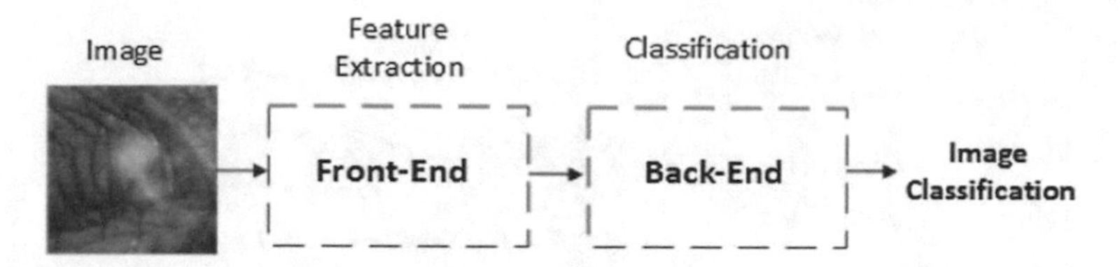

Fig. 1. Diagram of the proposed system. In the front end, the features of the image are extracted and then this information are used to determine whether the image contain or not vitiligo.

2.1 Front End

In the Front End, the relevant information that describes the principal characteristics of the image are extracted. Thus, from a RGB image, the MFCC and i-Vectors techniques are used. The procedure used to implement both techniques is the same as in the speech processing [13, 14].

Mel Frequency Cepstral Coefficients (MFCC). To obtain the MFCC, we use the HTK MFCC MATLAB [15] library which emphasizes obtaining a closely matching the MFCCs produced by HTK library [16] (widely used for MFCC extraction). The process is the following: each image is transformed from 2D to 1D concatenating their columns in a unique vector. Then, this one-dimensional vector is divided into overlapping frames of fixed duration that are multiplied by a windowing function hamming type.

Once we get the frames of the image, the next step has been obtaining the cepstrals coefficients. Thus, firstly a Fast Fourier Transform (FFT) algorithm for each frame has been applied, obtaining the corresponding magnitude and phase spectrum. Then, triangular filters banks spaced according to Mel scale have been applied to the magnitude spectrum values to produce filter bank energies (FBEs). Finally, the FBEs have been decorrelated using the discrete cosine transform to produce cepstral coefficients and the MFCC are obtained [17]. Additionally, applies sinusoidal lifter to produce liftered MFCCs [18] that closely match those produced by HTK. This additional step is based in the routines of Rastaman of Dan Ellis [19]. In the Fig. 2 it can be observed the full diagram of the procedure to obtain MFCC.

i-Vectors. To extract the i-Vectors, one must start off with the following mathematical relation:

$$M = m + Tw \tag{1}$$

Where M is a vector that contains the averages of the characteristics of a given model, in this case the model is composed with the MFCCs of each image. On the other hand, m is a gaussian mixtures model known as Universal Background Model (UBM) due to the fact that it contains all the information of the dataset (information of the healthy skin and skin with vitiligo), ergo it contains the global independent means of M [20]. T is the total variability matrix and is the one who allows to extract the i-Vectors. w contains the i-Vectors.

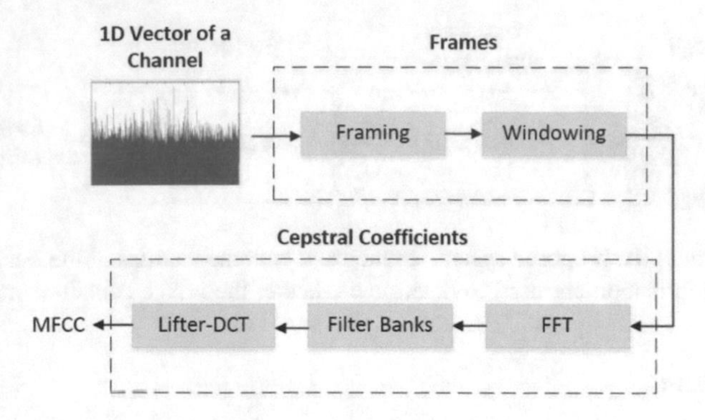

Fig. 2. Procedure diagram to obtain the MFCCs. It can be noted that from an RGB image, to each channel is obtained a one-dimensional vector that is break up in frames, then the cepstrals coefficients are obtained and consequently the MFCC.

I-Vectors process extraction is an iterative process in which first the i-Vectors are randomly initialized, then the values are used to estimate the T matrix. Using the obtained T matrix, the i-Vectors are re-estimated, then, using the i-Vectors the T matrix is estimated again, and so on, until in each iteration the global likelihood of the model does not decrease. When the iterative process ends, the w vector contains the i-Vectors of the image. To extract the i-Vectors we have used the MSR Identity Toolbox [21] developed by Microsoft.

2.2 Back End

Support Vector Machine. The Support Vector Machines (SVM) are binary classifiers that unlike other algorithms of machine learning is not based in statistical or probabilistic foundations, but through geometry minimize an error margin to obtain a separation hyperplane that equidistant two classes.

The SVM mathematically is a lineal classifier that minimize the classification margin and is defined as follows:

$$f(x) = sgn\left(w^T.x + b\right) \tag{2}$$

Where w^T e is a vector orthogonally to the hyperplane and "." is the dot product. In non-linearly separable data the lineal classification becomes restrictive. So, a Kernel function $\Phi(x)$ is used, this function takes the data to a high dimensionality (named feature space) where is more easy define a separation hyperplane. The main idea has been to find a separation hyperplane that is equidistant to the closest examples of each class, consequently a maximum margin on each side of the hyperplane. Considering this, the Eq. (2) would be rewritten in the following way:

$$f(x) = sgn\left(w^T \Phi(x) + b\right) \quad (3)$$

There is variety of kernel function, in this work we use three different types: Gaussian, lineal and polynomial.

Artificial Neural Network. The Artificial Neural Networks (ANN) are mathematical models that simulate the brain behavior to respond to a stimulus, both in functionality and how the information is processed. The ANN are formed by several interconnected artificial neurons and model the relationship between a set of input signals and a set of output signals, these characteristics allows it to be versatility to different tasks like the classification or numerical prediction. There are different topologies (architectures) that can be differentiate depending on the number of layers, how the data flows (forward, backward) and the number of neurons in each layer. However, the most commonly used is the Multilayer Perceptron of three layers: an input layer, hidden layer and the output layer [22, 23].

3 Materials and Methods

The study protocol was approved by the Research Coordination Center of the José Carrasco Arteaga Hospital in Cuenca, Ecuador. The patients who participated in the study signed an informed letter of consent, they have skin type II and type III according to the Fitzpatrick scale [4], their age varies between 18–65 years, and they did not have any type of burn (sunburn and phototherapy) near the photographed area. Eight hundred images were selected from different body areas; 400 images with vitiligo and another 400 that did not have it. All the images have a size of 200 × 200 pixels and due to are RGB images, each cannel was used as if it were different images. Thus, in total 2400 images (800 × 3 channels) were used for the modeling and validation. From all the dataset, 70% were used to train the classifiers of the Back End, 15% to adjust the model (Dev) and the rest 15% as test data (Eval) to validation. It should be noted that it was ensured that each sub sample of the dataset was composed by images of the both classes (with vitiligo and without vitiligo) equally.

3.1 Full Diagram

The feature vectors obtained in the Front End (using MFCC and i-Vectors) are use in the Back End and through a classifier it is determinate whether or not vitiligo is present in an image. For this, two classifiers were used: An Artificial Neural Network type Multilayer Perceptron (MLP) [22] and Support Vector Machines [23].

Subsequently, to determinate the differences between the above mentioned techniques, we proposed four combinations among the feature extraction techniques and the classifiers:

- The first combination (C1) corresponds to MFCC as feature extractor of the image and SVM as classifier.
- The second combination (C2) correspond to MFCC as feature extractor and MLP as classifier.

- The third combination (C3) correspond to i-Vector as feature extractor and SVM as classifier.
- Finally, the fourth combination (C4) correspond to i-Vectors as feature extractor and MLP as classifier.

In Fig. 3 it can be observed the diagram of the experiment proposed as well as the combinations.

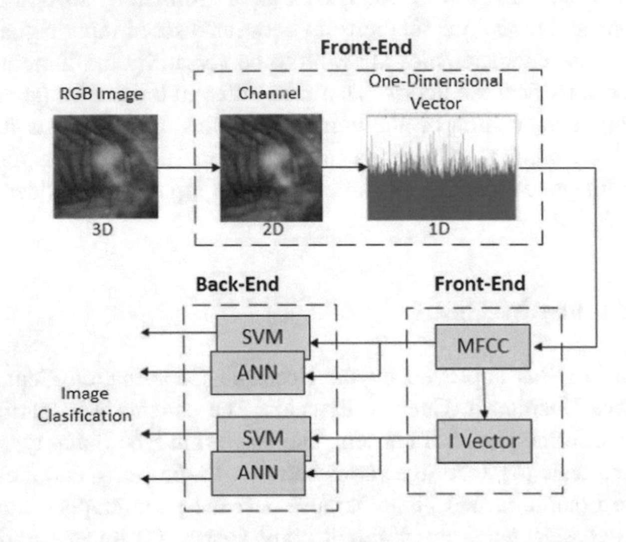

Fig. 3. Diagram of the experiments of proposed method. It can be noted that in the front end the features of the image are extracted by MFCC and i-Vectors then this information are used to train two classifier that determine whether the image contain or not vitiligo.

The performance is measured based on the detection rate of images with vitiligo and images without vitiligo. To do this, sensitivity, specificity and the Accuracy are used as metrics; the sensitivity shows the quantity of images with vitiligo correctly detected (True positives), if an image without vitiligo was detected as if had it the metric decrease their value. The specificity shows the quantity of images without vitiligo correctly detected (True Negatives), if an image with vitiligo was detected as if did not have it, the metric decrease their value. On the other hand, the Accuracy is defined as follows:

$$Accuracy = \frac{Images\ with\ Vitiligo + Images\ without\ vitiligo}{All\ Data} \tag{4}$$

This metric gives a global vision of the performance of the combination and shows if all the cases were detected correctly, the "All Data" parameter refers to the total amount of images used. In all the metrics, the value of 1 indicates 100% of correct answers, if errors are made, depending on the number of error that are committed this value decrease.

3.2 Front End Features

MFCC. The selection of parameters to extract the MFCC was made from configurations already used in speech recognition works [24], [25], after through a fine tuning this parameters were adapted. Then, each image has been separated in fixed duration frames of 24 ms every 19 ms and 18 triangular filters banks. Then, from each image 13 MFCC Coefficients with 719 characteristics each one are obtained. In the combination with MFCC as feature extractor, all the elements are concatenated in a one dimensional supervector of 9347 characteristics (13×719) of each image.

I Vector. Related to the i-Vectors technique, we have used the MFCC of each image to obtain an UBM that include whole the training corpus, then from there we have estimated the i-Vectors (each i-Vector is a one value) to each image. In the speech processing generally are obtained 400–600 i-Vectors, but in this case according to the test performed when we increasing the number towards those size of i-Vectors, their value tends to lower (close to zero) leading to the singularity, hence the most significant i-Vectors were used. Being 15 the total number of i-Vectors used.

3.3 Back End Configuration

Support Vector Machines. Several test have been performed using different kernels in the training phase. Nevertheless, the best result has been obtained using a second order polynomial kernel (scaling factor was selected using fine tuning). Table 1 are shown the results obtained using SVM as classifier, to the sub samples of Train, Dev and Eval.

Table 1. Results of the SVM classifier.

Back End	Train	Dev	Eval
C1	99.95%	92.22%	94.16%
C3	98.51%	91.94%	94.44%

Artificial Neural Networks. The artificial neural network used is the Multilayer Perceptron (MLP). The architecture of the MLP is the following: 50 neurons and an activation function type logarithmic tangential in the hidden layer. In the training process we have used a learning rate of 0.001 and 10 000 epochs. This configuration was selected due to was the one that obtained the best results during the tests carried out. Table 2 shown the results using MLP as classifier, to the sub samples of Train, Dev and Eval.

Table 2. Results of the MLP classifier

Back End	Train	Dev	Eval
C2	99.28%	92.50%	93.06%
C4	95.71%	90.83%	95.28%

4 Results

Table 3 shown the results obtained to the different combinations of the proposed system according to the set of test (Eval).

Table 3. Results of all the combinations in the proposed system regarding to the baseline model (GMM). In C1 and C2 the MFCC is the feature extracted and SVM and MLP are the classifiers, respectively. In C3 and C4 the i-Vectors are the feature extracted and SVM and MLP are the classifiers, respectively.

Back End	Sensibility	Specificity	Accuracy
GMM	85.85%	85.82%	85.84%
C1	93.00%	95.63%	94.16%
C2	95.34%	94.61%	93.06%
C3	94.36%	94.55%	94.44%
C4	96.34%	94.67%	95.28%

The MFCC and i-Vectors shows a significant improvement compared with the baseline technique. Nonetheless, i-Vectors got a better response both for the SVM (C2) and MLP (C4) as classifiers. In C4, based on the sensitivity obtains a better performance than the others combinations, but Specificity is lower than C1. In other words, better performance is obtained in images with vitiligo than images without vitiligo, this fact may be associated with the natural brightness that could be present in the skin, producing the detection of false positives by the algorithm.

5 Conclusions

In this article we have presented a novel method for the detection and recognition of the pathology of vitiligo, which makes use of the techniques used in the speech processing. The proposed method consists of two components, the Front End where the extraction of features is done, and the Back End where a classification task is performed. The best result was obtained when i-Vectors were used as an extractor of characteristics. Additionally, after contrasting the results between the two classification techniques, it is possible to state that the abstracted characteristics allow to describe the images and in turn to differentiate them independently if the technique is probabilistic (MLP) or if it is based on geometry (SVM). The techniques used to extract the characteristics are widely used in the speech processing, based on the results obtained in this experimental work, it is shown that these techniques can also be used in the field of images which give us great benefits as the reduction of dimensionality and in the case of i-Vectors allows us to represent in format of low dimensionality the most relevant information of the image without losing the characteristics of the high dimensionality; in this case the original image size of 200 × 200 (40 000 features) has been reduced to a vector containing 15 I-Vectors (being an i-Vector a single value). At the present time, in order to obtain a good training of a system use large amounts of data of which not all the information contained is relevant, in this context, the I-Vector technique could be used in image processing as an alternative option to the systems of traditional extraction of characteristics. It should be noted that none pre-processing was done on the images because we want to evaluate the performance of feature extraction algorithms by themselves, in future works it could be extended to the use of these techniques to enhance the relevant information (e.g. decrease the brightness) to improve the performance of the system.

References

1. Fadzil, M.H.A., Norashikin, S., Suraiya, H.H., Nugroho, H.: Independent component analysis for assessing therapeutic response in vitiligo skin disorder. J. Med. Eng. Technol. **33** (2), 101–109 (2009)
2. Alikhan, A., Felsten, L.M., Daly, M., Petronic-Rosic, V.: Vitiligo: a comprehensive overview. J. Am. Acad. Dermatol. **65**(3), 473–491 (2011)
3. Papadopoulos, L., Bor, R., Legg, C.: Coping with the disfiguring effects of vitiligo: a preliminary investigation into the effects of cognitive-behavioural therapy. Br. J. Med. Psychol. **72**(3), 385–396 (1999)
4. Fitzpatrick, T.B.: The validity and practicality of sun-reactive skin types I through VI. Arch. Dermatol. **124**(6), 869–871 (1988)
5. Das, N., Pal, A., Mazumder, S., Sarkar, S., Gangopadhyay, D., Nasipuri, M.: An SVM based skin disease identification using local binary patterns. In: 2013 Third International Conference on Advances in Computing and Communications, Cochin, India, pp. 208–211 (2013)
6. Alghamdi, K.M., Kumar, A., Taïeb, A., Ezzedine, K.: Assessment methods for the evaluation of vitiligo: Vitiligo assessment methods. J. Eur. Acad. Dermatol. Venereol. **26** (12), 1463–1471 (2012)
7. Nurhudatiana, A.: A Computer-aided diagnosis system for vitiligo assessment: a segmentation algorithm. In: Intan, R., Chi, C.-H., Palit, H.N., Santoso, L.W. (eds.) Intelligence in the Era of Big Data, vol. 516, pp. 323–331. Springer, Heidelberg (2015)
8. Hani, A.F.M., Nugroho, H., Shamsudin, N., Baba, R.: Melanin determination using optimised inverse monte carlo for skin—light interaction. In: 2012 4th International Conference on Intelligent and Advanced Systems (ICIAS), vol. 1, pp. 314–318 (2012)
9. Tiwari, V.: MFCC and its applications in speaker recognition. Int. J. Emerg. Technol. **1**(1), 19–22 (2011)
10. Nisar, S., Ashraf, M.W.: A new approach for toe recognition using mel frequency cepstral coefficients. In: 2016 13th International Bhurban Conference on Applied Sciences and Technology (IBCAST), Islamabad, Pakistan, pp. 291–294 (2016)
11. Gupta, S., Jaafar, J., wan Ahmad, W.F., Bansal, A.: Feature extraction using MFCC. Sig. Image Process. Int. J. **4**(4), 101–108 (2013)

12. Dehak, N., Kenny, P.J., Dehak, R., Dumouchel, P., Ouellet, P.: Front-End factor analysis for speaker verification. IEEE Transact. Audio Speech Lang. Process. **19**(4), 788–798 (2011)
13. Salamea, C., D'Haro, L.F., Córdoba, R., Caraballo, M.Á.: Incorporación de n-gramas discriminativos para mejorar un reconocedor de idioma fonotáctico basado en i-vectores. Procesamiento del lenguaje natural (51), 145–152 (2013)
14. Dehak, N., Torres-Carrasquillo, P.A., Reynolds, D., Dehak, R.: Language recognition via i-vectors and dimensionality reduction. In: Twelfth Annual Conference of the International Speech Communication Association (2011)
15. HTK MFCC MATLAB - File Exchange - MATLAB Central. https://www.mathworks.com/matlabcentral/fileexchange/32849. Accessed 27 Mar 2019
16. Young, S., Evermann, G., Gales, M., Hain, T.: The_HTK_book.pdf. Engineering Department (2006)
17. Hasan, R., Jamil, M., Rahman, G.R.S.: Speaker identification using mel frequency cepstral coefficients. Variations **1**(4) (2004)
18. Paliwal, K.K.: Decorrelated and liftered filter-bank energies for robust speech recognition. In: Sixth European Conference on Speech Communication and Technology (1999)
19. Ellis, D.: Reproducing the feature outputs of common programs in Matlab using melfcc.m (2005). http://www.ee.columbia.edu/~dpwe/LabROSA/matlab/rastamat/mfccs.html. Accessed 26 Mar 2019
20. Hasan, T., Hansen, J.H.L.: A study on universal background model training in speaker verification. IEEE Transact. Audio Speech Lang. Process. **19**(7), 1890–1899 (2011)
21. Sadjadi, S.O., Slaney, M., Heck, L.: MSR identity toolbox v1.0: a MATLAB toolbox for speaker-recognition research. Speech Lang. Process. Tech. Comm. Newsl. **1**(4), 1–32 (2013)
22. Orhan, U., Hekim, M., Ozer, M.: EEG signals classification using the K-means clustering and a multilayer perceptron neural network model. Expert Syst. Appl. **38**(10), 13475–13481 (2011)
23. Ceballos-Magaña, S.G., et al.: Characterisation of tequila according to their major volatile composition using multilayer perceptron neural networks. Food Chem. **136**(3–4), 1309–1315 (2013)
24. Koolagudi, S.G., Rastogi, D., Rao, K.S.: Identification of language using mel-frequency cepstral coefficients (MFCC). Proc. Eng. **38**, 3391–3398 (2012)
25. Hassan, F., Alam Kotwal, M.R., Rahman, M.M., Nasiruddin, M., Latif, M.A., Nurul Huda, M.: Local feature or Mel frequency cepstral coefficients - which one is better for MLN-based Bangla speech recognition? In: Abraham, A., Lloret Mauri, J., Buford, J.F., Suzuki, J., Thampi, S.M. (eds.) Advances in Computing and Communications, vol. 191, pp. 154–161. Springer, Heidelberg (2011)

Using Deep Neural Networks for Stock Market Data Forecasting: An Effectiveness Comparative Study

Carlos Montenegro(✉) and Marco Molina

Escuela Politécnica Nacional, Quito, Ecuador
{carlos.montenegro,marco.molinab}@epn.edu.ec

Abstract. Stock market value forecasting has been a challenge, because data are massive, complex, non-linear and noised. Nevertheless, some deep learning promising techniques can be reviewed in technical literature. Using S&P500 historical data as a case study, this work proposes the following approach: (i) NARX and Back Propagation Neural Networks are selected and trained for representing Index data; (ii) A sliding window technique for Index value forecasting is defined and tested; and, (iii) An effectiveness comparison is performed. The results suggest the best model for representing and forecasting S&P500 Index data. Thus, the academics can revise a new experience in data analysis; and practitioners will have an approach concerning the forecasting calculation in the stock market.

Keywords: Deep learning · Neural networks · S&P500 index · Forecasting · Stock market

1 Introduction

Predicting stock indices has been regarded as one of the most challenging tasks in econometrics. Measuring market risk requires quantitative techniques to analyze individual financial instruments and a portfolio of assets. This quantitative measure or model captures trends and behaviors in data which are then used to deduce future values [1].

The predictability of the stock market has been long a research topic. According to Fang [2], the overall stock indices are generally easier to work with than individual stocks, and longer-horizon (over a month) predictions make more sense than short-horizon ones. Although the researchers cannot agree on whether stock markets are predictable or not [3], studying whether one can predict them is an interesting issue [4].

Stock prediction can be made by using a statistical approach, which treats the stock data as a time series. Examples include Exponential Smoothing Models (ESM), ARIMA models, ARCH and GARCH models, among others [1]. The financial models are based on statistical proprieties and assumptions in data. In some instances, unrealistic assumptions are made to simplify the problem or to allow a mathematical derivation of the model. Given the complex behavior of financial markets, these models can potentially misrepresent or fail to represent critical features in data [1].

M. Botto-Tobar et al. (Eds.): ICAETT 2019, AISC 1066, pp. 399–408, 2020.
https://doi.org/10.1007/978-3-030-32022-5_37

On the other hand, feature-based machine learning approaches take advantage of economic data, as well as historical stock data. Examples include Support Vector Machines, Genetic Algorithms, and Artificial Neural Network (ANN). ANN has been one of the most successful applications [8] and, recently, there has been a great interest in reviving the study of deep network structures. Deep Neural Networks (DNN) has various successful reports in machine learning applications [2].

There are desirable features of neural networks, which makes them a suitable tool for market modeling. According to Mostafa *et al.* [1] and Qian [5], the achieved results show the superiority of neural networks over statistical models. Also, they are naturally suitable to model nonlinearities in the data.

The described scenario facilitates exploring alternatives to stock market forecasting, using artificial neural networks representative models, as is posed in this study. S&P500 index daily data of the last five years are used to illustrate the proposal.

2 Background and Related Work

2.1 S&P 500 Index

The Standard & Poor's 500 (S&P 500), is an American stock market index based on the information of 500 large companies having common stock listed on the NASDAQ or NYSE. S&P Dow Jones Indices determine the S&P 500 index parameters values. It is considered one of the best representations of the U.S. stock market.

Table 1 shows an extract of daily available data for the S&P500 Index and for each member enterprise. Open, High, Low and Close refer to stock value during a day; Volume refers to the number of shares of the stock market. The mentioned data can be obtained from URL: http://www.financeyahoo.com.

Table 1. S&P500 Index data (Extract)

Date	Open	High	Low	Close	Volume
08/05/2018	2670.26	2676.34	2655.20	2671.92	3717570000
09/05/2018	2678.12	2701.27	2674.14	2697.79	3909500000
10/05/2018	2705.02	2726.11	2704.54	2723.07	3333050000
11/05/2018	2722.70	2732.86	2717.45	2727.72	2862700000
14/05/2018	2738.47	2742.10	2725.47	2730.13	2972660000

2.2 Stock Forecasting Using Neural Networks for Deep Learning

Many artificial intelligence methods have been employed to predict stock market prices [6, 7]. ANN remains a popular choice for this task, it is widely studied and has shown to exhibit excellent performance [8]. Recent literature suggests that researchers are attempting to use deep learning for stock prediction [8, 9].

On the other hand, DNN is a machine-lcarning paradigm for modeling complex nonlinear mappings between input and output, in which the internal parameters are updated iteratively to make the given inputs fit with target outputs [10, 11]. Deep learning has more complex ways of connecting layers, also has more neurons count than previous networks to express complex models, and requires more computing power to train [12]. Standard DNN provides a universal framework for modeling complex and high-dimensional data. An especially attractive feature of DNN approach is the inherent capability of covering all stages of data-driven modeling (features selection, data transformation, and classification/regression) within a single framework, i.e., ideally, the practitioner can start with raw data from a domain of interest and get ready-to-use solution [13]. Nevertheless, Prastyo *et al.* [14] suggest that ANN models can present anomalies in extrapolation problems, thus that is a reason why it is recommendable define a strategy to control this behavior.

Deep Learning approaches consist in adding multiple layers that can be repeated to a neural network [15–18]. This study compares the effectiveness of two representative fundamental models: NARX, a dynamic model, and Back Propagation Neural Network (BPNN), a static model.

NARX Neural Networks

A recurrent neural network (RNN) is represented by a directed graph containing at least one directed cycle (feedback). Unlike the traditional artificial neural networks, the recurrent neural network is ideal for predicting sequential problems, because the learning technique keeps track of a sequence of events [19–21].

Nonlinear AutoRegressive models with eXogenous Inputs (NARX) have limited feedback which comes only from the output layer rather than from hidden ones. They are formalized by

$$y(t) = \Psi\left(u(t-n_u),\ldots,u(t-1),u(t),y(t-n_y),\ldots,y(t-1)\right)$$

Where $u(t)$ and $y(t)$ represent input and output of the network at time t; n_u and n_y are the input and output order; and, the function Ψ is a mapping performed by a Multilayer Perceptron [22]. Typically, this network architecture has been considered to model chaotic data [23]. Figure 1 presents a deep learning architecture based on the NARX model.

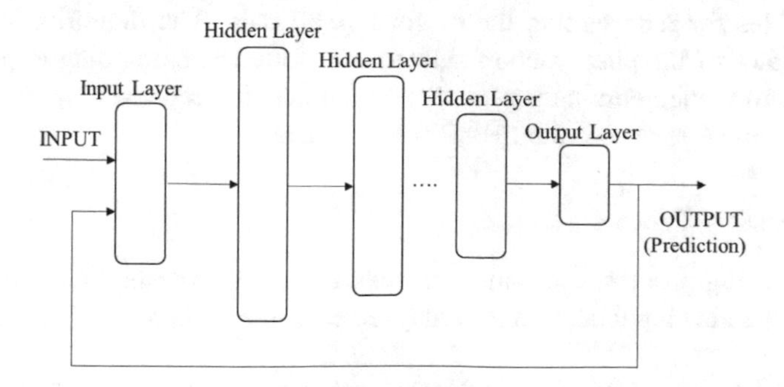

Fig. 1. NARX neural network for deep learning architecture

Back Propagation Neural Network
The BPNN architecture for Deep Learning has its layers in each fully connected stack, with fewer nodes than the preceding [9, 21, 24, 25]. The model is formalized by

$$y = h(A) = h(g(I)) = h(g(f(\Sigma x_{pi} \times w_{ji})))$$

Where x_{pi} is the input vector of dimension p; f is the input function; g is the activation function; and, h represents the training function. Weights w_{ji} are updated using a backpropagation process. Figure 2 presents a deep learning architecture based on the BPNN model.

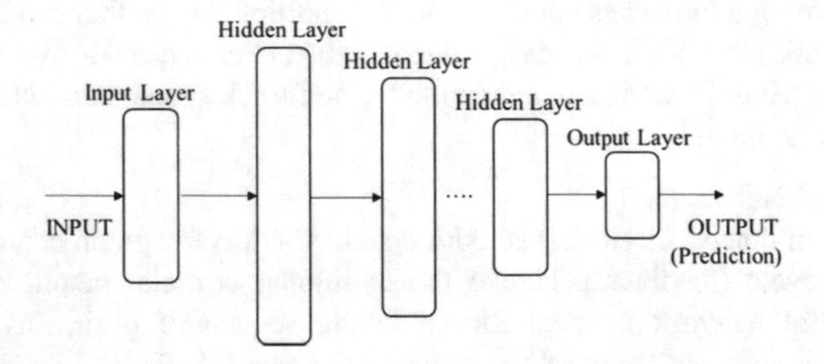

Fig. 2. BPNN for deep learning architecture

3 Methodology

As it was stated before, stock market data, as S&P500, are massive, complex, non-linear and noised. Thus, the forecasting process definition has been a challenge. The methods to achieve this goal are the following.

3.1 Representing Data

In this study, the previously described models are used: (i) NARX Neural Networks; (ii) Back Propagation Neural Network. Previous to training, it is necessary to define a set of variables for representing the original (real) data. The definition of input and output variables of the phenomenon aims to transform the extrapolation problem into an interpolation one, eliminating explicit variables for representing the date. The trained DNN must represent the SP&500 Index data.

3.2 Forecasting Process

At the forecasting process, the trained models must perform interpolation or extrapolation using the new input data. In the analyzed case, the trained DNN is simulated with new data.

ANN models can exhibit abnormal behavior in extrapolation cases, as some classical studies suggest [26, 27], as well as more recently investigations [28, 29]. In this study, an initial explorative work shows behavior anomalies in extrapolation forecasting. To overcome the problem, Barnard and Wessels [26] suggest simulating the ANN with values around the trained data.

The forecasting model must represent the data adequately, and produce future index values. According to the above considerations, only the next values in the time variable is calculated, using the data of the previous step (next-step prediction). The sliding window technique is showed in Fig. 3.

Data of n days is used to forecast the value of day $n + 1$. At the next step, the real data of the day $n + 1$ is added, and the data on the first day is deleted. Then, the value of the day $n + 2$ is forecasted and so on. Besides, as is showed later, new variables must be defined to obtain an implicit representation of time, and transform the extrapolation into an interpolation problem.

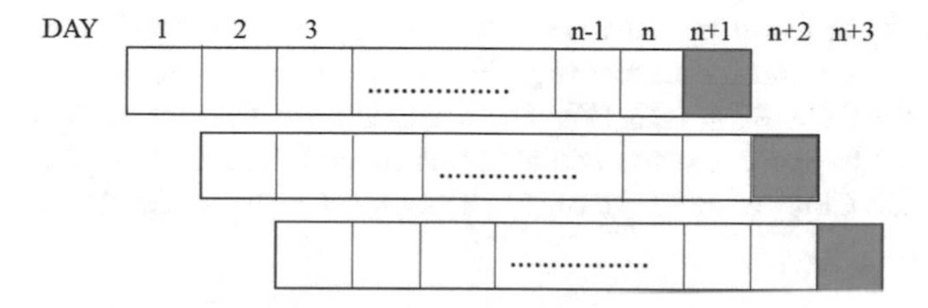

Fig. 3. Sliding window technique

3.3 The Comparison Criterion

The effectiveness of DNN models is evaluated using the R^2 of the modeled output. R^2 is determined as follows:

$$R^2 = 1 - \frac{\sum \left(y_{exp} - y_{pred}\right)^2}{\sum \left(y_{exp} - \bar{y}\right)^2}$$

Where y_{exp} is the experimental value, y_{pred} is the predicted value and $\bar{y}$ the mean value. This criterion is used to assess the effectiveness of the data representation and also, to assess the effectiveness of the forecasting process.

4 Results

4.1 Data Representation

The used data correspond to a period from June 7, 2013, to June 6, 2018. MATLAB software [30] is used for the process of modeling and calculations. The following ten variables are derived, for representing data and will be used as input in the forecasting process:

Input Variables

- Month: it refers to the month to which a given record belongs to.
- MonthDay: day of the month to which a given record belongs to.
- WeekDay: refers to the day of the week corresponding to a given stock record.
- LowDiff: For two consecutive slots S1 and S2. If L1 and L2 refer to the Low values for S1 and S2 respectively, then LowDiff for S2 is computed as (L2–L1).
- HighDiff: the difference between the High values of two successive slots. The computation is identical to LowDiff.
- CloseDiff: If two successive slots S1 and S2 have close values C1 and C2 respectively, then CloseDiff for S2 is calculated as (C2–C1).
- VolDiff: For two consecutive slots S1 and S2, if the mean values of Volume for both the slots are V1 and V2 respectively, the VolDiff for S2 is (V2–V1).
- RangeDiff: For two consecutive slots S1 and S2, suppose the High and Low values are H1, H2, L1 and L2 respectively. Hence, the Range value for S1 is R1 = (H1–L1) and for S2 is R2 = (H2–L2). The RangeDiff for the slot S2 is (R2–R1).
- OpenClose: Suppose two consecutive slots: S1 and S2. Let the Open price of S2 be X2, and the Close price of S1 be X1. The OpenClose for the slot S2 is (X2–X1).

Output Variable

- OpenPerc: Suppose two consecutive slots: S1 and S2. Let the Open price of the stock for the first record of S1 be X1, and that for S2 be X2, the OpenPerc for the slot S2 is computed as (X2–X1)/X1*100.

NARX Neural Network uses three hidden layers with sixty-six neurons and a closed loop with two delays (Fig. 4). The first hidden layer attempts to produce weights from the activations, to be used for regression. The second and third hidden layers act as pooling layers that perform downsampling along the spatial dimensionality of the given input, further reducing the number of parameters of the phenomenon. The linear transfer function on the output layer acts as a regression layer, as is required by the forecasting process.

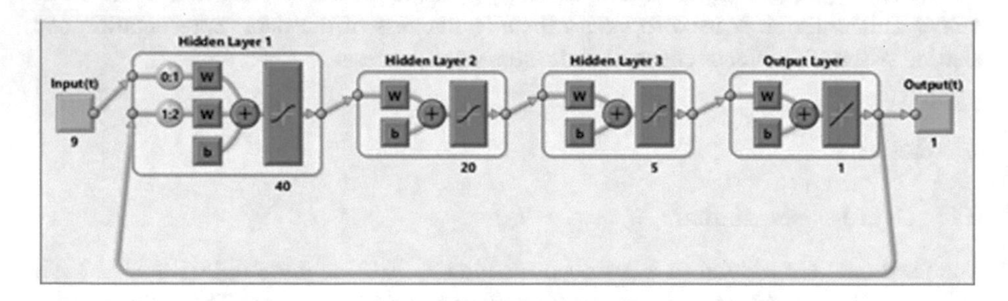

Fig. 4. NARX architecture

BPNN uses similar hidden layers as NARX (Fig. 5). This detail guarantees a better comparison.

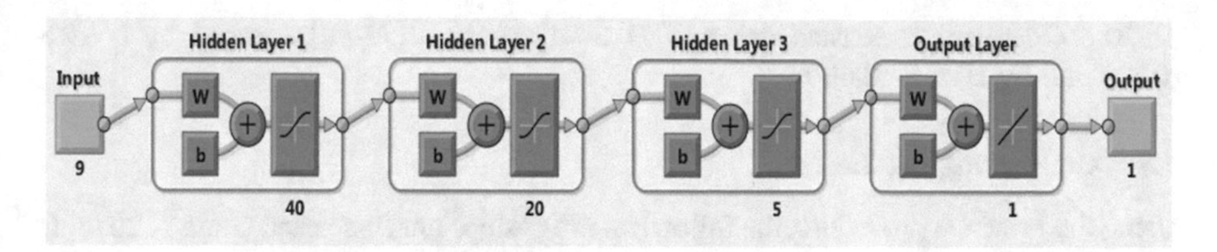

Fig. 5. BPNN architecture

Figure 6 shows the training adjustment, including R-value. The figures are drawn using an extract of the data. Besides, for each case, 70% of the data is used for training, 15% for validation, and 15% of the data is taken into account for testing. The network uses training data that is adjusted according to its error. Validation data are used to measure network generalization, and to halt training when generalization stops improving. Test data do not affect training and provide an independent measure of network performance during and after training.

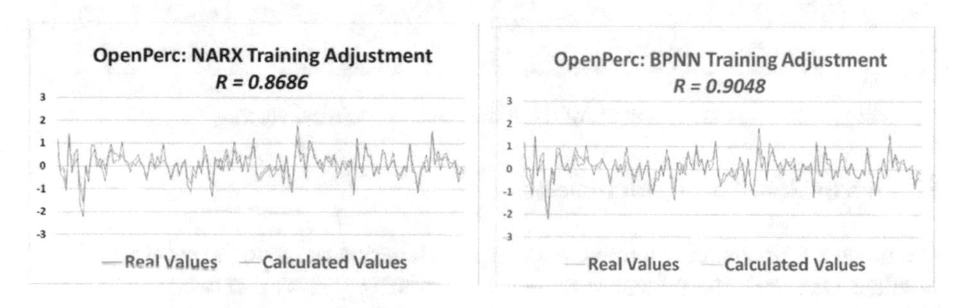

Fig. 6. NARX and BPNN training adjustment

4.2 Forecasting Process

Based on the historical movement of the stock prices it is predicted the stock price in the next slot. In other words, if the current time slot is S1, the technique will attempt to predict OpenPerc for the next slot S2. See Table 2, where the last line only varies the Weekday and the Month day values. Other values are the same as in the previous day.

Table 2. Data for OpenPerc variable prediction

Case	Month	Month day	Week day	Low diff	High diff	Close diff	Vol diff	Range diff	Open close	Open perc
1	6	10	2	13.99	4.29	−0.57	−393260000	−9.70	1.29	1.19
2	6	11	3	−16.34	−8.56	−16.68	456980000	7.78	−4.17	−0.37
3	6	12	4	−12.00	−2.42	−13.61	−233160000	9.58	3.81	-0.53
....	....	....	....	....	....	....	....	....	......	....
1258	6	6	4	8.95	19.78	23.55	133850000	10.83	4.45	0.17
1259	6	7	5	8.95	19.78	23.55	133850000	10.83	4.45	***FV***

The OpenPerc forecasted value (FV) for the case 1259 is generated by NARX (0.83) and by BPNN (0.6809).

4.3 Comparing the Results

Using the previous procedure, the following thirty steps are forecasted (June 7, 2018, to July 19, 2018) as a proof of the adjustment quality. Figure 7 shows the graph of the forecasted and real values of Open variable. BPNN shows the best effectiveness, according to R-value.

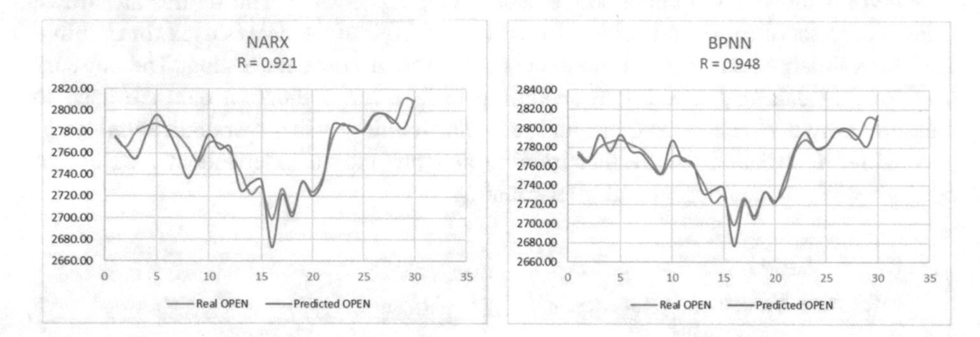

Fig. 7. Adjust of the Index Open forecasted values

5 Discussion and Conclusions

The process for computing a forecasting value is laborious. Due to the necessary data to simulate a network, the following values are successively less exact because the Neural Nets can exhibit abnormal behavior, even more in extrapolation cases. Here, a heuristic is used to control the possible anomalies: to learn about a topic it is necessary to know it; then, for the learning process is used some previous data, and is adopted a sliding window technique that updates the data, step to step (day to day), for the forecasting process. Besides, definitions of input variables of the phenomenon are designed, in order to transform the extrapolation into an interpolation problem, because Date does not appear as an explicit variable. In this controlled scenario, the BPNN model shows the best effectiveness.

The forecasted values generated by BPNN models are significant, as is probed by the adjustment criterion. While more detailed data are, more exact the forecasting will be, because more predictive variables can be defined; besides enables other periods for prediction (hour, daily, weekly, etc.). The number of cases for training is another parameter that can influence effectiveness.

Finally, as the theory and the results support it, this research describes a theoretical and a practical tool for academics and practitioners. The academics can revise a new experience of using neural networks models for forecasting a set of massive, complex, non-linear and noised data. The practitioners can expand the existing knowledge, concerning the criteria for representing and forecasting stock market data. One interesting use of the results is in the investment decision on the stock market.

References

1. Mostafa, F., Dillon, T., Chang, E.: Computational Intelligence Applications to Option Pricing, Volatility Forecasting and Value at Risk, Studies in Computational Intelligence, vol. 697. Springer, Heidelberg (2017)
2. Fang, Y.: Feature selection, deep neural network, and trend prediction. J. Shanghai Jiaotong Univ. (Sci.) **23**(2), 297–307 (2018)
3. Zheng, X., Chen, B.: Stock Market Modeling and Forecasting. LNCIS, vol. 442, p. 1–11. Springer, London (2013)
4. Olden, M.: Predicting stocks with machine learning (2016)
5. Qian, X.: Financial series prediction: comparison between precision of time series models and machine learning methods (2017)
6. Banik, S., Khan, A.: Forecasting US NASDAQ stock index values using hybrid forecasting systems. In: 18th International Conference on Computer and Information Technology (ICCIT) (2015)
7. Cao, Z., Wang, L., Melo, G.: Multiple-weight recurrent neural networks. In: Proceedings of the Twenty-Sixth International Joint Conference on Artificial Intelligence (IJCAI-17) (2017)
8. Singh, R., Srivastava, S.: Stock prediction using deep learning. Multimedia Tools Appl. **76**, 18569–18584 (2017)
9. Yong, B., Rahim, M., Abdullah, A.: A stock market trading system using deep neural network. In: Mohamed Ali, M.S., et al. (eds.): AsiaSim 2017, Part I, Singapore (2017)
10. Yang, B., Gong, Z., Yang, W.: Stock market index prediction using deep neural network ensemble. In: Proceedings of the 36th Chinese Control Conference, Dalian (2017)
11. Ma, J., Yu, M., Fong, S., Ono, K., Sage, E., Demchak, B., Sharan, R., Ideker, T.: Using deep learning to model the hierarchical structure and function of a cell. Nat. Methods **15**, 1–12 (2018)
12. Abiodun, O., Jantan, A., Omolara, A.E., Dada, K., Mohamed, N.A., Arshad, H.: State-of-the-art in artificial neural network applications: a survey. Heliyon **4**(e00938), 1–41 (2018)
13. Gavrishchaka, V., Yang, Z., Miao, R., Senyukova, O.: Advantages of hybrid deep learning frameworks in applications with limited data. Int. J. Mach. Learn. Comput. **8**(6), 549–558 (2018)
14. Prastyo, A., Junaedi, D., Sulistiyo, M.: Stock price forecasting using artificial neural network. In: Fifth International Conference on Information and Communication Technology (ICoICT) (2017)
15. Addo, P., Guegan, D., Hassani, B.: Credit risk analysis using machine and deep learning models. Documents de travail du Centre d'Economie de la Sorbonne 2018.03 (2018)
16. Fischer, T., Krauss, C.: Deep learning with long short-term memory networks. FAU Discussion Papers in Economics, No. 11/2017, Friedrich-Alexander-Universität Erlangen-Nürnberg, Institute for Economics, Erlangen (2017)
17. Abdou, H.: Prediction of financial strength ratings using machine learning and conventional techniques. Invest. Manag. Financ. Innov. **14**(4), 194–211 (2017)
18. Edet, S.: Recurrent neural networks in forecasting S&P 500 Index, 12 July 2017. SSRN: https://ssrn.com/abstract=3001046. (2017)
19. Unadkat, S., Ciocoiu, M., Medsker, L.: Introduction in Recurrent Neural Networks. Design and Applications. CRC Press, Boca Raton (2001)
20. Urban, G., Subrahmanya, N., Baldi, P.: Inner and outer recursive neural networks for chemoinformatics applications. J. Chem. Inf. Model. **58**, 1–13 (2018)
21. Cook, T., Hall, A.: Macroeconomic Indicator Forecasting with Deep Neural Networks (2017)

22. Siegelmann, H., Horne, B., Lee, C.: Computational capabilities of recurrent NARX neural networks. IEEE Trans. Syst. Man Cybern. **27**(2), 208–215 (1997)
23. Diaconescu, E.: The use of NARX neural networks to predict chaotic time series. Wseas Trans. Comput. Res. **3**(3), 182–191 (2008)
24. Abe, M., Nakayama, H.: Deep learning for forecasting stock returns in the cross-section (2017)
25. Moghaddama, A., Moghaddamb, M., Esfandyaric, M.: Stock market index prediction using the artificial neural network. J. Econ. Financ. Adm. Sci. **21**, 89–93 (2016)
26. Barnard, E., Wessels, L.: Extrapolation and interpolation in neural network classifiers. IEEE Control Syst. Mag. **12**(5), 50–53 (1992)
27. Haley, P., Soloway, D.: Extrapolation limitations of multilayer feedforward neural networks. In: IJCNN International Joint Conference on Neural Networks, Baltimore (1992)
28. Pektas, A., Cigizoglu, H.: Investigating the extrapolation performance of neural network models in suspended sediment data. Hydrol. Sci. J. **62**(10), 1694–1703 (2017)
29. Hettiarachchi, P., Hall, M., Minns, A.: The extrapolation of artificial neural networks for the modeling of rainfall-runoff relationships. J. Hydroinformatics **7**(4), 291–296 (2005)
30. MathWorks, Neural Network Toolbox™. User's Guide. R2014a, The MathWorks, Inc. (2014)

Online Action Recognition from Trajectory Occurrence Binary Patterns (ToBPs)

Gustavo Garzón and Fabio Martínez(✉)

Biomedical Imaging, Vision and Learning Laboratory (BIVL2ab), Universidad Industrial de Santander, Bucaramanga, Colombia
{gustavo.garzon, famarcar}@saber.uis.edu.co

Abstract. Online action recognition is nowadays a major challenge on computer vision due to uncontrolled scenarios, variability on dynamic action representations, unrestricted capture protocols among many other variations. This work introduces a very compact binary occurrence motion descriptor that allows to recognize actions on partial video-sequences. The proposed approach starts by computing a set of motion trajectories that represent the developed activity. On that regard, a local counting process is performed over bounded regions, and centered at each trajectory, to search for a minimal number of neighboring trajectories. This process is then codified in a vector of binary values (ToBPs) that will create a regional description, at any time of the video sequence, to represent actions. This regional description is obtained by determining the most recurrent binary descriptors in a particular video interval. The final regional descriptor is mapped to a machine learning algorithm to obtain a recognition. The proposed strategy was evaluated on three public datasets, achieving an average accuracy of 70% in tasks of action recognition by using a local descriptor of only 51 values and a regional descriptor of 400 values. This compact description constitute an ideal condition for real-time video applications. The proposed approach achieves a partial recognition above 70% on average accuracy using only the 40% of videos.

Keywords: Action recognition · Binary motion patterns · Occurrence patterns · Motion trajectories

1 Introduction

Human action recognition (AR) still holds as one of the most relevant and challenging tasks in computer vision, with potential applications such as video- surveillance, gesture recognition, video indexing, among much others [6]. In general terms, AR consists on automatically associating an action label, localized in space and time, to a particular video. Nevertheless, AR remains as an open research problem because several difficulties and challenges such as camera and background motion, variability in clothing and shapes, scale subject changes, among much others. Additionally, strong variations related to the particular dynamic that exhibit the object of interest, which perform the actions, could produce a wrong recognition.

M. Botto-Tobar et al. (Eds.): ICAETT 2019, AISC 1066, pp. 409–418, 2020.
https://doi.org/10.1007/978-3-030-32022-5_38

In the literature has been proposed several strategies ranged from global silhouettes characterization [2], to local patch representation that outperform problems related with occlusions and shape variability [7]. These approaches however remain dependent of appearance variability and limit the motion representation. To overcome such problems in the state-of-the-art has been reported some approaches that represent activities from kinematic primitives using optical flow representations and long motion trajectories [15]. Additionally, exhaustive learning methods and recurrent architectures to represent activities in videos have reported favorable accuracy but demanding large number of samples for training and the adjustment of hyper-parametric functions [12]. These approaches are robust to background appearance and motion variability, but a major limitation of such descriptors is that the spatio-temporal volumes are cut off to a fixed temporal length. Additionally, these strategies require a complete computation of spatio-temporal volumes to effectively compute the descriptor. Hence, a proper frame action representation could be achieved by describing local motion patterns at each frame, and modelling motion relationships into the codification.

The main contribution of this work is a very compact action descriptor that codes dense motion Trajectories as local Occurrence Binary Patterns (ToBPs). Firstly, a set of long trajectories are computed as motion primitives. Such primitives are naturally grouped on regional spatio-temporal patterns that serve to describe actions. Hence, a binary spatial point process is defined around of each trajectory to quantify the spatial distribution of neighboring trajectories. The local distribution of trajectories is measured by following a circular counting distribution scheme. Then, a bit-representation vector is defined on local counting process by a local occurrence threshold. The set of ToBPs is available at each frame along the sequence, which is therefore mapped to a previously trained dictionary. From such mapping is obtained a frame-occurrence histogram that represented the partial action that is developed during video streaming. This frame-level histogram is mapped, at each time, to a support vector machine to obtain a label action prediction. Each resulting ToBPs is very compact with an average size of 51. Also, the mid-level representation, from the mapping to the ToBPs dictionary, results on a dimension of 400 scalar values. This compact motion descriptor achieves partial recognition above 70% on average accuracy using only 40% of each video sequence.

2 Local Binary Patterns (LBP) on Action Recognition

One of the most interesting local descriptions for computer vision problems correspond to Local Binary Patterns (LBP) [10, 11]. The main idea behind LBP descriptors is to compute local differences around of a central point, and then generate a bit-vector quantification using a binary representation. If a neighborhood value is larger than central point then the bit is marked as one, otherwise the bit is fixed as zero. This compact representation allows a more stable description of local features to solve several problems, such as, texture modelling, pattern recognition, and even to model local dynamic patterns by extending the concept at consecutive frames [3].

Regarding action recognition, several strategies have exploited LBP coding to locally represent actions. For instance, Nanni et al. [8] extended the analysis of human shapes, over orthogonal video planes, by using a coding from local ternary patterns using gray scale differences. Nevertheless, this approach is sensible to occlusion and to a proper background computation. Yeffet et al. [17] proposed local trinary patterns by coding differences among consecutive frames during time. Then, in partial intervals of the video are computed occurrence histograms of the obtained local trinary patterns. This approach achieve interesting results but its computation is dense and requires rigid splitting of videos.

Nguyen et al. [9] also proposed to codify textures and motion using uniform LBP and Local Trinary Pattern (LTP) respectively, from neighboring image patches and a set of tracked points. Such technique, namely Spatial Motion Patterns (SMP), allowed to codify distinct velocities in local positions, but sequences with a significant amount of background motion might not perform well since only one spatial scale was considered for the analysis. The LBP scheme has also been extended on depth sequences [1, 16] for action classification and partial recognition. In such cases, the LBP descriptors are computed from Depth Motion Maps (DMMs) to complement the dynamic texture representation.

3 Proposed Method

Visual systems codify spatial distribution of key points that are moving in a coherent way alongside time to identify actions. Under this premise, in this work is presented an strategy that locally code a spatial occurrence process of key motion trajectories and then integrates them to represent activities. The pipeline of Fig. 1 illustrates the proposed strategy.

3.1 Motion Trajectories Representation

A set of key-motion trajectories was herein computed at each video to obtain a primary kinematic action representation. This work implements the salient motion trajectories proposed by Wang et al. [15], which densely follow a set of salient points with motion coherence along time. The tracking of such salient points underlies on Farneback dense optical flow field $\omega_t = (u_t, v_t)$. Then, each trajectory is obtained by concatenating points $(P_t, P_{t+1}, P_{t+2}, \ldots, P_{t+N})$ along time that follow velocity vectors from the dense optical flow. These trajectories are fixed during a number of N frames and smoothed according to a temporal median filter. Because video complexities, some trajectories could represent noisy or static performances, corresponding to sudden motion patterns and wrong detections. These are removed if motion magnitude change too fast and if there is no change in spatial position through time. Hence a primary kinematic representation of videos is given by a set of $\mathbf{P}$ trajectories, whose density is correlated with particular dynamic actions.

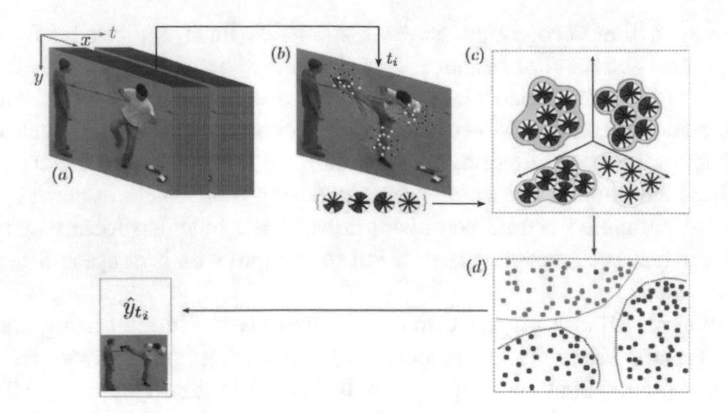

Fig. 1. Pipeline of the proposed method: (a) Trajectories are calculated for frame t_i. (b) A circular grid is defined to perform a LBP-based measurement resulting in a vector of motion density. (c) Such vector is used to create a regional dictionary representation (d) An activity classification process assign a label $\hat{y}_{ti}$ for each frame t_i of the video sequence.

3.2 Local Occurrence Binary Patterns (ToBPs)

The existence of spatial trajectory clusters imply significant kinematic patterns related with the developed activity. The proposed approach locally measured such patterns by considering regions with a fixed center around each trajectory. These atomic representations are modeled as a spatial point process composed of trajectories that fall inside a bounded local region at a particular frame t. That is, each active trajectory is identified by the spatial density of closely located trajectories, modeled as a random variable: $\mathbf{s} = \{s_1(\tau), s_2(\tau), \ldots, s_n(\tau)\}$.

In this work, the occurrences of neighboring trajectories are carried out into a bounded circular region. To achieve this, a counting process is defined around each trajectory with a layout of circles and angles. This creates sub-regions r_i bounded by γ concentric circles and α angular divisions. A compact and robust representation is carried out by computing occurrence motion patterns as a LBP- like operator (ToBPs). For doing so, a bit string vector is formed by considering binary values to represent local occurrences of neighboring trajectories if they comply with a given threshold. Our strategy considers the number of trajectories n_i passing through region r_i as a component of a dynamic texture distribution T such that: $T \approx t(m_\tau(n_1), m_\tau(n_2), \ldots, m_\tau(n_k))$ where, a function m_τ compares each value n_i with a Minimal Number of Trajectories (MNT) τ. The feature vector is then defined as $u := \sum_{1 \leq i \leq \gamma \cdot \alpha} 2^{i-1} m_\tau(n_i)$

Thus, we consider any spatio-temporal boundary $\mathbb{C}$ of an action sequence as a representation of a set of l-dimensional ToBPs points $\{u_1, u_2, \ldots, u_z\}$ with $u_i \in \mathbb{R}^l$. The boundary $\mathbb{C}$ could be considered as the total video sequence in classification tasks or as frame-level for online action recognition. These ToBPs locally represent actions by coding kinematic patterns occurring around motion trajectories. The bit-vector coding added robustness to occurrence disparities among motion patterns and achieved a compact description of actions (Fig. 2).

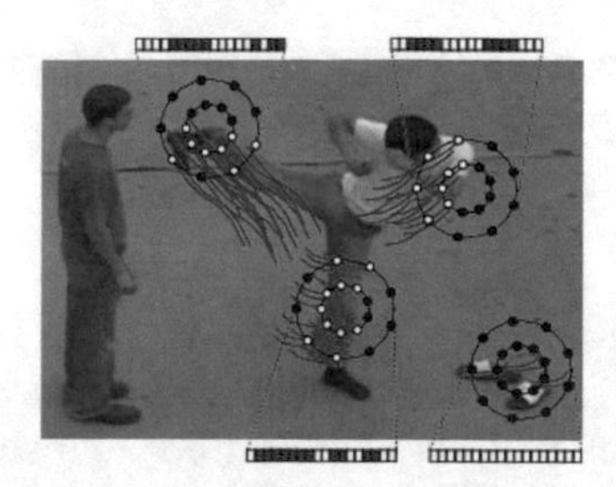
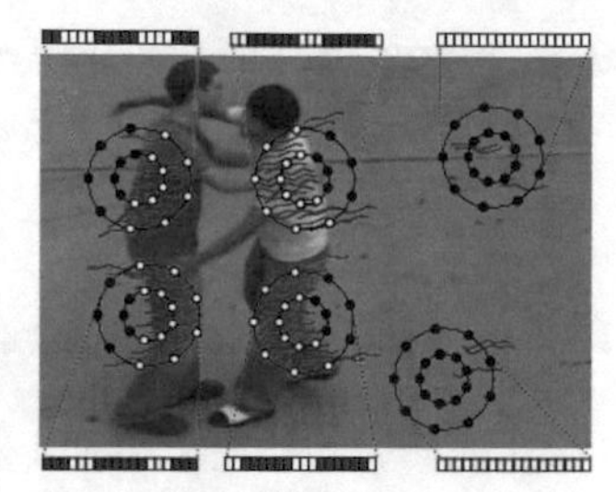

Fig. 2. Detail of our LBP-based scheme: bounded subregions r_i with a given motion density n_i are codified with a value of 1 (if $n_i \geq \tau$) or 0 otherwise. Subregions with not enough trajectories are excluded from the model.

3.3 Bag of Local Occurrence Binary Patterns (ToBPs)

A global descriptor of spatio-temporal regions is then attained by computing a global histogram representation of ToBPs patterns present on a specific temporal boundary interval C. In the first place, a dictionary of k representative ToBPs patterns is calculated from k-means function, computed inset of training videos. The set of k centroids $C = [c_1, c_2, \ldots c_k] \in \mathbf{R}^{l \times k}$ are recovered using an objective function: $C(k) = \min_{ck} = \sum_{m=1}^{M} \sum_{k=1}^{K} \left\| \mathbf{u}(\tau)_m - c_k \right\|_2^2$.

As soon as the dictionary is computed, a mid level action representation is generated by coding ToBPs as global histogram occurrences w.r.t. previously learned centroids. This representation can be obtained in any instant Δt of video sequence. For each instant, each obtained word $\mathbf{u_n}$ will contribute with the counted occurrence for centroid i using the minimal Euclidean distance s. Therefore, a normalized instance of the set of k centroid occurrences is considered the histogram representation of the current action in a given instant. This frame representation allows to describe partial sequences and it exhibits robustness to occlusion problems during capture. Compared with the previous step, this analysis is not directly associated to the neighboring trajectories over each region, which allows a certain scale invariance in the description of motion patterns. During the experimental stage, redundant information was observed and it added a convenient complexity to the model.

3.4 Action Classification and Recognition

At this point, the proposed motion descriptor is used as input for different machine learning methods in order to obtain action labels for any frame t_i. Our aim was to attain a favorable balance between accuracy and prediction time, thus, a random forest and kernel-based methods were herein implemented and evaluated. Particularly, Random Forest classification (RF) reaches favorable results by grouping decision tree classifiers (DT) and make inferences using a voting approach. Alternatively, a multi-class adaptation of Support Vector Machines (SVM) was evaluated using linear and Radial Basis Function (RBF) kernels [4]. While a linear kernel reported a fast performance on the prediction task, the RBF kernel dealt with non-linear class boundaries. In both kernel-based methods, the approach randomly selected, from training, support vectors that allowed to maximize distance between classes and defined hyper-parametric boundaries.

3.5 Data

To validate the proposed strategy, an evaluation was performed over 3 academic sets of labeled video sequences. The KTH dataset: [14] with six human actions executed by 25 distinct subjects. For validation, indications of KTH authors were followed: training (760 sequences), testing (863 sequences) and validation (768 sequences). The Weizmann dataset [5] has 10 actions distributed over 90 video sequences with an evaluation using a leave-one-out strategy. Finally, the UT-Interaction dataset [13] features surveillance sequences that were captured in uncontrolled scenarios with two sets of 60 video sequences, containing 6 actions. A 10-fold leave-one-out method was performed, following the dataset's authors indication.

4 Experimental Results

The proposed method generated a descriptor that was assessed two-fold: employing full video sequences in order to obtain a prediction, and using consecutive fragments of a sequence to obtain a cumulative prediction for online applications.

4.1 Full Sequence Action Labeling

Firstly, we evaluated the proposed approach to obtain an optimal value for the Minimal Number of Trajectories (MNT) threshold τ. Our circular grid was preconfigured with $\alpha = 9$ angles and MNT was set as $\tau = \{2, 3, 4, 5, 6, 7, 8\}$, increasing the amount of trajectories n_i admitted inside each region r_i around an averaged center, over all datasets.

In Fig. 3 is reported the performance of the proposed approach using different MNT (τ) values over all datasets. For the UT sequence 1 the maximum score of 75% was achieved with ($\gamma = 8, r = 8$), $\tau = 6$, as for sequence 2, the maximum score of 70% was obtained with ($\gamma = 4, r = 6$), $\tau = 2$. Sequence 2, which shows camera jitters and human interactions, require a smaller density of trajectories and a smaller reach ($\gamma = 4$) w.r.t. sequence 1 ($\gamma = 8$). Also, these features are explained because of the dynamic closeness of activities such as punching and pushing in sequence 1, and on the opposite, due to the uncontrolled scenarios of sequence 2.

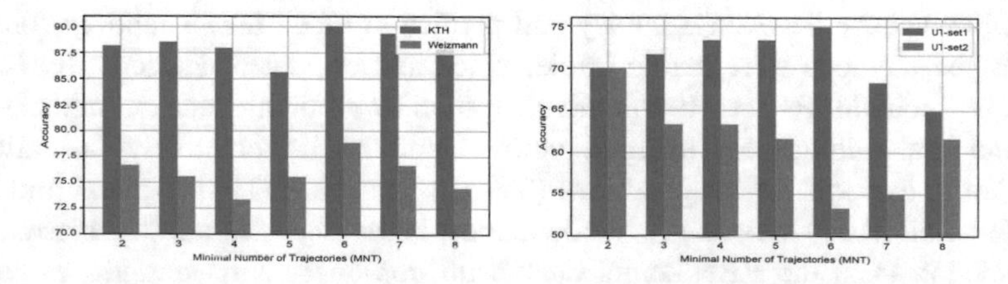

Fig. 3. Reported accuracy for different values of MNT occurring on each cell of the proposed circular grid over (left) KTH and Weizmann datasets, and (right) UT-Interaction dataset.

On the other hand, results for KTH are shown in Fig. 3 (left) featuring values of MNT from 2 to 8, achieving an accuracy of 90.03% for ($\gamma = 8, r = 6$) and $\tau = 6$. This is explained by the periodic featured actions and the constant spatial cropping of the sequences. Regarding Weizmann dataset, the best score (78.88%) was achieved with $\tau = 6$ and ($\gamma = 5$, $r = 10$). This dataset groups most action classes and therefore the classification is more challenging. Larger radius sizes have better performance, since the spatial frame resolution is much more larger than for KTH. In general, the ToBPs coding shows stable results for different configurations, with a mode above of 70% for very different and challenging datasets.

Secondly, we analyze the impact of trajectories length to describe actions. The experimental setup involved to increase the number of frames for trajectory tracking from 1 to 49 frames and a recognition using a SVM with a RBF kernel. Figure 4 (right) shows a crescent shape with a local maximum on l = 37 for UT- Interaction dataset, but resulted unstable for neighboring configurations (Fig. 6 green line). Best performance is then found at $l = 22$ because a favorable trade-off between descriptor size ($\alpha \times \gamma \times 22$) and online action recognition behaviour (Fig. 6 blue line). Sequence 2 obtained a local maxima on l = 49 but it implied a significant increase in descriptor size ($\alpha \times \gamma \times 49$) for obtaining only an extra 1%. Thus, the default length of l = 15 yields a sufficient performance, as shown in figure (Fig. 6 red line). As for Weizmann dataset, we used the same experimental configuration and obtained local maxima on l = 15 and l = 37 as shown in Fig. 4 (left). Therefore l = 15 yields a favorable trade-off with a descriptor of only ($\alpha \times \gamma \times 15$). Regarding KTH dataset, best performance was achieved with only l = 15 which yields a compact descriptor size of ($\alpha \times \gamma \times 15$).

In a third experiment, three classification strategies were evaluated over the best MNT configurations of ToBPs. These classification strategies adequately dealt with non linear spaces, and shown a reasonable trade-off in terms of computational time. Namely, we used: Random Forest (RaF), SVM (with linear kernel) and SVM (with RBF kernel). These methods were evaluated under a grid search parameter setup to obtain optimal configurations w.r.t to the proposed action descriptor.

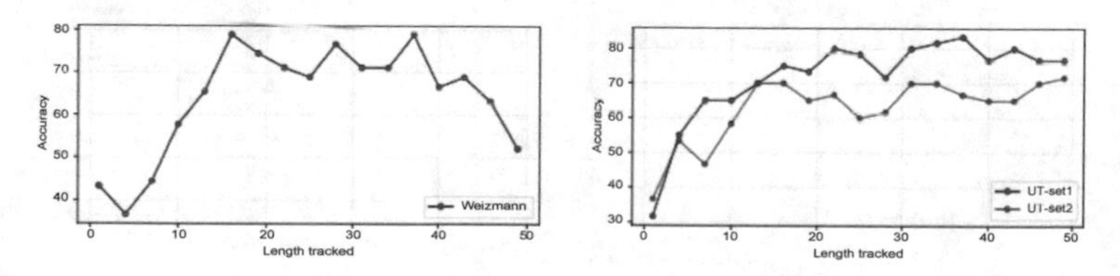

Fig. 4. Reported accuracy for different lengths of tracking over (left) Weizmann and (right) UT-Interaction datasets.

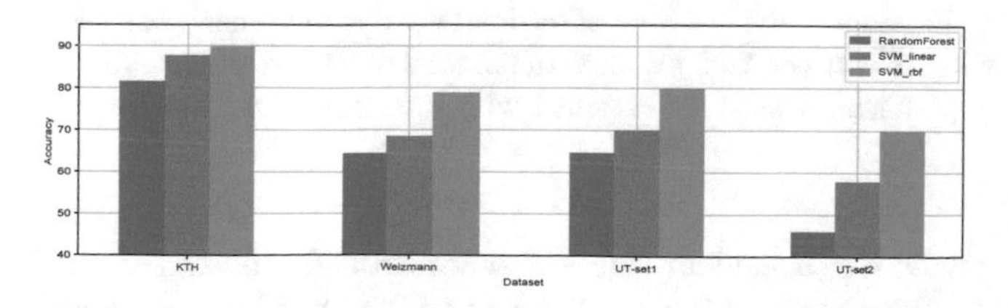

Fig. 5. Classification strategies for KTH, Weizmann and UT-Interaction datasets. Best overall scores are shown.

As shown in Fig. 5, the SVM + RBF kernel obtained the best performance in terms of accuracy and low computational complexity which is adequate for real-time classification tasks. We emphasize that the proposed descriptor achieves a favorable performance with only 400 scalar values for full video sequences, compared with state-of-the-art strategies that in general require thousands of scalar values. Moreover, our descriptor obtains a balanced ratio between accuracy and dimensionality, which makes it adequate for online classification tasks.

4.2 Action Recognition from Partial Sequences

Online action recognition requires an updated prediction for each frame t_i of the video sequence and then a evaluation was executed over partial sequences on a frame level representation. This strategy aims to verify the performance of the proposed descriptor when the dynamics of motion are incomplete. In order to calculate statistics for partial recognition, our method was proved using distinct temporal percentages of the sequences, and for all the datasets an averaged score was reported.

Online recognition performance for all datasets is shown in Fig. 6. Regarding KTH dataset, a competitive prediction is obtained with approximately 40% of the sequence for an accuracy of more than 74%. The periodic behaviour of the actions explain the sustained increase in the prediction score when more frames are employed for the prediction task. On the other hand, red dashed line illustrates online recognition accuracy for Weizmann dataset, where a promising score if acquired with only 20% of the total number of frames. Also, after 40% of the sequence the score shows stability and a moderate increase at the end of the sequence. Existent accuracy fluctuations are associated with a distinct duration between actions of the same class and the different scale cropping of the sequences. It is worth noting that a quick prediction is obtained with just 10% of the sequences.

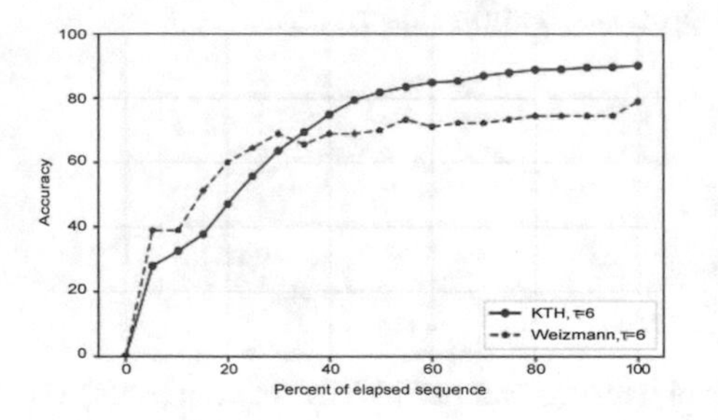

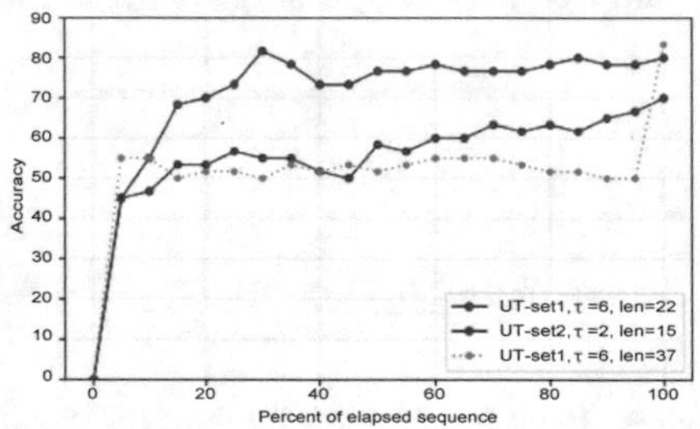

Fig. 6. Performance simulation for an online application over (left) KTH dataset: our method achieves promising results with just 30% of the total number of frames. (Left) Weizmann dataset: promising results are obtained with just 20% of the total number of frames. As for UT-Interaction dataset (right), promising results are obtained with just 15% of the total number of frames for sequence 1.

Lastly, predictive behaviour for UT-Interaction is illustrated in Fig. 6 (right) including set 1 (blue) and set 2 (red). Set 1 obtains promising results with just 15% of the elapsed sequences and sustained an increasing score with small variations. As for set 2, a reduced performance is explained by the noise introduced by additional featured subjects and background motion, in addition to changes in the spatial cropping of the sequences. Particularly, reported results correspond to a length of tracking $l = 22$ for set 1, $l = 15$ for set 2 and an extra test with $l = 37$ (green dotted) for set 1, which shown a final increase in prediction score for the entire sequences, but an inadequate performance for online recognition.

5 Conclusions

The proposed strategy features a compressed descriptor that is able to distinguish human actions on a frame-level by employing statistics generated from densities of active neighboring trajectories. Such ToBPs setup considered inside bounded regions whose occurrences were codified as sequences of binary values. The proposed ToBPs coding used in average only 51 scalar values, and 400 features for mid-level change of representation. Aforesaid characteristics are convenient for online applications that run over restricted hardware configurations and devices with low-cost specifications. The proposed approach showed competitive results for three different academic and public datasets, being robust to occlusion and above all the capability to recognize actions on continuous video streaming. Additional testing will be implemented with different features, as well as exhaustive validations over different datasets and new hardware architectures.

Acknowledgements. This work was partially funded by the Universidad Industrial de Santander. The authors acknowledge the Decanato de la Facultad de Ingenierías Fisicomecánicas and the Vicerrectoría de Investigación y Extensión (VIE) of the Universidad Industrial de Santander for supporting this research registered by the project: *Reconocimiento continuo de expresiones cortas del lenguaje de señas*, with SIVIE code 2430.

References

1. Al-Akam, R., Al-Darraji, S., Paulus, D.: Human action recognition from RGBD videos based on retina model and local binary pattern features. In: 26 Conference on Computer Graphics, Visualization and Computer Vision (WSCG), pp. 1–7 (2018)
2. Al-Ali, S., Milanova, M., Al-Rizzo, H., Fox, V.L.: Human action recognition: contour-based and silhouette-based approaches. In: Computer Vision in Control Systems-2, pp. 11–47. Springer, Heidelberg (2015)
3. Bouwmans, T., Silva, C., Marghes, C., Zitouni, M.S., Bhaskar, H., Frelicot, C.: On the role and the importance of features for background modeling and foreground detection. Comput. Sci. Rev. **28**, 26–91 (2018)
4. Chang, C.C., Lin, C.J.: Libsvm: a library for support vector machines. ACM Trans. Intell. Syst. Technol. (TIST) **2**(3), 27 (2011)
5. Gorelick, L., Blank, M., Shechtman, E., Irani, M., Basri, R.: Actions as space-time shapes. IEEE Trans. Pattern Anal. Mach. Intell. **29**(12), 2247–2253 (2007)

6. Herath, S., Harandi, M., Porikli, F.: Going deeper into action recognition: a survey. Image Vis. Comput. **60**, 4–21 (2017)
7. Laptev, I., Lindeberg, T.: Local descriptors for spatio-temporal recognition. Lect. Notes Comput. Sci. **3667**, 91–103 (2006)
8. Nanni, L., Brahnam, S., Lumini, A.: Local ternary patterns from three orthogonal planes for human action classification. Expert Syst. Appl. **38**(5), 5125–5128 (2011)
9. Nguyen, T.P., Manzanera, A., Vu, N.S., Garrigues, M.: Revisiting LBP-based texture models for human action recognition. In: Iberoamerican Congress on Pattern Recognition, pp. 286–293. Springer, Heidelberg (2013)
10. Ojala, T., Pietikäinen, M., Harwood, D.: A comparative study of texture measures with classification based on featured distributions. Pattern Recogn. **29**(1), 51–59 (1996)
11. Ojala, T., Pietikainen, M., Maenpaa, T.: Multiresolution gray-scale and rotation invariant texture classification with local binary patterns. IEEE Trans. Pattern Anal. Mach. Intell. **24** (7), 971–987 (2002)
12. Rahmani, H., Mian, A., Shah, M.: Learning a deep model for human action recognition from novel viewpoints. IEEE Trans. Pattern Anal. Mach. Intell. **40**(3), 667–681 (2018)
13. Ryoo, M.S., Aggarwal, J.K.: Spatio-temporal relationship match: video structure comparison for recognition of complex human activities. In: 2009 IEEE 12th international conference on Computer vision, pp. 1593–1600. IEEE (2009)
14. Schuldt, C., Laptev, I., Caputo, B.: Recognizing human actions: a local SVM approach. In: Proceedings of the 17th International Conference on Pattern Recognition 2004, ICPR 2004, vol. 3, pp. 32–36. IEEE (2004)
15. Wang, H., Kläser, A., Schmid, C., Liu, C.L.: Dense trajectories and motion boundary descriptors for action recognition. Int. J. Comput. Vis. **103**(1), 60–79 (2013)
16. Yang, Y., Zhang, B., Yang, L., Chen, C., Yang, W.: Action recognition using completed local binary patterns and multiple-class boosting classifier. In: 2015 3rd IAPR Asian Conference on Pattern Recognition (ACPR), pp. 336–340. IEEE (2015)
17. Yeffet, L., Wolf, L.: Local trinary patterns for human action recognition. In: 2009 IEEE 12th International Conference on Computer Vision, pp. 492–497. IEEE (2009)

Application of Data Mining and Data Visualization in Strategic Management Data at Israel Technological University of Ecuador

Paul Francisco Baldeon Egas(✉), Miguel Alfredo Gaibor Saltos, and Renato Toasa

Universidad Tecnológica Israel, Quito, Ecuador
{pbaldeon, canciller, rtoasa}@uisrael.edu.ec

Abstract. Currently, data analysis in higher education institutions is not a luxury, it is a necessity. The large amounts of data generated through university academic functions are the main reason for an analysis and representation of these; since they will allow an adequate decision making in the university academic processes. In this work we propose to perform an analysis of the data generated in the Israel Technological University from Ecuador in the period 2012–2018; for this we apply Data mining algorithms to make suitable predictions and by using data visualization techniques to represent this information allowing us to easily understand it; as a result, relevant information is obtained that will allow the personnel in charge to make the appropriate decisions and improve the processes that have low percentages.

Keywords: Data mining · Visualization · Strategic management · Higher education

1 Introduction

The data is on all sides, in essence, everything we do or say can be understood as the starting point to generate a data; and these data must be interpreted in an adequate way to achieve a better decision making in any field, currently, for correct data management, new and innovative disciplines are used, such as data mining and data visualization, which are two disciplines that allow the interpretation of data and improve the decision-making process in any area in which they are applied [1]. Data analysis and data visualization are very important tools for engineers, analysts, policy makers and decision makers. Originally developed for "small data", these techniques have had mixed success in recent centuries and decades [2].

Data, information and knowledge are three widely used terms, often in an interrelated context. In many cases, they are used to indicate different levels of abstraction, understanding or truthfulness. Information is data that has been given meaning through a relational connection, while knowledge is the appropriate collection of information, so that its intention is useful [3]. Knowledge is a deterministic process that must be handled correctly, to achieve this science of data must provide reliable and verified knowledge so that the visualization of data can represent them through clear graphics

M. Botto-Tobar et al. (Eds.): ICAETT 2019, AISC 1066, pp. 419–431, 2020.
https://doi.org/10.1007/978-3-030-32022-5_39

and allows the correct decision making in any type of institution. Data mining more than just understands the numbers used for the analysis; the data supports evidence-based practice and complements the data visualization.

Companies in all technological areas haven't realized that they need to hire data scientists to project themselves in the future and improve their processes, currently higher education institutions use data mining and data visualization to generate indicators that will be of support in the academic processes that are part of the Strategic Planning, which is a participatory, systematic, critical and self-critical, comprehensive and reflective process that allows formulating objectives and strategies in different time horizons, responds to the demands of the environment and the institution itself, and whose results they require monitoring and evaluation [4].

One of the reasons why Higher Education Institutions are focusing on the importance of data is that data mining and data visualization is closely related to other important concepts, such as data-driven decision making and big data, they are also growing in importance and attention.

In this research work, we propose to conduct a study of data mining and data visualization focused on the data of Strategic Planning at Israel Technological University from Ecuador, to determine the impact that the use of these disciplines have on the success of the daily university activities, for this we perform an analysis of academic data such as enrollments, schedules, students of Israel Technological University.

The rest of the document is organized as follows: the next section provides a description of the Literature Review. In Sect. 3 we describe in depth the case study for this research, in the next Sect. 4 we present the development of Data Mining for academic prediction, in Sect. 5 we mentioned the adequate visualization techniques for data mining. The tests and results are presented in Sect. 6. Finally, the conclusions are presented in Sect. 7.

2 Literature Review

Firstly, it is necessary to carry out a review of the literature of work related to the topic proposed in this research. For this we consider the following steps [5], (see Fig. 1).

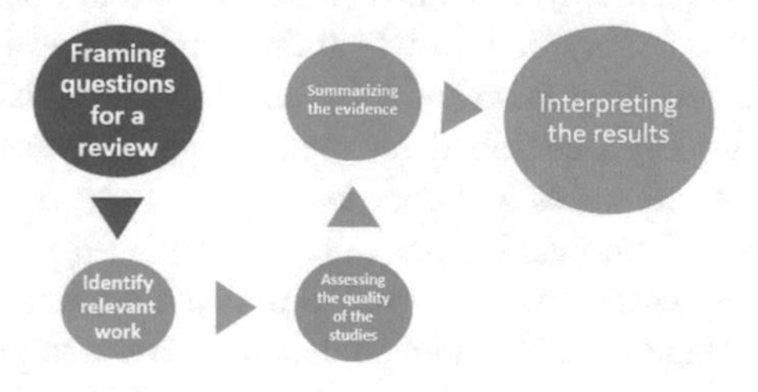

Fig. 1. Steps to literature review [5].

2.1 Framing Questions for a Review

In order to start with the current research, questions were posed that allowed us to identify related works: Who has already researched the Data Mining and Visualization of data in university processes? What has been done? What are the tools used?

2.2 Identify Relevant Work

To identify which works are relevant in the scientific world, criteria were used as the language of the research, the year of publication, the database in which the article is indexed, in this way to ensure that the articles found are of great help for the development of current research.

2.3 Assessing the Quality of the Studies

The quality of the works to which reference must be made is an important factor for the impact that this research will have. To determine the quality of the works, the number of citations that he had was verified and if he is responsible for initiating new related researches.

2.4 Summarizing the Evidence

Once several relevant works were found, a detailed review of each work is carried out, writing summaries with the main ideas, identifying which part could be useful for the current research.

2.5 Interpret the Results

The interpretation of results began with an analysis of the summary of the articles or works found, identifying those that can be referenced in the current research, see Table 1.

Document	Key features
Information Visualization and Visual Data Mining [6]	Information Visualization Visual Data Mining Visual Data Exploration
Data Science, Predictive Analytics, and Big Data: A Revolution That Will Transform Supply Chain Design and Management [7]	Data science Predictive analytics Big data Logistics Education

(continued)

(*continued*)

Document	Key features
Minería de Datos Educativa, Patrones de Comportamiento, Tecnologías de la Información y Comunicación, Agentes Inteligentes, Granularidad Educativa [8]	Minería de Datos Educativa Patrones de Comportamiento, Agentes Inteligentes Granularidad Educativa
Big Data for Education: Data Mining, Data Analytics, and Web Dashboards [9]	Data Mining Data Analytics Web Application Dashboard

The documents found were selected for their relevance and importance for the purposes of this research, several of these works focus on the importance of data science and data visualization, first in [10], develop a model based on data science in a doctoral program in nursing, this model contains domain, ethics, theory, technique, analysis and dissemination, with concepts compiled for use in the curriculum of scientific nursing and It provides important information in the decision making that is given to define the contents of the curriculum.

In the work [7], they use data science and data mining for jobs related to the supply chain, based on this they examine possible applications and skills to train personnel and develop new supply chain leaders. In [8] common trends in the analysis of chemical analytical data are presented, the common terms of data science are summarized and discussed, then systematic methodologies are presented for planning and executing big data analysis projects, finally, it is provided a general description of standard and new data analysis methods.

In relation to data visualization, at work [11] and describe that beauty in this context has four components; In addition to being aesthetically pleasing, it must also be novel, informative and efficient.

In the context of data science and data mining, additional restrictions arise, since the amount of data to be plotted is very large. Therefore, to be able to transmit meaningful messages, it is necessary to resort to pre-processing techniques, in order to extract a significant structure of the data [12]. In this way, data mining, visualization techniques are inherently related to data analysis; and this allows optimal decision making by organizations of any kind.

As it can be evidenced as described above, in large scientific databases there are research works that carry out similar studies but that do not focus on the University Strategic Management, which allows us to determine that current research will be useful throughout the scientific and academic community of the world.

3 Case Study

When dealing with an investigation on the application of Data Mining and Data Visualization of strategic management in Higher Education Institutions, it is important to systematically develop the different processes to achieve the proposed objective.

Therefore, as a case study, we have a predictive analysis of the academic information held in the Integrated System of Strategic Management (SIGE) of the Technological University of Israel (UISRAEL). This study has main information of grades, subjects, levels, academic periods and age.

For the development we worked with the technique of knowledge extraction based on the phases and stages of Data Mining (see Fig. 2) [13].

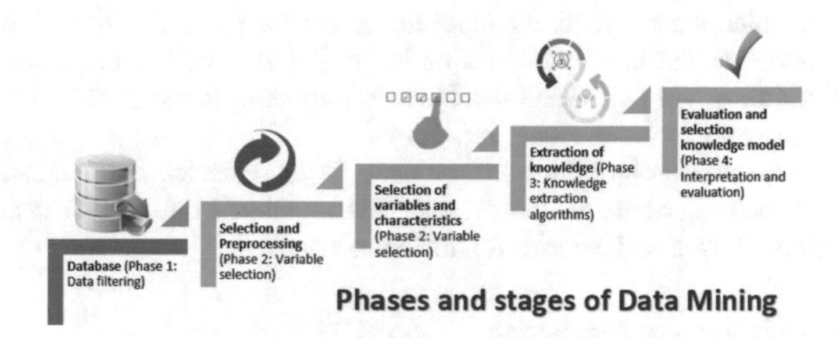

Fig. 2. Phases and stages of data mining

The four phases that were developed for predictive analysis using Data Mining are the data filtering, the Variable selection, Knowledge extraction algorithms, and finally interpretation and evaluation.

3.1 Data Filtering

For the realization of this phase, it was obtained from the source of data that in the present case study is in the DBMS SQL Server, where the academic information of the SIGE of the UISRAEL is housed.

In this phase, data processing was developed where the number of incorrect, invalid data and unknown data were reduced according to the need of the practical case. This processing was developed by creating a view, where it contains the necessary information.

Once the first phase is completed, it is necessary to have the necessary tools for the development of Data Mining itself, in which, after an evaluation, it was decided to work with SQL Server Data Tools (SSDT).

Through the tool, the data mining process was carried out through the filters of the three applied algorithms, which were through expansion, index of histogram, second plane level and determining the level.

3.2 Variable Selection

In this phase the SSDT tool was used, in which by creating a Business Intelligence project, we proceed to create the data source to SQL Server and the data source view, where you get the data from the view that was created in the data filtering phase in SQL Server. Once the origin and its data view is created, the selection and preprocessing of data is carried out, where the choice of influential characteristics and variables of the problem was established as a methodological strategy, through the analysis of better attributes. The variables determined for the present study can be visualized in Fig. 6.

3.3 Knowledge Extraction Algorithms

For the extraction of knowledge it is first necessary to create the different Data Mining structures, in which the key field, the input fields and the prediction fields are selected. The input fields are the input variables or inputs for the prediction process that the different algorithms will use. Additionally, it is important to establish the percentage and number of cases in the test data set.

To determine which algorithms will be used, it was selected in the creation of the Data Mining models, where in the present investigation Microsoft Decision Trees, Microsoft Naive Bayes and Microsoft Clusters were used.

3.4 Interpretation and Evaluation

For the interpretation of the data obtained through the different models of Data Mining, it is important to verify if the results are coherent, to validate and to verify the conclusions that the different established models show. To evaluate and determine which is the best model that fits the case study, it must be done using the SSDT Data Mining precision graph tool, where the three models that were used are selected and the elevation graph is generated, determining which is the most suitable model.

4 Develop for Academic Prediction SIGE UISRAEL

The academic prediction SIGE is an analysis through Data Mining, which shows the academic performance of students in the span of 7 years from 2012 to 2018. This application is based on the information received by the SIGE, which is developed through proprietary software platforms of Microsoft and DevExpress companies, which have a worldwide ranking for the versatility, effectiveness and safety of their products.

SIGE is composed of 23 modules, with a total of 41 sub-modules, included among the substantive and support functions, of which the information of the Academic Management Module (AMM) will be used in the present research.

The SIGE architecture is generated from the following platforms in Fig. 3:

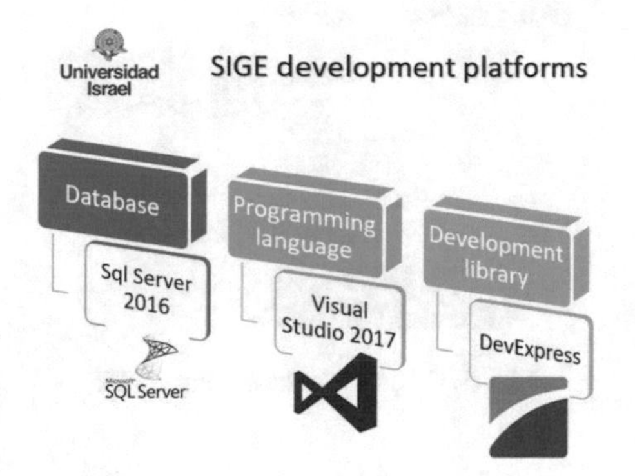

Fig. 3. SIGE development platforms

The SSDT software was used for the academic prediction analysis based on Data Mining, which allowed the development of the stages and phases without any problem.

In phase 1 (see Fig. 4), the data filtering was carried out from the base of the SIGE that is in SQL Server, and taking as an input for the next phase the view with the academic information needed for the case study.

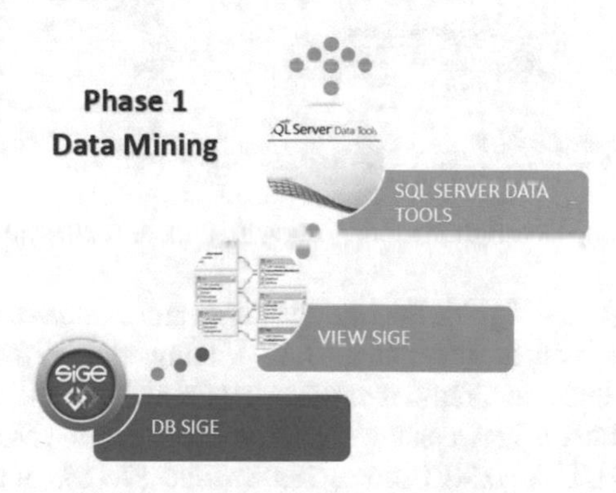

Fig. 4. Phase 1 data mining SIGE

In phase 2 (see Fig. 5), the variables were selected using the SSDT tool, which is based on the creation of a Business Intelligence project, following the following steps:

Fig. 5. Phase 2 data mining SSDT

In phase 3 (see Figs. 6 and 7), the knowledge extraction algorithm is selected, in which the present investigation was carried out with the models of Decision Trees of Microsoft, Microsoft Naive Bayes and Microsoft Clusters.

Phase 3 Data Mining – Algorithm models

Árboles de decisión de Microsoft Bayes naive de Microsoft Clústeres de Microsoft

Estructura	V MD NOTAS UISRAEL	MD_NOTAS_UISRAEL_CLUSTERES	MD_NOTAS_UISRAEL_BAYES
	Microsoft_Decision_Trees	Microsoft_Clustering	Microsoft_Naive_Bayes
ASI DESCRIPCION	Input	Input	Input
CARRERA	Input	Input	Input
EDAD	Input	Input	Input
ESTADO	PredictOnly	PredictOnly	PredictOnly
GEN DESCRIPCION	Input	Input	Input
MOD DESCRIPCION	Input	Input	Input
NIV DESCRIPCION	Input	Input	Input
PER CODIGO	Key	Key	Key
PRA ANIO LECTIVO	Input	Input	Input
PRA CODIGO	Input	Input	Input

Fig. 6. Phase 3 data mining algorithm models and variables

Finally, in phase 4 (see Fig. 7), the interpretation and evaluation of the data obtained through the different models of applied Data Mining algorithms was performed. In which, through the Elevation Graph, it can be determined that the best algorithm model for the present case study is between the Decision Trees and the Microsoft Naive Bayes, which present the 0.91/1 and 0.90/1 score, 58.08% and 57.18% of the target population, and finally 76.82% and 80.67% prediction probability respectively.

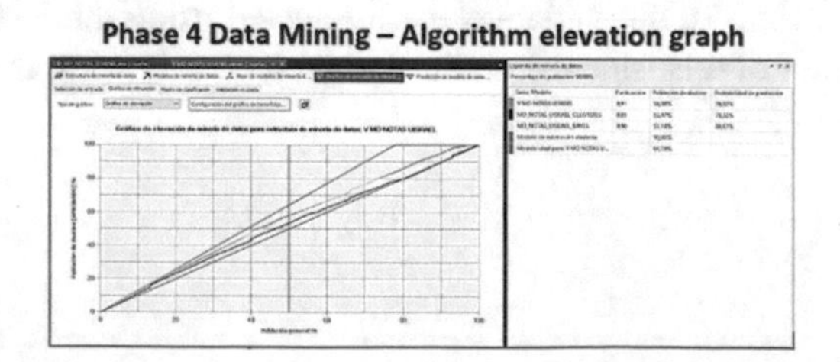

Fig. 7. Phase 4 data mining – algorithm elevation graph

5 Data Mining Visualization

The information shown in the previous section is very important to carry out the decision-making process, but only experts in the area can interpret this information; To avoid this, and to achieve that anyone can understand this data, it is necessary to represent the information by means of appropriate graphs, for this the data visualization techniques are used.

Currently, novel data visualization techniques have been developed, which allow the visualization of multidimensional data sets; Like those generated in Data Mining, these techniques can be classified according to three criteria: the data to be visualized, the visualization technique and the interaction and distortion technique used [14].

The next table provides some of the techniques that have proven to be of great value in the analysis of exploratory data, and also have a high potential to represent large databases.

Table 1. Data visualization techniques for data mining.

Technique	Description	Graphic
Autocharting	Autocharting produces the best visual based on what data you drag and drop onto the palette. It is important to note that auto charting may not always create the exact visualization the user has in mind [15].	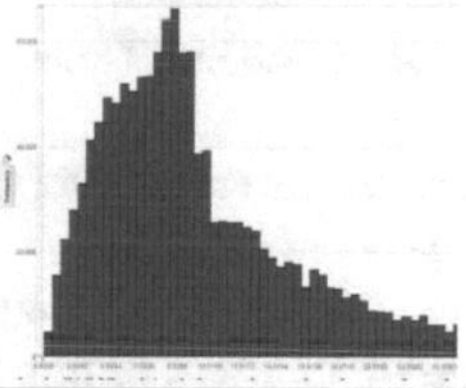
Correlation -Matrix	Shows a table which lists the correlation coefficients of the columns and rows and combines big data and fast response times to quickly identify which variables are related [15].	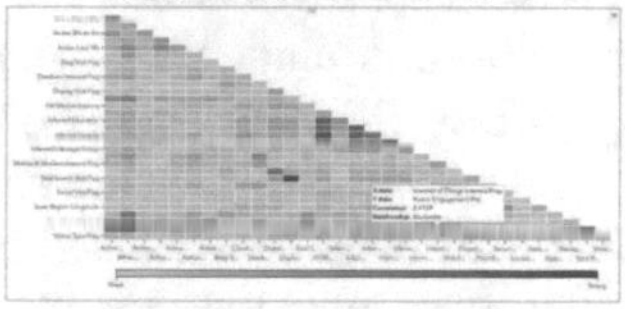
Network Diagram	Provides relationships in terms of nodes (representing individual actors within the network) and ties (which relationships between the individuals) [16].	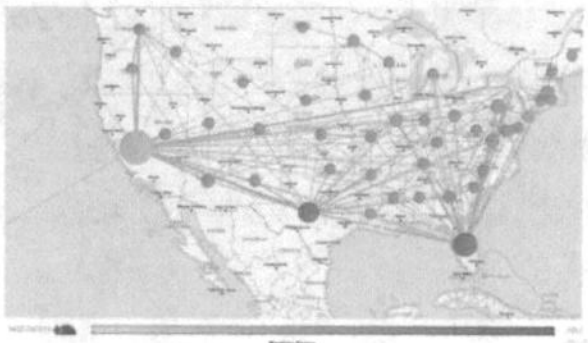
Sankey Diagrams	These diagrams use path analysis to show the dynamics of how transactions move [17].	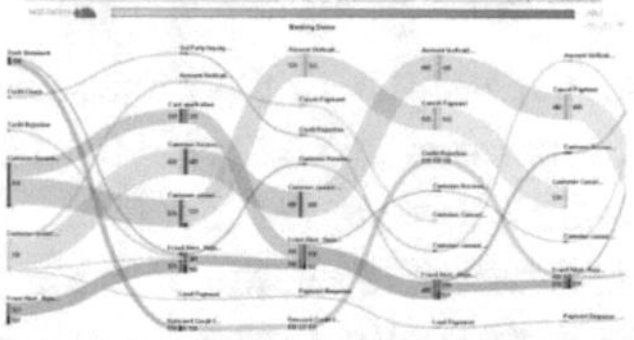

The use of any technique mentioned facilitates the understanding of the data and therefore the decision making.

6 Test and Results

From the Data Mining algorithms applied to the case study, it was possible to obtain significant results on the academic prediction of the possibility of the students approving the assignments in the different levels. From this scene we worked with a net view without discrimination of 99,193 records, giving an approximate of 1'388.702 data, in the lapse of 7 years (2012–2018). The variables that were used can be visualized in phase 3.

According to the results obtained from the case study, it is foreseen in the realization of ontologies for the representation of knowledge based on the semantic web of educational systems.

When executing the different phases of Data Mining with its three algorithms, applying a percentage of 30% for tests, with a maximum number of cases of 90,000, it was possible to carry out the predictive analysis of 4,361 real cases:

6.1 Microsoft Decision Trees Algorithm

- In their degree careers, generally as a University there is a 78.04% probability that students pass the subjects.
- According to the different input variables proposed for the analysis of the prediction that a student approves or not, is directly linked to the level. Therefore, according to the projection, in ninth level there is the highest degree of probability to pass the level, reaching 94.59%. On the other hand, it can be predicted that the second and third levels of all the races are the most complicated, where there is a probability that they will retest 60.82% and 50% respectively.
- There are some input variables that intervene in the prediction of a student approve a level. First, sixth, seventh, eighth and tenth have incidences of these variables.
- The modality affects the first level, in which probability they approve is 60.21%. In addition to the rest of the study modalities, there is a probability of 75.81%, having as another age incidence variable, where people who are currently 39 years old have a probability of 99.93% to pass, and those who are different in age 39 years have 75.54%.
- As antecedent in sixth level has as variable of incidence to the Academic period, where it is validated that the second semester of 2018 94.91% approved, and in the rest of periods from 2012 to the first semester of 2018, only 64.55% has the probability to pass the level.
- For the seventh level, the variable with the highest incidence is the academic period, where it generally exceeds the 80% probability of approve.
- The variable with the highest incidence in the eighth level is the career, where Business Administration prevails with 96.74% probability that students approve. In

the rest of the races, there is a probability of 12.22% that the students will reject the level.

- Finally in the tenth level prevails with 21 cases the entry variable subject, in which in the period proposed the subject of Fundamentals of Audiovisual Production, has the 99.99% probability of approved. In the rest of the subjects in this level, there is a probability of 15.52% that fail.

6.2 Microsoft Bayes Naive Algorithm

In general, it can be shown that the variable with the highest incidence in the prediction of a student's passed or failed status is the level at which it is enrolled (See Fig. 8).

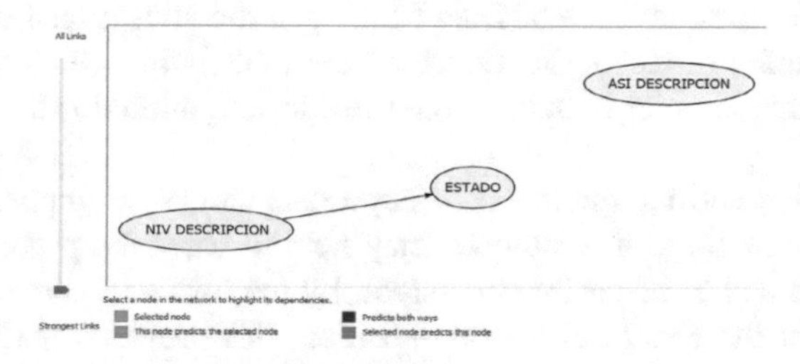

Fig. 8. Results Bayes Naive algorithm

6.3 Microsoft Clusters Algorithm

Based on this algorithm, clusters 7 and 9 prevail in all the variables, with which it can be predicted that in the coming years there will be an increase in the blended modality, where there is a high level of probability that the students approve the levels, and that the majority will be male (See Fig. 9).

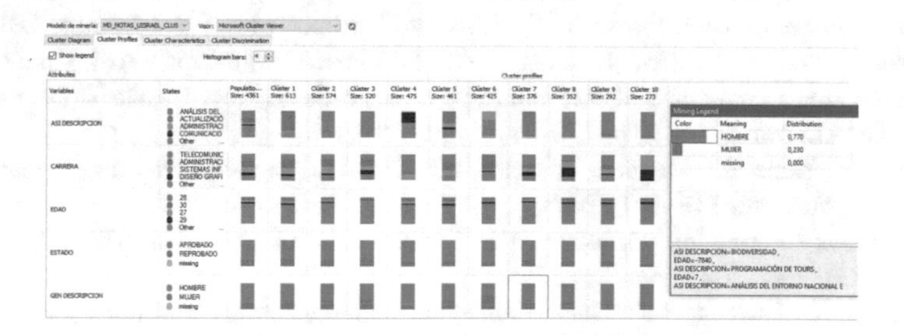

Fig. 9. Results clusters algorithm

7 Conclusions

A fundamental pillar for making correct decisions is the application of different techniques using Data Mining algorithms and data visualization, since it allows to have a large scale of hidden information patterns that the Computer Systems possess. Specifically, in the UISRAEL the results obtained were decisive in the decision making, the analysis was in the levels of the different careers that the students approve without inconvenience and the levels where the students have problems in approving, are the two extremes of the analysis.

Today in the world, Artificial Intelligence, Business Intelligence and Data Analytics are predominant, where Data Mining has become a powerful tool for Higher Education institutions to obtain efficiency in their value chain.

With the significant advance of Data Mining in the educational field, it is evident that it will continue to innovate its development tools, with which they will become fully parametrizable in order to obtain better results and minimize the risk of making a bad decision.

The exploration and interpretation of large data sets is an important problem; but the Information visualization techniques may help to solve the problem, allowing the information presented to be easily understood by the user and allow him to make the right decisions for the benefit of his organization, this provided that the correct visualization techniques are used.

References

1. Provost, F., Fawcett, T.: Data science and its relationship to big data and data-driven decision making. Big Data **1**(1), 51–59 (2013)
2. Xyntarakis, M., Antoniou, C.: Data science and data visualization. In: Mobility Patterns, Big Data and Transport Analytics, pp. 107–144. Elsevier (2019)
3. Toasa, R., Maximiano, M., Reis, C., Guevara, D.: Data visualization techniques for real-time information—a custom and dynamic dashboard for analyzing surveys' results. In: 2018 13th Iberian Conference on Information Systems and Technologies, pp. 1–7, June 2018
4. Rivero, J.L.A., López, J.G.: El proceso de planificación estratégica en las universidades: desencuentros y retos para el mejoramiento de su calidad. Rev. Gestão Univ. na América Lat. - GUAL, vol. 5, no. 2, pp. 72–97, August 2012
5. Khan, K.S., Kunz, R., Kleijnen, J., Antes, G.: Five steps to conducting a systematic review. J. R. Soc. Med. **96**, 118–121 (2003)
6. Keim, D.A.: Information visualization and visual data mining. IEEE Trans. Vis. Comput. Graph. **7**(1), 1–8 (2002)
7. Waller, M.A., Fawcett, S.E.: Data science, predictive analytics, and big data: a revolution that will transform supply chain design and management. J. Bus. Logist. **34**(2), 77–84 (2013)
8. Román, A.B., Sánchez-Guzmán, D., Salcedo, R.G.: Minería de datos educativa: Una herramienta para la investigación de patrones de aprendizaje sobre un contexto educativo (2014)
9. West, D.M.: Big data for education: data mining, data analytics, and web dashboards (2012)
10. Shea, K.D., Brewer, B.B., Carrington, J.M., Davis, M., Gephart, S., Rosenfeld, A.: A model to evaluate data science in nursing doctoral curricula. Nurs. Outlook **67**(1), 39–48 (2019)

11. Steele, J., Iliinsky, N.P.N.: Beautiful visualization : [looking at data through the eyes of experts] (2010)
12. Kotu, V., Deshpande, B., Kotu, V., Deshpande, B.: Data science process. In: Data Science, pp. 19–37. Morgan Kaufmann (2019)
13. SistemaEduca: Fases Data Mining. https://sistemeduca.com/fases-data-mining/. Accessed 28 Mar 2019
14. Keim, D.A.: Visual exploration of large data sets. Commun. ACM **44**(8), 38–44 (2001)
15. SAS: Data Visualization Techniques, White Paper, pp. 2–16 (2013)
16. Rose, S.: Return on information : the new ROI getting value from data. SAS Inst. Inc., USA (2014)
17. Sankey Flow Show - Attractive flow diagrams made in minutes! http://www.sankeyflowshow.com/. Accessed 28 Sept 2017

Path Planning for Mobile Robot Based on Cubic Bézier Curve and Adaptive Particle Swarm Optimization (A2PSO)

Daniel Soto[1] and Wilson Soto[2(✉)]

[1] Universidad de Antioquia, Medellín, Colombia
anderson.soto@udea.edu.co

[2] Universidad Católica de Pereira, Pereira, Colombia
wilson.soto@ucp.edu.co

Abstract. In this work, a new approach is proposed for getting a solution of path-planning for mobile robot based on cubic Bézier curve and adaptive particle swarm optimization (A2PSO). Paths generated using a cubic Bézier curve are optimized globally through the A2PSO algorithm. The A2PSO algorithm is significantly more powerful than conventional PSO algorithm. Our approach was successful in determining the shortest path in several environments full of obstacles compared to the performance of the conventional PSO.

Keywords: Path planning · Cubic Bézier curve · Bio-inspired algorithms · Particle swarm optimization

1 Introduction

In mobile robotics, one main problem is path planning. Path planning consists to find a collision-free path in an environment with obstacles for a mobile robot from start location to a target location.

There are four large groups of path planning techniques: graph search-based planners, sampling-based planners, interpolating curve planners and numerical optimization. In the group of interpolating curve planners the most frequently used parametric curves in robotics are, Béziers, B-splines, rational Bézier curves (RBCs), and non-uniform rational B-splines (NURBSs) [1,2]. The parametric curves are capable of enhancing the robot motion planning, because, the parametric curve is a continuous curve, smooth and easy to manipulate for avoiding obstacles, that help to move the robot gradually from start to goal.

The Bézier curves are specially more particular and simpler than B-splines, RBCs and NURBSs curves. Some other advantages of Bézier curves are: low computational cost, the control points can generate a curve of desired characteristics, the curves can be connected with each other to get the desired shape, convex hull property, interpolation of first and last control point, tangency to

M. Botto-Tobar et al. (Eds.): ICAETT 2019, AISC 1066, pp. 432–441, 2020.
https://doi.org/10.1007/978-3-030-32022-5_40

the control polygon first and last point, invariant under affine transformations and global control [1,3].

In [4] shows a summary of related work of parametric curves used in mobile robot applications.

The robot motion planning can be seen as an optimization problem. The commonly used optimization algorithms for solving the problem are particle swarm optimization (PSO), genetic algorithms (GA), ant colony optimization (ACO) and simulated annealing (SA). In recent years, many researchers have used the PSO due to its fast convergence, fewer parameters, strong robustness, and simple features compared with other algorithms. Although, it has some disadvantages, such as weak local search ability and poor speed updating formula, which seriously affect the efficiency and reliability of path planning. Many researchers nowadays have started to combine the existing path planning algorithms to achieve new innovations instead of spending time and effort to search for some new algorithms [5,6].

Several works are about PSO, but very few literature are found on adaptive PSO. Moreover, only 6% of implementation of PSO is related to mobile robot navigation of all the classical and heuristic algorithms [7]. We show in Table 1 a summary of the most important and recent works of optimal mobile robot path planning using PSO optimization.

Table 1. Optimal mobile robot path planning using PSO

Year	Author	Algorithm
2006	[8]	Ferguson Splines and PSO
2014	[9]	Ferguson Splines and PSO-Radial Basis Function (RBF)
2016	[5]	Simulated Annealing PSO
2018	[7]	Adaptive PSO (APSO) - Inertia weight
2018	[10]	Self-adaptive Learning PSO (SLPSO)
2018	[11]	Intelligent Bezier and Chaotic PSO (CPSO)

The rest of the paper is organized as follows. Overviews of the parametric curve Bézier and the PSO algorithm in Sect. 2. The approach is explained in detail in Sect. 3. The results and discussion are introduced in Sect. 4. Conclusions are given in Sect. 5.

2 Background

In this section, a brief introduction is given to cubic Bézier curve and Adaptive Particle Swarm Optimization (A2PSO) algorithm used in the proposed approach.

2.1 Cubic Bézier Curve

A Bézier curve is a polynomial curve (non-rational) for smoothing a graph defined by a set of points. Given a set of $n+1$ control points $P_0 \ldots P_n$, the Bézier curve is given by,

$$C(t) = \sum_{i=0}^{n} B_{i,n}(t) P_i \qquad t \in [0,1] \tag{1}$$

where, $B_{i,n}(t)$ is a Bernstein polynomial. A Bernstein polynomial of degree n is defined by,

$$B_{i,n}(t) = \frac{n!}{i!(n-i)!}(1-t)^{n-i}t^i \tag{2}$$

So a cubic Bézier curve could be generated by formulas (1) and (2) as follows:

$$C(t) = (1-t)^3 P_0 + 3(1-t)^2 t P_1 + 3(1-t)t^2 P_2 + t^3 P_3 \qquad t \in [0,1] \tag{3}$$

Bézier curves can represent complex curves by increasing the number of control points or several cubic Bézier curves end to end can form composite Bézier curves (Fig. 1).

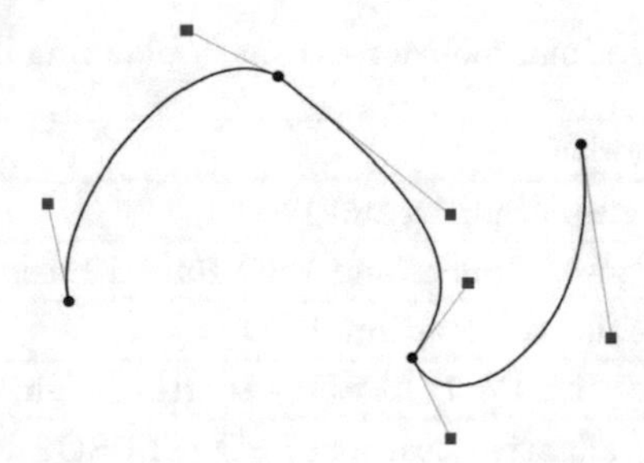

Fig. 1. Three cubic Bézier curve linked together. The blue points are control points. The black points are start/end points.

The two properties more important of Bezier curves [12] are:

1. Bézier curves begin at the start point and stop at the end point.
2. In the first derivative at the start/end of a Bézier curve the vector tangent to the Bézier curve at the start/stop is parallel to the line connecting the first two/last two control points.

2.2 Adaptive Particle Swarm Optimization (A2PSO)

The particle swarm optimization (PSO) is a population–based algorithm, in which swarm of particles explores a search space in order to find an optimal solution to a problem. Each particle located into a search space is a possible solution to the optimization problem. Each particle has three properties: coordinates, velocity, and the own memorized optimization. The trajectory of the particle is influenced by three components: the inertial component, the cognitive component, and the social component.

Formally, the particle swarm optimization is expressed through the mathematical model:

$$v_i^{t+1} = wv_i^t + c_1 r_1(x_i^{lbest} - x_i^t) + c_2 r_2(x^{gbest} - x_i^t) \tag{4}$$

$$x_i^{t+1} = x_i^t + v_i^{t+1} \tag{5}$$

where v is the velocity of particle, t is the iteration number, i is the particle index, x is location of the particle at time t in the search space, w is the inertial component, c_1 is the cognitive component, c_2 is the social component and the random variables r_1 and r_2 between 0 and 1.

If the optimization objective is the minimization, the variables *lbest* and *gbest* are given by:

$$lbest_i(t+1) = \begin{cases} lbest(t) & \text{if } f(x_i(t+1)) \geq lbest_i(t) \\ x_i(t+1) & \text{otherwise} \end{cases} \tag{6}$$

$$gbest(t+1) = argmin_{lbest_i} f(lbest(t+1)) \text{ where } 1 \leq i \leq N \tag{7}$$

In an adaptive PSO, literally, the word adaptive means which will capable to adjust the change continuously in every iteration for any parameter, such as velocity (4) (improving the coverage of the swarm location in the search space) or social component (social neighborhood) [7].

In [13] shows an adaptive PSO called A2PSO, where its performance is better than the conventional PSO.

In the A2PSO algorithm, the initial location is based on the concept of the electrostatic repulsion. The electrostatic repulsion consists in assign to each particle a repulsion area inside which cannot locate another particle. The PSO algorithm with electrostatic repulsion randomly places the particles inside the search space and iteratively corrects the positions with the calculate of the volume, radio and interaction area of the repulsion of each particle.

The hypervolume ν of electrostatic repulsion for each particle is calculated by (8):

$$\nu = \frac{L_1 * L_2 * ... * L_d}{2 * N} \tag{8}$$

where N is the population, d are the dimensions of the search space and L_i represents the length of the search space on the dimension i.

With the hypervolume ν calculated we can determine the repulsion radio $\mathcal{R}$ which is used in the location of particles. The Eq. (9) calculates the repulsion radio with the hypervolume of a hypersphere due to be a scheme n-dimensional.

$$\mathcal{R} = \left(\frac{\nu * \Gamma((d/2)+1)}{\pi^{d/2}} \right)^{1/d} \tag{9}$$

For avoiding that a particle is in a repulsion area of another particle, the particle must satisfy:

$$\sqrt{(x_1' - x_1'')^2 + (x_2' - x_2'')^2 + ... + (x_d' - x_d'')^2} > 2 * R \tag{10}$$

The A2PSO algorithm also has the characteristic of social neighborhood. The social neighborhood is a kind of neighborhood generally without fixed topology, adaptive and static or dynamic behavior. The social neighborhood does not have geographic or distance rules because a particle simply can have neighbors defined in anywhere of the search space. The concept of the social neighborhood can be represented by an interaction vector (11).

$$z_i^{rand(N)} = 0.2; \quad \forall n, |z_i^n(0) \neq 0| = 5 \tag{11}$$

The values of the interaction vector are iteratively updated under the concept of weakly symmetric interactions. The Eq. (12) calculates the updating rule depending on the nth neighbor selected.

$$z_i^n(t+1) = \begin{cases} z_i^n(t) + \dfrac{fit_i - fit_n}{4 * max_N(fit_i - fit_n)} & \text{if } 0.2 \leq z_i^n(t) < 1.0 \\ 1 & \text{if } z_i^n(t) \geq 1.0 \\ 0 & \text{if } z_i^n(t) < 0.2 \end{cases} \tag{12}$$

where fit is the fitness function.

The Eq. (13) evaluates the interactions between 6 or 7 nearest neighbors for each particle. If the condition is false, then a random position in the vector is selected assigning it the value 0. The process is iterative until to obtain the predefined number of individuals for creating the social neighborhood.

$$z_i^{rand(N)}(t+1) = 0.2 \text{ if } z_i^{rand(N)}(t) = 0; \forall n, |z_i^n(t+1) \neq 0| = 5 \tag{13}$$

After update the interaction vector, for each particle i are calculated the location and velocity with (14) and (15), respectively.

$$v_i^{t+1} = wv_i^t + r_1c_1(x_i^{lbest} - x_i^t) + r_2c_2\left(\sum_{n=0}^{N} z_i^N(t)(x_n^{lbest} - x_i^t)\right) \tag{14}$$

$$x_i^{t+1} = x_i^t + v_i^{t+1} \tag{15}$$

3 Robot Path Planning Based on Cubic Bézier Curve and A2PSO

The proposed algorithm (Algorithm 1) uses the A2PSO algorithm to optimize the control points of the cubic Bézier curve aim to get the minimum smooth path from the start point to the end point.

The Algorithm 1 computes the fitness function (fit) has two elements: the arc length of the Bézier curve (16), and a penalty based on the crosses of the Bézier curve and obstacles lines (19).

$$\mathcal{L} = \int_0^1 \sqrt{\left(\frac{d}{dt}x(t)\right)^2 + \left(\frac{d}{dt}y(t)\right)^2}\, dt \tag{16}$$

where

$$x(t) = Px_0(1-t)^3 + 3Px_1t(1-t)^2 + 3Px_2t^2(1-t) + Px_3t^3 \tag{17}$$

$$y(t) = Py_0(1-t)^3 + 3Py_1t(1-t)^2 + 3Py_2t^2(1-t) + Py_3t^3 \tag{18}$$

Algorithm 1. Proposed Algorithm Path-Planning Mobile Robot

```
Require
n : Number of individuals or particles
r : Repulsion radio
mbp : Max. bisection points
ibp : Iterations bisection points
c1, c2, c3 : Parameters for A2PSO
Ensure
gbest : Global optimum

1: Begin
2:   While not find solution and number bisection points < mbp do
3:     Locate the n particles (keep for each n particle radius > r)
4:     Evaluate the fitness of n particles
5:     Generate the neighborhood for each n particle
6:     Determine the local optimum of each n particle (lbest)
7:     Update the lbest value
8:     Calculate the neighborhood social values
9:     While number iterations < ibp do
10:        Move the particles (using the parameters c1,c2 and c3)
11:        Evaluate the fitness n particles
12:        Update the lbest for each n particle value
13:        Calculate the neighborhood social values
14:        Update gbest value
15:    End while
16:    Update gbest value
17:    New random intermediate point in the path (random bisection)
18:  End while
19:  Return gbest
20: End
```

The penalization is calculated through the sum of segments of all the crosses or intersection points (19). With Cardano's method (standard way to find a real root of a cubic polynomial function) we can get the set of intersection points between the Bézier curve and obstacle lines.

Two examples of crosses between curves and obstacles can see in Fig. 2. At left, a curve has 2 intersection points generating one segment. At right a curve has 4 intersection points generating two segments. The sum of the segments represents the total length of the cross.

$$\mathcal{X} = \sum_{a} \sqrt{ip_i^2 + ip_{i+1}^2}, \; a = 2 \times n, n \in N \wedge n < \frac{M}{2} \tag{19}$$

where ip_i is the i-th intersection point and M the number of intersection points.

The fitness function (*fit*) for each particle can be calculated as follows (20),

$$fit = \mathcal{L} + (\mathcal{X} \times k) \tag{20}$$

where, k is a constant.

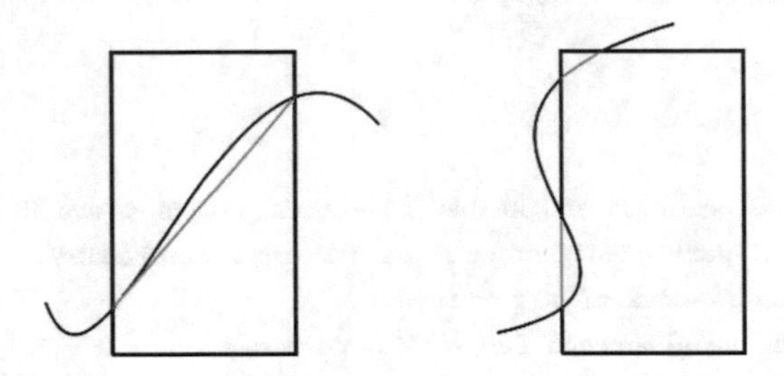

Fig. 2. Crosses between curves and obstacles

The Algorithm 1 includes random bisection (line 17). The random bisection is used to split the path into segments (random points) with the aim to obtain a curve that satisfies the optimization criteria. The optimization criteria of the curve are defined by do not cross obstacles and the shortest path from start point to end point.

4 Results

In this paper, the algorithms are implemented in C++ using the MPFR library (multiple-precision floating-point computation with exact rounding) on Quad-Core Intel Xeon, 2.4 GHz clock, 4 Gb memory running 64-bit GNU/Linux operating system. The parameters of the A2PSO algorithm are set in Table 2.

Table 2. Parameters used in the environments for the proposed algorithm

Parameter	Value
Iterations	200
Individuals	20
Neighborhood	5
Max bisection	10
Iterations by bisection	50
Constant k	2000
Random Parameters (c_1, c_2, c_3)	$[0.25, 0.5], [0.5, 1], [0.5, 1]$

The proposed algorithm is tested in several environments in order to show how the robot moves in scenarios full of obstacles and verify the effectiveness of Cubic Bézier with A2PSO for solving mobile robot path planning problem. The path planning environments were developed based on [14]. Moreover, we applied the method of obstacles inflated to guarantee that a collision does not occur in the path planning [15].

The best solution for each scenario is plotted in Fig. 3. The robot using cubic Bézier with A2PSO completely avoids the obstacles and follows the target path. In Fig. 3(a) and (b) the robot escaped from trapped condition. In Fig. 3(c) the robot has successfully finished the wall following. The mobile robot successfully navigates in a maze environment as shown in Fig. 3(d).

From Table 3, it is noticed that the A2PSO has greatly improved the performance results compared with the conventional PSO. The path length covered by the robot in PSO in the environments is more than A2PSO, so more energy is consumed. Therefore, using A2PSO the robot can achieve the goal more efficiently.

Table 3. Path length covered by robot using cubic Bézier with A2PSO and PSO

Figure	Algorithm	Min.	Max.	Average	Variance	Solutions
Figure 3(a)	A2PSO	48,805	93,214	68,560	89,784	950
	PSO	51,972	106,186	69,606	113,272	570
Figure 3(b)	A2PSO	32,257	96,862	64,222	182,116	62
	PSO	33,117	146,428	60,064	444,774	32
Figure 3(c)	A2PSO	50,656	75,567	55,172	41,552	410
	PSO	50,622	104,986	57,990	107,867	411
Figure 3(d)	A2PSO	116,959	147,196	131,601	82,076	24
	PSO	119,232	145,370	132,546	48,956	32

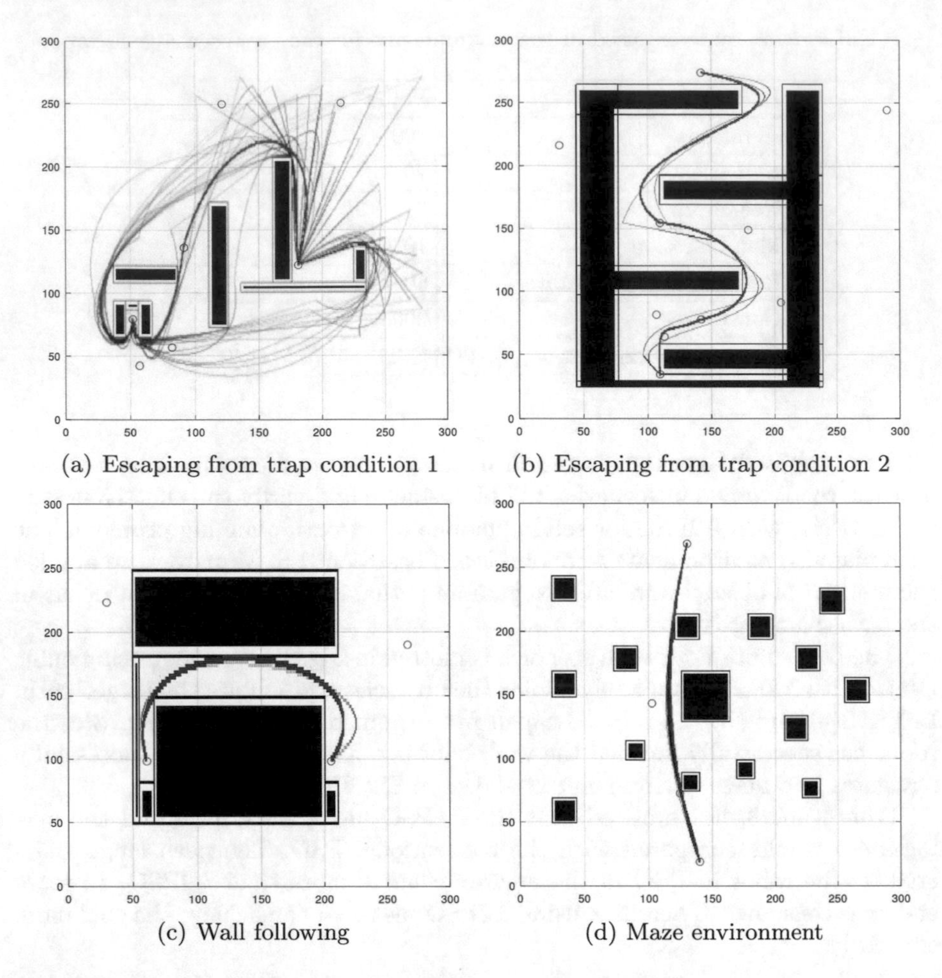

(a) Escaping from trap condition 1
(b) Escaping from trap condition 2
(c) Wall following
(d) Maze environment

Fig. 3. The simulation results of the path planning mobile robot using cubic Bézier with A2PSO

5 Conclusions

A new approach for getting a solution of path-planning for mobile robot was shown. The new approach combines the cubic Bézier curves and a variant of PSO algorithm called A2PSO. The Bézier curves provide an efficient way to generate the optimized path in mobile robot planning due to simplicity and easy handling. Using the Bézier curve as the swarm in the particle swarm optimization process meant that the optimized Bézier curve of the mobile robot was found by the A2PSO algorithm. The A2PSO is an adaptive PSO algorithm that improves the initial location of the particles and includes the concept of the social neighborhood. Using A2PSO in different types of environments the robot has successfully performance with shortest path and number of solutions than conventional PSO.

References

1. Pérez, L.H., Mora Aguilar, M.C., Sánchez, N.M., Montesinos, A.F.: Path planning based on parametric curves, chapter 7, pp. 125–143. InTech, September 2018
2. Costanzi, R., Fanelli, F., Meli, E., Ridolfi, A., Allotta, B.: Generic path planning algorithm for mobile robots based on Bézier curves. IFAC-PapersOnLine **49**(15), 145–150 (2016). 9th IFAC Symposium on Intelligent Autonomous Vehicles IAV 2016
3. Ravankar, A., Ravankar, A., Kobayashi, Y., Hoshino, Y., Peng, C.-C.: Path smoothing techniques in robot navigation: state-of-the-art, current and future challenges. Sensors **18**(9), 3170 (2018)
4. Elbanhawi, M., Simic, M., Jazar, R.N.: Continuous path smoothing for car-like robots using B-Spline curves. J. Intell. Robot. Syst. **80**(1), 23–56 (2015)
5. Nie, Z., Yang, X., Gao, S., Zheng, Y., Wang, J., Wang, Z.: Research on autonomous moving robot path planning based on improved particle swarm optimization. In: 2016 IEEE Congress on Evolutionary Computation (CEC), pp. 2532–2536, July 2016
6. Zafar, M.N., Mohanta, J.C.: Methodology for path planning and optimization of mobile robots: a review. Procedia Comput. Sci. **133**, 141–152 (2018). International Conference on Robotics and Smart Manufacturing (RoSMa 2018)
7. Dewang, H.S., Mohanty, P.K., Kundu, S.: A robust path planning for mobile robot using smart particle swarm optimization. Procedia Comput. Sci. **133**, 290–297 (2018)
8. Saska, M., Macas, M., Preucil, L., Lhotska, L.: Robot path planning using particle swarm optimization of Ferguson splines. In: 2006 IEEE Conference on Emerging Technologies and Factory Automation, pp. 833–839, September 2006
9. Arana-Daniel, N., Gallegos, A.A., López-Franco, C., Alanis, A.Y.: Smooth global and local path planning for mobile robot using particle swarm optimization, radial basis functions, splines and Bézier curves. In: 2014 IEEE Congress on Evolutionary Computation (CEC), pp. 175–182, July 2014
10. Li, G., Wusheng, C.: Path planning for mobile robot using self-adaptive learning particle swarm optimization. Sci. China Inf. Sci. **61**(5), 052204 (2018)
11. Tharwat, A., Elhoseny, M., Hassanien, A.E., Gabel, T., Kumar, A.: Intelligent Bézier curve-based path planning model using Chaotic Particle Swarm Optimization algorithm. Cluster Comput. 1–22 (2018)
12. Liang, J.J., Song, H., Qu, B.Y., Liu, Z.F.: Comparison of three different curves used in path planning problems based on particle swarm optimizer. Math. Prob. Eng. **2014**, 1–15 (2014)
13. Soto, D., Soto, W.: A parallel adaptive PSO algorithm with non-iterative electrostatic repulsion and social dynamic neighborhood. In: Madureira, A., Abraham, A., Gamboa, D., Novais, P. (eds.) Intelligent Systems Design and Applications, pp. 570–581. Springer, Cham (2017)
14. Gaillard, F.: Approche cognitive pour la planification de trajectoire sous contraintes. Ph.D. thesis, Université des Sciences et Technologie de Lille - Lille I (2012)
15. Connors, J., Elkaim, G.: Analysis of a spline based, obstacle avoiding path planning algorithm. In: 2007 IEEE 65th Vehicular Technology Conference - VTC2007-Spring, pp. 2565–2569, April 2007

ANN-Based Model for Simple Grammatical Cases Teaching in Spanish Language

Laura Márquez García(✉) and Adán Gómez Salgado

Córdoba University, Montería, Colombia
{lauramarquezg, aagomez, mflorezmadrigal}@correo.unicordoba.edu.co

Abstract. The Multilayer Perceptron (MLP) is one of the most powerful and popular network architectures, due to its ability to solve a large number of problems successfully. The objective of this work is to develop a model based on neural networks to solve simple grammar cases in the Spanish language, which in turn, allows the teaching of Artificial Neural Networks (ANN) in students. The presented model consists of 12 stages that allow to simulate simple grammatical cases. An illustrative example is presented with two programs that allow to simulate the grammatical case "Uppercase Identification" and "Infinitive Verbs" using MLP; each had a training scenario and a learning verification scenario.

Keywords: Artificial Neural Network · Gramatical cases · Multilayer perceptron · Simulator

1 Introduction

Artificial Neural Networks (ANN) are one of the most used and efficient practices in artificial intelligence [1, 2, 24, 25], ANN constitute one of the most important topics in science and computer engineering (Computer Machinery Association (ACM)) as in the development as in the teaching process of automatic learning techniques [3]. ANN are simplified models of information treatment, whose basic processing unit is inspired by the human nervous system basic cell: the neuron [4] and they are composed by complex mathematical algorithms, whose functioning depend on their interconnections and of the activation function [1]. ANN have been used for the development of multiple tasks in the artificial intelligence, such as: speech recognition, computer-based vision and gradually in the Natural Language Processing [5, 30].

Currently, in the literature exists tools such as R [27], WEKA [28] and SNNS [29] which have been used to facilitate the teaching, application and development of ANN models [24]. However, these require in depth knowledge of this type of software [25]. In this point, [24] considers that the implementation of the ANN in real cases requires to have a previous knowledge about data abstraction processes of a specific problem of the real world representing it in quantifiable terms which will be processed by the neural network. For this reason, [3] states that it is necessary to create alternative educational tools for the learning, design and development of ANN models.

M. Botto-Tobar et al. (Eds.): ICAETT 2019, AISC 1066, pp. 442–453, 2020.
https://doi.org/10.1007/978-3-030-32022-5_41

The ANN has been used to solve language processing tasks [6–9], affirm that the ANN (especially ANN with supervised learning) have been used in tasks that imply a degree of understanding of natural language processing such as: words and phrases categorization and paraphrase detection.

[10] states that the Multilayer Perceptron (MLP) developed by Rosenblatt is one of the types of ANN, most used [24], considering it as a linear binary classifier model for human decision making. The level of connectivity must be high, and its extension depends on the synaptic weights of the network [11].

[8] MLP can capture and understand the internal structures of sentences in natural language and patterns in their interactions by applying learning algorithms. [12] used MLP for examine the relation between reading comprehension and lexical and grammatical knowledge in students of English as a foreign language. [2] affirm that the application of ANN in different study areas have influenced in a superficial knowledge of their application to the real world. Therefore, they developed software that facilitate the learning about how to implement neural models in a simple cases. However, these simple cases are not related with grammatical components of a clause. [26] developed an ANN with multilayer perceptron to classify low and high capacity readers according to their grammar and vocabulary measures.

The implementation of techniques such as computer simulation allows the student improve the understanding of scientific concepts in a clear and dynamic. [3], developed a web-based tool which allows to learn ANN concepts and provide visualization capabilities that facilitate track the network step by step.

How is observed in the current literature, does not have been developed tools for teaching models based on ANN where a MLP is used to simulating simple grammatical cases, [1] affirms that in the design process of the ANN Architecture, the designer has to clearly identify the problem to solve for defining the ANN Topology and the other constitutive elements of the ANN. Otherwise, the Undergraduate Program Licenciatura en Informática of the Universidad de Córdoba, in the current curriculum [13] does not have a course that develops topics for automatic learning, especially ANN. For this reason, the students of this program present lack of knowledge about this kind of artificial intelligence techniques and its application in real cases.

Therefore, this model allows to detail and apply in the simple way the topology and architecture of an ANN with MLP to solve real cases, specifically simple cases of grammar in the Spanish language.

This document is structured as follows: The second chapter defines the multilayer perceptron; In the third chapter we explain the model based on neural networks to solve cases simple. The next chapter shows examples of this model mentioned above. Finally, the conclusions in the last chapter are presented.

2 Multilayer Perceptron

In the study field of neural networks, MLP implemented by [14] is one of the most powerful and popular network architectures, [15, 16] due to its ability to solve a large number of problems successfully [17]. This architecture is characterized by having the neurons grouped in layers of different levels: the input layer, continuous or discrete values (binary), the hidden layers and the output layer [21].

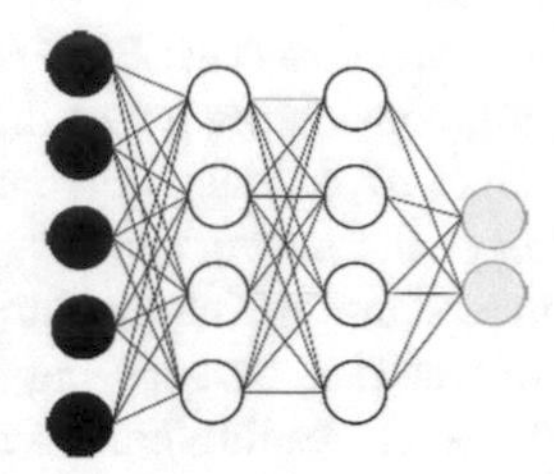

Fig. 1. Structure of multilayer perceptron.

As observed in the Fig. 1 the Input Layer functions as a receiver that only receive the signals or patterns from the outside and send them to all the neurons in the hidden layer [22]. In this layer, the neurons perform a non-linear processing of the received patterns (using an activation function) [23]. The output layer provides the external response of the network for each one of the input patterns [16].

MLP connections use a supervised learning model, they are always directed forward "feedforward" [1, 18]. The connections between the neurons of the network also have a threshold associated, which in this case is usually treated as a connection to the neuron, whose input is constant and equal to 1 [19].

This type of networks are very implemented in applications of recognition or classification of behavior patterns and even to predict the behavior of a system [1].

3 MLP-Based Model for Simple Grammatical Cases

The present model is adjusted a type of supervised learning, in this type of learning, a set of input patterns is presented to the network together with the expected output. In addition, specific weights are used to modify them proportionally to an error which occurs between the actual output of the network and the expected output.

This MLP-based model which allow to simulate simple grammatical cases have following stages:

1. Create the Input Matrix:
Identify the rules which characterize the grammatical case of the problem which to solve is wanted. The set of rules will become in a Vector of Bits (1 and 0) that will constitute the input layer of the network. In this Vector of Bits, a 1 represents if the rule is accomplished and a 0 if the rule is not accomplished. The final result of this step is an Input Matrix of bits structured by each Vector of Bits previously designed. In this matrix, each column represents an Input (a single bit) in the network and each row represents a whole pattern which enters to the network.

[31] affirms that the efficiency of a ANN depends as much on the input and output layers, such of its structure, learning algorithm, processing unit types and parameters.

So, in this model, the input layer of the ANN always is defined through a vector of bits, unlike other models, this allows way the abstraction of data of the case to solve in an easier way.

Add the Desired Output Column to the Input Matrix

The Desired Output (d) is a value that represents the response that the network waits for a certain question. This d is a numerical value that will represent the output of each pattern which will enters to the network reflecting in this way, all the possibilities of the matrix. To obtain this number, any mathematical calculation can be used. In this model, a simple formula will be performed (see formula 1):

$$d = \chi_i 2^i \tag{1}$$

Where i = 0 … n. In other words, multiply each bit of the Input Matrix by 2^i. This mathematical operation is used in order to convert the bit vector into an easy to understand and process data.

2. Convert d to a Number Between 0 and 1

In this step, with the purpose of improving the legibility of d, the variable Ψ will be used to convert this desired output to a number between 0 and 1. To know the value of Ψ, the following rule will be used (see formula 2 and 3):

$$\text{if } d \geq 0 \text{ and } d \leq 9 \text{ - > } \Psi = 10^{-1} \tag{2}$$

$$\text{if } d \geq 10 \text{ and } d \leq 99 \text{ - > } \Psi = 10^{-2} \tag{3}$$

3. Implement the Hidden Layer Activation Function

According to [1], the output of a neuron depends on both the input data and the activation function. This activation function is executed in a trial and error process, which is called the *Network Learning Mechanism*. ANN have the ability to learn from their environment and improve performance through learning. The procedure used for the self-learning process of a ANN is called learning algorithm [20].

In this model, the following mathematical function is proposed, which as *Activation Function* will be used (see formula 4):

$$y' = \sum_{i=1}^{n} xi.Wi + \theta \tag{4}$$

4. Develop the Learning Mechanism of the Hidden Layer

This process is calculated by performing cycles or iterations until a condition is accomplished. In this model, the condition consists in that the difference between the calculated output and d must be zero (0). This condition is called Error. While the error is not zero (0), the following formula will be used (see formula 5):

$$wi = wi + \eta(d - y')xi \tag{5}$$

The new weight will be equal to the sum between the current weight plus the learning rate multiplied by the difference between *d* and the calculated output multiplied by the current input.

5. Create the Graphic Model of the MLP-Based Model
Taking into account that the present model is oriented to teaching of the ANN functioning process, to design a graphic model of this process is proposed.

6. Create the Outputs of the Hidden Layer in a Vector of Bits that will Enter the Output Layer (Input Matrix for the Output Neuron)
In the same way, as in the first step of this model, an input matrix for the output neuron will be created. Taking into account that the Bits that will structure this input matrix will be obtained from the calculated output of the hidden layer.

7. Add the Desired Output Column to the Input Matrix of the Output Layer
In the same way, as in the second step of the model, the desired output is calculated for this input matrix of the output layer, using the same function.

8. Convert the Desired Output of the Input Matrix of the Output Layer to a Number Between 0 and 1
In the same way, as in the third step of the model, the *d* will be become to a number between 0 and 1, using the variable Ψ, with the following rule:

9. Implement the Output Layer Activation Function
In the same way, as in the fourth step of the model, will be used the following activation function, in this case for the output layer:

$$y' = \sum_{i=1}^{n} xi.Wi + \theta \tag{4}$$

10. Develop the Learning Mechanism of the Output Layer
In the same way, as in the fifth step of the model, will be used the following formula for implementing the learning mechanism of the output layer:

$$wi = wi + \eta(d - y')xi \tag{5}$$

11. Verification of Learning
To verify if the ANN learned, the Transfer Function must be applied to each of the calculated outputs (that is, to verify that the neuron must be activated or not).

$$\begin{aligned} f(x) &= 1\, if\, y' * \Psi = d * \Psi \\ &= 0\, else \end{aligned} \tag{6}$$

Finally, Apply the Transfer Function for the calculated outputs of the last neuron, that is to say, verify whether the neuron must be activated or not with the calculated output that was obtained.

4 Illustrative Example

Identify of infinitive verbs

1. Create the Input Matrix:

In spanish language a verb is a word ended in "ar" "er" or "ir" and this word must not be a noun ended in "ar" "er" or "ir" (for example, the word must not be in this list: alcazar, alfiler, alquiler, altar, amanecer, anochecer, atardecer, avatar, azúcar, bar, cancer, character, catéter, caviar, elixir, emir, esfinter, éter, faquir, impar, lunar, martir, master, mir, par, pilar, pinar, polar, tapir, telar, ureter, postrer, devenir, taller, pajar). For this reason, the input matrix will be created using words ended in "ar", "er" or "ir". In addition, will be used words which not be a noun ended in "ar", "er" or "ir" (Table 1).

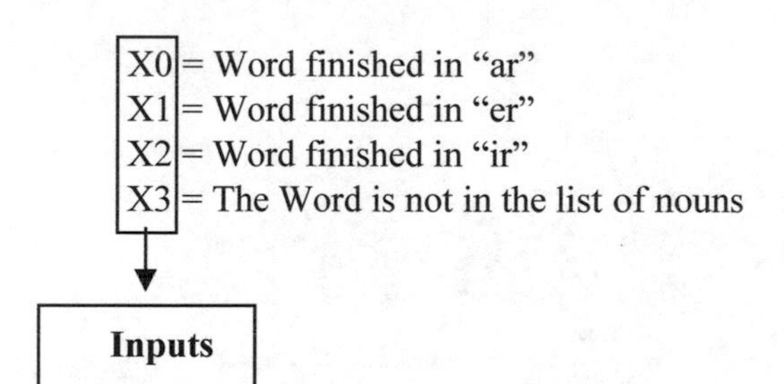

Table 1. Structure input matrix

x0	x1	x2	x3
1	0	0	1
0	1	0	1
0	0	1	1

2. Add the Desired Output (*d*) Column to the Input Matrix

(See Table 2).

$$2^0 *1 + 2^1 *0 + 2^2*0 + 2^3*1 = 9$$

$$2^0 *0 + 2^1 *1 + 2^2*0 + 2^3*1 = 10$$

$$2^0 *0 + 2^1 *0 + 2^2*1 + 2^3*1 = 12$$

Table 2. Structure *d* matrix

x0	x1	x2	x3	*d*
1	0	0	1	9
0	1	0	1	10
0	0	1	1	12

3. Convert *d* to a number between 0 and 1
(See Table 3).

Table 3. Structure for *d* matrix

	x1	x2	x3	x4	d	Ψ	d*Ψ	
P1	1	0	0	1	9	0,1	0,9	Finish in "ar" and it is not in the list
P2	0	1	0	1	10	0,01	0,10	Finish in "er" and it is not in the list
P3	0	0	1	1	12	0,01	0,12	Finish in "ir" and it is not in the list

4. Implement the Hidden Layer Activation Function
5. Create the outputs of the Hidden Layer in a Vector of Bits that will enter the Output Layer (Input Matrix for the Output Neuron)
6. Create the Graphic Model of the MLP-based model
(See Fig. 2).

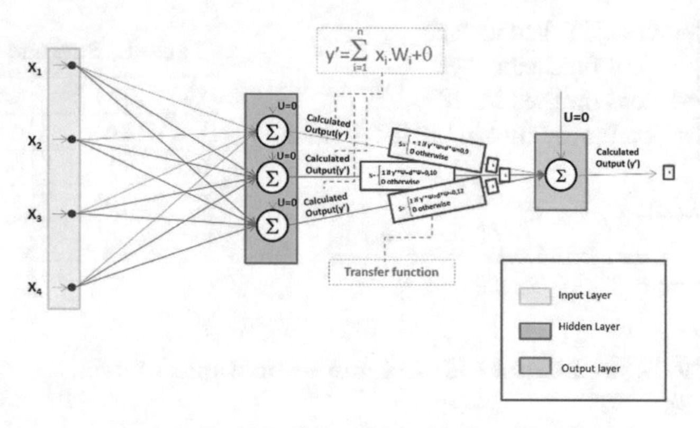

Fig. 2. Graphic model of the MLP-based model

7. Create the outputs of the hidden layer in a Vector of Bits that will enter the output layer (Input Matrix for the Output Neuron)
(See Table 4).

Table 4. Structure input matrix for the output neuron

x0	x1	x2
1	0	0
0	1	0
0	0	1

8. Add the *d* Column to the Input Matrix of the Output Layer

(See Table 5).

Table 5. Structure input matrix for the output neuron

x0	x1	x2	*d*
1	0	0	1
0	1	0	2
0	0	1	4

9. Convert the *d* of the Input Matrix of the Output Layer to a number between 0 and 1

(See Table 6).

Table 6. Structure input matrix for the output neuron

x0	x1	x2	*d*	Ψ	$d\Psi$
1	0	0	1	0,1	0,1
0	1	0	2	0,1	0,2
0	0	1	4	0,1	0,4

10. Implement the Output Layer Activation Function

11. Develop the Learning mechanism of the Output layer

12. Verification of Learning

(See Table 7).

Table 7. Matrix structure of verification of learning

	x0	x1	x2	x3	d	Ψ	d*Ψ	Transfer function
P1	1	0	0	1	9	0,1	0,9	$f(x) = 1$ if $y' * \Psi = d * \Psi = 0$ else
P2	0	1	0	1	10	0,01	0,10	$f(x) = 1$ if $y' * \Psi = d * \Psi = 0$ else
P3	0	0	1	1	12	0,01	0,12	$f(x) = 1$ if $y' * \Psi = d * \Psi = 0$ else

5 Computational Implementation

To computationally implement the model described above, two software were developed to validate the efficiency of the model. Below is the description of the interface of each of this software. It must be taken into account that the example and the grammatical cases are in Spanish language.

In each software, the first interface shows the input data with its respective d. To start the training of the network, the “Entrenar” button must be selected until the Error is zero (0) (see Figs. 3 and 4).

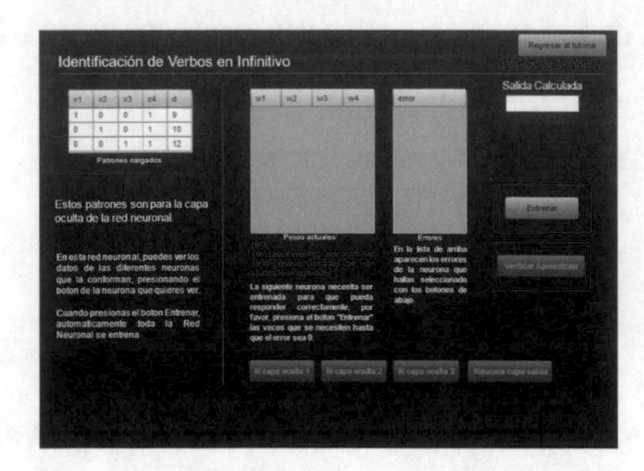

Fig. 3. ANN of "identification for infinitive verbs"

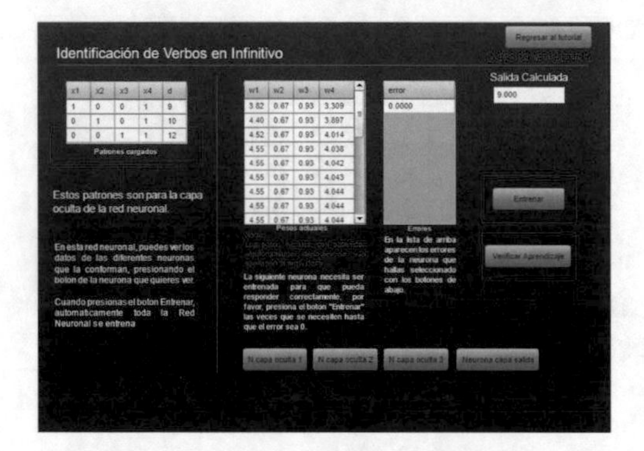

Fig. 4. ANN of "identification of infinitive verbs"

This model allows to verify the training process of each of the hidden layers y Output layer, for this, must press the "N capa oculta 1", "N capa oculta 2", "N capa oculta 3" and "Neurona capa de salida" buttons (see Figs. 5 and 6).

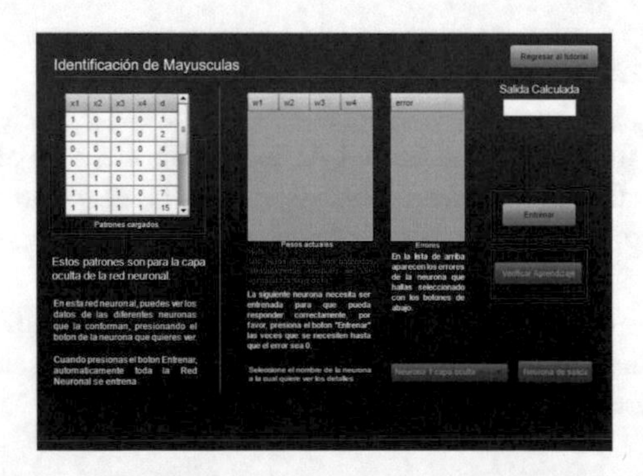

Fig. 5. Training of the ANN "uppercase identification"

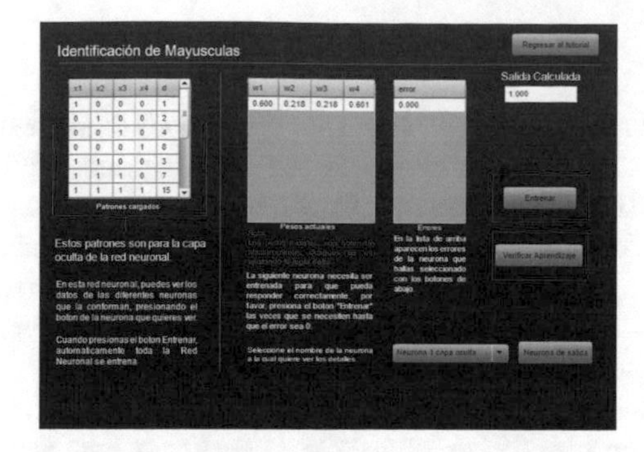

Fig. 6. Training of the ANN "uppercase identification"

After selecting the Verification of Learning you can observe the training process and the desired output (see Figs. 7 and 8).

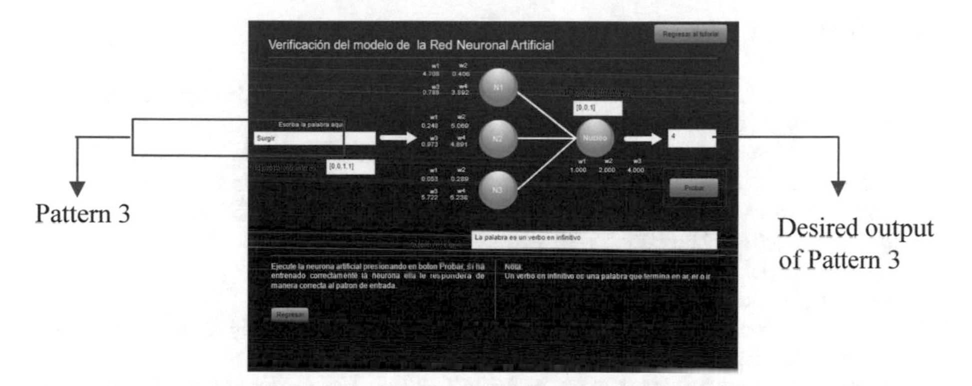

Fig. 7. Verify of learning of model of ANN for infinitive verbs.

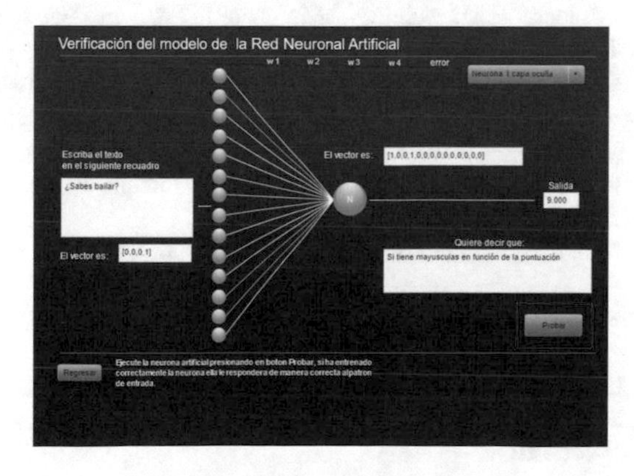

Fig. 8. Verify of learning of the ANN for uppercase identification

6 Conclusions

The elaboration of the ANN-based Model for Simple Grammatical Cases Teaching in Spanish language, is a first contribution to the teaching and application of automatic learning techniques for Undergraduate Program Licenciatura en Informática of the Universidad de Córdoba. The present model is adjusted a type of supervised learning, in this type of learning, a set of input patterns is presented to the network together with the expected output. This model is composed by twelve steps where a MLP is used to represent simple grammatical cases.

The first three steps of the model (*"Create the Input Matrix", "Add the Desired Output Column to the Input Matrix" y "Convert d to a number between 0 and 1"*), allowed to clearly define the input data, hidden layers and desired outputs, which simplified the application of the following steps of this model. The difference of this model regards to other models is the utilization of Bits Vector and decimal system data for representing in a easy way the information which will flow through of the ANN.

References

1. De Armas, T.: Redes Neuronales Artificiales. Fundamentos básicos de inteligencia artificial con Matlab, pp. 152–156. Educosta (2009)
2. Díaz, P.-C.J., Soria, O.E., Martínez, M.J., Escandell, M.P., Gómez, S.J.: Software educativo basado en las GUI de Matlab para cursos de redes neuronales. In: Psicología y salud mental: conceptos, metodologías, herramientas y aplicaciones, pp. 1730–1755. En I. Management Association (2016). http://doi.org/10.4018/978-1-5225-0159-6.ch075
3. Ugur, A., Kinaci, A.-C.A.: A web-based tool for teaching neural network concepts. Comput. Appl. Eng. Educ. **18**(3), 449–457 (2010)
4. Matich, D.J.: Neural Networks: Basic Concepts and Applications. Universidad Tecnológica Nacional, México (2001)
5. Goldberg, Y.: Neural network methods for natural language processing. Synth. Lect. Hum. Lang. Technol. **10**(1), 1–309 (2017)
6. Klein, M., Kamp, H., Palm, G., Doya, K.: A computational neural model of goal-directed utterance selection. Neural Netw. **23**(5), 592–606 (2010)
7. Carneiro, H.C., França, F.M., Lima, P.M.: Multilingual part-of-speech tagging with weightless neural networks. Neural Netw. **66**, 11–21 (2015)
8. Hu, B., Lu, Z., Li, H., Chen, Q.: Convolutional neural network architectures for matching natural language sentences. In: Advances in Neural Information Processing Systems, pp. 2042–2050 (2014)
9. Clarke, D.: A context-theoretic framework for compositionality in distributional semantics. Comput. Linguist. **38**(1), 41–71 (2012)
10. Nielsen, M.A.: Neural Networks and Deep Learning, vol. 25. Determination Press, USA (2015)
11. Haykin, S.S.: Rosenblatt's perceptron. In: Neural Networks and Learning Machines, pp. 48–62 (2009)
12. Barandela, R., Gasca, E., Alejo, R.: Corrección de la Muestra para el Aprendizaje del Perceptron Multicapa. Inteligencia Artificial. Revista Iberoamericana de Inteligencia Artificial **5**(13), 2–9 (2001)

13. Universidad de Córdoba. https://www.unicordoba.edu.co/index.php/facultad-educacion-y-ciencias-humanas/lic-informatica/plan-de-estudio-lic-informatica/
14. Rumelhart, D.E., Hinton, G.E., Williams, R.J.: Learning representations by back-propagating errors. Cogn. Model. **5**(3), 213–221 (1998)
15. García, N.B.: Implementación de técnicas de deep learning (2015). https://riull.ull.es/xmlui/handle/915/1409
16. Longoni, M.G., Porcel, E., López, M.V., Dapozo, G.N.: Modelos de Redes Neuronales Perceptrón Multicapa y de Base Radial para la predicción del rendimiento académico de alumnos universitarios. In: XVI Congreso Argentino de Ciencias de la Computación (2010)
17. Alonso, A.-D.C., Jara, E.A.: Visión por computadora: identificación, clasificación y seguimiento de objetos. FPUNE Scientific, vol. 10 (2016)
18. Cinca, C.S., del Brío, B.M.: Predicción de la quiebra bancaria mediante el empleo de redes neuronales artificiales. Revista española de Financiación y Contabilidad. **74**, 153–176 (1993)
19. Tabares, H., Branch, J., Valencia, J.: Generación dinámica de la topología de una red neuronal artificial del tipo Perceptron Multicapa. Revista Facultad de Ingeniería Universidad de Antioquia **38**, 146–162 (2006)
20. Yiğit, T., Işik, A.H., Bilen, M.: Web based educational software for artificial neural networks. In: ICEMST 2014, pp. 629–632 (2014)
21. Zhang, Y., Sun, Y., Phillips, P., Liu, G., Zhou, X., Wang, S.: A multilayer perceptron based smart pathological brain detection system by fractional Fourier entropy. J. Med. Syst. **40**(7), 173 (2016)
22. Parlos, A.G., Fernández, B., Atiya, A.F., Muthusami, J., Tsai, W.K.: Un algoritmo de aprendizaje acelerado para redes de perceptrón multicapa. IEEE Trans. Neural Netw. **5**(3), 493–497 (1994)
23. Tang, J., Deng, C., Huang, G.B.: Extreme learning machine for multilayer perceptron. IEEE Trans. Neural Netw. Learn. Syst. **27**(4), 809–821 (2016)
24. Deperlioglu, O., Kose, U.: An educational tool for artificial neural networks. Comput. Electr. Eng. **37**(3), 392–402 (2011)
25. Díaz, M.P., Carrasco, J.J., Soria, O.E., Martínez M.J., Escandell M.P., Gómez S.J.: Educational software based on Matlab GUIs for neural networks courses. In: Handbook of Research on Computational Simulation and Modeling in Engineering, pp. 333–358. IGI Global (2016)
26. Aryadoust, V., Baghaei, P.: Does EFL readers' lexical and grammatical knowledge predict their reading ability? Insights from a perceptron artificial neural network study. Educ. Assess. **21**(2), 135–156 (2016)
27. Gentleman, R., Ihaka, R.: R Project for Statistical Computing (3.6.0) (software) (1993). https://www.r-project.org/
28. University of Waikato: WEKA (3.6.6) (software) (2006). https://www.cs.waikato.ac.nz/ml/index.html
29. Eberhard Karls University of Tübingen.: SNNS (4.3) (software) (2008). http://www.ra.cs.uni-tuebingen.de/SNNS/welcome.html
30. Chartier, B.P., Mahler, K.: Machine Translation and Neural Networks for a multilingual EU. Zeitschrift für Europäische Rechtslinguistik (ZERL) (2018)
31. Vurkaç, M.: Prestructuring multilayer perceptrons based on information-theoretic modeling of a Partido-Alto-based Grammar for Afro-Brazilian music: enhanced generalization and principles of Parsimony, including an investigation of statistical paradigms. Portland State University (2011)

Automatic Categorization of Answers by Applying Supervised Classification Algorithms to the Analysis of Student Responses to a Series of Multiple Choice Questions

Lisset A. Neyra-Romero[1,3](✉), Oscar M. Cumbicus-Pineda[2,3], Basilio Sicrra[3], and Samanta Patricia Cueva-Carrion[1]

[1] Departamento de Ciencias de la Computacion y Electronica, Universidad Tecnica Particular de Loja, San Cayetano Alto, Loja, Ecuador
laneyra@utpl.edu.ec

[2] Facultad de Energia, Carrera de Ingenieria en Sistemas y Computacion, Universidad Nacional de Loja, Ave. Pio Jaramillo Alvarado, La Argelia, Loja, Ecuador

[3] Departamento de Ciencias de la Computacion e Inteligencia Artificial, Universidad del Pais Vasco (UPV/EHU), Donostia-San Sebastian, Spain

Abstract. In recent years there has been a growing interest in machine learning for the classification and categorization of documents, texts and questions. This allows automating processes that if done with the intervention of the human being could have a high cost in time, and opens the doors for its implementation with inclusive systems for students with physical disabilities. This article describes a research work that uses data mining techniques to obtain classifiers that automatically identify the correct answers expressed by students, and these answers are then associated with a question with different options that are part of the process of evaluating the knowledge acquired by students during their formative process. In view of these consideration, where each question had multiple feasible, where each question had multiple feasible options to be selected; however, each question had only one correct answer. The answers are given by the students of the Open and Distance Modality of the Universidad Tecnica Particular de Loja were transcribed, with a total of 12960 transcriptions of the verbal answers obtained from the students. The results obtained by means of different classification algorithms were presented, analyzed and compared; giving as a result that the neural networks and the support vector machine (SVM) were the best to classify with an average percentage of 97% of success.

Keywords: Supervised classification · Categorization of answers · Supervised classification algorithms · Data mining · Artificial intelligence · Machine learning

M. Botto-Tobar et al. (Eds.): ICAETT 2019, AISC 1066, pp. 454–463, 2020.
https://doi.org/10.1007/978-3-030-32022-5_42

1 Introduction

In higher education, assessment is a major process for the training of students, in broad terms, educational evaluation is a process through which the merit of a given object in the field of education is assessed in order to make particular decisions [10].

Distance education is a way of teaching and learning based on a didactic dialogue mediated between the teacher (institution) and the student who, located in a different space from the former, learns independently (cooperatively)" [4,9].

The problem of solving text categorization dates back to the '80s, where what was used as a knowledge engine and an expert based on a set of rules of the form if X then Y; this technique, however, was unsuccessful because the set of rules required constant manual updating, which was totally impractical. It wasn't until the '90s that a Machine Learning [6] approach began to be used for the automation of this task, where through a set of data the algorithm is able to learn a model that allows it to categorize documents. Computational learning techniques have often been used to solve problems where large amounts of information are handled and it is necessary to find a pattern to determine the behavior of that information. The objective of computational learning is to develop models that are capable of learning from the previous experience of the events that are presented, from data sets [13]. Among some of the advantages offered by this approach is its high degree of automation, since the same algorithm takes care of the training of the classifier. The only requirement is to send you the amount of data needed to build such a classifying model. In addition, the results obtained in terms of accuracy are comparable to those of human experts. It is for these reasons that the use of automatic learning has become more widespread. A clear sample can be observed when using the different search applications offered by the company Google [15]. Question classification is a key point in many applications, such as Question Answering (QA, e.g. Yahoo! Answers), Information Retrieval (IR, e.g. Google search engine) and E-learning systems (e.g. Blooms tax. classifiers) [12].

Educational data mining (EDM) is an emerging interdisciplinary research field concerned with developing methods for exploring the specific and diverse data encountered in the field of education. One of the most valuable data sources in the educational domain are repositories of exam queries, which are usually designed for evaluating how efficient the learning process was in transferring knowledge about certain taught concepts, but which commonly do not contain any additional information about concepts they are related to beyond the text of the query and offered answers [5].

Universidad Tecnica Particular de Loja has two types of studies: Face-to-face education and Open and Distance (MAD), the latter with greater impact and acceptance in Ecuador. The Open and Distance Modality is a system of higher education that aims to reach all corners of the country. It consists of a didactic dialogue between the teacher and the student who learns in an independent and collaborative way [1]. This modality of studies is offered in university centers in Ecuador and in international centers in New York, Rome, Madrid, and Bolivia.

At the Universidad Tecnica Particular de Loja (UTPL), evaluation is a continuous process and a central element in the teaching-learning process through which students receive feedback on their progress. In the same way, it allows evaluating the effectiveness of the didactic techniques used, the scientific and pedagogical capacity of the educator, the quality of the didactic materials, among others [2].

The educational model of the Universidad Tecnica Particular de Loja (UTPL) for the Open and Distance Modality contemplates three types of evaluation: self-evaluation, and co-evaluation. According to the UTPL, the hetero-evaluation is of a permanent nature, in which the teacher permanently evaluates the student in order to know his work, performance, performance and learning styles. This is done through distance evaluations, online activities, and face-to-face evaluation [2].

The face-to-face evaluation is carried out using a mobile system of random questions that the student must answer, however for students with special disabilities this activity cannot be carried out individually so they must be accompanied and assisted in the evaluation by a relative, friend or acquaintance.

The importance of this work lies in the principle that "everyone has the right to learn and to participate in standardized education plans and curricula" [2]. On the matter in Chapter VII learning of people with disability of article 49 of the regulation of Academic Regime of Ecuador, the following provision is given for Higher Education Institutions (IES). "In each career or program, higher education institutions must guarantee to persons with disabilities appropriate learning environments that allow their access, permanence, and degree within the educational process, propitiating the learning results defined in the respective career or program. As part of the learning resources, IES shall ensure that persons with disabilities have access to information and communication systems and technologies (TIC) adapted to their needs" [7].

Therefore, the present work covers a first scope where the students' answers to each question is transformed into text for later classification and through the application of supervised classification algorithms, it is determined if the answer is correct or not, in this way it is intended to automate the evaluation process through the online evaluation system of the Universidad Tecnica Particular de Loja.

Silva's study revealed that SVM is the main algorithm of the Machine Learning used, while BOW (Bag of Words) and TF-IDF (Term Frequency-Inverse Document Frequency) are the main techniques for feature extraction and selection, respectively [14], on the basis of this study, the BOW technique was used in the processing of responses and, it had been confirmed that one of the best algorithms for this type of problem was the SVM.

2 Methodology

For the development of this research work, the data mining process was used as a base, where 4 phases are identified. This process was applied to the case study carried out at the Universidad Tecnica Particular de Loja and the implementation will be carried out in the evaluation system (Fig. 1).

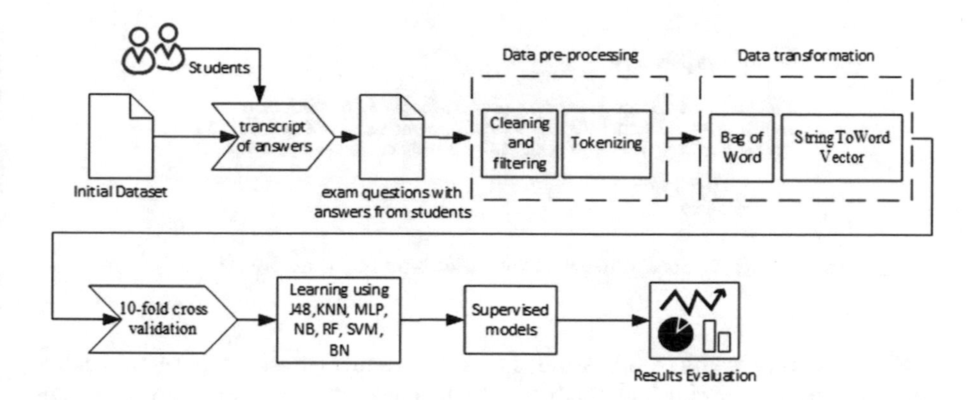

Fig. 1. Overview of the process

2.1 Phase 1: Data Selection

The objective of this stage is to select the data sources for the data mining process, as an internal source of the Universidad Tecnica Particular de Loja, evaluations were selected from the question banks of the Direction of Educational Technologies (DTE) referring to two educational components, which have the largest number of students. The academic components to be used are: Study Methodology and National and Environmental Reality, these components are common in all university degrees, in each academic period these academic components is attended by approximately 7,000 students. The backgammon was hosted in the university's evaluation system in HTML format and was transformed into XLS format for processing.

Subsequently, we proceeded to transcribe the answers of 20 students referring to 648 questions, obtaining a total of 12960 text-type answers Table 1; for each question, there is only one correct answer and it can be dichotomous or polytomous.

Table 1. Total questions and answers

Type of question	Total number of questions	Total answers
Dichotomous	456	9120
Polytomous	192	3840
Total	648	12960

Figure 2, there is a question of the educational component called Study Methodology where the answer choices are true or false, these are considered to be dichotomous. The correct answer is FALSE.

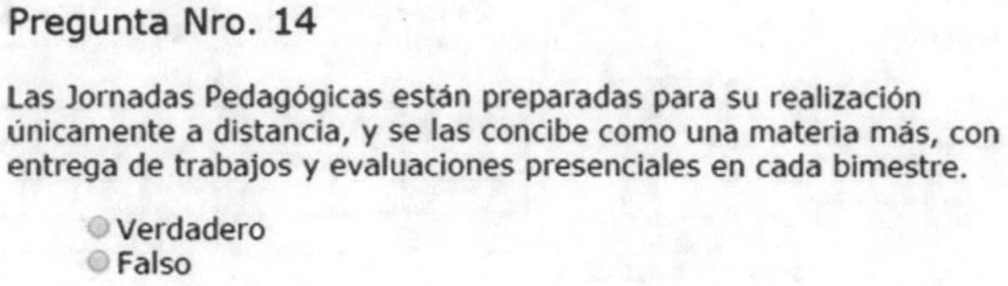

Pregunta Nro. 14

Las Jornadas Pedagógicas están preparadas para su realización únicamente a distancia, y se las concibe como una materia más, con entrega de trabajos y evaluaciones presenciales en cada bimestre.

- Verdadero
- Falso

Fig. 2. Study methodology dichotomous question [3]

Figure 3, represents a question from the educational component called National and Environmental Reality, which has multiple answer options. The correct answer is: They reject traditional parties.

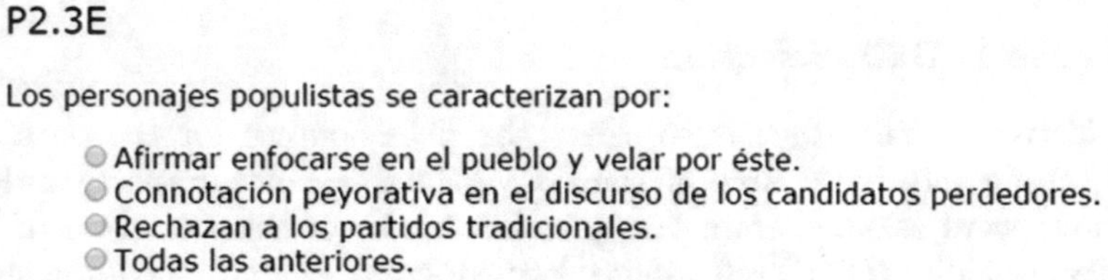

P2.3E

Los personajes populistas se caracterizan por:

- Afirmar enfocarse en el pueblo y velar por éste.
- Connotación peyorativa en el discurso de los candidatos perdedores.
- Rechazan a los partidos tradicionales.
- Todas las anteriores.

Fig. 3. Study methodology polytomous question [3]

2.2 Phase 2: Data Preprocessing

The objective of this stage was to obtain clean data, data without null or anomalous values that allow us to obtain quality patterns, that is to say, once the transcription of the answers given by the students was carried out, we proceeded to eliminate those that did not have a value that coincided with one of the possible values of these answers, for example those answers in which the student answered with the phrases: I do not know, I am not sure, I do not know the answer, among others.

Another activity that was carried out in this phase was tokenization which consisted of replacing in the transcription of the answers all the characters that the tool used for the training process WEKA (Waikato Environment for Knowledge Analysis) did not admit, characters such as letter, accents, double spaces. This process was done massively and automatically with the help of specific commands for this task.

2.3 Phase 3: Data Transformation

The objective of this phase is to transform the data source into a set ready to apply the different data mining techniques. In order to facilitate the training process, all the answers for questions were adapted to the ARFF format (Attribute-Relation File Format) required by WEKA (Waikato Environment for Knowledge Analysis) and the class was defined with YES or NO values, where for each correct answer transcribed by the students the YES value was set and the incorrect ones the NO value. Figure 4, question 52 is shown whose correct answer is: Conflict.

P52

Prédicas paternalistas y demagógicas han impedido que el pueblo adquiera conciencia de sus responsabilidades y, por ende ha impedido fortalecer la construcción de ciudadanía. Esta definición hace referencia a:

- Líderes populistas.
- Regionalismo.
- Conflictividad.

Fig. 4. Question 52 [3]

Figure 5, you can see the result of the transformation process of question 52 with their respective answers.

```
@relation Pregunta52

@attribute text string
@attribute class {YES, NO}

@data
"los Lideres populistas",NO
"seria los Lideres populistas",NO
"creo que son Lideres populistas",NO
"me parece que son los Lideres populistas",NO
"supongo que son Lideres populistas",NO
"imagino que son Lideres populistas",NO
"sin lugar a dudas son los Lideres populistas",NO
"el Regionalismo",NO
"Imagino el Regionalismo",NO
"me parece que el Regionalismo",NO
"podria ser el Regionalismo",NO
"imagino que es el Regionalismo",NO
"la Conflictividad",YES
"seria la Conflictividad",YES
"Imagino que la Conflictividad",YES
"considero que la Conflictividad",YES
"supongo la Conflictividad",YES
"podria ser la Conflictividad",YES
"definitivamente la Conflictividad",YES
"sin lugar a dudas la Conflictividad",YES
```

Fig. 5. The outcome of the transformation process of question 52

For the purposes of vectorization we have decided to use the Bag of Words using for this purpose the StringToWordVector filter, which converts the String attributes into a set of attributes representing the occurrence of the words in the text.

2.4 Phase 4: Experimentation

For this step we chose 7 supervised classification algorithms K-nearest neighbours (IBK), decision tree (J48), multilayer perceptron (MLP), Bayesian network (BN), random forest (RF), support vector machine (SVM), naive bayes (NB), which were applied to each bag of words of the answers to the 648 selected questions, which gave us a total of 4536 applications of the aforementioned algorithms, applications that were made with the help of a script.

To avoid overflowing, the models generated by the originally named classifiers were validated using the k-fold cross-validation technique (in our study we used 10-fold-CV). This technique randomly divides the original sample into 10 folds or sub-samples. One of the nine sub-samples is used to test the model, while the remaining nine are used for the algorithm training process. This process is repeated 10 times for each of the sub-samples of k. Thus, the 10 results are obtained which are then averaged to evaluate the performance of the classifier [8].

For the evaluation of the supervised models, the overall percentage of success of the algorithms in each of the data sets was taken into account, thus observing that for dichotomous questions the algorithms that obtained the best results were J48 and SVM (Fig. 6), for polytomous questions the algorithms MLP and SVM (Fig. 7) obtained the best results.

3 Evaluations and Results

Once the classifier is generated, it is necessary to measure its capacity to predict the new instances. To do so, it is necessary to have a set of examples (different from the set with which the classifier was generated), called the evaluation set. In the absence of an independent evaluation set of the training set, k-fold cross validation is performed. There are two methods to obtain a general evaluation of the performance of the classifier, macro average, and micro average. For both methods, it is necessary to obtain four values for each class [11]:

3.1 Results

Figure 6, shows the percentages obtained in each question for each algorithm executed. In the case of dichotomous questions, it was considered that having the same answer option for all the questions, it was included in a single execution both for when the correct answer is TRUE and also for when it is FALSE.

Figure 7 shows the percentage of success achieved by the algorithms when applied to the answers to the polytomous questions.

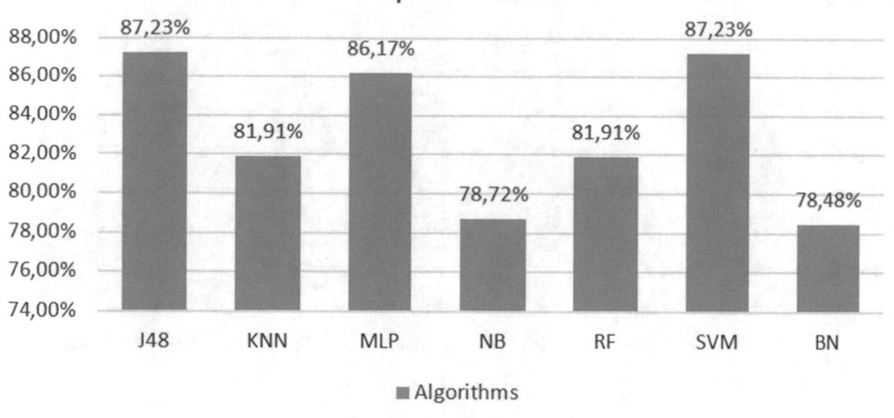

Fig. 6. Average percentage of successes in dichotomous question.

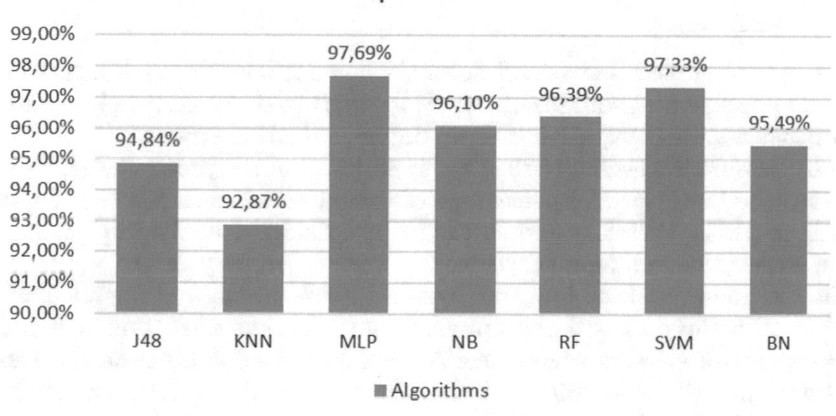

Fig. 7. Average percentage of successes in polytomous question.

Figure 8, it-is demonstrated that the algorithm that obtained a greater global percentage of successes in the classification process are the neural networks (MLP), followed by the algorithm of Support Vector Machine (SVM) obtaining also good results; on the other hand, the algorithm that obtained a lower global percentage of successes in comparison with the other algorithms executed is k-nearest neighbors (IBK).

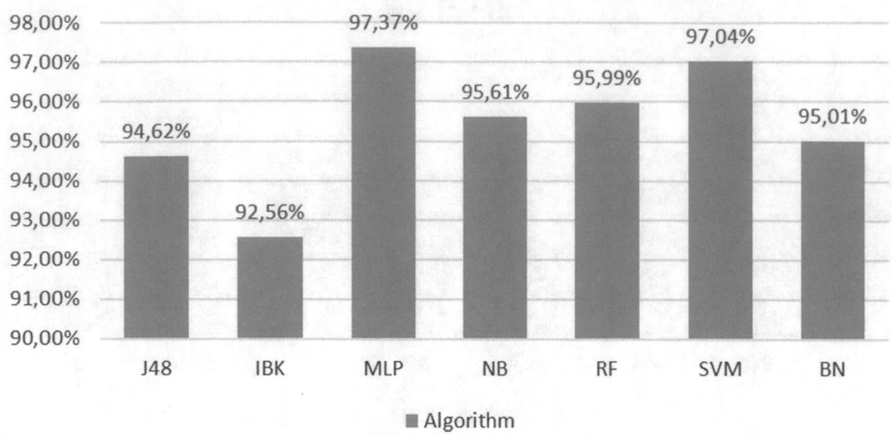

Fig. 8. Overall average of hit percentages.

4 Conclusion and Future Work

4.1 Conclusion

It has been identified that neural networks and Support Vector Machines have proven to be the most suitable classifiers for this classification problem, and can be implemented in integration with an online evaluation system.

The classifier according to the results obtained in the present work that has obtained the lowest overall percentage of correct classification is the K-nearest neighbors (IBK), therefore excludes it to be considered in the implementation phase in the online evaluation system.

In the questions that have not reached 100% in the classification of the answers with the proposed algorithms, it has been identified that it is due to the existence of answers whose classification should be focused on the textual semantic similarity that exists between the answers given. Where there is the challenge of finding the degree of similarity between a paragraph and a sentence, a sentence and a phrase, a phrase and a word and a word and a meaning.

4.2 Future Work

Implementation in the online assessment system that includes a voice recognition module so that the answers given by the students in a spoken form are transcribed.

Determine the algorithm that obtains 100% in the classification of the answers for which a total correct classification has not been obtained.

Expand the process of experimentation with other educational components whose level of difficulty in the classification process is greater, since it is required to interpret semantically the answers issued.

Implementation of the automatic learning process in the online assessment system in such a way as to correctly classify the text-type responses obtained by students who have been transcribed and to provide immediate qualification of the assessment applied, thus directly benefiting students with physical and visual disabilities.

References

1. Modalidad abierta y a distancia (2012). https://distancia.utpl.edu.ec/modalidad-abierta/descripcion
2. Sistema de evaluación de los aprendizajes (2012). https://distancia.utpl.edu.ec/modalidad-abierta/sistema-evaluacion
3. Banco de preguntas de metodología de estudio (2013)
4. Guía general mad (2014). http://www.utpl.edu.ec/sites/default/files/pregrado/guia-general-MAD.pdf
5. Begusic, D., Pintar, D., Skopljanac-Macina, F., Vranic, M.: Annotating exam questions through automatic learning concept classification, pp. 176–180 (2018)
6. Bishop, C.M.: Pattern Recognition and Machine Learning (2006). **60**(1), 78 (2012)
7. Ces, C.d.: Reglamento de regimen académico. Reglamento de Régimen Académico. Quito, Pichincha, Ecuador: Consejo de Educación Superior (2013)
8. Cumbicus-Pineda, O., Ordoñez-Ordoñez, P., Neyra-Romero, L., Figueroa-Diaz, R.: Automatic categorization of tweets on the political electoral theme using supervised classification algorithms. Commun. Comput. Inf. Sci. **895**, 671–682 (2019)
9. Lorenzo, G.A.: La educación a distancia: de la teoría a la práctica. Cap. IV, Ariel (2001)
10. Méndez Martínez, J., Ruiz Méndez, R.: Evaluación del aprendizaje y tecnologías de información y comunicación (tic): De la precensialidad a la educación a distancia. Revsta de Evauación Educativa REVALUE **4**, (2015)
11. de la Rosa, A.G.R.: Clasificacion de textos utilizando informaciøn inherente al conjunto a clasificar (2010)
12. Sangodiah, A., Muniandy, M., Heng, L.: Question classification using statistical approach: a complete review. J. Theor. Appl. Inf. Technol. **71**(3), 386–395 (2015)
13. Sierra Araujo, B.: Aprendizaje automático: conceptos básicos y avanzados: aspectos prácticos utilizando el software weka (2006)
14. Silva, V., Bittencourt, I., Maldonado, J.: Automatic question classifiers: a systematic review. IEEE Trans. Learn. Technol. (2018)
15. Varguez-Moo, M., Uc-Cetina, V., Brito-Loeza, C.: Clasificación de documentos usando máquinas de vectores de apoyo. Abstraction and Application Magazine **6**, (2014)

SMCS: Mobile Model Oriented to Cloud for the Automatic Classification of Environmental Sounds

María José Mora-Regalado, Omar Ruiz-Vivanco(✉), and Alexandra Gonzalez-Eras

Universidad Técnica Particular de Loja, Loja, Ecuador
{mjmora5, oaruiz}@utpl.edu.ec

Abstract. This paper presents SMCS, a cloud-oriented mobile system model that uses a Convolutional Neural Network for the automatic classification of environmental sounds in real time. The model comprises an architectural schema with its corresponding deployment scheme in Google cloud services provider. Finally, the validation protocol of SMCS is applied in two experiments using respectively the base of free sounds FSDkaggle2018 and a selection of warning sounds extracted from the same sound base. The results of the validation of the model are promising with high values of precision in the classification of sounds, demonstrating that the SMCS model is expected to be a point of reference for the development of sound analysis systems, contributing to improving the quality of life of people with Hearing Impairment.

Keywords: Hearing impairment · Cloud Computing · Mobile system · Convolutional neural network · FSDkaggle2018

1 Introduction

Hearing impairment refers to any decrease or loss of the functioning of the person's hearing system, which limits their ability to recognize specific sounds and requires adjustments in order to meet their special needs [1]. There are devices that increase hearing person however, do not allow processing of sounds in such a way to generate additional context information [2]; for example, the sound of a siren is not only interpreted as a loud sound but, is associated with an alert on the passage of a vehicle at high speed, a fact of which a person with hearing impairment would not have full consciousness, using the current hearing devices.

Intelligent systems are a useful tool for people with disabilities, because they adapt to the special needs of the individuals who suffer them [2]; in such a way that they receive the perceptions of the user, process them through automatic learning techniques and generate answers, which improve the performance of individuals in specific environments and events [1]. Thus arises SMCS, a proposal for a mobile system based on a neural machine learning model (ANN) and a knowledge base to identify and classify sounds, SMCS model provides a tool that enables people with disabilities easily incorporated into any environment [1, 23], providing response to user requests in

M. Botto-Tobar et al. (Eds.): ICAETT 2019, AISC 1066, pp. 464–472, 2020.
https://doi.org/10.1007/978-3-030-32022-5_43

real time and from any geographic location [3]; for this, the intelligent system captures sounds from a cell phone and analyzes them using machine learning techniques available in real time through Cloud services [6, 23]; which ensures the scalability and accessibility of the intelligent system, and at the same time, people with hearing disabilities have immediate services and assistance [2, 5].

In the following sections, we present a review of related works that demonstrate the relevance of our proposal, then the methodological approach about the architecture and the validation protocol. In addition, we propose a case study to evaluate the system using a knowledge base of previously tagged sounds and a measure of accuracy in the automatic classification of all sounds and those that represent threats for people with hearing disabilities. Finally, the analysis of these results, the scope, and conclusions of the proposal.

2 Related Works

There are commercial products including computer-based programs to support processes of auditory verbal therapy in order to increase people listening skills of people [3] and improve communication through computer-assisted lip-reading recognition [4], using self-organization neural networks (SOMNN) and extension techniques for sound phonemes recognition and pronunciation. However, few have been done in the automatic classification of environmental sounds, providing people with hearing disabilities semantic context information, which helps the normal development of their activities in different environments.

In addition, sound analysis by Machine Learning techniques has had greater development in other areas of health, for instance detecting pulmonary disorders using convolutional neural networks (CNN) that classify different types of breathing sounds [7], others instead use recurrent neural networks (RNN) to classify snoring and noise patterns [8]. These works highlight the importance of preprocessing sounds and propose schemes such as random spectrogram cuts and Vocal Tract Length Perturbation (VTLN), according to Mel Frequency Cepstrum Coefficient (MFCC: precision 92.12%) and Local Binary Patterns (LBP: precision 71.21%) [7]; also combine the Fourier Transform to extract length and frequency audio segments and MFCC to extract the patterns, with a sensitivity of 75.7% [8].

Besides, research has been developed with emphasizing in the classification of whale whistles using spectrograms, spectral subtraction methods and CNN [8]; others to classify free-purpose environment sounds with CNN but transform the logarithm of the spectrogram of the signals (logmel) and then normalize in batches (BN) [10]. In contrast, the variant of the CNN model in [11] is the use of densely connected networks (DenseNet Keras) with a softmax output layer with multiple heads, the normalization of the audio at a dB level of −0.1 and of the spectrograms by Rectified Linear Units (ReLUs), with a 92.61% accuracy.

It is noteworthy none of the works previously analyzed use Cloud Computing, although these platforms offer powerful algorithms for the analysis of natural language (voice), to speech analysis; leaving sound analysis as an unexplored gap. Therefore, this work is a proposal for the development of a sound classification system using Cloud Computing as indicated by each process defined in the model.

3 Methodology

This section shows the components of the SMCS model, which comprises two parts: 1. the development of the architectural scheme under the 4 + 1 paradigm, with its corresponding phase of deployment in the Cloud, and 2. the use and validation protocol to the classification system of environmental sounds.

3.1 Architectural Schema of SMCS Model

Figure 1 shows the deployment view of the SMCS system, which comprises three components: a user device, a mobile application, and the cloud services provider. These elements are grouped in the following layers:

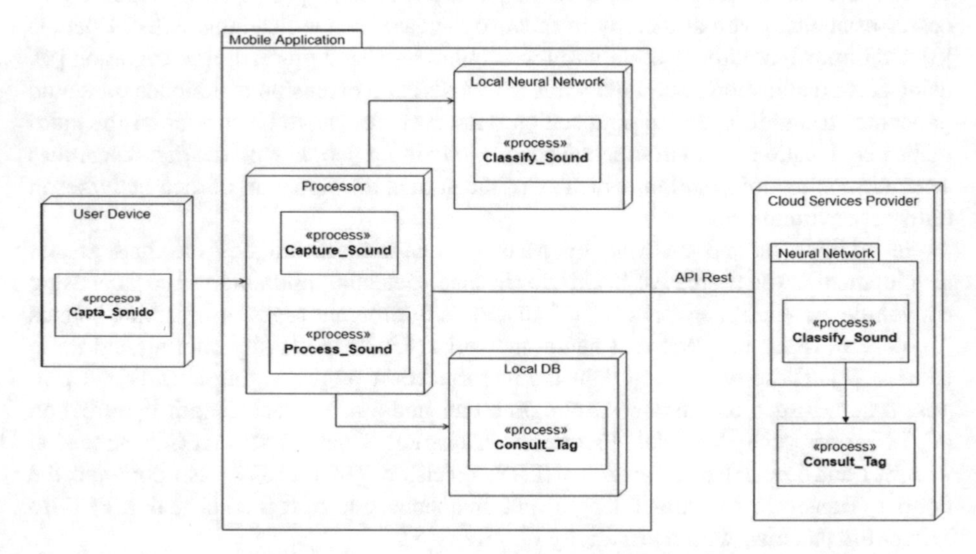

Fig. 1. Architectural model of the SMCS system

- The cloud-service provider layer, which offers services for IaaS, PaaS, and SaaS, also contains the knowledge base and the neural network model. Given the system requirements, the Google Cloud platform is chosen as a Cloud service provider [23]. Figure 2 shows the services required for the deployment of the system: App Engine for the management of processes in the cloud, REST for the management of sounds from the mobile device, ML Engine for the implementation of CNN and subsequent and Cloud Storage to store the knowledge base of sounds.

- The user layer comprises the user's device that contains the application for the capture and processing of sounds, as well as a light CNN, to classify priority sounds that constitute alerts, according to a reduced version of the knowledge base; In this way the SMCS system covers cases such as the lack of connection of the mobile device, and the non-availability of the service in the Cloud. The mobile application is developed in the IOS operating system that offers better features for capturing environmental sounds (five seconds in length), using the microphone of the cell phone and AVAudioRecorder [20]. In the implementation of the lighter CNN model, use the CoreML framework [22]. In addition, agile development methodologies are useful for the development of the application because of their increased quality of application, effective use of resources, lowered time-to-market and cost savings [23].

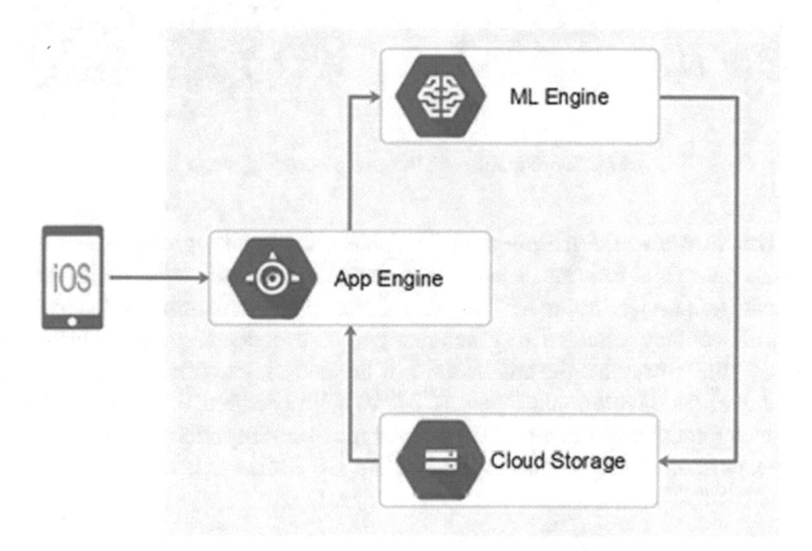

Fig. 2. Cloud architectural model of the SMCS system

3.2 Protocol of Validation of the Classification Model

The validation of the classification system of the SMCS includes the following steps:

- **Data processing:** begins with the selection of the set of general-purpose sounds of the Kaggle competition (FSDKaggle2018), which contains two datasets: the training set with 9473 audio records, classified into 41 categories (including human, animal, musical, environmental, among others), and the second test with 1600 audio records; each sound has a duration between 300 ms and 30 s [12, 13]. Then the audios enter a process of cleaning and normalization of environmental sounds with their respective labels [14, 15]. To improve training, fragments without audio sound are eliminated [11], and a spectrogram is obtained from the result, which is an image with the visual representation of the signal frequency spectrum of a sound [16] (Fig. 3).

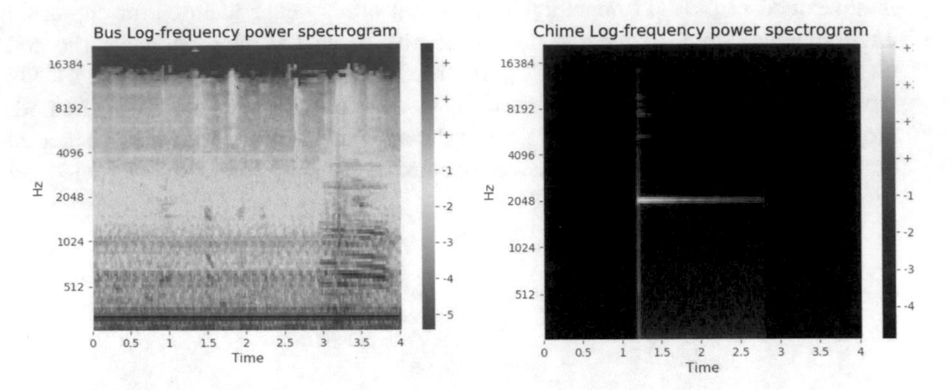

Fig. 3. Examples of the pre-processing stage

- **Experimentation and training:** in this phase, the spectrograms generated in the previous phase and their labels in the knowledge base are the input for the training of the model (supervised approach), with the CNN AlexNet, recommended in [8] by the values of accuracy achieved in speech recognition processes, computer vision among others. Table 1 presents the architecture of the neural network, which comprises an input layer, 5 convolutional layers (CONV) with rectified linear units (ReLU) in each, as an activation function [17] a layer max pooling after each CONV layer, a single layer completely connected and as the last SoftMax layer [18, 19].

Table 1. Convolutional neural network architecture

Input layer
Conv 1 – ReLU Max – Pooling
Conv 3 – ReLU Max – Pooling
Conv 2 – ReLU Max – Pooling
Conv 2 – ReLU Max – Pooling
Conv 2 – ReLU Max – Pooling
Fully connected layer
SoftMax

4 Experimentation

The experiment aims to carry out the evaluation of the SMCS system following the validation protocol presented in Sect. 3.2, to automatically classify sounds of events that happen in everyday environments (such as the sound of a horn in the city traffic) and that are difficult to recognize for people with hearing disabilities. For this purpose, we use a general-purpose sound collection available in the Kaggle competition in two experiments explained below, considering the previous classification of sounds dataset training to establish the degree of success of our model compared with the classification made by an expert on the Kaggle dataset.

4.1 First Experiment

Table 2 presents the results of the first experiment, where we randomly used all the Kaggle training dataset setting 80% for CNN training, and the remaining 20% for model validation; then, sounds preprocessing is performed using the spectrogram technique that finds relevant frequencies and unifies the duration times of each record. Then the CNN is configured according to Table 1 with two classification models: first, the MobileNets model [21], obtaining an accuracy of 73.34%; and second, the CNN Alexnet model [8], obtaining an accuracy of 71.1%, both measures in relation to the sounds original classification of the Kaggle dataset.

Table 2. Results with the complete Kaggle dataset

Experimentation	Neural model	Pre-processing	Precision
1	CNN Proposed based in Alexnet	Spectrogram	73,34%
2	MobileNets	Spectrogram	71,1%

4.2 Second Experiment

In this second experiment, 1583 sounds are extracted from the original dataset, distributed in 8 categories (Fig. 4) that, according to experts, contain sounds that represent a danger for a person with hearing disability [25, 26]. In the same way that in the first experiment, the collection is randomly divided into 80% for the training dataset and 20% for the test, just as the same techniques are used for the preprocessing of the sounds. By performing the classification with the two models CNN rated sounds reaching 73.2% accuracy with the MobileNets model and 79.41% with the Alexnet model for sounds previously classified in the dataset Kaggle is obtained (Table 3).

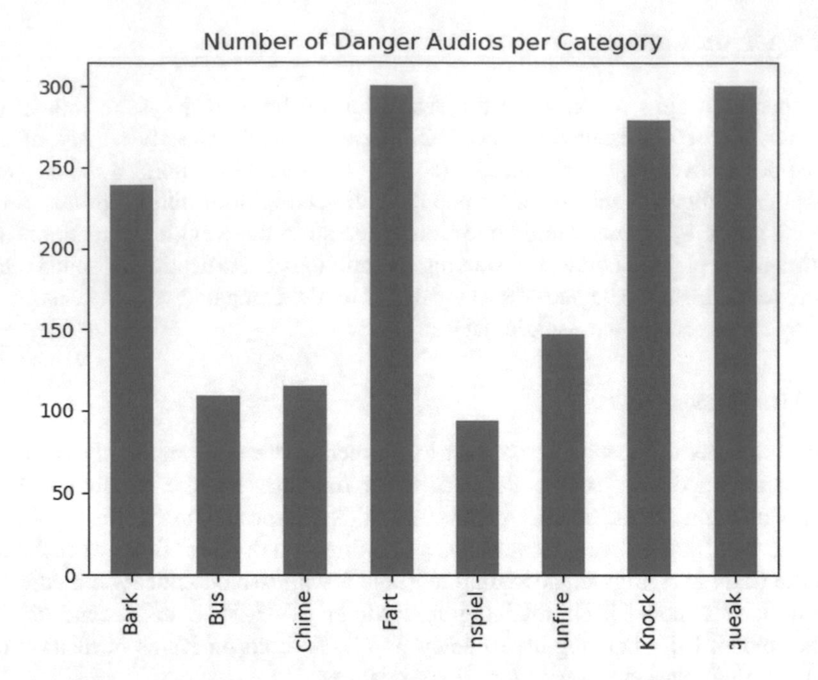

Fig. 4. Distribution of sounds into categories according to [25, 26]

Table 3. Results using the selected categories of the Kaggle dataset

Experimentation	Neural model	Pre-processing	Precision
1	CNN Proposed based in Alexnet	Spectrogram	79,41%
2	MobileNet	Spectrogram	73,2%

5 Conclusion

SMCS is a system that allows the automatic classification of environmental sounds in real time being a support for those with hearing impairment and who have serious problems to interact safely with the environment that surrounds them. Therefore, the objective of this proposal is to bring together in the SMCS model the best proposals and experiences in the field of sound analysis based on projects carried out to date.

Among the good practices obtained in the research is the pre-processing of the audio segments through normalization at a dB level of -0.1, cutting silence spaces and subsequently calculating the spectrograms. On the neuronal model, the most recommended is the CNN convolutional neural network based on AlexNet with five convolutional layers with rectified linear units and the use of the Softmax function in the output layer of the model.

For the development of the mobile application, it is important to consider the sound recording applications that exist, such as MediaRecorder (Android) and AVAudioRecorder (IOS). As well as Google Cloud services that among other features is compatible with the hardware available in the Medium range phones have a low complexity of implementation, offers several machine learning services, including the TensorFlow neural network model, as well as having a high-performance GPU instance in the cloud. In cases of limited or no connectivity, the application uses the local storage of a base of basic sounds. The architectural model of SMCS incorporates ML services to classify sounds in the cloud, enabling open implementation in other types of applications and platforms.

According to the experiments carried out, the CNN Proposed based on the Alexnet model obtains 73.34% precision in the first experiment and 79.41% in the second experiment, being the best implementation of the neuronal model obtained in the validation of the present work. The results obtained show that the proposed scheme is promising according to the precision values of the models reviewed compared to the accuracy achieved by the SMCS learning model.

Future work focuses on the improvement of the preprocessing phase of sounds, maintaining the representation based on spectrograms but emphasizing signal normalization for the reduction of environmental noise. Besides, we will increase the number of experimental scenarios considering the reduction of categories that contains alarm sounds, and thus clearly establishing knowledge bases for both the mobile application and the processing in Cloud services provider.

References

1. Vejarano, R., Alain, L.: Alternativas tecnológicas para mejorar la comunicación de personas con discapacidad auditiva en la educación superior panameña. Rev. Educ. la Univ. Granada **23**, 219–235 (2016)
2. Noor, T.H., Zeadally, S., Alfazi, A., Sheng, Q.Z.: Mobile cloud computing: Challenges and future research directions. J. Netw. Comput. Appl. **115**, 70–85 (2018). https://doi.org/10.1016/j.jnca.2018.04.018
3. Nanjundaswamy, M., Prabhu, P., Rajanna, R.K., et al.: Computer-based auditory training programs for children with HEARING impairment – a scoping review. Int. Arch. Otorhinolaryngol. **22**, 88–93 (2018). https://doi.org/10.1055/s-0037-1602797
4. Lay, Y.-L., Tsai, C.-H., Yang, H.-J., et al.: The application of extension neuro-network on computer-assisted lip-reading recognition for hearing impaired. Expert Syst. Appl. **34**, 1465–1473 (2008). https://doi.org/10.1016/j.eswa.2007.01.042
5. Chen, L., Tsai, C., Chang, W., et al.: A real-time mobile emergency assistance system for helping deaf-mute people/elderly singletons. In: 2016 IEEE International Conference on Consumer Electronics (ICCE), pp 45–46 (2016)
6. Jiang, J., Bu, L., Duan, F., et al.: Whistle detection and classification for whales based on convolutional neural networks. Appl. Acoust. **150**, 169–178 (2019). https://doi.org/10.1016/j.apacoust.2019.02.007
7. Bardou, D., Zhang, K., Ahmad, S.M.: Lung sounds classification using convolutional neural networks. Artif. Intell. Med. **88**, 58–69 (2018). https://doi.org/10.1016/j.artmed.2018.04.008
8. Lim, S.J., Jang, S.J., Lim, J.Y., Ko, J.H.: Classification of snoring sound based on a recurrent neural network. Expert Syst. Appl. **123**, 237–245 (2019). https://doi.org/10.1016/j.eswa.2019.01.020
9. Jeong, I.-Y., Lim, H.: Audio tagging system for DCASE 2018: focusing on label noise, data augmentation, and its efficient learning (2018)

10. Dorfer, M., Widmer, G.: Grating general -purpose audio tagging networks with noisy labels and interactive self-verification (2018)
11. Kaggle, F.: General-Purpose Audio Tagging Challenge
12. IEEE AASP Challenge on Detection and Classification of Acoustic Scenes and Events. http://dcase.community/challenge2018/
13. Shen, G., Nguyen, Q., Choi, J.: An environmental sound source classification system based on mel-frequency cepstral coefficients and gaussian mixture models. IFAC Proc. Vol. **45**, 1802–1807 (2012). https://doi.org/10.3182/20120523-3-RO-2023.00251
14. Fonseca, E., Plakal, M., Font, F., et al: General-purpose Tagging of Freesound Audio with AudioSet Labels: Task Description, Dataset, and Baseline (2018)
15. Ozer, I., Ozer, Z., Findik, O.: Noise robust sound event classification with the convolutional neural network. Neurocomputing **272**, 505–512 (2018). https://doi.org/10.1016/j.neucom.2017.07.021
16. Piczak, K.J.: Environmental sound classification with convolutional neural networks. In: 2015 IEEE 25th International Workshop on Machine Learning for Signal Processing (MLSP), pp. 1–6 (2015)
17. Peréz Peréz, E.: Diseño de una metodología para el procesamiento de imágenes mamográficas basada en técnicas de aprendizaje profundo. Universidad Politécnica de Madrid (2017)
18. Picazo Montoya, Ó.: Redes Neuronales Convolucionales Profundas para el reconocimiento de emociones en imágenes. Universidad Politécnica de Madrid (2018)
19. Developers G MediaRecorder. https://developer.android.com/reference/android/media/MediaRecorder
20. Developer A AVAudioRecorder
21. Howard, A.G., Zhu, M., Chen, B., et al.: MobileNets: Efficient Convolutional Neural Networks for Mobile Vision Applications. CoRR abs/1704.0 (2017)
22. Developer A CoreML. https://developer.apple.com/documentation/coreml
23. Marozzo, F.: Infrastructures for High-Performance Computing: Cloud Infrastructures. In: Ranganathan, S., Gribskov, M., Nakai, K., Schönbach, C. (eds.) Encyclopedia of Bioinformatics and Computational Biology, pp. 240–246. Academic Press, Oxford (2019)
24. Wasserman, T. Software engineering issues for mobile application development. In: FoSER (2010)
25. Łopatka, K., Zwan, P., Czyżewski, A.: Dangerous sound event recognition using support vector machine classifiers. In: Advances in Intelligent and Soft Computing, pp. 49–57 (2010)
26. Almaadeed, N., Asim, M., Al-Maadeed, S., Bouridane, A., Beghdadi, A.: Automatic detection and classification of audio events for road surveillance applications. Sensors **18**(6), 1858 (2018)

User-Based Cognitive Model in NGOMS-L for the Towers of Hanoi Algorithm in the Metacognitive Architecture CARINA

Yenny P. Flórez, Alba J. Jerónimo(✉), Mónica E. Castillo, and Adán A. Gómez

Universidad de Córdoba, Montería 230001, Colombia
{ajeronimomontiell4,mecastillo, aagomez}@correo.unicordoba.edu.co

Abstract. The Towers of Hanoi is a mathematical problem, which consists of three pegs, and a number "n" of disks of distinct sizes which can slide onto any peg. A cognitive model is a theoretical, empirical and computational representation of mental processes which belong to a cognitive function. A cognitive model generates a human-like performance for developing tasks, correcting errors, using strategies, and acquiring knowledge. A cognitive model constructed in a cognitive architecture is characterized to be runnable and producing specific behaviors. A cognitive architecture is a general-purpose control framework based on scientific theories to specify computational models of human cognitive performance. CARINA is a metacognitive architecture for artificial intelligent agents, obtained from the MISM Metacognitive Meta model. The objective of this paper is to present a cognitive model based on NGOMS-L to solve the algorithm of the Towers of Hanoi that can be runnable in the metacognitive architecture CARINA. The methodology used for the analysis of the cognitive task was: *pre-processing stage*, *processing stage*, *classification of subjects*, *description of cognitive task in natural language* and finally *description of the cognitive task in NGOMS-L*. The results obtained showed that of the four subjects originally selected, three of them were able to solve the problem and only one abandoned the problem. In the classification made, a successful and an unsuccessful subject was selected, to represent the cognitive task in natural language and finally express in the NGOMS-L notation the different Goals, Operators, Methods, and Selection Rules.

Keywords: Towers of Hanoi · Cognitive model · Cognitive architecture · Metacognitive architecture · CARINA

1 Introduction

The Towers of Hanoi is a mathematical problem, which consists of three pegs, and a number "n" of disks of distinct sizes which can slide onto any peg [1]. Towers of Hanoi is a well-structured test [2, 3] which allows the memory content of a person to be seen when solving problems [4, 5]. Cognitive modeling (CM) is a research procedure commonly used in cognitive sciences where theoretical assumptions are represented as

M. Botto-Tobar et al. (Eds.): ICAETT 2019, AISC 1066, pp. 473–484, 2020.
https://doi.org/10.1007/978-3-030-32022-5_44

computer programs [6]. The main objectives of cognitive modeling are: (a) to describe (b) to predict, and (c) to prescribe human conduct [7, 8] using computational representations of cognitive tasks denominated cognitive models [9]. A cognitive model is a theoretical, empirical and computational representation of mental processes which belong to a cognitive function [10, 11]. A cognitive model generates a human-like performance for developing tasks, correcting errors, using strategies, and acquiring knowledge [12].

Expert-based cognitive models are computational representations made by experts in different domains of experience (e.g., different disciplines, specialties, or cultures) who observe the data and respond to specific patterns of each cognitive model [13, 14]. User-based cognitive models are computational representations in which the subject solves a certain cognitive task to monitor user behavior and make generalizations and predictions based on observations, using well-structured problems such as the Towers of Hanoi [15]. User-based cognitive models can be used to predict a user's behavior, obtain knowledge, or improve existing computational models [15]. Analyzing the structure of the cognitive task is required to build a cognitive model [16]. This analysis process uses techniques which facilitate the description of the specific order of the events that structure of the cognitive task. One of these techniques is GOMS. GOMS (Goals, Operators, Methods and Selection Rules) is a specification of the knowledge that a system needs for completing a cognitive task [17]. A GOMS model describes the methods required to achieve specific goals [18]. NGOMS-L is a structured natural language notation for representing GOMS models and specifies processes for constructing them [19].

Cognitive architectures can be used in the building process of cognitive models [20]. A cognitive model constructed in a cognitive architecture is characterized by being runnable and producing specific behaviors [11]. A cognitive architecture is a general-purpose control framework based on scientific theories to specify computational models of human cognitive performance [21–23]. A metacognitive architecture according to [9–27] is a framework that integrates two levels of information processing: the Object Level (Cognition) and the Meta Level (Metacognition). Furthermore, in between levels an introspective monitoring process which facilitates exchange of information must exist [24] and self-reasoning on the implemented computational models. CARINA is a metacognitive architecture for artificial intelligent agents, obtained from the MISM Metacognitive Meta model [24–27]. Currently, the metacognitive architecture CARINA only has developed cognitive models based on experts [22, 23] and it has not used the user-based cognitive model creation method. For this reason, the present study describes the first stages of the creation process of this type of cognitive models. These stages are the following: the selection of the cognitive task, its representation in natural language and its representation in NGOMS-L [23]. The objective of this paper is to present a user-based cognitive model in NGOMS-L to solve the algorithm of the Hanoi towers in the metacognitive architecture CARINA.

The structure of this paper is as follows: the basic conceptualization of the Towers of Hanoi algorithm is shown in Sect. 2. The main principles of the CARINA metacognitive architecture are presented in Sect. 3. Section 4 points to the formal representation of cognitive models in the metacognitive architecture CARINA. Section 5 describes an overview of cognitive architectures and cognitive models for the

Towers of Hanoi algorithm. The methodology of this research is described in Sect. 6. And finally, the cognitive model of the Towers of Hanoi algorithm is presented in the NGOMS-L notation.

2 The Towers of Hanoi

The Towers of Hanoi is a mathematical problem, which consists of three pegs, and a number "n" of disks of distinct sizes which can slide onto any peg [1] (Fig. 1).

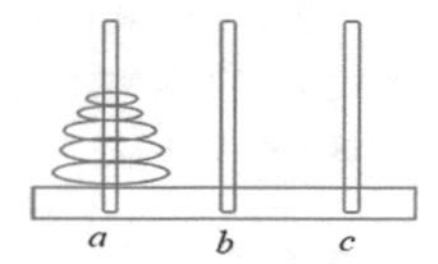

Fig. 1. Towers of Hanoi with five disk [own source].

The standard version of the Hanoi Towers consists of a series of a number "n" of disks, organized according to their size with the first being the smallest, and three pegs called <A>, <B> and <C>. In the initial state, the "n" disks are located on the "A" peg in order of size, with the biggest disk at the base and the smallest at the top. The goal of the problem is to put all the disks in "C", leaving the disks in the same location as the initial state. The constraints to solve the problem indicate that only one disk can be moved at a time (the one at the top), a bigger disk cannot be placed on top of a smaller one, and the subjects must come to the solution with the least number of movements [4–28].

3 CARINA Metacognitive Architecture

CARINA is a metacognitive architecture integrate for artificial intelligent agents. CARINA is concept from the MISM Metacognitive Metamodel [27]. CARINA has self-regulation and metamemory with capacity for metacognitive mechanisms of introspective monitoring and meta-level control. In this form CARINA assumes a functional focus to philosophy of mind, due to [23–29]. In accordance with [30], the mean "mechanism" includes instances and activities involving instances. The instances in CARINA are named "cognitive elements". In accordance with MISM, CARINA is shaped by three kinds of cognitive elements: structural elements, functional elements and basic elements. Structural elements are storage into which the functional and basic elements are embedded; the primal structural element is the "Cognitive level". The functional elements are tasks that stock on reasoning and decision-making. Basic elements conformed of the collection of elements that cooperate and interact in reasoning and meta-reasoning processes. The main functional elements in CARINA are: reasoning task and meta-reasoning task. Reasoning tasks (RT) are actions that stock on the processing (transformation, reduction, elaboration, storage and retrieval) of information by applying knowledge and decision making in order to meet the objectives of the system. A meta-reasoning task (MT) can be used to illustrate errors in some reasoning task. It can also be utilized to choose between "cognitive algorithms" to perform the reasoning [22–32].

4 Representation of a Cognitive Models in CARINA

This part officially explains a cognitive model in CARINA. A cognitive model is composed of declarative knowledge, denoting facts, likewise procedural knowledge, and denoting reasoning rules. Also, for modeling how to reason, in addition it is required to model the control to reason in a specific environment.

Given a cognitive model in CARINA with:

$$\boldsymbol{CM} \triangleq (\boldsymbol{P}, \boldsymbol{G}, \boldsymbol{S}, \boldsymbol{MS}, \boldsymbol{PK}, \boldsymbol{SK}) \tag{1}$$

Where:

P is the problem to resolve in CARINA. ***G*** is the objectives that intelligent system must accomplish in the resolution of a cognitive task. ***S*** is a sensor that has the objective of monitoring the profiles of cognitive tasks to detect failures that may occur in the reasoning of the cognitive system. ***MS*** are Mental States, which are true or false variables in which an intelligent agent can be found. ***PK*** constitutes the Procedural Knowledge that the system needs to solve the cognitive task. Procedural Knowledge is also called production rules and represent the processing knowledge of CARINA.

Let $\boldsymbol{r} \in \boldsymbol{R}$ is a Production Rule, with

$$\boldsymbol{r} \triangleq (\boldsymbol{condition}, \boldsymbol{conclusion}) \quad \text{with,} \tag{2}$$

$$\boldsymbol{condition} \triangleq (\boldsymbol{AS}, \boldsymbol{C}) \quad \text{with,} \tag{3}$$

AS is a Set of Attributes used to activate the values of a rule ***(r)***. Rules ***(r)*** are a conditional establishment to match the Set of Attributes used.

$$\boldsymbol{c} \in \boldsymbol{C} \text{ and } \boldsymbol{c} \neq \boldsymbol{C} \text{ and } \boldsymbol{c} \neq \boldsymbol{AS}$$

c represents an optional constraint inserted by the cognitive designer. A constraint ***(c)*** is a specific condition to match, in some conditions $\boldsymbol{c} = \{\}$. ***C*** is the conclusions. Conclusions are also executed when rules ***(r)*** are matched. In this context $\boldsymbol{a} \in \boldsymbol{A}$, with ***a*** as an action. ***SK*** is Semantic Knowledge. Semantic Knowledge is also the knowledge requirements to achieve the cognitive task. Semantic Knowledge is represented in the form of beliefs.

Let: $\boldsymbol{f} \in \boldsymbol{F}$ with f is a Field. Where:

$$\boldsymbol{F_D} \subset \boldsymbol{F} \wedge \boldsymbol{F_B} \subset \boldsymbol{F} \tag{4}$$

And F is a set of properties of the SMU. ***SMU*** is the Semantic Memory Unit. Structured as follows:

F_D are data types like string. F_B are Fields that point to a Beliefs of Semantic Memory. And β is Beliefs, that is, the minimum unit of declarative knowledge that structure the Semantic Memory in CARINA. Beliefs are elements based on facts and notions that are stored.

5 Overview of Cognitive Architectures and Cognitive Models for the Towers of Hanoi Algorithm

In the literature there is research which developed cognitive models for representing the Towers of Hanoi algorithm. For example, a symbolic computational model was developed by [33] to represent perceptual strategies in the resolution, like the time employed, the quantity of movements and the role of working memory plays to correctly solve problems of different ambit. Another implementation of the cognitive task of the Towers of Hanoi was done by [34] who presented a computational model able to solve problems in any order of the Towers of Hanoi through the use of 150,000 neurons called a neural engineering framework. In addition [35] implements a computational model in the ACT-R architecture that captures the performance of participants using strategies to solve this algorithm and gives a novel response using simple strategies to solve the problem. Also [36] examine the constraints of “modelling idioms” in SOAR and ACT-R cognitive architectures in the programming patterns that commonly appear in implemented models, analyzing the cognitive model the Towers of Hanoi and evaluating aspects such as decision timing, memory access, and error rates. Another study was developed by [37] that reviewed the evolution of ICARUS cognitive architecture through the elaboration of intelligent agents in a variety of domains that simulate classic problems such as Towers of Hanoi and challenging problems such as driving a vehicle in an urban environment simulated by an intelligent agent. And recently [38] discuss the use of an intelligent decision-making system based on cognitive psychology, motivation theory and theories that refer to memory, categorization, perception and decision making for the control of autonomous agents. These use the systemic ideas proposed in some cognitive architectures such as SOAR, CLARION, OpenCog, DUAL, MANIC, and LID as foundations.

6 Methodology

The methodology used is based on the protocol implemented by [28] where: *(i)* The environment was organized to make the subject comfortable; *(ii)* the expert gives the game instructions: move one disk at a time, move the top disk of a peg and cannot move a bigger disk on top of a smaller one; *(iii)* the main instruction given to the subjects to develop the cognitive task was “solve the task and verbalize what is happening in your mind”; *(iv)* during the process of solving the puzzle the expert participated only when the subjects stopped talking, telling the subject “continue talking”; and *(v)* the stages in which the research was developed are explained below:

1. Pre-processing stage: *(i)* processing of the report begins with the recording of the verbalization made by the subject and *(ii)* then recorded material is transcribed, so that the record will be as accurate as possible. Voids, repetitions, etc. are not omitted.
2. Processing stage: *(i)* after, the non-verbal information is eliminated from the transcription; *(ii)* the material obtained is segmented, where each segment has a sense; *(iii)* each protocol segment is coded to interpret contextual information contained in other verbalization segments and *(iv)* to determine the codification, the categories of analysis and the names of the different mental actions that the subject develops during the realization of the cognitive task and that constitute the meanings of the symbols are used in the codification.
3. Classification of subjects: (i) after codification, subjects that solved the cognitive task were classified as: successful and unsuccessful, where three of the subjects had a successful result and only one deserted the problem. For a later representation in natural language of the codification, a successful and an unsuccessful subject were taken to discuss the research results.
4. Description of the cognitive task in natural language: *(i)* after the classification of the subjects the successful subject with the least number of movements was selected and an unsuccessful one was also selected to describe the processes in natural language.
5. Finally, task description in NGOMS-L: all the Goals, sub-goals (steps), Actions and Mental States in the solution of the Towers of Hanoi are specified in the notation NGOMS-L to analyze the cognitive task and to be represented in a formal specification of cognitive models in CARINA in the future.

7 Cognitive Model for the Towers of Hanoi Algorithm in NGOMS-L

Process of construction of computational cognitive models starts with the analysis of the cognitive task [39]. The methodologies used to analyze cognitive tasks require a sequence of goals and sub-goals that are interpreted as the sequence of steps executed to solve the cognitive task that will be represented computationally. The commonly used method for the analysis of user-based cognitive problems is NGOMS-L, defined as a well-defined notation in natural language for the specification of GOMS models and a process for constructing them [18]. GOMS are Goals, Operators, Methods, and Selection Rules. Operators are specific steps that users develop in an established time. In case a goal is not executed using more than one method, selection rules are used to establish the correct method [19].

The NGOMS-L-based cognitive model of the Towers of Hanoi algorithm is presented (see Table 3).

8 Results/Discussion

This section describes the results obtained during the research.

The first stage (see Table 1) in which the investigation was developed begins with the recording of the verbalization made by the subject and the transcription of the recorded material without omitting voids and repetitions.

Table 1. Pre-processing stage.

1. Pre-processing stage		
Subject 1	E 46 I am going to write … eh … the orange wheel goes to the "A".	E: Each expression … pauses or silences
Subject 2	E12… I am thinking, I am thinking, I am thinking.	
Subject 3	E1 I move the green piece to stake "B"	
Subject 4	E2…I'm doin…, I'm writing the instruction…move the stake…the green hoop to stake "B"…	

According to the second stage of the analysis "processing stage" it was observed that three of the four subjects managed to solve the problem in approximately one hour, with an approximate of 80 movements. Subject 1 solved the problem with 120 movements. Subject 2 solved the problem with 60 movements and subject 3 solved the problem with 67 movements. And subject 4 quit the problem with 173 movements (see Table 2).

Table 2. Processing stage: production system and decision tree.

N°. Steps	Initial state A{G,Y, O, R, P}, B{}, C{}	Final state A{}, B{}, C{G,Y, O, R, P}	Operators
Subject 1: 122 (Successful)	A{G,Y, O, R, P}, B{}, C{}	A{}, B{}, C{G,Y, O, R, P}	Move (G,C)
Subject 2: 60 (Successful)	A{G,Y, O, R, P}, B{}, C{}	A{}, B{}, C{G,Y, O, R, P}	Move (G,C)
Subject 3: 67 (Successful)	A{G,Y, O, R, P}, B{}, C{}	A{}, B{}, C{G,Y, O, R, P}	Move (G,C)
Subject 4: 173 (Unsuccessful)	A{Y,O}, B{G,R,P},C{}	A{O}, B{G, R, P}, C{Y}	Move{Y,C}

Continuing with the third stage "classification of subjects", it was observed that in performing the cognitive task three of the subjects understood the problem (i.e., performing appropriate mental actions) and the rules of the problem. A subject also declined to finish the problem by explaining that he did not understand the problem.

Table 3. The NGOMS-L-based cognitive model of the Towers of Hanoi algorithm.

NGOMS-L notation

Method for goal γ_{700}: Complete the game
Step 1. (α^{ς}_{700}) Accomplish goal:γ_{701}
Step 2. (α^{ς}_{701}) Accomplish goal:γ_{702}
Step n. (α^{ς}_{702}) Return with goal accomplished.

Method for goal γ_{701}: Subject moves the green disk to peg B.
Step 1. (α^{ς}_{703}) Choose green disk
Step 2. (α^{ς}_{704}) Select peg
Step 3. (α^{ς}_{705}) Put green disk in the selected peg
Step n. (α^{ς}_{702}) Return with goal accomplished.

Method for goal γ_{702}: Subject moves the yellow disk to peg C.
Step 1. (α^{ς}_{706}) Choose yellow disk
Step 2. (α^{ς}_{707}) Select peg
Step 3. (α^{ς}_{708}) Put yellow disk in the selected peg
Step n. (α^{ς}_{702}) Return with goal accomplished.

Method for goal γ_{703}: Subject moves the green disk to peg C.
Step 1. (α^{ς}_{709}) Choose green disk
Step 2. (α^{ς}_{710}) Select peg
Step 3. (α^{ς}_{711}) Put green disk in the selected peg
Step n. (α^{ς}_{702}) Return with goal accomplished.

Method for goal γ_{704}: Subject moves the orange disk to peg B.
Step 1. (α^{ς}_{712}) Choose orange disk
Step 2. (α^{ς}_{713}) Select peg
Step 3. (α^{ς}_{714}) Put orange disk in the selected peg
Step 4. (α^{ς}_{715}) Accomplish goal:γ_{701}
Step n. (α^{ς}_{702}) Return with goal accomplished.

Method for goal γ_{705}: Subject moves the yellow disk to peg A.
Step 1. (α^{ς}_{716}) Choose yellow disk
Step 2. (α^{ς}_{717}) Select peg
Step 3. (α^{ς}_{718}) Put yellow disk in the selected peg
Step 4. (α^{ς}_{719}) Accomplish goal:γ_{703}
Step n. (α^{ς}_{702}) Return with goal accomplished.

Method for goal γ_{706}: Subject moves the yellow disk to peg B.
Step 1. (α^{ς}_{720}) Choose yellow disk
Step 2. (α^{ς}_{721}) Select column
Step 3. (α^{ς}_{722}) Put yellow disk in the selected peg
Step 4. (α^{ς}_{723}) Accomplish goal:γ_{701}
Step n. (α^{ς}_{702}) Return with goal accomplished.

Method for goal γ_{707}: Subject moves the red disk to peg C.
Step 1. (α^{ς}_{724}) Choose red disk
Step 2. (α^{ς}_{725}) Select peg
Step 3. (α^{ς}_{726}) Put red disk in the selected peg
Step n. (α^{ς}_{702}) Return with goal accomplished.

Method for goal γ_{708}: Subject moves the green disk to peg A.
Step 1. (α^{ς}_{727}) Choose green disk
Step 2. (α^{ς}_{728}) Select peg
Step 3. (α^{ς}_{729}) Put green disk in the selected peg
Step 4. (α^{ς}_{730}) Accomplish goal:γ_{702}
Step 5. (α^{ς}_{731}) Accomplish goal:γ_{701}
Step 6. (α^{ς}_{732}) Accomplish goal:γ_{703}
Step n. (α^{ς}_{702}) Return with goal accomplished.

Method for goal γ_{709}: Subject moves the Orange disk to peg A.
Step 1. (α^{ς}_{733}) Choose orange disk
Step 2. (α^{ς}_{734}) Select peg
Step 3. (α^{ς}_{735}) Put orange disk in the selected peg.
Step 4. (α^{ς}_{736}) Accomplish goal:γ_{701}
Step 5. (α^{ς}_{737}) Accomplish goal:γ_{705}
Step 6. (α^{ς}_{738}) Accomplish goal:γ_{703}
Step 7. (α^{ς}_{739}) Accomplish goal:γ_{706}
Step 8. (α^{ς}_{740}) Accomplish goal:γ_{701}
Step n. (α^{ς}_{702}) Return with goal accomplished.

Method for goal γ_{710}: Subject moves the Orange disk to pet C.
Step 1. (α^{ς}_{741}) Choose Orange disk
Step 2. (α^{ς}_{742}) Select pet
Step 3. (α^{ς}_{743}) Put orange disk in the selected pet
Step 4. (α^{ς}_{744}) Accomplish goal:γ_{701}
Step 5. (α^{ς}_{745}) Accomplish goal:γ_{706}
Step 6. (α^{ς}_{746}) Accomplish goal:γ_{703}
Step n. (α^{ς}_{702}) Return with goal accomplished.

Method for goal γ_{711}: Subject moves the purple disk to pet B.
Step 1. (α^{ς}_{741}) Choose purple disk
Step 2. (α^{ς}_{742}) Select pet
Step 3. (α^{ς}_{743}) Put purple disk in the selected pet.
Step 4. (α^{ς}_{744}) Accomplish goal:γ_{701}
Step 5. (α^{ς}_{745}) Accomplish goal:γ_{705}
Step 6. (α^{ς}_{746}) Accomplish goal:γ_{708}
Step 7. (α^{ς}_{747}) Accomplish goal:γ_{704}
Step 8. (α^{ς}_{748}) Accomplish goal:γ_{703}
Step 9. (α^{ς}_{749}) Accomplish goal:γ_{706}
Step 10. (α^{ς}_{750}) Accomplish goal:γ_{701}
Step n. (α^{ς}_{702}) Return with goal accomplished.

Method for goal γ_{712}: Subject moves the red disk to pet A.
Step 1. (α^{ς}_{751}) Choose red disk
Step 2. (α^{ς}_{752}) Select pet
Step 3. (α^{ς}_{753}) Put red disk in the selected pet
Step 4. (α^{ς}_{754}) Accomplish goal:γ_{703}
Step 5. (α^{ς}_{755}) Accomplish goal:γ_{705}
Step 6. (α^{ς}_{756}) Accomplish goal:γ_{708}
Step 7. (α^{ς}_{757}) Accomplish goal:γ_{710}
Step 8. (α^{ς}_{758}) Accomplish goal:γ_{703}
Step 9. (α^{ς}_{759}) Accomplish goal:γ_{706}
Step 10. (α^{ς}_{760}) Accomplish goal:γ_{701}
Step 11. (α^{ς}_{761}) Accomplish goal:γ_{709}
Step 12. (α^{ς}_{762}) Accomplish goal:γ_{703}
Step 13. (α^{ς}_{763}) Accomplish goal:γ_{705}
Step 14. (α^{ς}_{764}) Accomplish goal:γ_{708}
Step n. (α^{ς}_{702}) Return with goal accomplished.

Method for goal γ_{713}: Subject moves the purple disk to pet C.
Step 1. (α^{ς}_{765}) Choose purple disk
Step 2. (α^{ς}_{766}) Select pet
Step 3. (α^{ς}_{767}) Put purple disk in the selected pet.
Step 4. (α^{ς}_{768}) Accomplish goal:γ_{701}
Step 5. (α^{ς}_{769}) Accomplish goal:γ_{702}
Step 6. (α^{ς}_{770}) Accomplish goal:γ_{703}
Step 7. (α^{ς}_{771}) Accomplish goal:γ_{704}
Step 8. (α^{ς}_{772}) Accomplish goal:γ_{708}
Step 9. (α^{ς}_{773}) Accomplish goal:γ_{706}
Step 10. (α^{ς}_{774}) Accomplish goal:γ_{701}
Step 11. (α^{ς}_{775}) Accomplish goal:γ_{707}
Step 12. (α^{ς}_{776}) Accomplish goal:γ_{703}
Step 13. (α^{ς}_{777}) Accomplish goal:γ_{705}
Step 14. (α^{ς}_{778}) Accomplish goal:γ_{708}
Step 15. (α^{ς}_{779}) Accomplish goal:γ_{710}
Step 16. (α^{ς}_{780}) Accomplish goal:γ_{701}
Step 17. (α^{ς}_{781}) Accomplish goal:γ_{702}
Step 18. (α^{ς}_{782}) Accomplish goal:γ_{703}
Step n. (α^{ς}_{702}) Return with goal accomplished.

After the third stage, the cognitive task was described in natural language. The description was made according to the segmentation (second stage) and the classification of the subjects (third stage). For the formulation of the cognitive task in natural language, a successful subject with the least number of movements and an unsuccessful subject were taken. The representation of the cognitive task in natural language is presented (see Table 4).

Table 4. Description of the cognitive task in natural language.

Successful subject	Unsuccessful subject
Move green disk to peg B	Move green disk to peg B
Move yellow disk to peg C	Move yellow disk to peg C
Move green disk to peg C	Move green disk to peg A
Move orange disk to peg B	Move yellow disk to peg B

Finally, the cognitive task the Towers of Hanoi algorithm is described in NGOMS-L according to the Goals, Actions, and Mental States with the respective inventories. Below is a fragment of the cognitive model in NGOMS-L (see Table 5).

According to [33] the five disk algorithm is resolved with a number of 16 steps. However the successful subject resolved the algorithm in one hour with approximately 60 steps, and the unsuccessful subject used 173 steps in approximately two hours. This research is a contribution in the scientific world where the algorithm of the Towers of Hanoi is modeled with NGMOS-L, in a cognitive model based-user using a metacognitive architecture, in this case CARINA.

In addition, this paper constitutes a contribution to scientific knowledge in the area of cognitive informatics, because it provides a novel contribution in a current knowledge gap with respect to the Towers of Hanoi modeling in a user-based model, using NGOMS-L.

Table 5. Inventory of goals, mental states and actions.

Goals Inventory	Mental States Inventory	Actions Inventory
Goalγ_{701}: Subject moves green disk to peg B.	**Mental Stateσ_{702} : Green disk is moved to peg C.**	α^{c}_{700}) Accomplish goal γ_{700}
Goalγ_{702}: Subject moves yellow disk to peg C.	**Mental Stateσ_{704} : Orange disk is moved to peg B.**	α^{c}_{702}: Return with goal accomplished.
Goalγ_{703}: Subject moves green disk to peg C.	**Mental Stateσ_{705} : Yellow disk is moved to peg A.**	α^{c}_{703}: Choose green disk
Goalγ_{704}: Subject moves orange disk to peg B.	**Mental Stateσ_{706} : Yellow disk is moved to peg B.**	α^{c}_{704}: Select peg
Goalγ_{705}: Subject moves yellow disk to peg A.	**Mental Stateσ_{707} : Red disk is moved to peg C.**	α^{c}_{705}: Put green disk in selected peg
Goalγ_{706}: Subject moves yellow disk to peg B.	**Mental Stateσ_{708} : Green disk is moved to peg A.**	α^{c}_{706}: Choose yellow disk
Goalγ_{707}: Subject moves red disk to peg C.	**Mental Stateσ_{709} : Orange disk is moved to peg A.**	α^{c}_{708}: Put yellow disk in selected peg
Goalγ_{708}: Subject moves green disk to peg A.	**Mental Stateσ_{710} : Orange disk is moved to peg C.**	α^{c}_{712}: Choose orange disk
Goalγ_{709}: Subject moves orange disk to peg A.	**Mental Stateσ_{711} : Purple disk is moved to peg B.**	α^{c}_{714}: Put orange disk in selected peg
Goalγ_{710}: Subject moves orange disk to peg C.	**Mental Stateσ_{712} : Red disk is moved to peg A.**	α^{c}_{724}: Choose red disk
Goalγ_{711}: Subject moves purple disk to peg B.	**Mental Stateσ_{713} : Purple disk is moved to peg C.**	α^{c}_{726}: Put red disk in selected peg
Goalγ_{712}: Subject moves red disk to peg A.		α^{c}_{741}: Choose purple disk
Goalγ_{713}: Subject moves purple disk to peg C.		α^{c}_{743}: Put purple disk in selected peg

9 Conclusion/Future Work

This paper presents an approach to cognitively model the algorithm of the Towers of Hanoi using the NGOMS-L user-based method, which can be runnable in metacognitive architecture CARINA. The cognitive model in NGOMS-L was developed with one successful and one unsuccessful subject, where the successful subject had an average of 60 movements and the unsuccessful an average of 173 movements. According to the solution made by the successful subject, a cognitive model based on GOMS-L was made with an inventory of Goals, sub-goals (steps), Mental States and cognitive Actions that need to be achieved (see Table 5). This paper adds a new contribution to filling a knowledge gap with respect thc creation of cognitive models using NGOMS-L method of language for metacognitive architectures.

In future works it will be possible to model the algorithm of the Towers of Hanoi using the visual language M++ and that the cognitive model can be executed in the metacognitive architecture CARINA. In addition, it will also be possible to make a comparison between a cognitive model based on experts and a cognitive model based on users, where CARINA can understand the different reasoning failures that happen when a subject tries to solve the algorithm of the Towers of Hanoi.

References

1. Crowley, K., Sliney, A., Pitt, I., Murphy, D.: Evaluating a brain-computer interface to categorise human emotional response. In: 2010 10th IEEE International Conference on Advanced Learning Technologies, pp. 276–278 (2010)
2. Simon, H.A.: Information-processing theory of human problem solving. Handb. Learn. Cogn. Process. **5**, 271–295 (1978)
3. Reed, S.K.: The structure of ill-structured (and well-structured) problems revisited. Educ. Psychol. Rev. **28**(4), 691–716 (2016)
4. García, A.B.: Evaluación de estrategias de resolución de problemas. Rev. Educ. **287**, 275–286 (1988)
5. Ericsson, K.A., Simon, H.A.: Protocol Analysis: Verbal Reports as Data. The MIT Press, Cambridge (1984)
6. Strube, G.: Cognitive modeling: research logic in cognitive science (2001)
7. Marewski, J.N., Link, D.: Strategy selection: an introduction to the modeling challenge. Wiley Interdiscip. Rev. Cogn. Sci. **5**(1), 39–59 (2014)
8. Ziefle, M, et al.: A multi-disciplinary approach to ambient assisted living. In: E-Health, Assistive Technologies and Applications for Assisted Living: Challenges and Solutions, pp. 76–93. IGI Global (2011)
9. Fum, D., Del Missier, F., Stocco, A.: The cognitive modeling of human behavior: why a model is (sometimes) better than 10,000 words. Elsevier (2007)
10. Zacarias, M., Magalhaes, R., Caetano, A., Pinto, H.S., Tribolet, J.: Towards organizational self-awareness: an initial architecture and ontology. In: Handbook of Ontologies for Business Interaction, pp. 101–121. IGI Global (2008)
11. Jacko, J.A.: Human Computer Interaction Handbook: Fundamentals, Evolving Technologies, and Emerging Applications. CRC Press, Boca Raton (2012)
12. Salvucci, D.D.: Modeling driver behavior in a cognitive architecture. Hum. Factors **48**(2), 362–380 (2006)

13. Forsythe, C., Bernard, M.L., Goldsmith, T.E.: Cognitive Systems: Human Cognitive Models in Systems Design. Psychology Press, Hove (2006)
14. Hoffman, R.R.: How can expertise be defined? Implications of research from cognitive psychology. In: Exploring Expertise, pp. 81–100. Springer, Heidelberg (1998)
15. Van Rijn, H., Johnson, A., Taatgen, N.: Cognitive user modeling. In: Handbook Human Factors Web Design, 2nd edn. pp. 527–542. CRC Press, Boca Raton (2011)
16. Taylor, B.A., Harris, S.L.: Teaching children with autism to seek information-acquisition of novel information and generalization of responding. J. Appl. Behav. Anal. **28**(1), 3–14 (1995)
17. Carrillo, A.L., Guevara, A., Gálvez, S., Caro, J.L.: Interacción Guiada por Objetivos. In: Actas de INTERACCION, pp. 68–75 (2002)
18. Kieras, D.E.: A guide to GOMS model usability evaluation using GOMSL and GLEAN3, vol. 313. University of Michigan (1999)
19. John, B.E., Kieras, D.E.: The GOMS family of user interface analysis techniques: comparison and contrast. ACM Trans. Comput. Interact. **3**(4), 320–351 (1996)
20. Forstmann, B.U., Wagenmakers, E.J.: An Introduction to Model-Based Cognitive Neuroscience. Springer, Heidelberg (2015)
21. Langley, P., Laird, J.E., Rogers, S.: Cognitive architectures: research issues and challenges. Cogn. Syst. Res. **10**(2), 141–160 (2009)
22. Jerónimo, A.J., Caro, M.F., Gómez, A.A.: Formal Specification of cognitive models in CARINA. In: 2018 IEEE 17th International Conference on Cognitive Informatics & Cognitive Computing (ICCI* CC), pp. 614–619 (2018)
23. Olier, A.J., Gómez, A.A., Caro, M.F.: Cognitive modeling process in metacognitive architecture CARINA. In: 2018 IEEE 17th International Conference on Cognitive Informatics & Cognitive Computing (ICCI* CC), pp. 579–585 (2018)
24. Cox, M.T.: Field review: metacognition in computation: a selected research review. Artif. Intell. **169**(2), 104–141 (2005)
25. Cox, M.T., Oates, T., Perlis D.: Toward an integrated metacognitive architecture. In: 2011 AAAI Fall Symposium Series (2011)
26. Samsonovich, A.V., De Jong, K.A.: Metacognitive architecture for team agents. In: Proceedings of the 25th Annual Meeting of the Cognitive Science Society, pp. 1029–1034 (2003)
27. Caro, M.F., Josvula, D.P., Gómez, A.A., Kennedy, C.M.: Introduction to the CARINA metacognitive architecture. In: 2018 IEEE 17th International Conference on Cognitive Informatics & Cognitive Computing (ICCI* CC), pp. 530–540 (2018)
28. Jerónimo, A.J., Acosta, K., Caro, M.F., Rodriguez, R.: Protocolos verbales para el análisis del uso de estrategias metacognitivas en la elaboración de algoritmos por estudiantes de grado sexto de la institución educativa Mercedes Ábrego. In: V congreso Internacional y XIII encuentro nacional de educación en tecnología e informática (2017)
29. Caro, M.F., Josyula, D.P., Jiménez, J.A., Kennedy, C.M., Cox, M.T.: A domain-specific visual language for modeling metacognition in intelligent systems. Biol. Inspired Cogn. Archit. **13**, 75–90 (2015)
30. Huet, N., Mariné, C.: Memory strategies and metamemory knowledge under memory demands change in waiters learners. Eur. J. Psychol. Educ. **12**(1), 23–35 (1997)
31. Cox, M., Oates, T., Perlis, D.: Toward an integrated metacognitive infrastructure. In: 2011 AAAI Fall Symposium, pp. 74–81 (2011)
32. Florez, M.A., Gomez, A.A., Caro, M.F.: Formal representation of introspective reasoning trace of a cognitive function in CARINA. In: 2018 IEEE 17th International Conference on Cognitive Informatics & Cognitive Computing (ICCI*CC), pp. 620–628 (2018)

33. Goela, V., Pullara, D., Grafman, J.: A computational model of frontal lobe dysfunction: working memory and the Tower of Hanoi task. Cogn. Sci. **25**(2), 287–313 (2001)
34. Stewart, T.C. Eliasmith, C.: Neural Cognitive Modelling: A Biologically Constrained Spiking Neuron Model of the Tower of Hanoi Task The Tower of Hanoi (2011)
35. Gunzelmann, G., Anderson, J.R.: An ACT-R model of the evolution of strategy use and problem difficulty. In: Proceedings of the Fourth International Conference on Cognitive Modeling, pp. 109–114 (2001)
36. Jones, R.M., Lebiere, C., Crossman, J.A.: Comparing modeling idioms in ACT-R and Soar. In: Proceedings of the 8th International Conference on Cognitive Modeling, pp. 49–54 (2007)
37. Choi, D., Langley, P.: Evolution of the ICARUS cognitive architecture. Cogn. Syst. Res. **48**, 25–38 (2018)
38. Zdzisław, K., Michał, C.: Embodying intelligence in autonomous and robotic systems with the use of cognitive psychology and motivation theories. In: Advances in Data Analysis with Computational Intelligence Methods, pp. 335–352. Springer, Heidelberg (2018)
39. Wong, J.H., Kirschenbaum, S.S., Peters, S.: Developing a cognitive model of expert performance for ship navigation maneuvers in an intelligent tutoring system (2010)

Application of Knowledge Discovery in Data Bases Analysis to Predict the Academic Performance of University Students Based on Their Admissions Test

María Isabel Uvidia Fassler(✉),
Andrés Santiago Cisneros Barahona,
Gabriela Jimena Dumancela Nina,
Gonzalo Nicolay Samaniego Erazo,
and Edison Patricio Villacrés Cevallos

Universidad Nacional de Chimborazo, Riobamba, Ecuador
muvidia@unach.edu.ec

Abstract. In 2012, the Ecuadorian Higher Education System implemented a standardized test called "Ser Bachiller" as a compulsory requirement to be admitted on an undergraduate program at any public university of the country. Based on the test, applicants receive a score that allows them to apply to the academic program of their choice. On a meritocratic basis, students with the highest grades from the examination process have greater opportunities to be admitted. The present study focuses on the admission process at Universidad Nacional de Chimborazo (Unach), a public university located in Riobamba-Ecuador. Through the application of knowledge discovery in database (KDD), this analysis generated a data warehouse (DW) after collecting the information of the tests scores from the admitted students and the relation with their academic performance once they were enrolled in the leveling courses. In addition, through Data Mining (DM) processes, it was possible to identify patterns that provide information and knowledge about the relationship between the admission score and the academic performance of the students. In parallel, through Business Intelligence (BI) reports, higher education institutions can inform their decisions in order to generate strategies for strengthening the processes of admission and knowledge leveling. This can allow them to guarantee a high academic performance among the admitted students, which additionally supports the general development of institutions and their undergraduate programs.

Keywords: Higher Education · Knowledge discovery from data bases · Data Mining · Universidad Nacional de Chimborazo

1 Introduction

Education is a key component in the development of nations [1]; during the past decades, Ecuador has keenly sought to improve its academic growth in all stages of education. For instance, as of 2012, the government through the National Secretariat of

M. Botto-Tobar et al. (Eds.): ICAETT 2019, AISC 1066, pp. 485–497, 2020.
https://doi.org/10.1007/978-3-030-32022-5_45

Higher Education, Science, Technology and Innovation (Senescyt) updated the policies and mechanisms of application and admission to a third level degree at public universities and polytechnic schools. Preliminarily, the objectives for this change included to increase the access to third level education, as well as the relevance and academic quality of the programs offered by universities, among others. With this logic, Senescyt created an instrument of selection, used until the present, which consists in a standardized test that aims to measure the following fields on the applicants: mathematical reasoning, social and scientific knowledge, language application, and abstract thinking. Students are graded over 1,000 points and this evaluation is the only requirement to apply to any public higher education institution in the country.

Currently, the literature indicates that the test has strongly reduced student desertion at universities, significantly improving the graduation rates; in addition, as stated by the Ministry of Education, the waste of resources has been avoided [2]. Similarly, it has been possible to guarantee a meritocratic access to education. The Organic Law of Higher Education (LOES), in its article 63, section "On Admission and Leveling", allows Senescyt through its National Leveling and Admission System (SNNA) to regulate the national admission process, guaranteeing that the most qualified students obtain a space to continue their education. In addition, since 2012, all public universities and polytechnic schools are required to open a minimum number of places for every cohort they create; this must be approved by SNNA in order to guarantee that all institutions receive a certain number of new students.

Under this framework, the present document centers on the system of quotas applied at the National University of Chimborazo (Unach), in relation to the creation of its leveling courses, compulsory for all applicants admitted, and their performance after their enrollment. Unach has been part of the leveling and admissions process for six years, which has made possible to gather enough information about students' admission scores and analyze them in relation to their performance at the leveling courses. This analysis uses the process of Knowledge Discovery in Database (KDD), where Data Mining (DM) is the core process that integrates data from different sources to subsequently extract important knowledge; meaning to identify transcendent, valuable and useful information, from which Unach and other higher education institutions could base their decision making processes [3], in order to increase the quality of their academic programs as well as the attraction and retention of students.

2 Process of Application to Universities and Polytechnical Schools in Ecuador

Since 2012, the national higher education system, based on the Organic Law of Higher Education, in its Article 63 "Admission and Leveling", states:

> The admission process to higher education institutions is regulated through the Leveling and Admission System, for all applicants. The system is governed by the principles of merit, equal opportunities and freedom of choosing academic programs and institutions by the applicants. […] The best candidates, who obtain the best scores will access the academic program of their choice based on the available offer at the higher education institutions.

The country [4] has an admissions process that depends on a test graded under 1,000 points. Once the applicant receives a score from the test "Ser Bachiller", she or he can complete the application process, which means choosing an academic program at any public university or polytechnic school in the country. The students can choose up to five options in an order of preference; they will be accepted depending on the number of applicants at a specific institution. If the grade is high enough as to compete with the rest of candidates, they will receive an admission offer, that they can accept or reject (Fig. 1).

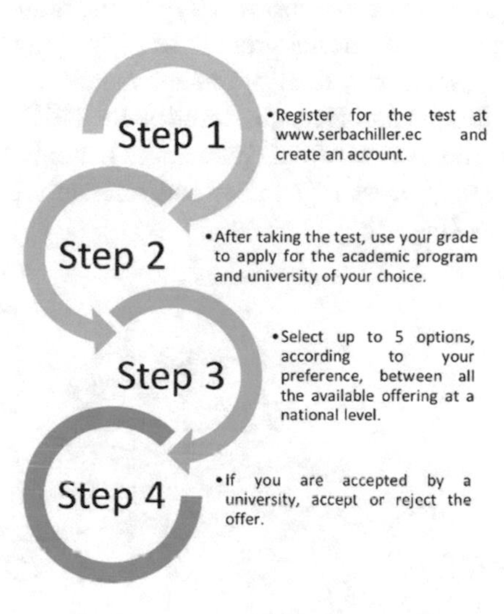

¡APPLY NOW!
Step 1
•Register for the test at www.serbachiller.ec and create an account.
Step 2
•After taking the test, use your grade to apply for the academic program and university of your choice.
Step 3
•Select up to 5 options, according to your preference, between all the available offering at a national level.
Step 4
•If you are accepted by a university, accept or reject the offer.

Fig. 1. Application process

Since the changes of the admission system to public higher education, Universidad Nacional de Chimborazo has offered about 1800 places in each semester distributed within its current academic offering, which includes 31 academic programs, associated to 4 schools: Health Sciences, Political and Administrative Sciences, Education Sciences and Engineering. The rate of effective enrollment in the leveling courses is over 92%.

3 Leveling Courses

The objectives of the leveling courses are various, as detailed in the Regulation for Leveling and Admission for students admitted at Universidad Nacional de Chimborazo, its article 11th states: "The courses aim to prepare the candidates who obtained a place in a program offered by the University in order to guarantee their academic performance during their studies, based on the development and strengthening of specific skills" [5]. In addition, the University proposes several strategies to guarantee the effective performance of its admitted students; for example, during these past 8 years Universidad Nacional de Chimborazo planned and executed Career Leveling Courses, according to its four academic areas, allocating the necessary number of hours in order to allow an efficient leveling of knowledge and the adaptation of students to the higher education system.

4 Discovery of Knowledge in Databases

The Discovery of Knowledge in Databases (KDD), is an automatic, exploratory and modeling analysis of large data repositories. Fayyad mentions that "the discovery of knowledge in databases is the non-trivial process of identifying patterns in data that are valid, novel, potentially useful and, finally, understandable" [6]. Complementing this definition, the authors Maimon and Rokach, describe the KDD as an organized process of valid, novel, useful identification of understandable patterns from large and complex data sets; in addition KDD process is iterative and interactive, the author defines it in nine steps [7], as follows (Fig. 2):

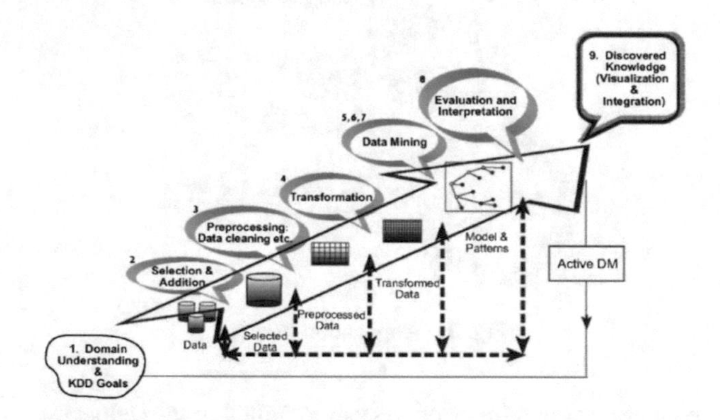

Fig. 2. The process of knowledge discovery in databases [7].

The KDD process of this study, uses the methodology of the KDD – ESPOCH analysis, applied to the Leveling and Admission Unit at the Polytechnic School of Chimborazo (Espoch), described in the article "Moving Towards a Methodology Employing Knowledge Discovery in Databases to Assist in Decision Making Regarding Academic Placement and Student Admissions for Universities", published in Communications in Computer and Information Science, vol. 798. Springer [8]. In demonstrating the usefulness and benefit of this former analysis system, it was applied again in the Admission and Leveling Office at Unach. It is employed to integrate data from the performance of the students currently and previously enrolled on leveling courses (from 2012 to 2019), in relation to their score on the admissions test. The present study, mainly focuses on data mining; within this method, it is centered on the application of new Data Mining (DM) grouping techniques and the visualization of data through Business Intelligence tools.

The foundation methodology that is used within the Data Warehouse (DW) subprocess includes: learning the domain of the application, selecting and creating a set of data on which the discovery, pre-processing and cleaning, and data transformation can be carried out. The DW product is created with validated and consistent data; in the DM subprocess, the selection of the appropriate DM task is generated by choosing the DM algorithm, which later is used to develop patterns as an outcome. Finally, in the stage of evaluation and usage of the knowledge discovered, it is possible to encounter the information useful to inform the decision making process at higher education institutions.

5 Data Mining

"DM refers to the extraction of patterns or models from the data" [6]. Data mining is the core of the KDD process (Knowledge Discovery in Database, Knowledge Discovery in Data Bases) that allows the identification of valid, useful and understandable patterns, as well as structures that convert data into useful information to achieve an organized work. In addition, it allows the implementation of relevant further actions [9]. Within KDD there are two main objectives in data mining: prediction and description; prediction refers to the monitoring of data mining (proven techniques), while descriptive data mining includes unsupervised aspects (techniques not tested and validated) and data mining visualization [10].

Additionally, DM is defined as the process of extracting useful and understandable knowledge, previously unknown, from large amounts of data stored in different formats. In this sense, the fundamental task of data mining is to find intelligible models from the collected data [11]. According to Han and Kamber, (2006) DM refers to the extraction of knowledge or mining of large amounts of data. By linking the ideas of data and extraction, data mining has become a more popular option. [12]. Parallelly, according to González (2005), DM is the process by which a useful model for prediction is generated. This model is built from the data found in a database, to which an algorithm has been applied to propose a model [13]. For this reason, knowledge management is a topic addressed by practitioners and scholars as a trend. It also has become a relevant aspect in today's organizations, as it contributes to make them more competitive and at the same time maintain their internal knowledge through time [14].

5.1 DM Grouping Techniques - Clustering

"The analysis of grouping techniques takes ungrouped data and uses automatic techniques to cluster them out; the grouping is not supervised and does not require a learning set" [15]. "Currently, only a few clustering algorithms are included in standard clusters: KMeans, EM for Bayes simple models, farthest-first clustering and Cobweb" [7].

The goal of grouping or clustering in a database is the division of it into segments or clusters of similar records, that share a number of properties and are considered homogeneous. By heterogeneity it is understood that the records in different segments are not similar, according to a measure of similarity [16].

6 Business Intelligence

According to Bernabeu (2010), Business Intelligence (BI) can be described as a concept that integrates storage on the one hand and the processing of large amounts of data on the other, with the main objective of transforming this information into knowledge and real-time decisions, through a simple analysis and exploration. This knowledge must be well-timed, relevant, useful and must be adapted to the context of the organization. Popularly, Business Intelligence is defined as: "The process of turning data into knowledge and knowledge into action, for a decision-making process" [10]. Having accurate, real time information allows to identify and correct situations before they become problems as well as to minimize risks within a company.

7 Results of the Case Study

During the application of the KDD methodology in the Admissions and Leveling Office at Unach, the following results were obtained:

- In the subprocess of DWH it was possible to consolidate the information of the students who were admitted at the University, and who were part of the Career Leveling Courses from 2012 to the second academic period of the year 2018, which ended in February 2019. The admission grade and the average leveling grade were taken into account, as well as student's background information, including their province of origin, gender, careers and schools by academic period, on which several analyzes were performed as shown in the next sub process of the KDD methodology.
- In the DM sub process, Weka is one of a collection of machine learning algorithms for data mining tasks. Several important results were obtained:

(a) Data Mining Technique

Within the DM techniques of grouping, we used the partitioning technique with the Kmeans algorithm, implemented in Weka, as SimpleKmeans, "in which the number of groups (NumClusters) to be formed and the seed (seed) is configured, which is used in the generation of a random number, which is used to make the initial allocation of instances to the groups" [17] (Fig. 3).

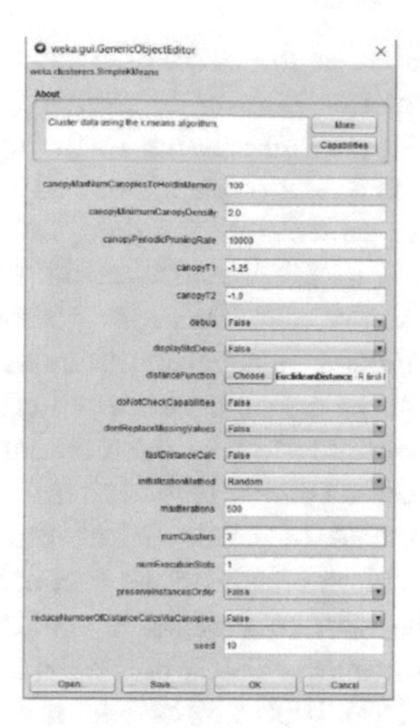

Fig. 3. Definition of the parameters of the DM-K Means technique

In the present case study, two important data were considered, such as: application grade from the standardized test "Ser bachiller" and the GPA from the Career Leveling Courses to demonstrate their relationship. The configuration of the number of clusters in the SimpleKmeans algorithm was 3 clusters, obtaining the following result (Fig. 4):

```
=== Run information ===
Scheme:       weka.clusterers.SimpleKMeans -init 0 -max-candidates 100 -periodic-pruning 10000 -min-density 2.0 -t1 -1.25 -t2 -1.0 -N 3 -A
"weka.core.EuclideanDistance -R first-last" -I 500 -num-slots 1 -S 10
Relation:    data DM March 2019
Instances:   18158
Attributes:  4
             Academic Program
             Period
             Final GPA
             Application grade
Test mode:   evaluate on training data

=== Clustering model (full training set) ===

kMeans
======
Number of iterations: 8
Within cluster sum of squared errors: 30672.67720363749

Initial starting points (random):

Cluster 0: 'BASIC EDUCATION','SECOND SEMESTER 2015',6.6,724
Cluster 1: 'CLINIC LABORATORY ','FIRST SEMESTER 2016',8.3,814
Cluster 2: ARCHITECTURE,'FIRST SEMESTER 2018',7.3,898

Missing values globally replaced with mean/mode

Final cluster centroids:
                                        Cluster#
Attribute                  Full Data               0                   1                   2
                           (18158.0)            (2936.0)            (8209.0)            (7013.0)
================================================================================================
ACADEMIC PROGRAM           MEDICINE             BASIC EDUCATION     LAW                 ARCHITECTURE
PERIOD                     SECOND SEMESTER 2018 SECOND SEMESTER 2015 FIRST SEMESTER 2016 FIRST SEMESTER 2018
FINAL GPA                  7.3766               4.823               7.8424              7.9004
APPLICATION GRADE          795.5195             777.4283            758.7328            846.1537

Time taken to build model (full training data) : 0.07 seconds

=== Model and evaluation on training set ===
Clustered Instances

0     2936 ( 16%)
1     8209 ( 45%)
2     7013 ( 39%)
```

Fig. 4. Results – SimpleKmeans WEKA

The result of the application of the SimpleKmeans technique in 18,158 records, shows that the generated patterns identify 3 groups of information with centroids (reference information of the group) in relation to the application grade and the final GPA of the leveling courses; this relationship is directly proportional. The first cluster considered has a concentration of 16% of the data, being the lowest one with a test score of 777.4283/1000 points, its GPA at the leveling stage is 4.823/10 points; this rates are below the acceptable average. This group shows the lowest academic performance and the highest number of academic difficulties; the group admitted on the second semester of 2015 displays the highest concentration of data with this particularity. The second group concentrates 45% of the total data, being the largest amount of data concentrated on a cluster, at a general level. It has a score of 758.7328/1000 points on the admissions test and a GPA of 7.8424/10 on the leveling courses. This group concentrates its data with greater emphasis on the first semester of 2016, it also gathers the largest number of students with an acceptable academic performance.

Finally, the third group concentrates 39% of the data from the analyzed population, with an entry score of 846.1537/1000 points and a GPA on the leveling stage of 7.9004/10. This third group has the greatest academic performance, with the largest concentration of data in the first semester of 2018.

This analysis allows the identification of the academic performance of all the students who were admitted to Universidad Nacional de Chimborazo from 2012 until the present. It also grants the possibility to stablish a direct relationship between the admissions test and the students' final GPA. Figure 5 shows the concentration of the data, where each color represents one of the 3 clusters; X axis corresponds to the admissions test grade and Y axis to the final GPA.

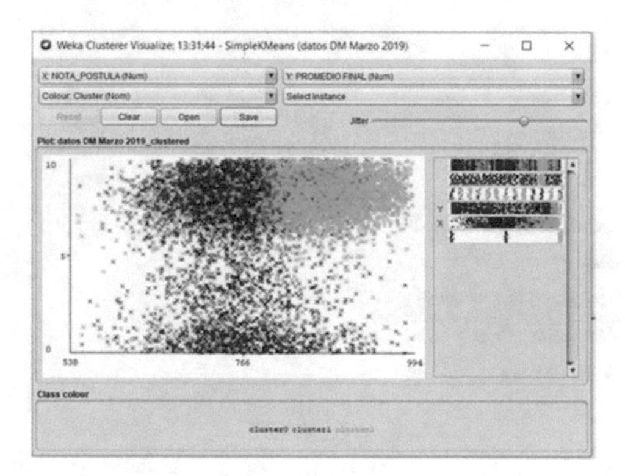

Fig. 5. Results – SimpleKmeans WEKA clusterer.

(b) Business Intelligence

Through the use of the Qlik Sense Desktop application, several data analysis of the academic periods ranging from 2012 and 2018 was also generated, obtaining dynamic results about the institutional academic reality, that will allow the university to take decisions for its improvement. In this sense, it is necessary that higher education institutions place special attention to their admission process as it constitutes an important element to guarantee an appropriate academic performance among the students.

The first analysis carried out, once again proves the direct relationship between the admissions test and the final GPA from the leveling courses. Universidad Nacional de Chimborazo has 31 undergraduate academic programs; the highest grades of the test belongs to the ones most demanded. Additionally, students with the highest entry grades maintain their high academic performance on the leveling classes. From the periods analyzed, the admission score at Universidad Nacional de Chimborazo is 795.52/1000 points and the GPA in the leveling process is 7.36/10 (Table 1).

Table 1. Relationship between "Ser Bachiller" test and final GPA at the preparatory courses "Nivelación".

School	Academic Program	Minimum grade to apply	Average grade to apply	Final GPA
Total		538,00	795,52	7,3651612
Health School	Medicine	627,00	902,52	8,378262
Health School	Dentistry	610,00	879,67	8,0877249
Engineering School	Civil Engineering	647,00	862,61	7,6814633
Engineering School	Architecture	619,00	849,60	7,616063
Health School	Clinic Psychology	580,00	848,96	7,7812618
Health School	Nursing	637,00	838,21	7,8603043
Health School	Physiotherapy	616,00	826,78	7,5261397
Health School	Clinical laboratory	547,00	825,07	7,6530219
Political and Administrative Sciences School	Law	561,00	810,88	7,7691528
Engineering School	Environmental Engineering	580,00	798,59	7,009663
Political and Administrative Sciences School	Economics	610,00	777,70	7,6096933
Education School	Sports and Pedagogy of Physical activities	700,00	775,12	7,3785919
Education School	Pedagogy of the national and international languages	700,00	774,41	7,335021
Engineering School	Industrial Engineering	586,00	772,31	6,6300093
Engineering School	Telecommunications	553,00	768,64	6,124045
Education School	Mechanical, Industrial and Automotive Education	700,00	763,57	6,879359

(*continued*)

Table 1. (*continued*)

School	Academic Program	Minimum grade to apply	Average grade to apply	Final GPA
Total		538,00	795,52	7,3651612
Education School	Basic education	616,00	762,37	7,2896603
Engineering School	Information technologies	541,00	761,97	6,1245596
Education School	Psychopedagogy	547,00	760,85	7,6374564
Political and Administrative Sciences School	Communications	595,00	758,95	7,5130634
Political and Administrative Sciences School	Accounting and auditing	562,00	758,06	7,3404085
Political and Administrative Sciences School	Business administration	538,00	757,93	7,2410119
Education School	Pedagogy of the Experimental Sciences	562,00	757,49	7,4191488
Education School	Initial education	604,00	754,82	7,311561
Education School	Technical education in electricity and electronic	700,00	753,05	7,1944731
Political and Administrative Sciences School	Turism	565,00	752,96	7,273393
Education School	Graphic Design	617,00	750,19	7,5078137
Engineering School	Turism	700,00	748,33	8,3171429
Education School	Pedagogy of the Experimental Sciences	622,00	748,15	7,2299681
Engineering School	Agroindustrial Engineering	559,00	744,04	6,3356526
Education School	Pedagogy of the Language and Literature	604,00	742,19	7,3682796
Education School	Pedagogy of the history and Social Sciences	610,00	734,57	7,0901586
Education School	Pedagogy of Informatics	592,00	732,52	6,2042975
Education School	Pedagogy of the Arts and Humanities	580,00	730,35	7,5894652

At Universidad Nacional de Chimborazo, from the year 2012 to 2018, applicants show an average grade of 710 < = x < 860 on their admissions test (Fig. 6).

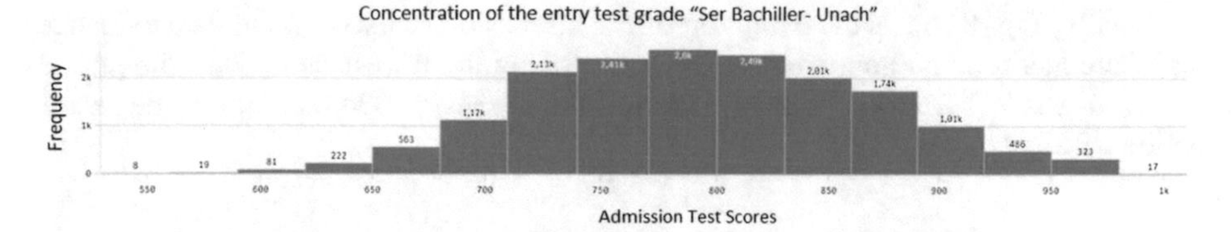

Fig. 6. Results. Concentration of the entry test grade "Ser Bachiller- Unach"

In addition to the concentration of data on the admission test score, this study also considers the final GPA of the leveling courses in the four schools of the university. The results obtained in the School of Political and Administrative Sciences show an average score of 772.72/1000 and the final grade point average is 7.49/10, which is considered as an acceptable academic performance, showing direct relation between both variables. In the School of Education, the average score is 754.5/1000 and the final GPA is 7.31/10, this is considered also as a proportional score.

In the School of Engineering, the average score on the admission test is 796.57/1000 and the final average grade point is 6.84/10 points. This is an exception case, that does not show a direct relationship between the variables analyzed on this study. The difference calls us to think about external factors that may cause a poor performance of students once they enroll in the preparatory classes. For example, high demands to manage a numerical and logical thinking that is not well explored during their high school years, evidencing the shortcomings of the national secondary education system.

The fourth case analyzed is the School of Health Sciences, which displays the best scores on the admission test 858.31/1000 and a final GPA of 7.93/10 points. This academic unit is characterized for having the most demanded academic programs of the university such as medicine, dentistry and nursing. Due to the demand, the minimum score to be accepted in the School of Education tends to be higher than in the other schools. According to the data, students from this academic unit show the best scores at the entry level and maintain a high academic performance once they enroll in the university (Fig. 7).

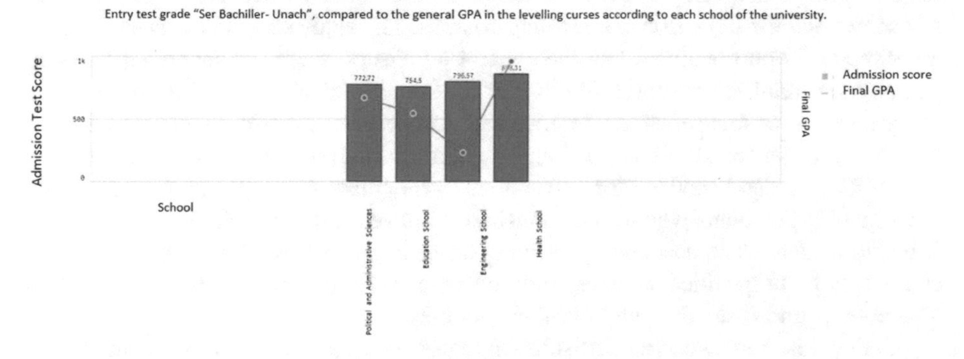

Fig. 7. Admission test score "Ser Bachiller- Unach", compared to the general GPA in the levelling curses according to each school of the university.

Finally, Fig. 8 shows the evolution of the admission test scores; the data evidences that there has been an improvement in recent years; the lowest score from the periods analyzed was 747.27/1000 in 2013, whereas the highest is 834.88/1000 in the second period of 2018.

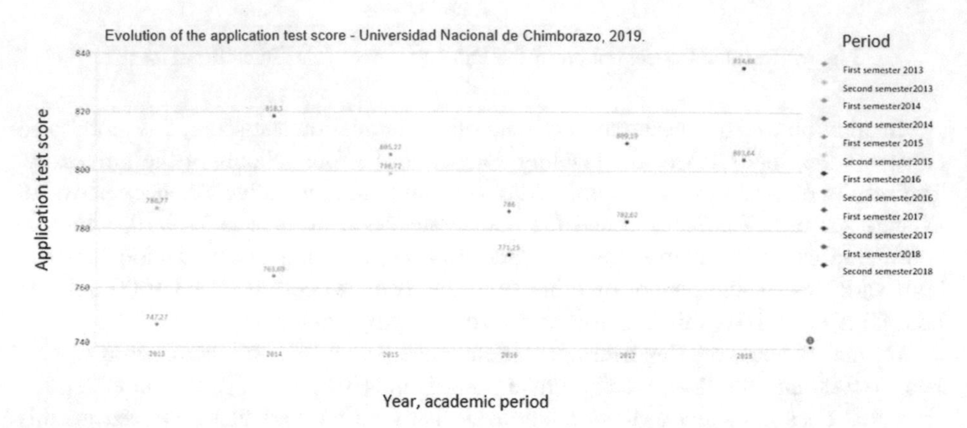

Fig. 8. Evolution of the application test score - Universidad Nacional de Chimborazo, 2019.

8 Conclusions

Through experimentation methods in this research work, it has been possible to apply a proposal of the KDD at Universidad Nacional de Chimborazo, obtaining within the DW subprocess a corporate database with historical information about applicants from the year 2012 until 2018. The study also allowed us the acquisition of data from the subprocess of DM that also evidences information and knowledge in the academic area of the institution. Within this framework, it is clear the importance of secondary education as a provider of the tools and knowledge that the students will require for an adequate development during their third level education.

The techniques of DM for grouping and the analysis of BI made possible to show the data groups that Universidad Nacional de Chimborazo displays from the admission test scores and the GPA from the leveling courses. The study shows that these variables are directly proportional and that the general admission score has increased in recent years. As a trend, entry grades are higher on the second semester of each year. The study also demonstrates that greater levels of competition motivate students to achieve higher scores on the admission test and maintain a high academic performance. This competition is also beneficial for universities as they have the opportunity to select the most qualified students who also demonstrate to have a greater commitment with their education. As a result, academic programs can meet new goals of development as they count with more qualified students, with the right skills to cope with pressure, who are also able to find a suitable job after their graduation.

Under this framework, advertising strategies also appear as an important tool to guarantee competition among students. From 2016 to the present, the Admissions

Office at Universidad Nacional de Chimborazo has increased the communication and promotion of the University's undergraduate programs, on different media, in order to increase the attractiveness of the institution. The knowledge generated from the study allows to consider "competition" as a key element for selection, retention and higher academic performance. New levels of competition can be achieved by offering greater amounts of information that attract students and that motivate them to apply to the university. Furthermore, the institution should strive to maintain this internal competition as a strategy for guaranteeing high academic levels in all its academic programs.

References

1. Gaviria, A.: Los que suben y los que bajan : educación y movilidad social en Colombia, p. 2012 (2012)
2. Guadagni, A.A.: Ingreso a la universidad en ecuador, cuba y argentina (2016)
3. Daza Vergaray, A.: Data Mining, Minería de Datos, Primera. Editorial Macro, Lima (2016)
4. P. de la R. Ecuador, Ley Organica de Educacion Superior, Loes, pp. 1–92 (2018)
5. U. N. de C. UNACH, Reglamento de Nivelación y Admisión para estudiantes de grado de la Universidad Nacional de Chimborazo, no. 0052, pp. 1–12 (2018)
6. Fayyad, U., Piatetsky-Shapiro, G., Smyth, P.: From data mining to knowledge discovery in databases. AI Mag. **17**(3), 37 (1996)
7. Maimon, O., Rokach, L.: Data Mining and Knowledge Discovery Handbook, pp. 22–38 (2015)
8. Uvidia Fassler, M.I., Cisneros Barahona, A.S., Ávila-Pesántez, D.F., Rodríguez Flores, I.E.: Moving towards a methodology employing knowledge discovery in databases to assist in decision making regarding academic placement and student admissions for universities. In: Technology Trends, pp. 215–229 (2018)
9. Moine, M., Haedo, S., Gordillo, D.S.: Estudio comparativo de metodologías para minería de datos. In: CACIC 2011 - XVII Congreso Argentino de Ciencias de la Computación (2011). http://sedici.unlp.edu.ar/bitstream/handle/10915/20034/Documento_completo.pdf?sequence=1. Accessed 03 Dec 2015
10. Bernabeu, R.: Hefesto, p. 146 (2010)
11. Hernández, J., Ferrari, C., Ramírez, M.: Introducción a la Minería de Datos. Pearson Educación, Madrid (2004)
12. Han, J., Kamber, M.: Data Mining Concepts and Techniques, 2nd Edn. Elsevier Inc., New York (2006)
13. González, L.: Una arquitectura para el análisis de información que integra procesamiento análitico en línea con minería de datos. Universidad de las Américas Puebla (2005)
14. Alzate Molina, C.A., Gallego Álvarez, G.A.: Gestión del Conocimiento, pp. 1–15 (2012)
15. Olson, D.L., Delen, D.: Advanced Data Mining Techniques (2008)
16. Cabena, P., et al.: Discovering Data Mining: from Concept to Implementation. Prentice Hall PTR, New Jersey (1997)
17. Jiménez, A., Timarán, S.: Caracterización de la deserción estudiantil en educación superior con minería de datos, vol. 28, no. Diciembre, pp. 447–463 (2015)

Deep Learning Model for Forecasting Financial Sales Based on Long Short-Term Memory Networks

Pablo F. Ordoñez-Ordoñez[1,2](✉), Martha C. Suntaxi Sarango[1], Cristian Narváez[1], Maria del Cisne Ruilova Sánchez[1], and Mario Enrique Cueva-Hurtado[1]

[1] Facultad de Energía, Universidad Nacional de Loja, Av. Pío Jaramillo Alvarado, La Argelia, Loja, Ecuador
pfordonez@unl.edu.ec

[2] Escuela Técnica Superior de Ingeniería de Sistemas Informáticos, Universidad Politécnica de Madrid, Calle Alan Turing s/n, 28031 Madrid, Spain

Abstract. The present article presents a model LSTM for the forecast of product sales, an alternative in deep learning for this type of dilemmas and not frequent in the area of financial knowledge. It was approached as a time series and following the steps for the construction of models with machine learning. The ILE company of Ecuador provided the data, between 2011 and 2018. The results showed this model has a minimum RMSE error of 2.20 compared to another two models: ARIMA and Single Perceptron.

Keywords: LSTM · Deep learning · Machine learning · Sales forecast

1 Introduction

Long Short-Term Memory Networks (LSTM) are the type of recurring networks, and arise to solve the problem of the enormous growth of the gradient and its fading over time, which occurs when training recurrent networks and makes it difficult to memorize long-term dependencies [6]. LSTM incorporated a series of steps to decide what information will be stored and eliminated through three floodgates that control the way information flows: **Input gate** used to control the new information input to the memory. **Forget gate** decides which data is important and which is discarded to make room for new data. **Output gate** used to control the result of the memories stored in the cell, that has optimization mechanisms based on the resulting network output error. The LSTM selection for these types of problems is based on a literature review [11], in particular, LSTM is the most widely used by other authors and is recommended for the area of finance.

An additional, several technologies were used to achieve this purpose, these were:

M. Botto-Tobar et al. (Eds.): ICAETT 2019, AISC 1066, pp. 498–506, 2020.
https://doi.org/10.1007/978-3-030-32022-5_46

1. **Keras** [1] is a library of neural networks developed in Python, can be run on Tensorflow or Theano, is characterized by modularity, minimalism and extensibility that allow an easy and fast prototyping, supports convolutional neural architectures, recurrent, in addition to arbitrary connectivity schemes. In [1] Python is currently compatible in versions 2.7 to 3.6 and can run algorithms on GPUs and CPUs. The Keras library consists of:
   (a) *Models*, Keras have two types of models available for implementation: the sequential model and the class *Model*, Both present several methods in common the $summary()$, $get_config()$ o $to_json()$ that return basic information and the models *getter* y *setter* to establish his weights.
   (b) *Layers*, neural network is designed based on the functions: *Dense*, which are totally connected regular layers. *Activation*, activating function for the output of the layers, these are: $softmax$, $softplus$, $softsign$, $relu$, $tanh$, $sigmoid$, $hard_sigmoid$, and $linear$. The following function is *Dropout*, randomly sets a fraction of input units to 0 for each update during the training phase, to avoid over-adjustment. *Reshape*, it transforms the output into a concrete form. Convolutional layers have different functions to develop according to the dimensions of their input, and recurrent layers exist in 3 types, the $simpleRNN$, $LSTM$ y *GRU*.
   (c) *Initializers*, determine how to initialize the weights of the layers through the argument *init*, the most common ones are: $uniform$, $normal$, *identity*, *zero*, $glorot_normal$, $glorot_uniform$.
   (d) *Missing functions*, is a parameter necessary to compile a model, Keras provides the following functions: $mean\_absolute\_error()$, $mean\_squared\_error()$, and $binary\_crossentropy()$.
   (e) *Optimizers*, is another of the values needed to compile a model, the principals are: *SGD*, a stochastic approximation method of the gradient descent method. *Adagrad*, a gradient algorithm that adapts the learning ratio to the parameters. *Adadelta*, an extension of the algorithm *Adagrad*; and *Adam*, based on the first-order gradient of objective stochastic functions, it is easy to implement, efficient in the consumption of computational resources.
2. **Python**, a programming language with efficient, high-level data structures, version 3.6 was used, standard library is freely available for major platforms from the Python website in [5]. This website contains distributions, links to free Python modules, programs, tools and documentation.
3. **Jupyter**, an interactive work environment that allowed Python code to be developed dynamically. This tool integrates blocks of code, text, graphics and images in the same document. Jupyter is frequently used in numerical analysis, statistics and ML, among other fields of computer science and mathematics [9].

To prove the efficiency of DL LSTM model, it was compared with Single Perceptron [7] and ARIMA [2,4], which results are illustrated in this section.

2 Methodology

The methodology used is based on [8] whose phases are the following: *(a) Data extraction* of the ILE Company, where sales records are stored with a total of 443 products between the dates 01-03-2011 and 28-02-2018. *(b) Hardware and software* for the execution of the model was an 8 GB memory, Intel Core i7 processor (2.2 GHz), Intel HD graphics 6000, MacOS Mojave operating system, and 500 GB solid disk. Keras was used, along with Tensorflow, Python 3 and the Jupyter work environment. To compare the DL model, two traditional models were implemented: a Single-layer Neural Networks (Perceptrons) in Keras and the regression model ARIMA for which the tool R version 1.1.423 was used. *(c) Preprocessing and exploratory analysis*, the data were expressed in time series [2], the variables of the data set of the ILE company is characterized by: **date, article** and **sale**, where the time series has temporary observations for different articles. As seen in Table 1 the data set was adjusted to represent a particular article in order to address a sales prediction model. The predictive model was made for total daily sales in the period from 01-03-2011 to 28-02-2018 giving a total of 2257 observations for training and comparison with other models.

Table 1. Fragment of data set adjusted for total sale

Date	Total_sale
01/03/2011	447,29
02/03/2011	0,00
03/03/2011	52,96
04/03/2011	230,80
05/03/2011	18,59
06/03/2011	0,00
07/03/2011	0,00
08/03/2011	0,00
09/03/2011	0,00
10/03/2011	13,25
11/03/2011	162,16
....	....

Next were applied transformations that allowed:

- Stabilize variance with function *ln* all data.
- Stationary the data with the variance, mean and covariance with the technique of the differences.
- Delete trends, because there is one instance per day, even if on that day, there is no sale.
- Apply the Dickey-Fuller Test [3] with R to validate if the series is stationary.

To train the deep neural network, it was necessary to normalize the input variables, the method used was the scaling of variables according to the Eq. 1.

$$X_std = \frac{(X - X.min)}{(X.max - X.min)} \tag{1}$$

Finally Fig. 1 shows the scaled and stationary observations. The first data were taken for training and the last for testing, with a distribution of 80% and 20%. That is, the first 2,044 data for training and the rest for testing.

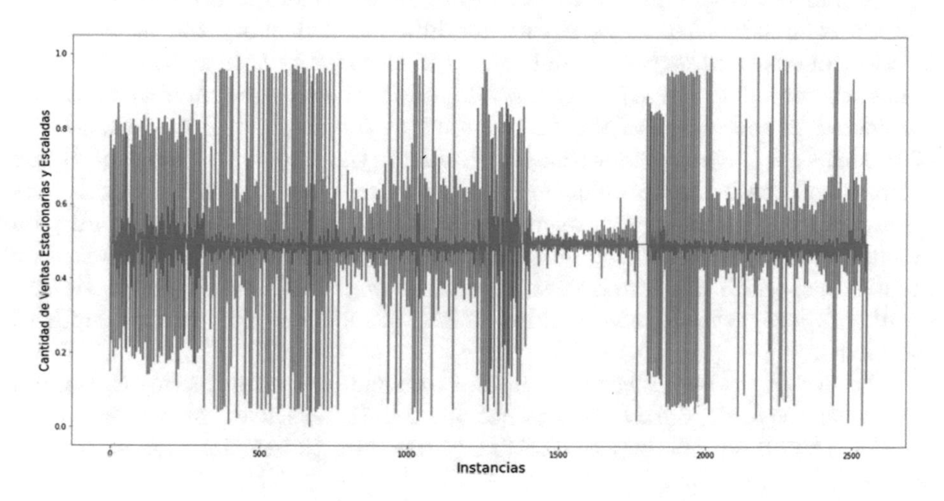

Fig. 1. Time series of escalated and stationary ILE sales

2.1 Selection of Models

To demonstrate the potency in the use of DL versus other ML techniques, three models were considered: ARIMA, Single-layer Neural Networks (Perceptrons) and the DL LSTM model [2,4]. For the comparison, all three models used the same data set and the error measure RMSE (square root of the mean square error) which is expressed in the Eq. 2.

$$RMSE = \sqrt{MSE} = \sqrt{E[(y - y)^2)]}) \tag{2}$$

To apply the perceptron and LSTM models in a time series, it was transformed to a regression, for this was added lagged variables, using the Eq. 3 in Python, where n is the size of lag or number of delayed variables:

$$y_t =_1 y_(t-1) +_2 y_(t-2) + ... +_n y_(t-n) \tag{3}$$

Then and finally Keras was used in Python to build the DL model and as a result of the SLR [11] an LSTM network was selected.

3 Results and Discussion

3.1 LSTM Model

The Table 2 summarizes the test values for LSTM: $Lag(n)$, measures the variables to consider to predict the target. $Dropout(p)$, avoids overfitting that occurs in deep nets. The parameter $0 \leq p \leq 1$ refers to the percentage of discarded units. It looks for different values of the extremes, if $p = 0$ nothing is discarded (no dropout is used); and if $p = 1$ all units are being discarded, that is, the layer would not have units. $Stateful(True/False)$, refers to the use of the last state of each sample, that is, it dictates how much memory the network has between batches. $Stateful$ is boolean, so if $stateful$ is $true$ it indicates that it has memory. $Number\_of\_blocks\_per\_layer$, it refers to the number of blocks considered in each layer of the LSTM. Size of the *batch*, refers to the size B of the lot $i_1, i_2, ...i_B$ of the stochastic gradient, the higher the value of B the network will reach the optimum in fewer iterations, but the cost for each iteration is higher. $number\_few$, refers to the number of times the data will pass through the model. Activation function refers to the activation function used by the neurons, which was tested were: *relu*, *tanh*, *sigmoid*, *hard_sigmoid*, *linear*. $number\_layers$, refers to the stacking of LSTM model to add greater depth to the model.

The size of the *batch* was kept in one to avoid an over adjustment, besides due to the size of the dataset it does not affect in the execution of the algorithm.

The Table 3 summarizes the LSTM parameters to get the best fit for the model.

3.2 ARIMA Model

In R [10] auto.arima() was used, this algorithm automatically calculates the parameters and returns the best possible model. The results are shown in the Fig. 2.

```
> modelo<-auto.arima(data_estacionaria)
> summary(modelo)
Series: data_estacionaria
ARIMA(1,0,2) with zero mean

Coefficients:
          ar1      ma1      ma2
      -0.7897  -0.0133  -0.8161
s.e.   0.0231   0.0177   0.0152

sigma^2 estimated as 17.6:  log likelihood=-7291.44
AIC=14590.88   AICc=14590.9   BIC=14614.27

Training set error measures:
                     ME     RMSE      MAE  MPE MAPE      MASE         ACF1
Training set 0.01017815 4.192876 2.520297 -Inf  Inf 0.4841581 -0.002365047
```

Fig. 2. R ARIMA results

Table 2. Experimental LSTM results for model optimization

Lag	Batch	Block	Layers	Stateful	Dropout	Periods	Activation Function	RMSE train	RMSE test	Parameters	Time
Base Parameters											
2	1	4	1	no	no	100	tanh	4,22761	**3,29912**	101	0:25:30
Dropout											
2	1	4	1	no	si, p = 0,5	100	tanh	3,87864	3,45729	101	0:19:25
2	1	4	1	no	si, p = 0,2	100	tanh	3,73345	3,30554	101	0:21:00
2	1	4	1	no	si, p = 0,8	100	tanh	4,1299	3,81689	101	0:18:04
2	1	4	1	no	si, p = 0,1	100	tanh	3,72496	**3,20202**	101	0:20:30
Lag											
3	1	4	1	no	si, p = 0,1	100	tanh	3,66505	3,23155	101	0:28:45
6	1	4	1	no	si, p = 0,1	100	tanh	2,79555	2,80478	101	0:30:03
5	1	4	1	no	si, p = 0,1	100	tanh	3,45550	3,20180	101	0:32:05
12	1	4	1	no	si, p = 0,1	100	tanh	2,74367	2,79802	101	0:38:00
13	1	4	1	no	si, p = 0,1	100	tanh	3,50090	2,87960	101	0:40:00
10	1	4	1	no	si, p = 0,1	100	tanh	2,90161	2,69407	101	0:38:46
8	1	4	1	no	si, p = 0,1	100	tanh	2,73881	**2,67092**	101	0:36:06
Block											
8	1	2	1	no	si, p = 0,1	100	tanh	5,55509	3,11815	35	0:30:00
8	1	5	1	no	si, p = 0,1	100	tanh	3,87060	3,28180	146	0:31:25
8	1	10	1	no	si, p = 0,1	100	tanh	2,69603	**2,62997**	491	0:32:18
8	1	6	1	no	si, p = 0,1	100	tanh	2,75467	2,84563	199	0:29:56
8	1	8	1	no	si, p = 0,1	100	tanh	2,62983	2,71030	329	0:30:35
Stateful											
8	1	10	1	si	si, p = 0,1	100	tanh	2,69664	**2,63961**	491	0:45:06
Periods											
8	1	10	1	no	si, p = 0,1	80	tanh	3,77974	3,80333	491	0:29:45
8	1	10	1	no	si, p = 0,1	200	tanh	3,11539	**2,61237**	491	0:38:47
Activation Function											
8	1	10	1	no	si, p = 0,1	200	relu	4,16812	3,20667	491	0:38:50
8	1	10	1	no	si, p = 0,1	200	sigmoid	2,19801	**2,20418**	491	0:45:32
8	1	10	1	no	si, p = 0,1	200	hard_sigmoid	3,10938	2,81715	491	0:40:00
8	1	10	1	no	si, p = 0,1	200	lineal	3,04838	2,83978	491	0:50:00
Layers											
8	1	10	2	no	si, p = 0,1	200	sigmoid	2,03534	**2,38636**	1401	1:17:09
10	1	10	3	no	si, p = 0,1	100	relu	3,04838	2,46667	491	1:20:07
10	1	10	4	no	si, p = 0,1	100	sigmoid	3,10938	2,61715	491	1:25:08
10	1	10	5	no	si, p = 0,1	200	sigmoid	2,90658	2,67963	491	1:04:06

Table 3. Optimum parameter for LSTM

Parameter	Optimum
Lag size	8
Batch	1
Blocks	10
Layers	1
Stateful	False
Dropout	0,1
Periods	200
Activation Functions	Sigmoide
Evaluation in training and testing	
Train score	2.19801 RMSE
Test score	2.20418 RMSE

This predictive model obtained an RMSE error of 4.192876.

3.3 Single-Layer Neural Networks (Perceptrons)

With this neural network of the one-layer perceptron type, the parameters were obtained based on an experimental bank of required values: size of *lag*, number

Table 4. Experimental ARIMA results for model optimization

Lag	Blocks	Periods	Activation Function	RMSE train	RMSE test	Parameters	hour
Base Parameters							
2	4	100	tanh	4,22761	4,29912	41	0:14:10
Lag							
3	4	100	tanh	3,66505	3,23155	41	0:15:00
6	4	100	tanh	3,66505	3,23155	41	0:15:03
5	4	100	tanh	3,66505	3,23155	41	0:15:07
12	4	100	tanh	2,94367	2,99802	41	0:22:00
13	4	100	tanh	3,72305	3,83455	41	0:15:04
10	4	100	tanh	2,90161	**2,67407**	41	0:15:06
8	4	100		2,73881	2,69092	41	0:14:09
Blocks							
10	2	100	tanh	3,55509	3,11815	35	0:09:34
10	5	100	tanh	3,66505	3,23155	51	0:12:02
10	10	100	tanh	2,69603	**2,62997**	101	0:15:54
10	6	100	tanh	3,66505	3,23155	61	0:13:04
10	8	100	tanh	2,62983	2,71030	81	0:13:56
Periods							
10	10	80	tanh	3,77974	**3,60333**	101	0:11:56
10	10	200	tanh	2,90003	2,61501	101	0:12:34
Activation Function							
10	10	200	relu	3,18181	**2,60783**	101	0:10:23
10	10	200	sigmoid	4,19295	3,32873	101	0:12:34
10	10	200	hard_sigmoid	3,17539	2,91237	101	0:11:32
10	10	200	lineal	3,97313	2,82845	101	0:16:07
LSTM Parameters							
8	10	200	sigmoid	3,26215	2,82495	101	0:09:13

of *blocks*, number of *periods* and *activationfunction*. The Table 4 illustrates the experiments carried out in the search of the optimal parameters for the model, the error of this network does not move away from the error of the DL model, the execution times are short and better results were obtained with the *ReLU* activation function. When using the parameters of the DL model the error increased.

The Fig. 3 shows the final results of the neural network.

```
Evaluando modelo en entrenamiento y prueba:
Train Score: 3.18181 RMSE
Test Score: 2.60783 RMSE

El numero de parametros en cada capa es:
_________________________________________________________________
Layer (type)                 Output Shape              Param #
=================================================================
Capa_oculta (Dense)          (None, 10)                90
_________________________________________________________________
Capa_salida (Dense)          (None, 1)                 11
=================================================================
Total params: 101
Trainable params: 101
Non-trainable params: 0
_________________________________________________________________
```

Fig. 3. Optimal network design

The results obtained indicate that the RMSE error among the LSTM models was 2.20 and the error in the single-layer perceptron was 2.61 were close, unlike the ARIMA model that had an error of 4.19.

It is considered that the performance of the DL model is acceptable considering that only 2257 instances were used, this can improve if the size of the data set is increased, because in this way the architecture will have the opportunity to model much more complicated time dependencies, characteristic of DL models for abstract and hierarchical concepts. It was also possible to analyze and improve the noise in the data generated by the presence of values equal to zero or very small.

The prediction of sales for the company ILE was based on the historical years 2011 to 2018, and results were achieved with the preprocessing of data and exploratory analysis, the product was a set of stationary data and scaled between (0, 1) to run the model. For this particular case, the trial and error method was used to estimate the parameters of the LSTM model and demonstrate its effectiveness versus the ARIMA model and the one-layer perceptron network.

4 Conclusions and Future Work

The success of the DL models is due to difficult problems, the depth of the network allowed to learn abstract and hierarchical representations of the data, complex concepts were constructed from simple concepts. A sales prediction model was achieved using exploratory analysis and the heuristic method of trial and error, in the first case through the analysis of the data to adapt to the model, and in the second case to find the optimal parameters that reduce the error in the LSTM model, demonstrating results in the area of finance that does not frequent these models for the resolution of their problems. For future works it is advisable to add to this model of sales, the variables: place, products, and high seasons, and to adapt the use of the model in other fields of study, for example, prediction of energy consumption, indices of the price of petroleum, population growth, rate of mortality, etc.

References

1. Chollet, F.: keras (2015). https://github.com/fchollet/keras
2. Corres, G., Passoni, L.I., Zárate, C., Esteban, A.: Estúdio comparativo de modelos de pronóstico de ventas. Iberoamerican J. Ind. Eng. **6**(11), 113–134 (2014)
3. Dickey, D.A., Fuller, W.A.: Distribution of the estimators for autoregressive time series with a unit root. J. Am. Stat. Assoc. **74**(366a), 427–431 (1979)
4. Ferreira, R., Braga, M., Alves, V.: Forecast in the pharmaceutical area–statistic models vs deep learning. In: World Conference on Information Systems and Technologies, pp. 165–175. Springer, Heidelberg (2018)
5. Foundation, P.S.: Python language reference, version 3.6 (2019). http://www.python.org
6. Hochreiter, S., Schmidhuber, J.: Long short-term memory. Neural Comput. **9**(8), 1735–1780 (1997)
7. Moral Algaba, F.: Redes completamente convolucionales en la segmentación semántica de lesiones melanocíticas (2017)
8. Moreno, A.: Aprendizaje automático (1994)
9. Pérez, F., Granger, B.E.: IPython: a system for interactive scientific computing. Comput. Sci. Eng. **9**(3), 21–29 (2007). https://doi.org/10.1109/MCSE.2007.53, https://ipython.org
10. R Core Team: R: A Language and Environment for Statistical Computing. R Foundation for Statistical Computing, Vienna, Austria (2013). http://www.R-project.org/
11. Suntaxi-Sarango, M.C., Ordoñez-Ordoñez, P.F., Pesantez-González, M.A.: Applications of deep learning in financial intermediation: a systematic literature review. KnE Eng. **3**(9), 47 (2018). https://doi.org/10.18502/keg.v3i9.3645, https://knepublishing.com/index.php/KnE-Engineering/article/view/3645

Cellular Automata Based Method for Territories Stratification in Geographic Information Systems

Yadian Guillermo Pérez Betancourt(✉), Liset González Polanco, Juan Pedro Febles Rodríguez, and Alcides Cabrera Campos

University of Informatics Sciences, Havana, Cuba
{ygbetancourt,lgpolanco,febles,alcides}@uci.cu
http://www.uci.cu

Abstract. The stratification of territories is a powerful tool for the analysis and trend in health studies. An important element in health studies is the relationship established between geographical location and health indicators in correspondence with the first law of Geography. From this approach, the formation of compact strata allows the identification of local and global trends. This paper presents method for territories stratification in Geographic Information Systems. A clustering algorithm based on cellular automata theory is proposed to incorporate the treatment to heterogeneity and spatial dependence. The results obtained from the evaluation of validation indices demonstrates the utility and applicability of the proposal.

Keywords: Cellular learning automata · Compact groups · Geospatial data · Graph clustering · Spatial stratification

1 Introduction

Stratification is a procedure that allows classifying objects in homogeneous classes [15,21] based on analogies or relationships that are established between their characteristics [5]. This procedure allows the construction of subsets of aggregated units called strata, to concretize efforts and resources in the areas most in need. In health studies, it is often called epidemiological stratification and is part of the integrated diagnostic-intervention-evaluation process [18]. Epidemiological stratification uses several approaches to contribute to the selection of sites or areas with health problems and to plan intervention strategies [1,3,17].

Among the approaches used, we can mention risk stratification and absolute risk [2], by weighted indicators, frequency distribution patterns of the main associated risks and multivariate analysis techniques, mainly those based on group analysis or clustering. The weighted indicators have been widely used in health studies. The weighted indicators method uses a set of indicators or risk factors associated with the study. The indicators are weighted by criteria of experts in

M. Botto-Tobar et al. (Eds.): ICAETT 2019, AISC 1066, pp. 507–517, 2020.
https://doi.org/10.1007/978-3-030-32022-5_47

the field and values are obtained for each territory used to build the groups from established ranges. The main limitation of this strategy lies in the definition of the weights and ranges to build the groups because of the bias that can be introduced. The frequency distribution patterns of the main associated risks are mainly used when the risk indicators can not be determined or are not identified. Its objective is to identify areas where a certain factor has a greater frequency of appearance and then undertake intervention actions.

When it is possible to quantify the risk indicators and their influence on the area, the epidemiological stratification of risk has been used. This strategy allows obtaining an evaluation on the level of reduction of the problem if the risk factors are acted upon. Its main limitation is that it does not allow to identify changes on the groups because it constitutes a photograph of the problem under study, and it is usually complex due to the need of studies to determine the relative risk and the population attributable risk. The absolute risk distribution is based on the use of a single indicator to build the strata from distribution ranges according to the incidence rate or characteristic of the study region. It is widely used for short-term epidemiological surveillance because it allows to monitor certain indicators in priority areas in a quick, practical and timely manner.

Multivariate analysis techniques allow for more complex studies because many risk factors and other variables associated with the problem can be incorporated. From this approach, the use of clustering techniques have been widely used [19]. Methods based on clustering within the context of health studies have limitations. The measures of similarity used are focused on the thematic data, without taking into account the spatial component; this violates the first law of Geography, *"everything is related, but nearby objects are more related than distant objects"*. Spatial auto-correlation is usually ignored in the stratification; it is based on assumptions of an independent distribution when analyzing the data, which can produce hypotheses or inaccurate and inconsistent models. The treatment of spatial heterogeneity is insufficient, which limits the identification of local phenomena.

In this paper proposes an approach based on clustering techniques for territories stratification in Geographic Information Systems (GIS). A clustering algorithm based on cellular automata theory is proposed to incorporate the treatment to heterogeneity and spatial dependence.

This article is structured as follows. A section of related works is presented where the main existing references in the literature that are related to the approach proposed in this study are presented. Subsequently, the proposed method and the stages that comprise it are detailed. Then the application is presented to a case study that allowed to evaluate the applicability of the proposed approach and the treatment to spatial dependence in stratification. Finally, the conclusions and the lines of future work are presented.

2 Related Works

In the literature reviewed, the authors agree that the approach to be used in the stratification process depends on the urgency of the results to make decisions, the availability of information associated with the problem being studied, the resources available, the training of the specialists who will undertake the study and its purpose. From the geospatial perspective, training in techniques of exploratory data analysis or data mining is required, just to mention two examples.

The geospatial clustering from data mining has been used in the stratification [7] such a way that the members of a group are closely related from some measure of similarity. The measures of similarity used for such purposes depend on the context of application and the type of geospatial entity. In [4] the authors propose to incorporate spatial nature from a transformation on point objects or lines, in which they are then treated as thematic. This proposal does not allow to describe spatial relationships on objects and therefore hinders the incorporation of space in the process.

Clustering based on graphs is a branch within these techniques that tries to find partitions of nodes within the graph in such a way that the connections within a same partition are denser than those of different partitions [12,13]. This approach incorporates the advantages in scalability and interpretation provided by graph-based representation models. It should also be noted that it is possible to transform a problem of grouping in the vector space to one based on graph from the construction of a graph of similarity.

The spectral clustering algorithm has been one of the most used and studied due to its solid theoretical base and the optimal global solution [9,23]. Given a graph of n nodes, the method executes a decomposition of eigenvalues of the Laplacian matrix. Its main disadvantage is that the direct implementation of decomposition of eigenvalues has a complexity $O(n^3)$. Several data clustering algorithms based on social segregation models have been proposed [10]. In particular models based on Cellular Automata Theory have been used for classifiers [11].

2.1 Cellular Automata

Cellular automata (CA) are a mathematical idealization of complex systems constructed from identical components with local interaction. An AC is a non-linear dynamic model with discrete space and time. These characteristics make CAs suitable for modeling complex processes described as massive collections of objects that interact locally with each other.

Definition 1. *An CA is a d-dimensional structure $C = (Z^d, \Phi, N, F)$ where: Z^d represents the cellular space with dimension d. Φ is the finite set of states through which a cell can transit. N is a subset of Z^d called neighborhood vector. The neighborhood of a cell u is the set of cells that are obtained from the neighborhood function. $F : \Phi^{\bar{m}} \rightarrow \Phi$ is the local rule giving the new state for each cell from the current state of the cells in your neighborhood.*

2.2 Cellular Learning Automata

Cellular Learning Automata (CLA) is a combination of Cellular Automata with Learning Automata for modeling problems and decentralized phenomena. It is classified within the family of Stochastic Cellular Automata with the feature of using an automaton for probabilistic state transition. A d-dimensional CLA is formally defined by 2.

Definition 2. *An CLA is a d-dimensional structure* $C = (Z^d, \Phi, A, N, F)$ *where: A is the set of learning automata, each assigned to a space cell* Z^d, Φ *is the finite set of states, N is neighborhood definition and* $F : \Phi^{\bar{m}} \rightarrow \beta$ *where* β *is the set of values that the reinforcement signal can take.*

2.3 Irregular Cellular Learning Automata

To remove the restriction in terms of efficiency and scalability in the presence of large-scale data and take advantage of features based models have been reported investigations graphs representing the cellular space by a graph. This approach is known as Irregular Cellular Learning Automata (ICLA).

An ICLA is considered an undirected graph where each node is a cell of the automaton. Each node has a learning automaton and the nodes in the neighborhood are their local environment. The local environment is called non-stationary because the action probability vectors in the neighboring learning automatons change during the evolution of the ICLA [6].

Definition 3. *ICLA is a structure* $C = (G = \, <V, E>, \Phi, A, F)$ *where:*
G is an undirected graph, with V as the set of vertices and E as the set of edges. Φ *is a finite set of states. The state of cell* c_i *is defined by* φ_i. *A is the set of learning automata and* $F^i : \underline{\varphi_i} \rightarrow \underline{\beta}$ *is the local transition rule of ICLA in the cell* c_i, *where* $\underline{\varphi_i} = \{\varphi_j | \overline{\{i, j\} \in E\} \cup} \{\varphi_i\}$ *is the set of states of all neighbors of* c_i *and* $\underline{\beta}$ *is the set of values that the reinforcement signal can assume.*

This approach has been used in spatial analysis problems as sampling social networks [8] and vertex coloring problem [22]. From this approach, the theoretical bases for the development of proposal are established.

3 Proposed Method

This research is framed in the discipline of information systems (IS). IS are essentially artifacts that capture and represent knowledge about certain domains. Paradigms that support research discipline if identified as behavioral sciences and design sciences. Based on the nature of the problem addressed in this research and the relationship between its field of action and the IS discipline, the proposal is developed under the paradigm of design sciences.

The research was planned and executed from the process defined in [14] for research under the paradigm of design sciences and that defines the stages: problem identification and motivation, solution objectives, design and development, demonstration, evaluation and communication.

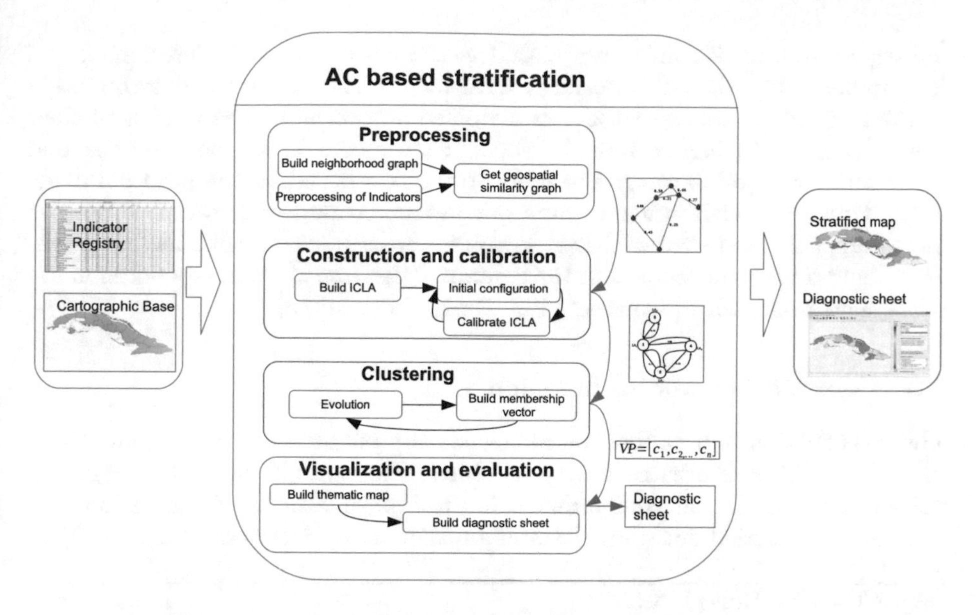

Fig. 1. Cellular automata based method for territories stratification in GIS

3.1 Preprocessing of the Indicator Register and the Cartographic Base

This stage aims to prepare the cartographic base and the indicators for stratification. It also ensures the creation of an intermediate structure as defined geospatial graph similarity is the basis for the work of the following method stages. The cartographic base must be formed by n layers with $n \geq 1$, for the case where $n = 1$ then, the layer must be polygons that represent the territories are studied. To construct the neighborhood graph, the definition base layer of the territories under study is used. From this we define a graph $G = (V, E)$ where V is the set of nodes of the graph and each node represents a polygon of the layer. Then the edge set E is built by adding an edge between each pair of nodes whose polygons belong to the neighborhood. The definition of neighborhood N is given by the objectives of the research and the available data.

The indicators for this stage according to their nature are classified into geospatial indicators and thematic indicators. Geospatial indicators are obtained from layers of the cartographic base. Each layer represents an indicator and, in correspondence with the geospatial object type of the layer, the layer indicator is obtained. To obtain the layer indicator in correspondence to the geospatial type, for points and lines the Eq. 1 and for polygons the 2.

$$I_{ij} = \frac{n}{A(p_i)} \forall p_i \in P \tag{1}$$

$$I_{ij} = \frac{\sum_{k=1}^{n} A(p_i k)}{A(p_i)} \forall p_i \in P \tag{2}$$

Subsequently, the risk index associated with each indicator is determined and used to determine the risk associated with each territory. For the determination of the risk index, the indicators are classified according to the impact of their value. Those with higher value of the indicator contribute more to risk and those with lower value contribute more to the risk based on the proposal made in [9]. Once the neighborhood graph is constructed and with the preprocessed indicators, the geospatial similarity graph is constructed. In this task the edges are weighted with the value that the similarity between the corresponding nodes takes. This approach was presented by the authors in [16].

3.2 Construction and Calibration

The construction and calibration stage has the purpose of building an ICLA. Initially an ICLA is created having as starting the geospatial similarity graph. Subsequently an initial configuration for the automaton is obtained and the calibration is carried out until the configuration is not the desired one.

Algorithm 1. Build ICLA

Require: A similarity graph $G = (V, E)$
Ensure: A ICLA *icla*
  $icla = ICLA(G)$
  for all $v_i \in V$ **do**
    $p_i = []$
    $action_i = actionsNeighborhood(v_i)$
    for all $a_j \in action_i$ **do**
      $p_i.add(weightEdge(v_i, a_j))$
    end for
    for all $j \in [0, len(p_i)]$ **do**
      $p_{ij} = \frac{p_{ij}}{\sum_{p_i}}$
    end for
    $LA_{v_i} = buildLearningAutomata(action_i, p_i)$
    $icla.addLearningAutomata(LA_{v_i})$
  end for
  return *icla*

The probability vector is calculated from the similarity between the nodes. Each action is assigned a probability according to the similarity.

3.3 Clustering

Once the neighborhood graph is constructed and with the preprocessed indicators, the geospatial similarity graph is constructed. In this task the edges are weighted with the value that the similarity between the corresponding nodes takes.

Algorithm 2. ICLASC

Require: $icla = (G, \Phi, LA, F)$
Ensure: A membership vector PV
  repeat
    for all $la_i \in LA$ **do**
      $S[i] = max(la_i.probVector)$
    end for
    $PV(t) = buildPVector(S)$
    for all $la_i` \in LA$ **do**
      if $coefIntra(la_i) < coefInter(la_i)$ **then**
        $\beta = 0$
      else
        $\beta = 1$
      end if
      $Z_i(t) = Z_i(t-1) + 1$
      $W_i(t) = W_i(t-1) + 1 - \beta$
      $\hat{D}_i(t) = \frac{W_i(t)}{Z_i(t)}$
      $a_i = max(\hat{D}_i)$
      for all $a_j \in la_i.actions$ **do**
        $la_i.probVector[j] = \begin{cases} la_i.probVector[j] + \alpha * (1 - la_i.probVector[j]) & \text{if } a_j = a_i \\ (1-\alpha) * la_i.probVector[j] & \text{in other case} \end{cases}$
      end for
    end for
  until $PV(t) \neq PV(t-1)$
  return VP

This algorithm results in the membership vector. With this vector, the groups or strata that are visualized in the thematic map are constructed.

The estimator learning algorithm used is defined as $\langle PV(t), \hat{D}(t)\rangle$ at moment t, according to [20] $PV(t)$ denotes the action probability vector and $\hat{D}(t) = [\hat{d_1}(t), \hat{d_2}(t), ..., \hat{d_r}(t)]^T$ represents the set of reward estimations.

Algorithm 3. BuildPVector

Require: $S = [a_1, a_2, \ldots a_n]$ Solution vector
Ensure: A membership vector pv
  $class = 0$
  $pv = [0, 0, \ldots, 0]$
  $visit = [$**false**, **false**, $\ldots$, **false**$]$
  for all $a_i \in S$ **do**
    if not$visit[i]$ **then**
      if $visit[S[i]]$ **then**
        $pv[i] = pv[S[i]]$
        $visit[i] =$ **true**
      else
        $visit(i, class, visit, pv, S)$
        $class = class + 1$
      end if
    end if
  end for
  return pv

3.4 Visualization and Evaluation

This stage has the function of evaluating the results obtained in the grouping stage and constructing the diagnostic sheet. During the creation of the thematic map, a color is assigned to each stratum in correspondence with the average risk index of the stratum. The diagnosis sheet is constructed with the most affected strata, the most affected territories and the indicators that most affect. The indicators that most affect the strata are obtained from those with positive auto-correlation and risk index higher than the expected value.

4 Stratification of Territories Method Application

Our novel stratification method can be seamlessly used in data set where you can establish a relationship of spatial neighborhood between objects. Here we show the application of the proposed method to perform a stratification of territories. The spatial dependence between the territories of a group is also evaluated.

4.1 Main Causes of Death in Cuba in 2016 Year

The 15 provinces of Cuba were selected, according to the current administrative political division. The vector layer with the polygons that represent each territory chosen for the analysis was obtained from the Spatial Data Infrastructure of the Republic of Cuba.

From the papers reported in the literature on the relation of diseases to space, it was decided to choose as variables the main causes of death in Cuba in 2016 year. The indicators of these variables by territories are accessible in the Statistical Yearbook of health.

The main goal of this experiment is to evaluate if the strata have spatial dependence. First, spatial autocorrelation is evaluated for the data set. This allows examining if there is a tendency to form groups in space over the data set. Then the groups are built and the internal group spatial dependence is evaluated.

From the literature review about stratification, was identified the K-Means algorithm as one of the most used and with the best results. It also Spectral clustering stands out for its solid theoretical base. From the selected data set and taking as reference the stratification presented in [17], the performance of the proposed method is evaluated. We evaluate the performance of each clustering algorithm, ICLA, KMeans and Spectral.

4.2 Stratification Results

When evaluating spatial autocorrelation, the MORAN index for data set obtained a positive value (0.67), so there is spatial dependence. The hypothesis test showed a p_value of 0.008, so the hypothesis of spatial dependence is approved. With the proposed method, three groups obtain positive values of auto-correlation [0.27, 0.81, 0.42] and for all three it is significant [0.001, 0.007, 0.01]. The other group has a single element and it is not possible to evaluate the spatial autocorrelation.

Spectral obtains a one stratum with a single element and the other three obtain the following values of autocorrelation [0.66, 0.50, −0.07]. Spectral obtains two groups with positive auto-correlation but not significant [0.18, 0.12, 0.41]. In the case of Kmeans, three groups obtain negative values of auto-correlation [−0.46, −0.34, −0.42] and the other group has a single element. The results of the evaluation of spatial dependence show that the proposed method obtains better performance in correspondence with the first law of geography.

Clustering techniques required different evaluation criteria for performance evaluation according to internal and external validation indices. Validation indices are used to evaluate the performance of the proposed method with respect to other clustering algorithms. For evaluate the quality of a data partition using quantities and feature inherited from the datasets includes Silhouette Score, Calinski Harabaz and Davies-Bouldin Index.

The results of different cluster validation indices reported in the literature show in Fig. 2.

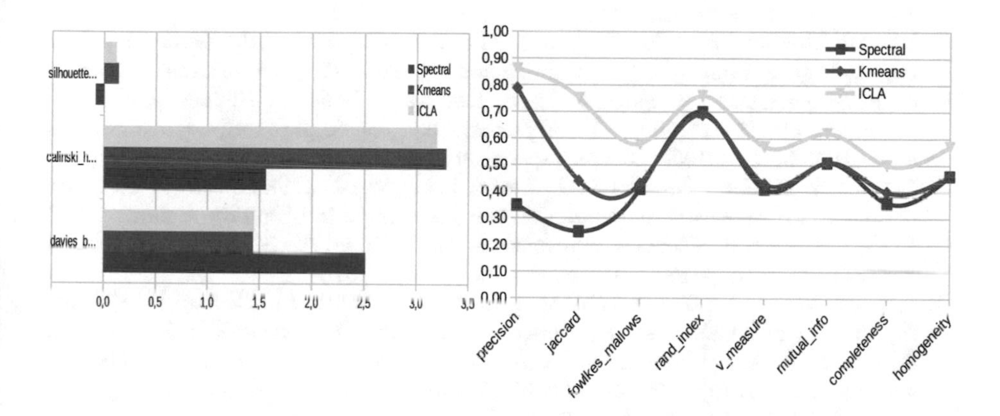

Fig. 2. Results of internal and external validation indices

For the internal validation indexes, the best results are obtained by KMeans and the proposed algorithm. Silhouette and Calinski-Harabasz show the validity of the clustering results. Davies Bouldin index results indicates that the clusters are more compact and far from each other for KMeans and ICLA. The evaluation of the external validation indices show the competitiveness of the proposal for this data set with the other two algorithms.

5 Conclusions and Future Work

The stratification of territories with the proposed method allows obtaining more compact strata in correspondence with the first law of geography and in the same way detects spatial structures. It also guarantees scalability with respect to the number of territories over which the analysis is carried out. By making use of automatons with irregular cellular learning, the time in the conformation of the strata is considerably reduced, so efficiency is guaranteed, which is evidenced in the experimental results obtained. The results obtained from the evaluation of validation indices demonstrate the utility and applicability of the proposal. As future lines of work, we intend to define a mechanism for evaluating the geosocial perception of the indicators that most affect the strata and territories. Similarly, when the level of detail of the stratified studies leads to the growth in the number of nodes and edges of the graphs, introduce the use of reduced graphs for the analysis. It is also intended to take advantage of the dynamics of cellular automata to incorporate the temporal component in this type of study.

References

1. Alegret Rodríguez, M.: Propuestas metodológicas para la incorporación más efectiva del análisis espacial en Ciencias de la Salud. dcmed, Universidad de Ciencias Médicas de Villa Clara (2007). http://tesis.repo.sld.cu/213/
2. de Araújo Nobre, M., Ferro, A., Maló, P.: Adult patient risk stratification using a risk score for periodontitis. J. Clin. Med. **8**(3), 307 (2019)
3. Batista Moliner, R., Coutin Marie, G., Feal Cañizares, P., González Cruz, R., Rodríguez Milord, D.: Determinación de estratos para priorizar intervenciones y evaluación en Salud Pública. Revista Cubana de Higiene y Epidemiología **39**(1), 32–41 (2001). http://scielo.sld.cu/scielo.php?script=sci_abstract&pid=S1561-30032001000100005&lng=es&nrm=iso&tlng=es
4. Betancourt, Y.G.P., Polanco, L.G., Pérez, R.M., Vega, Y.T.: Stratification of territories based on health indicators on the geographic information systems QGiS. Revista Cubana de Ciencias Informáticas **10**(0), 163–175 (2016). http://rcci.uci.cu/?journal=rcci&page=article&op=view&path[]=1374
5. Delgado Acosta, H., González Moreno, L., Valdés Gómez, M., Hernández Malpica, S., Montenegro Calderón, T., Rodríguez Buergo, D.: Estratificación de riesgo de tuberculosis pulmonar en consejos populares del municipio Cienfuegos. MediSur **13**(2), 275–284 (2015). http://scielo.sld.cu/scielo.php?script=sci_arttext&pid=S1727-897X2015000200005
6. Esnaashari, M., Meybodi, M.R.: Irregular cellular learning automata. IEEE Trans. Cybern. **45**(8), 1622–1632 (2015). http://ieeexplore.ieee.org/abstract/document/6914602/, 00009
7. Gewali, L.P., Manandhar, S.: Approaches for clustering polygonal obstacles. In: Latifi, S. (ed.) Information Technology - New Generations. Advances in Intelligent Systems and Computing, vol. 558, pp. 887–892. Springer, Heidelberg (2018)
8. Ghavipour, M., Meybodi, M.R.: Irregular cellular learning automata-based algorithm for sampling social networks. Eng. Appl. Artif. Intell. **59**, 244–259 (2017)
9. Langone, R., Mauricio Agudelo, O., De Moor, B., Suykens, J.A.K.: Incremental kernel spectral clustering for online learning of non-stationary data. Neurocomputing **139**, 246–260 (2014). http://www.sciencedirect.com/science/article/pii/S0925231214004433
10. Li, Z., Guan, X., Wu, H., Gong, J.: A novel k-means clustering based task decomposition method for distributed vector-based CA models. ISPRS Int. J. Geo-Inf. **6**(4), 93 (2017)

11. de Lope, J., Maravall, D.: Data clustering using a linear cellular automata-based algorithm. Neurocomputing **114**, 86–91 (2013). http://www.sciencedirect.com/science/article/pii/S0925231212007904
12. Miasnikof, P., Shestopaloff, A.Y., Bonner, A.J., Lawryshyn, Y.: A statistical performance analysis of graph clustering algorithms. In: Bonato, A., Prałat, P., Raigorodskii, A. (eds.) Algorithms and Models for the Web Graph. LNCS, pp. 170–184. Springer, Heidelberg (2018)
13. Moradi, P., Rostami, M.: Integration of graph clustering with ant colony optimization for feature selection. Knowl.-Based Syst. **84**, 144–161 (2015). http://www.sciencedirect.com/science/article/pii/S0950705115001458
14. Peffers, K., Tuunanen, T., Gengler, C.E., Rossi, M., Hui, W., Virtanen, V., Bragge, J.: The design science research process: a model for producing and presenting information systems research. In: Proceedings of the First International Conference on Design Science Research in Information Systems and Technology (DESRIST 2006), pp. 83–106. sn (2006)
15. Pérez, C.G., Aguilar, P.A.: Estratificación epidemiológica de riesgo. Revista Archivo Médico de Camagüey **17**(6), 762–783 (2013). http://www.medigraphic.com/pdfs/medicocamaguey/amc-2013/amc136l.pdf
16. Pérez Betancourt, Y.G., González Polanco, L., Febles Rodríguez, J.P.: Geospatial data preprocessing algorithm for the stratification of territories. In: Science and Technological Innovation, vol. 2, Chap. Technical sciences. EDACUN—Opuntia Brava (2018)
17. Pérez Betancourt, Y.G., González Polanco, L., Febles Rodríguez, J.P., Cabrera Campos, A.: Proposals for geospatial analysis in health studies. Revista Cubana de Ciencias Informáticas **12**(2), 44–57 (2018)
18. Quesada Aguilera, J.A., Quesada Aguilera, E., Rodríguez Socarras, N.: Diferentes enfoques para la estratificación epidemiológica del dengue. Revista Archivo Médico de Camagüey **16**(1), 109–123 (2012). http://scielo.sld.cu/scielo.php?script=sci_abstract&pid=S1025-02552012000100014&lng=es&nrm=iso&tlng=es
19. da Costa Resendes, A.P., da Silveira, N.A.P.R., Sabroza, P.C., Souza-Santos, R.: Determination of priority areas for dengue control actions. Revista de saude publica **44**(2), 274–282 (2010)
20. Rezvanian, A., Moradabadi, B., Ghavipour, M., Khomami, M.M.D., Meybodi, M.R.: Learning Automata Approach for Social Networks. Springer, Heidelberg (2019)
21. Santos-Garcia, A., Jacob, M.M., Jones, W.L.: SMOS near-surface salinity stratification under rainy conditions. IEEE J. Sel. Topics Appl. Earth Observ. Remote Sens. **9**(6), 2493–2499 (2016)
22. Vahidipour, S.M., Meybodi, M.R., Esnaashari, M.: Adaptive Petri net based on irregular cellular learning automata with an application to vertex coloring problem. Appl. Intell. **46**(2), 272–284 (2017)
23. Wang, S., Lu, J., Gu, X., Weyori, B.A., Yang, J.Y.: Unsupervised discriminant canonical correlation analysis based on spectral clustering. Neurocomputing **171**, 425–433 (2016). http://www.sciencedirect.com/science/article/pii/S0925231215008899

Determination of the Central Pattern Generator Parameters by a Neuro-Fuzzy Evolutionary Algorithm

Edgar Mario Rico Mesa[1]
and Jesús-Antonio Hernández-Riveros[2(✉)]

[1] SENA, Medellín, Colombia
emrico@sena.edu.co
[2] Universidad Nacional de Colombia, Medellín, Colombia
jahernan@unal.edu.co

Abstract. The selection of appropriate parameters of a Central Pattern Generator (CPG) facilitates its use on applications in robotics. A Central Pattern Generator is class of oscillator represented by a system of *n* differential equations of first order with m parameters, finding those *m* parameters to match a specific behavior is a complex problem to solve. For the development of applications in robotics, such as locomotion, the CPG have been applied as a decentralized control system. In this work, a Neuro-Fuzzy Evolutionary procedure is proposed to determine the parameters of a CPG for a specific cases of 2, 3, 4, 5 and 6 neurons. The developed methodology is presented in detail. The proposed algorithm guarantees a given response in the frequency and amplitude into the ranges required.

Keywords: CPG · Digital filter · Fuzzy classifier · Recurrent Neural Network · Genetic algorithms

1 Introduction

The Central Generators of Patterns (CPG) are a twentieth-century theory proposed by biologists, such as Sten Grillner, to explain the mechanical, involuntary and reflex movements that usually occur in living beings such as breathing, digestion, heartbeat and even locomotion [1]. The way to demonstrate this theory was through experiments performed with living beings such as mollusks, fish, snails, mice, and even in cats obtaining electrical signals from the central nervous system [2]. In the 21st century, this theory has been taken to robotics to optimize the control of the locomotor system in bipeds, quadrupeds, and hexapods using the oscillators as electrical signals emulator [3]. At the moment, there are many classes of oscillators that from the theory of chaos and bifurcations [4] were built for to experiment viability of CPG in robotics [5]. The CPG proposed have been built from the model proposed by the biologists for oscillators of two and four neurons represented by a system of *2 and 4* differential equations of the first order [6]. It has been developed for systems of differential equations of 2, 3, 4, 5 and 6 neurons. Each neuron controls each joint for the movement

M. Botto-Tobar et al. (Eds.): ICAETT 2019, AISC 1066, pp. 518–530, 2020.
https://doi.org/10.1007/978-3-030-32022-5_48

of a robotic system. Note that to find the value of the parameters of a recurrent neural network for a stable frequency is, in general terms, an activity of great complexity. Therefore, a Neuro-Fuzzy Evolutionary procedure to determine the parameters of the system of differential equations for 2, 3, 4, 5, 6 neurons is proposed in this paper. Each system has a frequency of operation for locomotion process in robotic. This paper has the following units. State of art: the technologies that in various parts of the world are developed with CPG. Conceptual Framework: to define and explain key theoretical concepts. Results: to present the scope achieved in the development of the algorithm. Data analysis: to make an interpretation of the data produced in the simulation.

2 State of Art

Following is a series of works from Europe, Asia and North America made in the last 5 years focused to solutions of CPG. It presents proposals used to find the CPG parameters as in [7] that the Gauss-Newton method is used that determines the values of the neuronal coupling parameters Morris - Lecar taking into account that a CPG has a cycle behavior limit exponentially stable. In [8] a procedure consisting of three steps is used, initially classical methods are used using an application called Feedback design to find the parameter values that allow a possible stable behavior, this set of values is evaluated according to the theory of chaos and bifurcations using the CPG Design application and finally the Simscan application is used to determine the sensitivity of the parameters. In [9] Hopfield networks analyzed from the spatiotemporal aspect of bifurcated periodic oscillations considering the stability criteria of the theory of chaos and symmetric bifurcation of equations and the representation theory of Lie groups. In [10] the Van Der Pol oscillators are used for the development of CPG and for the attainment of the optimal parameters the genetic algorithms are used, initially, the values of the parameters of the Van Der Pol oscillator are searched. The frequency and amplitude should be periodic, and then the values of the parameters whose synaptic weight allow a phase delay according to the type of CPG required are adjusted. In [11] determines the parameters of the oscillator in the chimera state that arise during an instant of local chaos. In this type of state, it is analyzed to determine the degree of stability. In [12] the Van Der Pol equation with periodic excitation proposed for the emulation of action potentials in the FitzHugh-Nagumo model. The intention is the study of the various dynamic nonlinear behaviors of the nervous system. In [13], a CPG of 4 neurons is synthesized directly with an evolutionary algorithm. The parameters are found by minimizing the error in the magnitude and the phase shift of the signals.

3 Conceptual Framework

3.1 CPG Models

In the development of movement coordination in mobile robotics, various algorithms based on the concepts and theories about CPG by biologists and researchers such as

Avis Cohen, Serge Rossignol, and Sten Grillner. These neural networks are not based on knowledge of the positions of the joints in space but act directly on the joints. The architecture Recurrent Neural Networks in continuous time (CTRAAN) consist of neural networks that have bidirectional connections between the nodes [14]. The mathematical representation of recurrent neural networks based on the Chiel-Beer Model [15] is a system of first-order differential equations, it is expressed in Eq. 1.

$$\tau_i \frac{dy}{dt} = -y_i + \sum\nolimits_{j=1}^{M} w_{ji}\sigma(y_j + \theta_j) + I_i, \text{ i} = 1\ldots.\text{M} \tag{1}$$

Where time constant (τ) is associated with the permittivity of the cell membrane, weights of synaptic connections between neurons (w), point of operation of each node (θ), transfer function of each neuron (σ), and external input to the system (I).

The typical models most used for the generation of CPG are the models of 2 and 4 neurons [16], see Fig. 1.

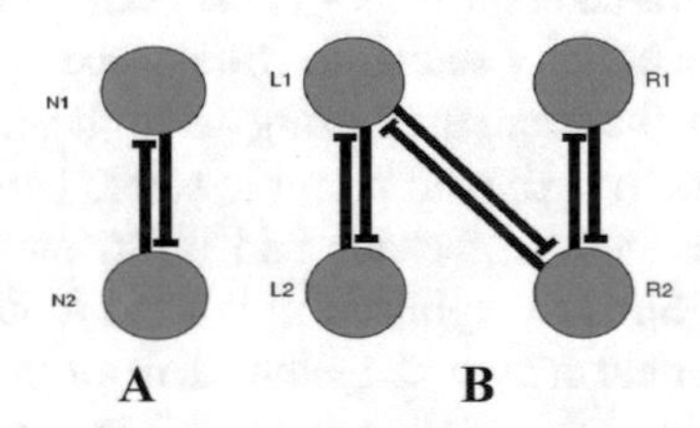

Fig. 1. Recurrent network of (A) two neurons, (B) four neurons.

4 Methodology

The next proposed methodology were defined to meet the objective of obtaining the parameters of a CPG for the locomotion of an articulated robot.

4.1 Design and Implementation of CPG

For the development of the central generator of patterns (CPG) the Rand proposal with their systems of differential equations of first order for 2 and 4 neurons are used of [16], the response of each neuron is the reproduction of cyclic signals (oscillator). Starting from the generic expression for 2 and 4 of Rand neurons, we have extrapolated for 3, 5 and 6 neurons presented in [17]. All cases are analyzed, starting for 2 neurons.

$$\tau_1 \frac{dy_1}{dt} = -y_1 + \frac{A(k_1 - Dy_2)^2}{(k_1 + Dy_2)^2 + (k_1 + Dy_1)^4} \tag{2}$$

$$\tau_2 \frac{dy_2}{dt} = -y_2 + \frac{A(k_2 - Dy_1)^2}{(k_2 + Dy_1)^2 + (k_2 + Dy_2)^4} \tag{3}$$

4.2 Obtaining the Parameters of Each System of Differential Equations

In the search for the solution, for example, for the Eqs. 2 and 3, it is fundamental to start from the initial conditions of the systems of differential equations, therefore, in order to comply with the initial conditions, an analysis has been made of the most important characteristics that the signals obtained from an oscillating system must possess.

Cyclical Oscillation
The frequency is of the order of milli-hertz, since in the different experiments carried out by the biologists it has been found that in repetitive activities such as locomotion the operating frequencies are extremely low.

Amplitude of Homogeneous Oscillation
The signals that have been captured in the alive beings by the biologists have had amplitudes of signals that tend to be uniform.

4.3 Data Analysis

To determine the values of the parameters of the central pattern generator it is necessary a procedure that allows reproducing the expected behavior. However, looking at the systems of equations you have the big problem that each parameter is a degree of freedom and each system of equations can have at least 6 parameters. Therefore, the manual selection of values of the parameters can be difficult and inappropriate for a robotic system that requires changing its type of locomotion or changing the speed of displacement immediately.

4.4 Selection of CPG Parameters

The procedure proposed to select the parameter values consists of the following steps:

(1) Select the values of the parameters in a random way. The values oscillate between 0 and 10 that will be assigned respectively to the parameters of the system of differential equations. The number of systems of differential equations reproduced is 100.
(2) Solve the system of differential equations by the Euler method. The Euler method is applied to the 100 systems of equations. The Euler method consists of a numerical integration procedure to solve ordinary differential equations from a given initial value.
(3) Use criteria of digital signal processing as digital passband filters by means of the convolution method. For each system of equations, the convolution is used multiplying each data obtained by the Euler method by the passband filter coefficients. The result of the convolution procedure will determine if the system of equations operates at the required frequency.
(4) Use computational intelligence technique as a diffuse classifier taking into account the experience of the designer. In the diffuse classifier, rules are used related to the amplitude of the signal and its periodicity. The signals that fulfilled the frequency

requirements are submitted to the fuzzy classifier. The diffuse classifier allows recognizing the periodic signals of the non-periodic ones.

(5) Select the most optimal family of parameters from the obtained score in the classification process. The classifier assigns a value between zero and one that indicates the periodicity of the signal reproduced by the hundred systems of differential equations of the first order.

(6) Creation of 2n systems of differential equations using genetic algorithm techniques. The new population will be of 2n and will be obtained from the last two generations with the mutation techniques with n/2 systems of difference equation with worse qualification and will apply the crossing between the first n/2 systems of differential equations with better score. The Euler method and then the diffuse classifier will be applied to the new population of n systems of differential equations.

(7) Ordination of the population of the last 2 generations. The fuzzy classifier will be used as a performance function for the classification with all the systems of differential equations, where those with the highest score will have a higher periodicity behavior and those with the lowest score will behave toward non-periodicity.

(8) Elimination of systems of differential equations of first order. From the result obtained when evaluating the n families of equations by the fuzzy classifier, the two systems of differential equations of the first order with the worst score will be eliminated. The process will continue from point 6 and will only end when there are only 2 systems of differential equations of first order, a situation in which the system of equations with the best score will be selected as the best family.

5 Results

Below are the results obtained taking into account the proposed methodology.

5.1 Algorithm for the Selection of the Initial Population

The flow diagram (Figure 2) of the algorithm implemented for shaping the initial population is presented below (steps 1–5).

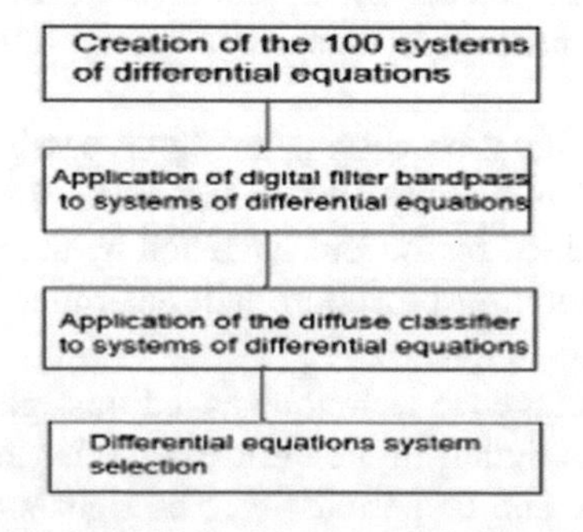

Fig. 2. Flow diagram of the algorithm

The filter is FIR type with a Kaiser Passband window with the following characteristics.

Cut frequency: 1, rejection band: 0, Cutting frequency: 2 rejection band: 0.03, Cutting frequency: 1, pass band: 0.004, Cutting frequency: 2, pass band: 0.005, Magnitude band rejection: 0, Bandwidth of the band: 1.

In obtaining the order of the filter the KaiserOrd function is used and to obtain the filter coefficients the function fir1 is used.

The recursive equation used for the implementation of the filter with the signal obtained from the differential equations system 4 is the following [18].

$$y[n] = a0 \cdot x[n] + a1 \cdot x[n-1] + a2 \cdot x[n-2] + \ldots + aN \cdot x[n-N] \quad (4)$$

Initially defining the cut and step frequency, the magnitude and oscillation of the passband to obtain the order of the filter through the Kaiser window, is shown below

```
fs = 1;
fc = [0.000001 0.004 0.005 0.03];
mag = [0 1 0];
dev = [0.34 0.33 0.34];
[n,Wn,beta,ftype] = kaiserord(fc,mag,dev,fs);
```

To find the 30 coefficients the function fir1 is applied as shown below.

```
b1 = fir1(n,Wn,ftype,kaiser(n + 1,beta),'noscale');
```

Having the filter coefficients, the convolution is applied with the data obtained with the execution of the Euler method in the systems of differential equations of the first order.

The bandpass filter is applied to the one hundred systems of first-order equations, eliminating those whose result of convolution tends to zero at most points.

Once the curves that do not meet the frequency range between 15 and 25 mHz are discarded, the diffuse classifier is used for the remaining curves.

The classification of the curves is done with two characteristics of the signal:

- **The periodicity of the signal (PS)**
  A comparison is made between two consecutive periods of the signal called the absolute relative error of the period, which is the ratio between the differential of the average of two periods and the average of the first period.

The periodicity of the signal is an input of the fuzzy classifier that presents the next fuzzy set (see Fig. 3).

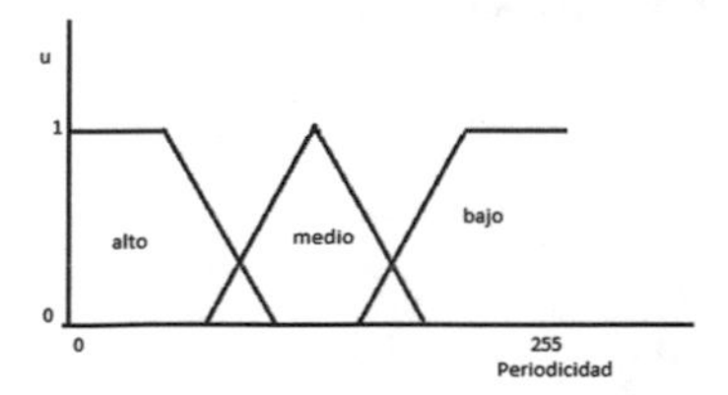

Fig. 3. Fuzzy set of periodicity

- **Signal Amplitude Uniformity**
  The comparison of two amplitudes is carried out as the absolute relative error of the amplitude as the ratio between the differential of two consecutive amplitudes and the first amplitude.

The uniformity of the amplitude of the signal is an input of the fuzzy classifier that presents the next fuzzy set (see Fig. 4).

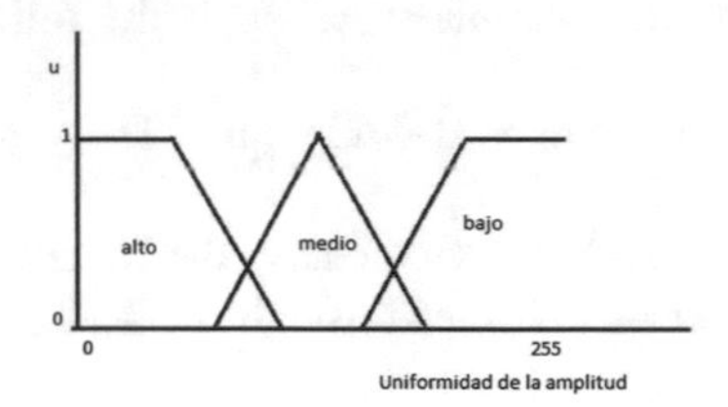

Fig. 4. Fuzzy set of uniformity of the amplitude

The set of rules that are formed from the inputs of the system is made up of 9 rules that are presented below

If UA is low and P is low then the signal is non-cyclical
If UA is low and P is medium then the signal is non-cyclical
If UA is low and P is high then the signal is non-cyclical
If UA is medium and P is low then the signal is non-cyclical
If UA is medium and P is medium then the signal is non-cyclical
If UA is medium and P is high then the signal is non-cyclical
If UA is high and P is low then the signal is non-cyclical
If UA is high and P is medium then the signal is non-cyclical
If UA is high and P is high then the signal is cyclical

The class of output for the system is composed of 2 according to its waveform, the signals are classified as cyclical and not cyclical [19]. In the inference machine, the diffuse intersection is used for each rule and the diffuse union in the results between rules obtaining a total value between zero and one [20], in which the values that tend to 1 mean that the signal is cyclical and the values that tend to zero means that the signal is non-cyclical. In order to define the most relevant parameters, the highest output value of the fuzzy classifier is sought between the systems of differential equations. With the system of differential equations, the previously described algorithm was incorporated, obtaining, as a result, the following curve (see Fig. 5).

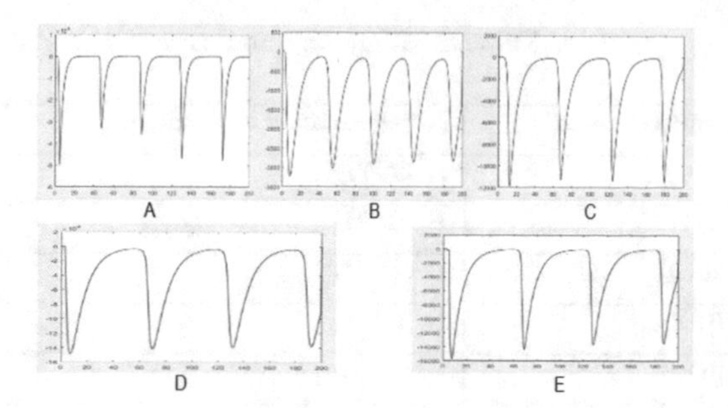

Fig. 5. Graphic representation (A) Two neurons, (B) Three neurons, (C) Four neurons, (D) Five neurons, (E) Six neurons.

5.2 Interaction Algorithm to Obtain the Parameters of the Differential Equations System

The flow diagram (Fig. 6) of the algorithm implemented to obtain the system of equations with the best periodicity performance is presented below (steps 6–8).

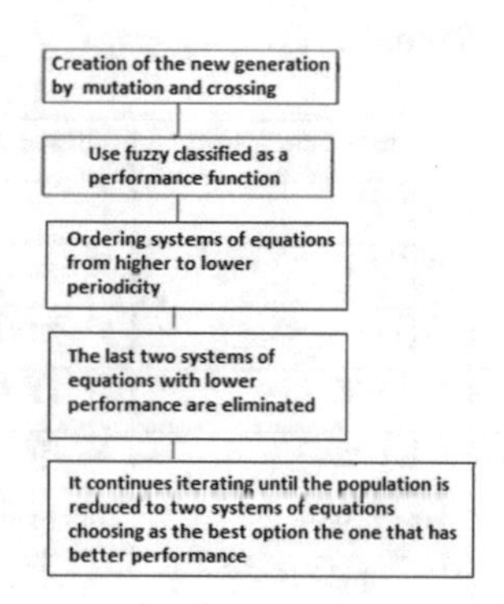

Fig. 6. Flow diagram steps 6–8

5.3 Creation of the New Population

The genetic operators are used with the aim of providing greater heterogeneity to the population.

Mutation: The parameters of the worst-rated differential equations systems are binarized and the 2 intermediate bits of each parameter that will provide a new set of parameters of the differential equation system will be randomly modified (Table 1).

Table 1. Mutation process

Current generation	New generation by mutation
A [0110]	A [0000]
B [1111]	B [1001]
C [1010]	C [1000]
D [0101]	D [0111]
K1 [0000]	K1 [0010]
K2 [1110]	K2 [1010]
b [0111]	b [0101]

Crossing: The parameters of the systems with the best qualification equations are binarized and the heavier bits are exchanged with the lower weight bits between parameters of equations to configure a new set of parameters of the system of differential equations (Table 2).

Table 2. Crossing process

Current generation	Current generation	New generation by crossing
A [1100]	A [0011]	A [1111]
B [0011]	B [1101]	B [0001]
C [0000]	C [1111]	C [0011]
D [1010]	D [0101]	D [1001]
K1 [0111]	K1 [0001]	K1 [0101]
K2 [1110]	K2 [1010]	K2 [1110]
b [0011]	b [0100]	b [0000]

Classification of the Population: Once the population of systems of equations is constituted, each system is evaluated with the performance function of periodicity by means of the diffuse classifier. After the classification of the population from highest to lowest, the two systems of equations of minor valuation are eliminated. The algorithm obtains the definitive parameters of the system of differential equations, as in Fig. 7.

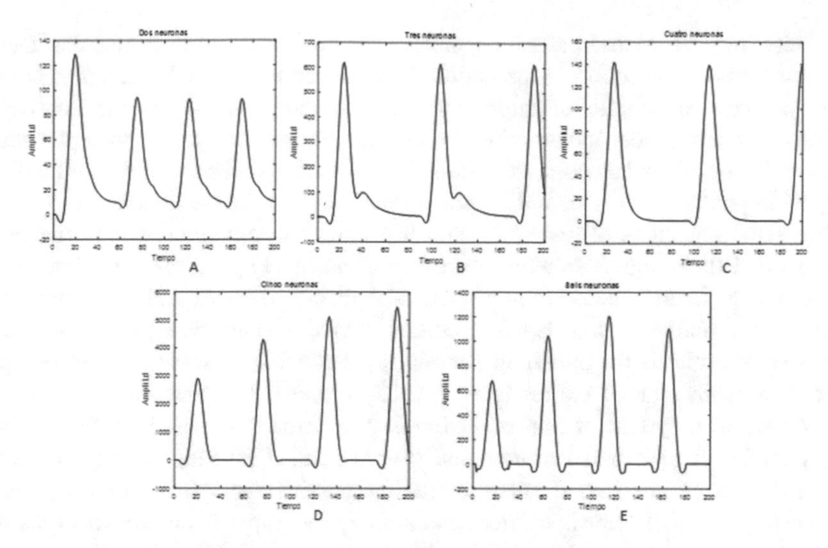

Fig. 7. Graphical representation (A) Two neurons, (B) Three neurons, (C) Four neurons, (D) Five neurons, (E) Six neurons.

5.4 Discussion

In the search for parameters for systems of differential equations of the first order. A series of works were analyzed. All the works have as main relation their application as central generators of patterns. These works are presented below.

In [7], the Gauss-Newton iterative method is presented to solve the neuronal coupling parameters that approximate the characteristics of a reference signal or initial estimation. The validation of the system is based on the theory of chaos and bifurcations. An oscillating system has an exponentially stable limit cycle which guarantees a cyclic signal. The paper applies solutions through simulations to central pattern generators. The solution is applied to two specific cases of systems of differential equations of the first order. The performance of the method depends on the initial seed or initial estimate proposed by the designer. In [8], three applications are socialized as a basis for the development of simulations of a two-joint robot for locomotion. The FEEDBACKDESIGN application is responsible for the design of servo motor control with feedback. The CPGDESIGN is responsible for the design of the CPG structure. The SIMSCAN function handles the feedback of CPG and the movement of a leg in function of the locomotion of the walk. However, the paper does not present a concrete methodology on the choice of CPG parameters. Although results are presented starting from certain values assigned to specific parameters and a particular parameter is varied that is related to the behavior of the CPG in a process such as locomotion. In [9] present an analysis under the theory of chaos and bifurcations for Hopf oscillators with four neurons per layer. The stability and periodicity are determined from the variation of the parameters in their differential equations of first order. Although its potential for implementation in locomotion in quadruped robots is demonstrated, no procedure is presented for the search of the values of the parameters as a function of frequency and periodicity. In [10], central pattern generators are proposed based on the Van Der Pol oscillator for hexapod robots. A procedure is proposed to obtain the parameters of the system that consists initially of finding the values of the parameters of the Van Der Pol oscillator for amplitude and specific frequency. Subsequently, the synaptic weight values of the oscillator network and initial data of each oscillator are defined to obtain the specific phase delay. It should be noted that Van Der Pol oscillators are composed of differential equations of first order and trigonometric functions, which makes calculations and their implementation more complex. In [11] the mathematical expressions of the "chimera" states are presented. The theory of chaos and bifurcation raises in these states oscillatory signals with a certain degree of stability. However, it does not address search criteria for obtaining parameters that achieve stable oscillatory signals under their mathematical expressions. In [12] propose the Van Der Pol oscillator incorporated in a system of first order differential equations based on the Hodgkin-Huxley model. A new model of neuronal network called FitzHugh-Nagumo has been generated. The author is dedicated to finding possible analytical solutions for the parameters whose behavior is cyclical and stable. The paper is an analytical study on the behavior of the proposed model. In [13] present a model that reproduces a mathematical expression from images of the generated oscillations of the hip, knee and ankle. These signals are required to develop the locomotion of a quadruped. The method used is an evolutionary algorithm called "MAGO". Although pertinent mathematical solutions are obtained to implement in a robotic system its dependence on digital image processing implies a higher computational cost.

Finally, unpublished procedure for the achievement of the parameters of a specific recurrent neural network as oscillator for 2, 3, 4, 5, 6 neurons has been presented. The equations have simple mathematical expressions based on elementary arithmetic operations. The purpose of the algorithm is to deliver the parameters that meet the frequency and periodicity. The proposed procedure is presented as a solution of a group of families between 2 and 6 differential equations of the first order. While the papers presented in the state of art [7–13] show their procedures to obtain the parameters of oscillators developed based on the theory of chaos and bifurcation. The vast majority of the works establish fixed values in certain parameters and vary one or two parameters. In other works such as [7] and [10] it is only presented as a specific solution for 1 or 2 cases. Their equations are constituted by mathematical expressions of greater complexity and therefore higher computational cost. For all the oscillators mentioned above, finding the parameters of the recurrent neural networks by the proposed method, in this work, is feasible to achieve in robotic applications.

6 Conclusions

The techniques and methods based on hybridize digital signal processing and computational intelligence to obtain the parameters of oscillators of several order is appropriate and standardized for all cases.

The success of the procedure outlined in this paper has been largely achieved by the experience on the management of systems of first order differential equations that has allowed to use the most suitable tools in the proposed solution.

The procedure described to obtain the oscillator parameters on systems of differential equations for 2, 3, 4, 5 and 6 neurons according to given operating criteria is a fundamental tool for the development of central pattern generators in robots locomotion.

References

1. Kassim, M.H., Zainal, N., Arshad, M.R.: Central pattern generator in bio-inspired robot: simulation using MATLAB. In: 2nd International Conference Underwater System Technology, 4–5 November 2008, Bali, Indonesia (2008)
2. Cohen, A.H., Rossignol, S., Grillner, S.: Neural Control of Rhythmic Movements in Vertebrate, pp. 1–500. Wiley, Hoboken (1998)
3. Buchli, J., Ijspeert, A.J.: Distributed central pattern generator model for robotics application based on phase sensitivity analysis. In: 1st International Workshop Bio-ADIT. Lecture Notes in Computer Science, vol. 3141, pp. 333–349 (2004)
4. Serón, M.M.: Sistemas no lineales, Notas de clase. Universidad Nacional de Rosario, Lab. de Sistemas Dinámicos y Procesamiento de Señales, pp. 1–161 (2000)
5. Rico, E.M., Hernandez, J.A.: Analysis and application of a displacement CPG-based method on articulated frames. In: Solano, A., Ordoñez, H. (eds.) CCC 2017. CCIS. Advances in Computing, vol. 735, pp. 495–510. Springer (2017). https://doi.org/10.1007/978-3-319-66562-7_33
6. Ramírez Moreno, D.F.: Modelo computacional de la modulación de la transformación sensorial motora, tesis para optar título de doctor presentada a la Universidad del Valle (2006)
7. Consolini, L., Lini, G.: A Gauss-Newton method for the synthesis of periodic outputs with central pattern generators. IEEE Trans. Neural Netw. Learn. Syst. **25**(7), 1394–1400 (2014)
8. Szczecinski, N.S., Hunt, A.J., Quinn, R.D.: Design process and tools for dynamic neuromechanical models and robot controllers. Biol. Cybern. **111**, 105–127 (2017)
9. Zhang, C., Sui, Z., Li, H.: Equivariant bifurcation in a coupled complex-valued neural network rings. Chaos, Solitons Fractals **98**, 22–30 (2017)
10. Li, W., Chen, W., Wu, X., Wang, J.: Parameter tuning of CPGs for hexapod gaits based on genetic algorithm. In: IEEE 10th Conference on Industrial Electronics and Applications (ICIEA), pp. 45–50 (2015)
11. Maxim, B., Lev, S., Grigory, O., Arkady, P.: Complex chimera states in a nonlinearly coupled oscillatory medium. In: 2nd School on Dynamics of Complex Networks and Their Application in Intellectual Robotics (DCNAIR), pp. 17–20 (2018)
12. Tabi, C.B.: Dynamical analysis of the FitzHugh–Nagumo oscillations through a modified Van Der Pol equation with fractional-order derivative term. Int. J. Non-Linear Mech. **105**, 173–178 (2018)
13. Balarezo Gallardo, F., Hernández-Riveros, J.-A.: Evolutionary parameter estimation of coupled non-linear oscillators. In: Solano, A., Ordoñez, H. (eds.) CCC 2017, CCIS. Advances in Computing, vol. 735, pp. 457–471. Springer (2017). https://doi.org/10.1007/978-3-319-66562-7_33
14. Cappelletto, J., Estévez, P., Grieco, J.C., Fernández-López, G., Armada, M.: A CPG-based model for gait synthesis in legged robot locomotion. In: 9th International conference on climbing and walking robots, pp. 56–64 (2006)

15. Cappelletto J., Estevez P., Grieco J.C., Medina–Melendez, W., Fernandez–Lopez, G.: Gait synthesis in legged robot locomotion using a CPG based model. In: Bioinspiraction and Robotics: Walking and Climbing Robots, Viena, pp. 227–246 (2007)
16. Bower, J.M., Beeman, D.: The Book of Genesis. Springer, Berlin (2003). (Chapter 8)
17. Rico, E.M., Hernandez, J.A.: Modulation of central pattern generators (CPG) for the locomotion planning of an articulated robot. In: Florez, H., Diaz, C., Chavarriaga, J. (eds.) ICAI 2018. CCIS. Advances in Computing, vol. 942, pp. 321–334. Springer (2018). https://doi.org/10.1007/978-3-030-01535-0_24
18. Capellini, V., Constantinides, A., Emiliani, P.L.: Digital Filters and Their Applications. Techniques of Physics. Academic Press, London (1978)
19. Robert, C., Michal, J., Jacek, L.: Introduction to fuzzy systems. In: Theory and Applications of Ordered Fuzzy Numbers (2017). https://doi.org/10.1007/978-3-319-59614-3_2
20. Nozaki, K., Ishibuchi, H., Tanaka, H.: Adaptive fuzzy rule-based classification systems. IEEE Trans. Fuzzy Syst. **4**(3), 238–250 (1996)

Improved ITU Model for Rainfall Attenuation Prediction of in Terrestrial Links

Angel D. Pinto-Mangones[1], Juan M. Torres-Tovio[1], Nelson A. Pérez-García[2](✉), Luiz A. R. da Silva Mello[3], Alejandro F. Ruiz-Garcés[1], and Joffre León-Acurio[4]

[1] Universidad del Sinú, Montería, Córdoba 230001, Colombia
{angelpinto,juantorrest,alejandroruiz}@unisinu.edu.co
[2] Universidad de Los Andes, Mérida 5101, Venezuela
perezn@ula.ve
[3] Pontifícia Universidade do Rio de Janeiro, Rio de Janeiro, RJ 22453-900, Brazil
smello@cetuc.puc-rio.br
[4] Universidad Técnica de Babahoyo, Babahoyo 120102, Ecuador
jvleon@utb.edu.ec

Abstract. Rain attenuation is one of the main detrimental effects on the performance of wireless telecommunications systems operating in frequencies above 10 GHz. Mitigation of its impacts requires, among other things, the use of rain attenuation models adequate to local climatic characteristics in which the communications systems will be implemented. In this paper, a modified version of the ITU-R Recommendation ITU-R.P.530-17 prediction method is proposed. The model maintains the concept of the distance correction factor used in the ITU-R model, but considers the full rainfall rate distribution. To derive the model, a nonlinear regression adjustment is performed based on results from measurements carried out in temperate and tropical climate. Subsequently, a fine-tuning of the model parameters is carried out using the computational intelligence technique, Particle Swarm Optimization (PSO). The accuracy of the proposed model, evaluated by the root mean square error (RMSE), is higher than that of several tested models for unavailability percentages of time in the range from 0.001% to 0.1%.

Keywords: Rain attenuation · Terrestrial links · Full rainfall rate distribution · Nonlinear regression · Particle swarm optimization

1 Introduction

For frequencies above 10 GHz the wavelength of propagating radio signals in a radio link is of the same order of magnitude of raindrop sizes. When rain cells crosses the propagation path, part of energy of the electromagnetic wave is lost due to absorption and scattering and does not reach the receiver. Besides frequency, the path attenuation depends on the rainfall rate, path length, radio wave polarization, etc.

A good estimation of rain attenuation in a particular link is one the keys to increase the accuracy of the planning and design process, so that the of link availability can be

M. Botto-Tobar et al. (Eds.): ICAETT 2019, AISC 1066, pp. 531–541, 2020.
https://doi.org/10.1007/978-3-030-32022-5_49

guaranteed even in the presence of rain and, on the other hand, avoiding overestimation of the required transmission power that may cause interference in other systems, at the same time taking care that, for example, there will be not oversizing or undersizing in the amount of actually required equipments [1].

Several models are proposed in the technical literature for the prediction of rain attenuation in terrestrial links as, for example, ITU model (Recommendation ITU-R P.530-17) [2, 3], Silva-Pontes-Souza-Perez (SPSP) method [4, 5], Crane model [6, 7], Differential Equations Approach (DEA) method [8] and Ghiani-Luini-Fanti (GLF) model [9]. In this work, a new model for the estimation of rain attenuation in terrestrial links is developed based on the general form of widely recognized and accepted Recommendation ITU-R P.530-17. The ITU-R method considers only the value of rainfall rate exceeded at 0.1% of time ($R_{0.01}$). As a result, if different climatic zones with different cumulative rainfall rate distributions have very similar values of $R_{0.01}$ the method will produce very similar cumulative attenuation distributions. In this sense, the main advantage of the proposed method in this paper is using complete rainfall rate distribution. This is achieved by using the concept equivalent rainfall rate introduced in the SPSP model [6], which avoids the extrapolation process used in ITU-R model for percentages of time other than 0.01%.

It is important to note that, although SPSP model considers the distance factor concept used in the ITU-R model to compensate for the non-homogeneity of rainfall rate along link, the form of this factor does not correspond to the current content of Recommendation ITU-R P.530-17. The model proposed in this paper is obtained by adjusting the parameters of ITU-R method distance factor through the combination of the quasi-Newton multiple nonlinear regression method [10] and the bioinspired technique particle swarm optimization (PSO) [11, 12], using rain attenuation measurements performed in temperate climates [13] and tropical regions [14, 15].

In this paper, Sect. 2 refers to the main aspects of the database used to obtain the improved model. In Sect. 3, a brief review of rainfall attenuation models considered in this work is presented. Section 4 explains the optimization methods employed in this paper for the development of the improved model. The details about development of the model are explained in Sect. 5, as well as the results obtained and the comparison of its performance with that of the other models reviewed in Sect. 3. Finally, in Sect. 6 presents the conclusions.

2 Measurements Database

The measurements used to improve on the prediction model given in Recommendation ITU-R P.530-17 include the long-term probability distributions of rainfall rate and rain attenuation currently available in the ITU-R database [13] plus additional data from measurements recently performed in a tropical climate region in Malaysia [14, 15]. The ITU-R databank includes measurements from 87 links, mainly located in regions with temperate climate (e.g., United Kingdom, Sweden, France, Italy, USA, etc.) with operating frequencies between 7 GHz and 137 GHz and path lengths range from 0.5 km to 58 km. Cumulative probability distributions of point rainfall rate and rain attenuation are given for time percentages between 0.001% and 0.1%. The additional data from Malaysia were obtained in six line-of-sights links, four of them operating at 14.8 GHz, one operating at 15.3 GHz and one operating at 26 GHz. The path length of the six links varies between 1.3 km and 11.3 km.

3 Rain Attenuation Prediction Models for Terrestrial Links

In this section, the main details of the models implemented in this paper for the evaluation of the improved model are succinctly explained.

3.1 Recommendation ITU-R P.530-17

Recommendation ITU-R P.530-17 is based on the prediction of rainfall attenuation $A_{0.01}$, exceeded during 0.01% of time, using as input data the point rainfall rate $R_{0.01}$ exceeded at the same percentages of time. For percentages of time $p(\%)$ different from 0.01% and between 0.001% and 1%, an extrapolation procedure that does not depend on rainfall distribution is applied [2, 3]. This is the most important limitation in the model, as the method will provide the same results for links with the same path length and frequency located in different climatic regions but have in common the input parameter $R_{0.01}$.

The ITU model considers a distance factor r, that also depends on $R_{0.01}$, operation frequency f, the path length d and the parameter α, that is one of the two power-law parameters of the specific attenuation and is function of frequency and wave polarization [16]. The attenuation $A_{0.01}$ is given by [2, 3]:

$$A_{0.01} = kR_{0.01}^{\alpha} d\, r \tag{1}$$

where k, that is the other power-law parameters of the specific attenuation, also depends of frequency and wave polarization [16]. The distance factor r is defined as [2];

$$r = \frac{1}{0.477 d^{0.633} R_{0.01}^{0.073\alpha} f^{0.123} - 10.579\left(1 - e^{-0.024d}\right)} \tag{2}$$

The rain attenuation A_p, exceeded for a percentage of time p other than 0.01% is determined by extrapolating $A_{0.01}$ using a function of p and f.

3.2 Silva-Pontes-Souza-Perez (SPSP) Model

In the SPSP model, the rain rate $R_{0.01}$ in Eq. (1) is replaced by an equivalent rainfall rate R_{eff}, which is a function of the full rain rate distribution, R_p (rain rate exceeded for the relevant percentage of time p) and the path length d [4, 5]:

$$R_{eff} = 1.763 R_p^{\left(0.753 + \frac{0.197}{d}\right)} \tag{3}$$

The SPSP model also uses a distance factor concept, as does the ITU-R model. Nevertheless, the factor in the SPSP model differs from expression (2) and includes a frequency dependence.

3.3 Crane Global Model

Crane method is a model used mainly in the USA for the prediction of rain attenuation in terrestrial links [6, 7]. In addition to also being based on full rainfall rate distribution, the Crane model results from the use of two exponential functions for the attenuation dependence on the path length, one for the interval between 0 km and a distance *d* (km) and the other from *d* to 22.5 km. Thus, rainfall attenuation for a percentage of time *p* is given by:

$$A_p = \begin{cases} kR_p^{\alpha}\left(\frac{e^{u\alpha D}-1}{u\alpha}\right);\ 0 \le D \le d \\ kR_p^{\alpha}\left[\frac{e^{u\alpha d}-1}{u\alpha} + \frac{b^{\alpha}}{c\alpha}\left(e^{c\alpha D} - e^{c\alpha d}\right)\right];\ d \le D \le 22.5 \end{cases} \tag{4}$$

where parameters *u*, *b*, *c* and *d* depend of the rain rate.

For paths greater than 22.5 km A_p is calculated considering *D* equal to 22.5 km and the resulting rainfall interruption time is multiplied by a factor of D/22.5 [7].

3.4 Differential Equations Approach (DEA) Model

Abdulrahman *et al.* [8] developed a model for rain attenuation prediction by the differentiation, in relation to precipitation rate, of rainfall attenuation measured in terrestrial links of Malaysia. The attenuation is given as a function of slope *S* resulting from aforementioned differentiation, i.e., $S(R_p) = dA_p/dR_p$. The rain attenuation A_p, exceeded for percentage of time *p*, is calculated as:

$$A_p = \mu S(R_p) \tag{5}$$

where $S_p = \beta R_p^{\alpha-1}$, indicating that the slope S and rain rate are related by a power law.

In the DEA model, the parameters μ and β are dependent on the rainfall rate, the path length *d*, the parameters *k* and α, and two adjustment parameters *a* and *b*.

3.5 Ghiani-Luini-Fanti (GLF) Model

In GLF model, a distance factor is obtained from more than 500 rainfall maps generated with numerical model MultiEXCELL [17]. The approach consists of obtaining, based on the aforementioned maps, an analytical expression for the distance factor that depends on rain rate and path length, as follows [9]:

$$r = \left(ae^{-bR_p} + c\right) \tag{6}$$

where the parameters *a*, *b* and *c* depend on path length *d*, with exponential tendency.

4 Optimization Methods

In this paper, we resort to two optimization methods to tune the ITU-R model to the rainfall and rain attenuation measurements considered in Sect. 2, and obtain an improved model. A first adjustment is achieved using the quasi-Newton method, that belongs to the family of non-linear regression methods. Next, a fine adjustment is implemented this time using one of the bioinspired techniques, specifically the particle swarm optimization (PSO).

4.1 Quasi-Newton Method

Considered among the methods based on gradient information most appropriate for the solution for nonlinear problems, the quasi-Newton technique is based on Newton's method and calculates an approximation of Hessian matrix inverse or variants of it, by recurrence formulas. The aforementioned approximation is constructed using a quadratic function, $f(x)$, of the form [10, 18]:

$$f(x) = f(x_c) + [\nabla f(x_c)]^T p + \frac{1}{2} p^T H(x_c) p \tag{7}$$

where T refers to "transpose", $\nabla f(x_c)$ and $H(x_c) = \nabla^2 f(x_c)$ represent the gradient vector and Hessian matrix de $f(x)$, respectively, both evaluated in x_c, for all the independent variables present in f. In turn, p refers to the search direction that follows the iteration process to estimate the minimum of function f.

It is important to point out that, the elements gradient vector and Hessian matrix are the first-derivates and second-derivates of f, respectively.

Then, using information obtained during the descent process, the direction vector, p_k, in iteration k, is computed as follows [10, 18]:

$$p_k = -H_k^{-1} \nabla f(x_k) \tag{8}$$

where H_k^{-1} is a positive matrix that approximates p_k to Newton's direction and is calculated in an approximate way, being updated in each iteration, without losing the advantage of rapid convergence depending on the chosen x_o initial value. Recurrence formulas such as Davidon-Fletcher-Powell (DFP) or Broyden-Fletcher-Goldfarb-Shanno (BFGS) algorithm, among others, are mainly used for this purpose [10].

The point new in next iteration $k + 1$ is given by [19]:

$$x_{k+1} = x_k + \alpha_k p_k \tag{9}$$

Equation (9) satisfies the condition $f(x_{k+1}) \leq f(x_k)$. The step-size, α_k, can be found using methods such as Armijo rule [20], step halving [21] or Weng-Mamat-Mohd-Dasril method [22], among others.

4.2 Particle Swarm Optimization (PSO)

Initially proposed by Kennedy and Eberhart, PSO incorporates aspects related to the social behavior of groups of individuals (fish schooling, bird flocking, bee swarms, etc.). The individuals, also known as particles or agents, represent the solution candidates and their movements in a *n*-dimensional space represent the solution set of optimization problem [11].

Each individual has a position x_i, and a velocity v_i. In addition, each individual knows the best position visited by it (p_{best}) in the past and the best position of the leader (g_{best}) in the cumulus until the previous iteration. The p_{best} value works as an autobiographical memory (each individual remembers their best explored places). On the other hand, g_{best} represents the tendency of the *n* group particles to visit places where it obtained good results in the past. The selection of the leader is made in each iteration. The leader is the particle that presents, in each iteration, the best value of the selected cost function. The velocity of each individual v_i^{k+1} in the iteration $k + 1$ is given by [11, 12]:

$$v_i^{k+1} = \omega v_i^k + C_1 R_1 \left(p_{best} - x_i^k\right) + C_2 R_2 \left(g_{best} - x_i^k\right) \tag{10}$$

where ω denote the inertia or moment (represents the degree of influence of the current velocity on the future velocity), C_1 is the cognitive parameter or "nostalgia" of the agent (representing the influence of the best position of the individual in the new velocity), C_2 is the social parameter (represent the influence of the best position of the leader in the future velocity) and R_1 and R_2 are random numbers belonging to a uniform distribution $U(0,1)$. Additionally, x_i^k and v_i^k are the current velocity and actual position of the agent, respectively.

Finally, the position of the individual is updated from [11, 12]

$$x_i^{k+1} = x_i^k + v_i^{k+1} \tag{11}$$

5 Improved Model and Comparative Tests

To obtain the improved model as a modification of the one given in Recommendation ITU-R P.530-17, the parameters of a generalized expression derived of this model together with the equivalent rainfall rate $R_{eff,}$ proposed by SPSP model, are adjusted. Thus, from (1), (2) and (3), the general expression for predicting the attenuation exceeded at $p\%$ of the time is given by:

$$A_p = kR_{eff}^{\alpha} dr = k\left[a_1 R_p^{\left(a_2 + \frac{a_3}{d}\right)}\right]^{\alpha} d \frac{1}{a_4 d^{a_5} R_p^{a_6} f^{a_7} + a_8\left(1 - e^{a_9 d}\right)} \tag{12}$$

The values of the parameters a_1 to a_9 were obtained, in a first phase using the quasi-Newton multiple nonlinear regression algorithm with the global root mean square error

(*RMSE*) between the measured attenuation values for all measurements and the values estimated by the model as objective function. The *RMSE* value is given by calculated [23]:

$$RMSE = \sqrt{\frac{\sum_{i=1}^{N} (V_{mi} - V_{ci})^2}{N - 1}} \quad (13)$$

where V_{mi} is the measured value (in this case, attenuation measured at each site for different time percentages), V_{ci} represents the calculated value by considered model (in this case, attenuation calculated at each site for corresponding time percentages) and N is total number of rain attenuation measurements considered.

Figure 1 shows the flowchart implemented for the use of quasi-Newton method in this letter.

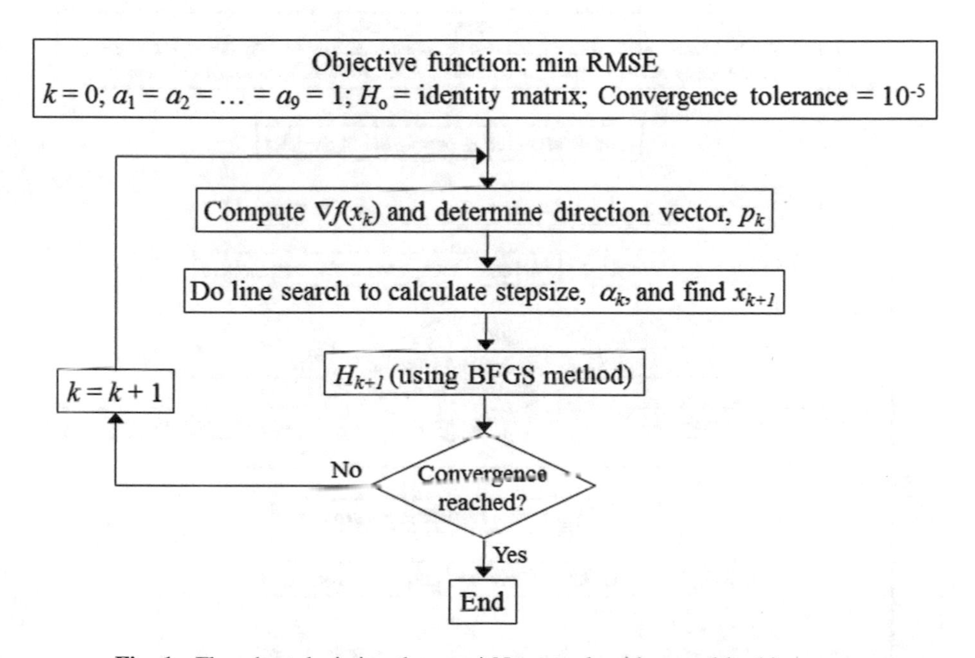

Fig. 1. Flowchart depicting the quasi-Newton algorithm used in this paper.

The best result was achieved with 200 iterations and a convergence tolerance equal to 10^{-5}. The coefficient of determination R^2 was equal to 81.89% and a global *RMSE* value was 5.713 dB. Table 1 summarizes global *RMSE* values obtained by the new model as well as those obtained with Recommendation ITU-R P.530-17, SPSP model, Crane model, DEA model and GLF model.

Table 1 shows that new model is more accurate in the estimation of rain attenuation than others models considered in this comparison. The model that has the poorest performance in this evaluation is the DEA model, probably due to the fact that this model was developed essentially for a tropical climate [8].

Table 1. Comparison of global *RMSE* values for the rain attenuation prediction models.

RMSE (dB)					
ITU-R P.530-17	SPSP	Crane	DEA	GLF	Improved ITU model
6.86	6.55	9.83	17.85	10.45	5.71

Then, using the computational intelligence technique PSO a fine adjustment was made to three parameters of the set a_1 to a_9 resulting from the adjustment carried out using the quasi-Newton method. For the implementation of the PSO, the minimization of the global *RMSE* was again considered as the objective. In Fig. 2, the flowchart implemented for the use of PSO technique in this work is shown, with the fixed parameters values, as well as the variables initial values that gave the best result.

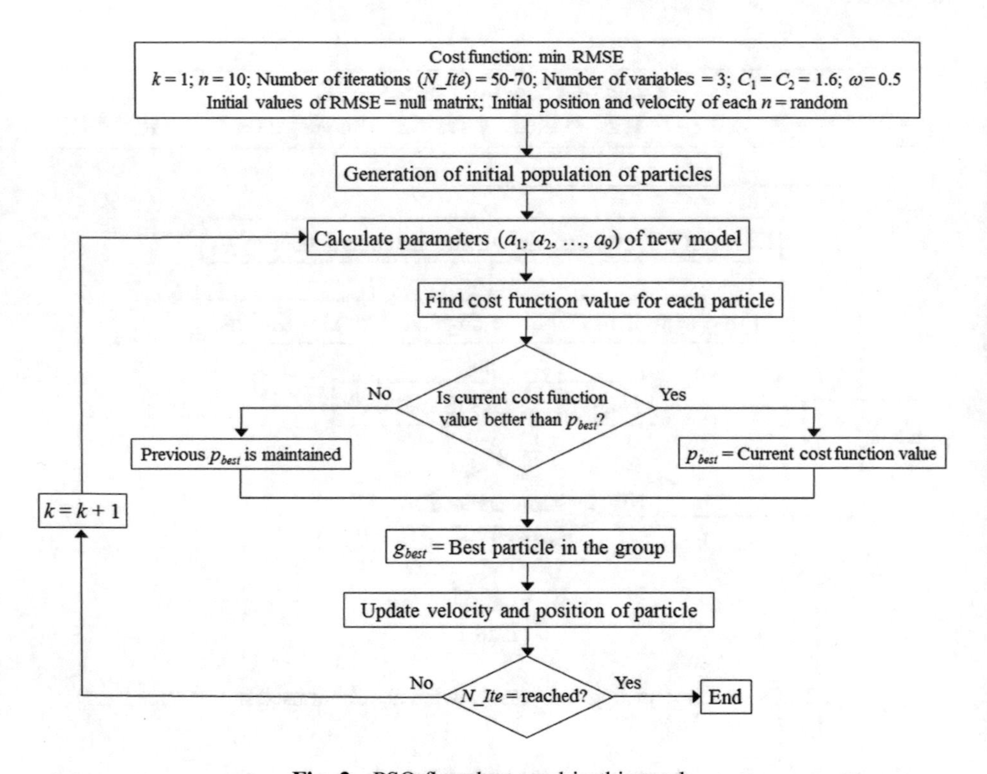

Fig. 2. PSO flowchart used in this work.

A coefficient R^2 equal to 81.96% was obtained, similar to that achieved with quasi-Newton algorithm, but with a global *RMSE* equal to 5.343 dB, better than the previous one. Table 2 shows the final values obtained for the parameters a_1 to a_9 of modified model.

The scatterplot of the measured values of rain attenuation from database and those corresponding values estimated by the new model can be observed in Fig. 3.

Table 2. Values of the adjustment parameters.

a_1	a_2	a_3	a_4	a_5	a_6	a_7	a_8	a_9
1.499	1.0397	–0.0903	0.1602	0.6784	0.2559	0.2656	81.3873	0.0012

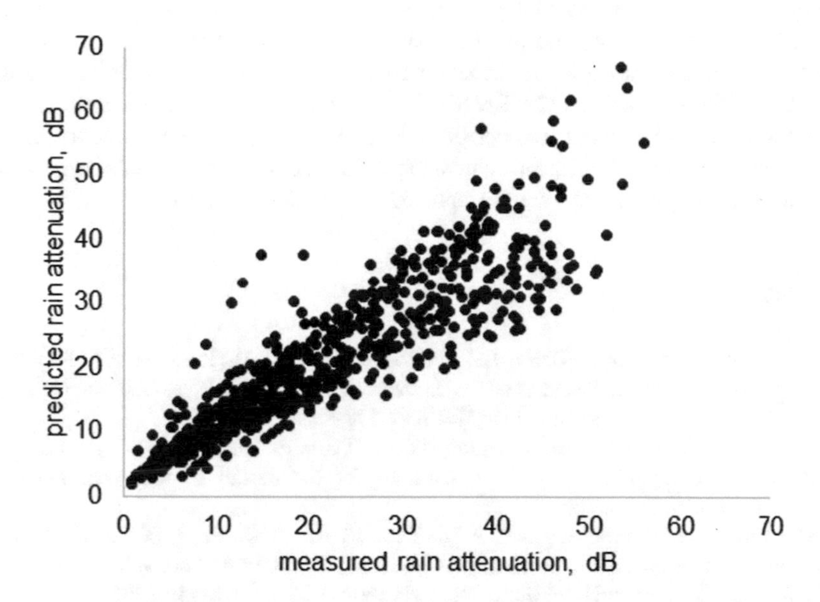

Fig. 3. Scatterplot of measured and predicted attenuation values (by new model).

Finally, Table 3 summarizes *RMSE* values for all considered models discriminated by percentages of time $p(\%)$ between 0.001 and 0.1%, the same interval of $p(\%)$ of rain attenuation measurements used for optimization. A best performance of the new model can be observed for all exceedance probabilities considered.

Table 3. Comparison of RMSE values for several percentages of time.

Propagation model	$p(\%)$						
	0.001	0.003	0.006	0.01	0.03	0.06	0.1
	RMSE (dB)						
ITU-R P.530-17	12.02	8.11	5.75	6.12	5.52	4.88	4.23
SPSP	9.07	7.21	6.79	6.94	6.09	4.66	3.84
Crane	14.90	11.00	9.84	10.36	8.72	6.64	5.33
DEA	32.59	22.33	17.80	18.02	12.37	7.05	4.76
GLF	18.77	12.31	9.40	10.56	8.44	6.13	4.69
Improved ITU model	7.63	6.03	5.34	5.48	4.70	3.81	3.43

6 Conclusions

The improved rain attenuation prediction model for terrestrial links proposed in this work, based on method Recommendation ITU-R P.530-17 and Silva-Pontes-Souza-Perez (SPSP) model, was derived using current ITU-R database with the addition of measurements carried out in Malaysia. It not only overcomes the main limitation of original ITU-R model, i.e., the use of a single value of the rainfall rate cumulative distribution, but also improves on the accuracy of the attenuation estimation in terms of the global RMSE as well as the RMSE for all percentages of time considered. The model parameters adjustment was performed by means of a first adjustment made with quasi-Newton multiple nonlinear regression followed by a fine adjustment through bioinspired technique particle swarm optimization (PSO) technique.

References

1. Perez-Garcia, N.A., Torres-Tovio, J.M., Pinto-Mangones, A.D., Ramirez, E.J.: Planificación y Dimensionamiento de Sistemas de Comunicación Vía Satélite. Sello Editorial Corporación Universidad del Sinú, Montería (2018). ISBN 978-958-8553-52-8
2. International Telecommunication Union (ITU): ITU-R Recommendation P.530-17. Propagation data and prediction methods required for the design of terrestrial line-of-sight systems, Geneva, Switzerland (2017)
3. Perez-Garcia, N., Pinto-Mangones, A.D., Torres-Tovio, J.M., Perez-Di Santis, T.: Planificación y Dimensionamiento de Sistemas Celulares y de Radio Acceso, vol. 1. Sello Editorial Corporación Universitaria del Sinú, Montería (2017). ISBN 978-958-8553-41-2
4. Silva Mello, L.A.R., Pontes, M.S., Souza, R.M., Perez-Garcia, N.A.: Prediction of rain attenuation in terrestrial links using full rainfall rate distribution. Electron. Lett. **43**(25), 1442–1443 (2007). https://doi.org/10.1049/el:20072410. ISSN 0013-5194
5. Silva Mello, L.A.R., Pontes, M.S.: Unified method for the prediction of rain attenuation in satellite and terrestrial links. J. Microwaves Optoelectron. Electromagnet. Appl. **11**(1), 1–14 (2012). https://doi.org/10.1590/S2179-10742012000010001. ISSN 2179-1074
6. Crane, R.K.: Prediction of attenuation by rain. IEEE Trans. Commun. **28**(9), 1717–1733 (1980). https://doi.org/10.1109/TCOM.1980.1094844. ISSN 0090-6778
7. Koryu-Ishii, T.: Handbook of Microwave Technology, vol. 2. Academic Press, San Diego (1995). Applications
8. Abdulrahman, A.Y., Rahman, T.A., Rahim, K.A., Islam, M.R., Abdulrahman, M.K.A.: Rain attenuation predictions on terrestrial radio links: differential equations approach. Trans. Emerg. Telecommun. Technol. **23**(3), 293–301 (2012). https://doi.org/10.1002/ett.1531. ISSN 2161-3915
9. Ghiani, R., Luini, L., Fanti, A.: A physically based rain attenuation model for terrestrial links. Radio Sci. **52**(8), 972–980 (2017). https://doi.org/10.1002/2017RS006320. ISSN 1944-799X
10. Bonnans, J.F., Gilbert, J.C., Lemaréchal, C., Sagastizábal, C.A.: Numerical Optimization Theoretical and Practical Aspects, 2nd edn. Springer, Heidelberg (2006)
11. Cheng, S., Lu, H., Lei, X., Shi, Y.: A quarter century of particle swarm optimization. Complex & Intelligent Systems **4**(3), 227–239 (2018). https://doi.org/10.1007/s40747-018-0071-2. ISSN 2199-4536

12. Sun, J., Lai, C.H., Wu, X.J.: Particle Swarm Optimisation: Classical and Quantum Perspectives. CRC Press, Boca Raton (2012)
13. International Telecommunication Union (ITU): Databank DBGS3. http://www.itu.int/pub/R-SOFT-SG3/en. Accessed 24 January 2019
14. Abdulrahman, A.Y., Falade, A., Olufeagba, B.J., Mohammed, O.O., Rahman, T.A.: Statistical evaluation of measured rain attenuation in tropical climate and comparison with prediction models. J. Microwaves Optoelectron. Electromagnet. Appl. **15**(2), 123–134 (2016). https://doi.org/10.1590/2179-10742016v15i2624. ISSN 2179-1074
15. Ulaganthen, K., Rahman, T.A., Islam, M.R.: Complementary cumulative distribution function for rain rate and rain attenuation for tropical region Malaysia. Int. J. Manag. Appl. Sci. **3**(1), 54–57 (2017). Special Issue-1, ISSN 2394-7926
16. International Telecommunication Union (ITU): ITU-R Recommendation P.838-3. Specific attenuation model for rain for use in prediction methods, Geneva, Switzerland (2005)
17. Luini, L., Capsoni, C.: MultiEXCELL: a new rainfall model for the analysis of the millimetre wave propagation through the atmosphere. In: 3rd European Conference on Antennas and Propagation, March 2009, Berlin, Germany, pp. 1946–1950 (2009). ISBN 978-1-4244-4753-4
18. Dennis Jr., J.E., Schnabel, R.B.: Numerical Methods for Unconstrained Optimization and Nonlinear Equations. Society for Industrial and Applied Mathematics (SIAM), Philadelphia (1996)
19. Lange, K.: Optimization, 2nd edn. Springer, New York (2013)
20. Armijo, L.: Minimization of functions having Lipschitz-continuous first partial derivatives. Pac. J. Math. **16**(1), 1–3 (1966). https://doi.org/10.2140/pjm.1966.16-1. ISSN 0030-8730
21. Lange, K.: Numerical Analysis for Statisticians, 2nd edn. Springer, New York (2010)
22. Wen, G.K., Mamat, M., Mohd, I.B., Dasril, Y.: A novel of step size selection procedures for steepest descent method. Applied Mathematical Sciences **6**(51), 2507–2518 (2012). ISSN 1314-7552
23. Rujano-Molina, L.M., Perez-Garcia, N.A., Nariño, T.: Distribuciones acumulativas de la tasa de lluvias con tiempo de integración de 1-minuto en Venezuela. Ingenieria Al Dia **3**(1), 24–44 (2017). ISSN 2389-7309

Novel Lee Model for Prediction of Propagation Path Loss in Digital Terrestrial Television Systems in Montevideo City, Uruguay

Juan M. Torres-Tovio[1], Nelson A. Pérez-García[2(✉)], Angel D. Pinto-Mangones[1], Mario R. Macea-Anaya[3], Samir O. Castaño-Rivera[3], and Enrique I. Delgado Cuadro[4]

[1] Universidad del Sinú, Montería, Córdoba 230001, Colombia
{juantorrest, angelpinto}@unisinu.edu.co
[2] Universidad de Los Andes, Mérida 5101, Venezuela
perezn@ula.ve
[3] Universidad de Córdoba, Montería, Córdoba 230001, Colombia
{mariomacea, sacastano}@correo.unicordoba.edu.co
[4] Universidad Técnica de Babahoyo, Babahoyo 120102, Ecuador
edelgado@utb.edu.ec

Abstract. During the planning and dimensioning (P&D) of a Digital Terrestrial Television (DTT) system plays a very important role to use a suitable model that allows estimating the propagation path loss with the most possible precision according to the typical propagation characteristics of the site in which the system will be implemented. The imprecision in that estimation will lead to oversize or undersize of the system. In this sense, in this paper, one of the propagation most used models for the PyD process in ultra high frequency (UHF) band, in which the DTT systems operate, as the Lee model, is optimized using measurements carried out in Montevideo city, Uruguay, and also using Ant Colony Optimization (ACO) computational intelligence technique. The performance presented by new Lee model, compared with the performance shown by propagation models such as Okumura-Hata, Hata-Davidson, TDT-Uruguay and original Lee model, was the best, with a root mean square error (RMSE) of 9.43 dB.

Keywords: Digital Terrestrial Television · Planning and dimensioning · Lee model propagation · Ant Colony Optimization

1 Introduction

The success in planning and dimensioning (P&D) and subsequent implementation of a Digital Terrestrial Television (DTT) system and, in general, of a wireless telecommunications system, depends on, among other things, on the precision with which the coverage distance for a given transmitter of the system is estimated. This is closely related to the accuracy in which the propagation path loss of the RF signal between transmitter and receiver are estimated. The imprecision in the aforementioned prediction will result in an increase in the percentage of undersize or oversize of the system

M. Botto-Tobar et al. (Eds.): ICAETT 2019, AISC 1066, pp. 542–553, 2020.
https://doi.org/10.1007/978-3-030-32022-5_50

under study [1, 2]. Hence, it is essential to use propagation models that have been developed based on own propagation characteristics of the environment (climate, topography, morphology, orography, etc.) in which the DTT system will be implemented [3, 4].

There are several propagation models that can be used for propagation path loss prediction in DTT systems, e.g., Longley and Rice, also known as Irregular Terrain Model (ITM) [5, 6], Fernandez [7], Okumura-Hata [8, 9], Lee [10–12], Hata-Davidson [13], Recommendation ITU-R P.1546-5 [14], Recommendation ITU-R P.1812-4 [15], TDA-Venezuela-I [16], TDA-Venezuela-II [16], TDA-Venezuela-III [17], TDT-Uruguay [18, 19], TDT-Ecuador-I [20], TDT-Ecuador [20] and TDT-Ecuador-III [9]. The only model, of those mentioned, that was obtained directly from measurements in a TDT system carried out in Uruguay, specifically, in Montevideo city, is TDT-Uruguay model. In this model, the least squares method was used to find the values of the adjustment parameters of classical equation for propagation path loss varying only with the logarithmic dependence on transmitter-receiver distance. However, the aforementioned model has the disadvantage of depending only on this distance, without taking into account other parameters that impact loss, such as, operating frequency, transmitter and receiver antenna heights and the environment surrounding the reception site under consideration.

Thus, in this paper a new model is developed based on DTT measurements performed in that city, by adjusting the Lee model, which is one of those used for the UHF band, with relatively simple implementation. Lee model optimization for the indicated DTT measurements is performed using the intelligent computing technique known as Ant Colony Optimization (ACO) [21–24].

2 Propagation Path Loss for DTV Terrestrial Systems

In this section, a review of Digital Terrestrial Television (DTT) propagation models system that will be used in this paper to compare the performance of new Lee models, in the estimation of propagation path loss in Montevideo city, is accomplished.

2.1 Lee Model

It was developed by William C. Y. Lee for the first time in 1977, for point-to-point wireless links, flat terrain and systems operating at 900 MHz, based on measurements carried out in the United States [10]. Later, Lee developed a new version of his model, this time for area-to-area systems, originally also valid for flat terrain and 900 MHz operating frequency, transmitter power of 10 W, transmitter antenna height of 30.48 m, receiver antenna height of 3 m and transmitter and receiver antenna gains of 6 dBd and 0 dBd, respectively [10–12]. For other conditions, i.e., other values for frequency, f, or antenna heights or antenna gains, Lee incorporated correction factors. The propagation path loss, L, is given by [11, 12]:

$$L(dB) = L_o + 10\gamma \log\left(\frac{d}{d_o}\right) - 10 \log F_A \tag{1}$$

where L_o is the propagation path loss measured at 1.6 km from the transmitter (see Table 7.2 in [12]), γ is the power loss coefficient with distance (see Table 7.2 in [12]), d represents the distance between transmitter-receiver [km] and d_o is a reference distance equal to 1.6 km.

F_A factor is determined from [11, 12] and it consists of corrections due to transmitting antenna height (h_T), receiving antenna height (h_R), operating frequency (f), transmitting antenna gain (G_T) and receiving antenna gain (G_R).

The Lee model has been the subject of some adjustments to improve its performance in propagation path loss estimation for certain telecommunication systems, frequency bands and localities [25–27]. However, there are no known adjustment results for that model specifically for Digital Terrestrial Television (DTT) systems.

2.2 Okumura-Hata Model

Based on measurements made by Yoshihisa Okumura in Tokyo city, Japan, the model was developed by Masaharu Hata in 1980 [8] and is valid for frequency (f) between150 MHz and 1500 MHz, effective transmitting antenna height (h_1) from 30 m to 200 m, effective receiving antenna height (h_2) between 1 m and 10 m, and transmitter-receiver distance (d) from 1 km to 20 km.

The application environments or types of service area considered by the Okumura-Hata model are urban areas (large cities and medium or small cities), suburban areas, and rural or open areas.

The mean value of the basic propagation loss, L, is given by [9]:

$$\begin{aligned} L(dB) = {} & 69.55 + 26.16 \log f - 13.82 \log h_1 \\ & - a(h_2) + (44.9 - 6.55 \log h_1) \log d + C \end{aligned} \tag{2}$$

where $a(h_2)$ is a correction factor that depends on the receiver antenna height or mobile station height and the type of service area; and C depends on the frequency and also on the environment in which the receiver is located [8, 9].

2.3 Hata-Davidson Model

Published by the Telecommunications Industry Association (TIA) in 1997, the Hata-Davidson model is derived from the Okumura-Hata model to extend its use for transmitter-receiver distances, d (up to 300 km) and effective transmitter height, h_1 (up to 2500 m) [13].

The model consists of six (6) factors:

$$L(dB) = L_{OH} + A - S_1 - S_2 - S_3 - S_4 \tag{3}$$

where L_{OH} corresponds to the propagation path loss given by Eq. (2), while the remaining factors are given by [13] and are adjustment factors as function of the distance d (A, S_1, S_2 and S_4), height h_1 (A and S_2) and frequency (S_3 and S_4).

Hata-Davidson model is valid for frequencies between 150 MHz and 1500 MHz, and effective receiving antenna heights from 1 m to 10 m.

2.4 TDT-Uruguay Model

As mentioned above, this model also consists of the classical equation of propagation path loss varying only with the logarithm of the transmitter-receiver distance, using of DTT measurements performed in Montevideo city, Uruguay, for equation adjustment [18, 19].

The propagation path loss given by the mentioned model is:

$$L(dB) = 100.6 + 36.22 \log d \quad (4)$$

where distance, d, is given in kilometers.

TDT-Uruguay model presented a standard deviation (SD) of the mean error of 6.99 dB, improving the SD obtained with original Okumura-Hata model, which was 9.3 dB.

It should be noted that this review does not include the models recently developed in Venezuela and Ecuador for assessment of the model to be developed, because those models were adjusted for DTT measurements specific to both countries and not to Uruguay.

3 Ant Colony Optimization

The computational intelligence (CI) technique known as Ant Colony Optimization (ACO) is part of the family of bioinspired techniques. It was proposed for the first time by Marco Dorigo and collaborators at the beginning of 1990 with the purpose, initially, of solving combinatorial optimization problems [21]. ACO is based on stigmergy, science that explains the simple collective work of individuals, through the modification of the environment in order to communicate messages indirectly. This type of communication is exhibited, for example, by insects such as ants, termites, bees and wasps [23]. Specifically, ACO emulates the ant foraging process [21, 28]. Initially, the ants randomly explore the area around their nest. When an ant obtains a source of food, it moves a part to the nest, having previously evaluated the amount of food found and its quality. During his journey to the nest, the ant segregates in the ground a chemical substance known as pheromone, whose quantity depends on the amount and quality of the food. Pheromones attract other ants from the colony, increasing the probability that they obtain their food.

ACO is a simple and computationally efficient optimization method that uses information that evaluates the heuristic preference of artificial ants to move from one node to another in a graph established by the search algorithm. ACO is also based on the information related with the traces of artificial pheromones, to evaluate the desirability learned in the movement from one node to another. The heuristic information remains constant during the execution of the algorithm, while the information related to the traces of artificial pheromones is modified depending on the solutions found by the ants [21, 22, 28].

ACO technique, similarly to other evolutionary techniques, uses an adjustment function, also known as cost, aptitude or fitness function.

3.1 Search Process in Ant Colony Optimization

Figure 1 shows the essence of Ant Colony Optimization [21]. The search of solutions by each artificial ant is developed in a graph, *G*, conformed by *n* nodes and *A* arcs connected to each other. The objective is to find the path that represents the lowest cost (represented by the red arrows in Fig. 1). Each ant, *j*, of the colony of *m* ants, walks between the nodes of the input set and deposits pheromone in each one of those nodes to interlace them. After each ant has built its way in iteration *i*, the selection of the next node by each ant is carried out in a probabilistic way, taking into account the amount of pheromone in the node, in combination with heuristic information. It is important to emphasize that: (a) the mentioned lowest cost route also includes the return to the node of origin; (b) each ant has a stigmergetic memory of its journey, so that it does not pass through nodes which it had already passed.

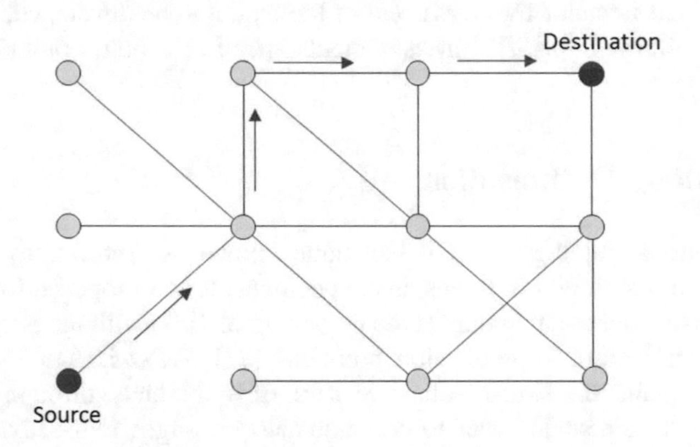

Fig. 1. Search for solutions in the ACO algorithm

In the first ACO algorithms, the set of probabilities related to the points selection of the graph by the ant was given by a discrete distribution function, such is the case of the well-known Ant System (AS) [21]. Later, continuous functions began to be used, for example, the normal or Guassian probability density function (*pdf*), which establishes that the probability, $(p_l)_i$, of nodes selection in iteration *i*, by each ant, is given by the weight normalization, $(\omega_l)_i$, of each solution $(S_l)_i$, by the sum of their respective weights [24]:

$$(p_l)_i = \frac{(\omega_l)_i}{\sum_{e=1}^{k} (\omega_e)_i} \tag{5a}$$

$$(\omega_l)_i = \frac{1}{qk\sqrt{2\pi}} e^{-\left(\frac{l-1}{qk\sqrt{2}}\right)} \tag{5b}$$

where l is the argument and k represents the population size of solutions. The weight ω_l, as can be seen in Eq. (5b), is given by a Gaussian probability density function with mean equal to 1 [24]. In this equation, q is a parameter of the ACO algorithm that impacts on the robustness (in terms of the solution found) and algorithm convergence speed.

The pheromone trace update is implicit in the standard deviation, $(\sigma_l)_i$, in the solution $(S_l)_i$, (solution in iteration i) of Gaussian pdf's set established for the nodes selection of the graph by each ant [24]:

$$(\sigma_l)_i = \xi \sum_{e=1}^{k} \frac{|(S_e)_i - (S_l)_i|}{k-1} \tag{6}$$

where ξ represents the evaporation rate of the pheromone and $(S_e)_i$ refers to all solutions stored up to step i.

In this paper, to select the probability-based solution candidate, nonlinear probability weighting is incorporated into the cost function, as follows:

$$Candidate(i) = w(p_i) \times V(FC_i) \tag{7}$$

where *Candidate(i)* represents the best candidate in iteration i, $w(p_i)$ is the probability weighting function (PWF) of the probability given in Eq. (5a) in step i and $V(FC_i)$ is the value function (VF) of the cost function (FC) in iteration i. Both PWF and VF are given by [29]:

$$w(p_i) = \frac{p_i^\gamma}{[p_i^\gamma + (1-p_i)^\gamma]^{1/\gamma}} \tag{8a}$$

$$V(FC_i) = \begin{cases} FC_i^\alpha; & \text{for } x \geq 0 \\ -\lambda(-FC_i)^\beta; & \text{for } x < 0 \end{cases} \tag{8b}$$

where γ is known as weighting parameter, α (0,1) is the adjustment parameter for gain and λ (>1) and β (0,1) are the adjustment parameters for loss.

Lastly, the run of the ACO algorithm ends after reaching a certain stop criterion or, failing that, a maximum number of iterations.

4 Measurements in DTV Terrestrial in Montevideo City

The measurements carried out in Montevideo city, Uruguay, which will be the basis for the Lee model optimization object of this paper, are part of the campaign of power level received measurements in outdoor environments performed in the aforementioned country in order to implement the deployment of Digital Terrestrial Television (DTT) during 2014 and 2015, with Integrated Services Digital Broadcasting - Terrestrial Built-in (ISDB-Tb) standard.

4.1 Experimental Setup

The transmitter was installed under the administration of the Televisión Nacional de Uruguay (TNU), in the building of the Administración Nacional de Telecomunicaciones (ANTEL), Central Aguada, Montevideo, at geographic coordinates 34.870112° South and 56.180102° West, operating at 569 MHz (channel 30 of the UHF band) and with a transmission power, P_T, equal to 1 kW [18].

The transmitting antenna was located at 112 m height (h_T), with gain, G_T, in the direction of maximum radiation equal to 15.1 dBi. The loss in the transmitter's transmission line, L_T, was approximately equal to 6.35 dB [18].

Therefore, equivalent isotropic radiated power (*EIRP*), in the direction of maximum radiation, is determined from [20] and is equal to 68.75 dBm, i.e., approximately the 7.5 kW referred to by the TNU on its web portal [30].

On the other hand, the effective radiated power, *ERP*, that is the information reported in [18], is given by [12] and is equal to 66.6 dBm.

The reception system consisted of a portable spectrum analyzer for the measurement of the received power level, with an antenna deployed on a 6 meters height mast, with a gain, G_R, equal to 9 dBi in the direction of the radiation pattern main lobe. The transmission line loss, L_R, in the receiver system was 9.53 dB [18].

Measurement Sites in Montevideo. For the received power level measurement, seven (7) radials were selected and a procedure called as the "Four Corners Method" by the researchers responsible for the measurements [18, 19] was used, which consisted of carrying out, whenever possible, one measurement in each of the four corners of the block to which each of the selected sites belonged and averaging at least two of these values depending on the particularity of each point in terms of propagation. In [18, 19] it is possible to observe the 22 sites that were selected.

5 Propagation Path Loss Measured

From the measured level of received power, P_R, at each of the 22 sites above mentioned, propagation path loss, L, is obtained at the above-mentioned points using the classical link budget equation for wireless communications systems [18]:

$$L(\mathrm{dB}) = EIRP_{\theta}(\mathrm{dBm}) + G_R(\mathrm{dBi}) - L_R(\mathrm{dB}) - P_{\mathrm{R}}(\mathrm{dBm}) \tag{9}$$

where $EIRP_{\theta}$ is the *EIRP* considering the receiving antenna gain in azimuth direction, θ, to each measurement site.

Figure 2 shows propagation path loss measured as a function of distance from the transmitter, together with the trend line of this loss. It is observed that, as expected, as the transmitter-receiver distance increases, the loss is greater in a way that, in a first approximation, the exponent of power loss, γ, in the case of Montevideo city and DTT systems, is equal to 2.494.

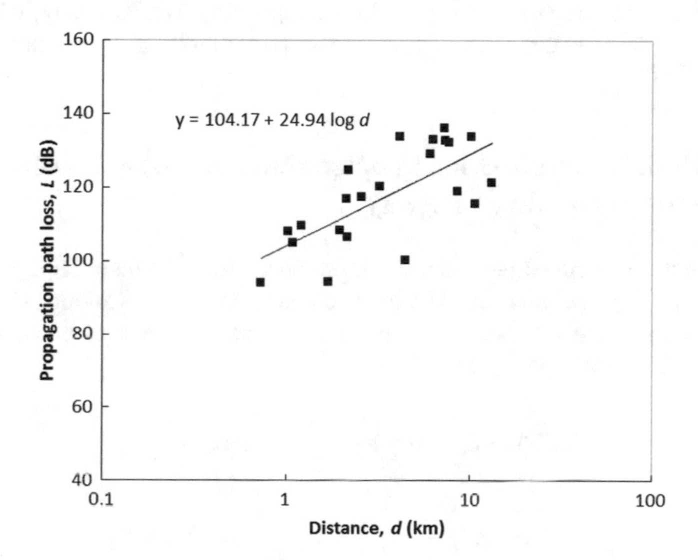

Fig. 2. Measured path loss propagation in DTV terrestrial, Montevideo, Uruguay

5.1 Performance Evaluation of Propagation Models

Considering technical parameters previously provided by the transmitter and receiver systems of the experimental configuration used for measurements, propagation models discussed in this paper are implemented and estimated propagation path loss are compared with those that were obtained experimentally.

To quantitatively and objectively evaluate the performance of these models in estimating propagation path loss, root mean square error (*RMSE*) is used, given by [17]. The *RMSE* consists of the square root of the sum of the squared errors and amplifies and penalizes with greater force those errors of greater magnitude.

Table 1 summarizes *RMSE* values obtained for the four propagation models implemented in this paper for the prediction of losses in DTT systems in Montevideo city, Uruguay.

It is observed in Table 1 that indeed the Hata-Davidson, TDA-Uruguay and Okumura-Hata models are the ones that show the lowest *RMSE* in their estimation of

Table 1. *RMSE* value of propagation models compared to the measurements

RMSE value			
Lee	Okumura-Hata	Hata-Davidson	TDT-Uruguay
34.69	11.97	9.84	10.65

propagation path loss for measurements under consideration. On the other hand, Lee model is the one with the worst performance, with a high *RMSE* of 34.69 dB.

It is worth highlighting that, in Table 1, for the Okumura-Hata and Hata-Davidson models, it was considered the environment corresponding to a large city, while for the Lee model it was assumed the environment corresponding to an urban area of Philadelphia [12].

6 Lee Model Optimized for Propagation Path Loss Prediction in Montevideo City, Uruguay

To improve the Lee model performance in propagation path loss in Digital Terrestrial Television (DTT) systems in Montevideo city, the Ant Colony Optimization (ACO) technique will be used. For this purpose, the functions to be optimized are firstly defined, which in this case will be:

$$L(dB) = L_o + 10\gamma\ \log\left(\frac{d}{d_o}\right) - 10x_1\ \log F_A \tag{10}$$

$$L(dB) = x_2 L_o + 10x_3\ \gamma\ \log\left(\frac{d}{d_o}\right) - 10\ \log F_A \tag{11}$$

Equations (10) and (11) represent variations of Eq. (1), which corresponds to the original Lee model. In this paper, these variations will be referred to as Lee-V1 and Lee-V2 models, respectively.

Next, the cost or fitness function is defined, that in this case is the *RMSE* value, so that this value is the minimum possible. In this way, the values of x_1, x_2 and x_3 that will enforce such condition will be determined.

To implement the ACO algorithm that will tune Lee model to the DTT measurements performed in Montevideo, it was considered, after several tests: (a) number of variables to be optimized equal to 1, for the cost function given by Eq. (10); and 2, for the cost function given by Eq. (11); (b) population size solutions, $k = 10$; (c) number of ants, $m = 5$; (d) pheromone evaporation rate, $\xi = 0.85$; (e) maximum number of iterations, $i_{max} = 100$; (f) minimal error $= 1 \times 10^{-4}$ (stopping criteria). Also, $q = 0.1$; $\gamma = 0.6$ [29]; $\alpha = \beta = 0.8$ [29]; $\lambda = 2.25$ [29] and the initial value of the optimal cost is equal to 0.

The values obtained for the adjustment parameters x_1, x_2 and x_3 were 0.1282, 1.3304 and 0.6779, respectively. Substituting these values in Eqs. (10) and (11), the new versions of the Lee model (Lee-V1 model and Lee-V2 model) are obtained, which give *RMSE* values in their estimate of propagation path loss equal to 10.41 dB and 9.43 dB, respectively. Therefore, the errors obtained with these new versions are significantly lower than that obtained when the original Lee model was implemented, i.e., 34.69 dB. Even, in the case of the Lee-V1 model the *RMSE* obtained was better than that resulting with the application of TDA-Uruguay model, while with the Lee-V2 model, the resulting *RMSE* was better than that obtained with any of the four remaining models.

Lastly, in Fig. 3 the success values of the propagation path loss in the last iteration compared to the first iteration, can be clearly observed.

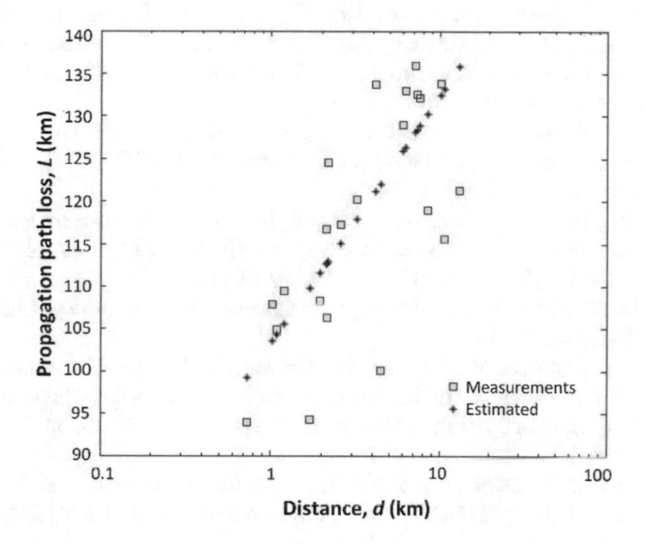

Fig. 3. Performance of ACO optimization for Lee-V2 model in last iteration.

7 Conclusions

In this paper, two new versions of the Lee model were developed, that is one of the typically used models for predicting propagation path loss in UHF band, which is commonly used by Digital Terrestrial Television (DTT) systems. For the development of both versions, named in this paper as Lee-V1 model and Lee-V2 model, the intelligent computing technique Ant Colony Optimization (ACO) was used, which allowed to adjust the Lee model to measurements carried out in DTT systems Montevideo city, Uruguay.

The Lee model modified versions considerably improved its performance in propagation path loss predicting compared to the performance shown by the original version of the model, reducing the *RMSE* value from 34.69 dB to 10.41 dB (in case of Lee-V1 model) and 9.43 dB (in case of Lee-V2 model). In fact, in the case of Lee-V2 model, in which two of the factors from original Lee model were adjusted, its performance in the mentioned propagation loss estimation was better than the performance not only of the original model, but also of the remaining models implemented for comparison, i.e. Okumura-Hata, Hata-Davidson and TDT-Uruguay.

References

1. DTV Status: Map - Snapshot of January 2016. http://www.dtvstatus.net/map201601/map.html. Accessed 15 Jan 2019
2. Augustyniak, L.: Terrestrial television reception disturbances. Przegląd Elektrotechniczny (Electr. Rev.) **2012**(9), 296–297 (2012). ISSN 0033-2097
3. Azpilicueta, L., López-Iturri, P., Aguirre, E., Mateo, I., Astrain, J.J., Villadangos, J., Falcone, F.: Analysis of radio wave propagation for ISM 2.4 GHz Wireless Sensor Networks in inhomogeneous vegetation environments. Sensors **14**(12), 23650–23672 (2014). https://doi.org/10.3390/s141223650. ISSN 1424-8220
4. Farjow, W., Raahemifar, K., Fernando, X.: Novel wireless channels characterization model for underground mines. Appl. Math. Model. **39**(19), 5997–6007 (2015). https://doi.org/10.1016/j.apm.2015.01.043. ISSN 0307-904X
5. Longley, A.G., Rice, P.L.: Prediction of tropospheric radio transmission loss over irregular terrain. A computer method. ESSA Technical Report ERL 79-ITS 67, National Technical Information Service (NTIS), Washington, DC, USA (1968)
6. Parsons, J.D.: The Mobile Radio Propagation Channel, 2nd edn. Wiley, Chichester (2000). ISBN 978-0471988571
7. Fernández, J.R., Quispe, M., Kemper, G., Samaniego, J., Díaz, D.: Adjustments of log-distance path loss model for digital television in Lima. In: XXX Simpósio Brasileiro de Telecomunicações (SBrT 2012), Setembro 2012, Brasília, DF, Brasil, pp. 1–4 (2012). ISBN 978-85-89748-07-0
8. Hata, M.: Empirical formula for propagation loss in land mobile radio services. IEEE Trans. Veh. Technol. **VT-29**(3), 317–325 (1980). https://doi.org/10.1109/T-VT.1980.23859. ISSN 0018-9545
9. Vargas-Zambrano, N.Y., Pérez-García, N.A., Ramírez, J.E., Pérez-Cabrera, M.L., Uzcátegui, J.R.: Nuevo modelo para predicción de pérdidas de propagación en sistemas de televisión digital abierta. Ingeniería Al Día **4**(1), 38–62 (2018). ISSN 2389-7309
10. Lee, W.C.Y.: Wireless and Cellular Telecommunications, 3rd edn. McGraw-Hill, New York (2006). ISBN 978-0071436861
11. Stuber, G.L.: Principles of Mobile Communication, 2nd edn. Kluwer Academic Publishers, New York (2002). ISBN 0-792-37998-5
12. Seybold, M.: Introduction to RF Propagation. Wiley, Hoboken (2005). ISBN 978-0-471-65596-1
13. Faruk, N., Adediran, Y.A., Ayeni, A.A.: Optimization of Davidson model based on RF measurement conducted in UHF/VHF bands. In: 6th International Conference on Information Technology, May 2013, Amman, Jordan, pp. 1–7. ISBN 978-9957-8583-1-5
14. International Telecommunication Union (ITU): Recommendation ITU-R P.1546-5. Method for point-to-area predictions for terrestrial services in the frequency range 30 MHz to 3000 MHz. Geneva, Switzerland (2013)
15. International Telecommunications Union (ITU): Recommendation ITUR P.1812-4. A path-specific propagation prediction method for point-to-area terrestrial services in the VHF and UHF bands, Geneva, Switzerland (2015)
16. Pinto-Mangones, A.D., Torres Tovio, J.M., García-Bello, A.S., Pérez-García, N.A., Uzcátegui, J.R.: Modelo para estimación de pérdidas de propagación en sistemas de

televisión digital abierta. Revista Ingeniería Electrónica, Automática y Comunicaciones (RIELAC) **XXXVII**(2), 67–81 (2016). ISSN 1815-5928
17. Perez-Garcia, N.A., Pinto, A.D., Torres, J.M., Rengel, J.E., Rujano, L.M., Robles-Camargo, N., Donoso, Y.: Improved ITU-R model for digital terrestrial television propagation path loss prediction. Electron. Lett. **53**(1), 832–834 (2017). https://doi.org/10.1049/el.2017.1033. ISSN 0013-5194
18. Gómez-Caram, A., Labandera, A., Marín, G.: Mediciones y modelo de cobertura para televisión digital terrestre. Tesis para optar al grado de Ingeniero Electricista, Universidad de La República, Montevideo, Uruguay (2014)
19. Guridi, P.F., Gómez-Caram, A., Labandera, A., Marín, G., Simon, M.: Studying digital terrestrial TV coverage. In: 11th International Symposium on Broadband Multimedia Systems and Broadcasting (BMSB 2014), June 2014, Beijing, China, pp. 1–5 (2014). https://doi.org/10.1109/BMSB.2014.6873498. ISBN 978-1-4799-1654-2
20. Delgado-Cuadro, E.I., León-Acurio, J.: Optimization of recommendation ITU-R P.1812-3 for the propagation losses prediction in digital terrestrial television system. In: Botto-Tobar, M., Esparza-Cruz, N., León-Acurio, J., Crespo-Torres, N., Beltrán-Mora, M. (Eds.): Technology Trends, CITT 2017. CCIS 798, pp. 3–17. Springer, Heidelberg (2017). https://doi.org/10.1007/978-3-319-72727-1. ISSN 1865-0929. ISBN 978-3-319-72726-4
21. Dorigo, M., Stutzle, T.: Ant Colony Optimization. Massachusetts Institute of Technology (MIT), Cambridge (2004)
22. Sumathi, S., Paneerselvam, S.: Computational Intelligence Paradigms: Theory & Applications Using MATLAB. CRC Press, Boca Raton (2010)
23. Abraham, A., Grosan, C., Ramos, V.: Stigmergic Optimization. Springer, Heidelberg (2006)
24. Socha, K., Dorigo, M.: Ant colony optimization for continuous domains. Eur. J. Oper. Res. **185**(3), 1155–1173 (2008). https://doi.org/10.1016/j.ejor.2006.06.046. ISSN 0377-2217
25. Chebil, J., Lwas, A.K., Islam, R., Zyoud, A.: Adjustment of Lee path loss model for suburban area in Kuala Lumpur-Malaysia. In: International Proceedings of Computer Science and Information Technology (IPCSIT), vol. 5, Article 45, pp. 252–257 (2011). ISSN 2010-460X
26. Nissirat, L., Ismail, M., Nisirat, M.A.: Macro-cell path loss prediction, calibration, and optimization by Lee's model for south of Amman city, Jordan at 900 and 1800 MHz. J. Theor. Appl. Inf. Technol. **41**(2), 253–258 (2012). ISSN 1992-8645
27. Al-Salameh, M.S.H., Al-Zubi, M.M.: Prediction of radiowave propagation for wireless cellular networks in Jordan. In: 7th International Conference on Knowledge and Smart Technology (KST 2015), January 2015, Chonburi, Thailand, pp. 149–154. https://doi.org/10.1109/KST.2015.7051452. ISBN 978-1-4799-6048-4
28. Dorigo, M., Socha, K.: An Introduction to Ant Colony Optimization. In: Gonzalez, T.F. (ed.): Handbook of Approximation Algorithms and Metaheuristics. Methodologies and Traditional Applications, 2nd edn., vol. I, pp. 395–407. CRC Press, Boca Raton (2018). ISBN 978-1-4987-7011-8
29. Tversky, A., Kahneman, D.: Advances in prospect theory: cumulative representation of uncertainty. J. Risk Uncertainty **5**(4), 297–323 (1992). https://doi.org/10.1007/BF00122574. ISSN 0895-5646
30. Televisión Nacional de Uruguay (TNU): Televisión digital abierta. http://www.tnu.com.uy/institucional. Accessed 6 Feb 2019

Author Index

M. Botto-Tobar et al. (Eds.): ICAETT 2019, AISC 1066, pp. 555–557, 2020.
https://doi.org/10.1007/978-3-030-32022-5

Printed in the United States
By Bookmasters